S0-AUW-417

THE MYCOLOGICAL SOCIETY OF AMERICA
MYCOLOGIA MEMOIR NO. 19

Lignicolous Corticioid Fungi
(Basidiomycota)
of North America
Systematics, Distribution, and Ecology

J. Ginns and M. N. L. Lefebvre

APS PRESS
The American Phytopathological Society
St. Paul, Minnesota

Authors' Address
Centre for Land and Biological Resources Research
Saunders Building
Central Experimental Farm
Agriculture Canada
Ottawa, Ontario, Canada K1A 0C6

This book has been reproduced directly from computer-generated copy submitted in final form to APS Press by the authors. No editing or proof-reading has been done by the Press.

Reference in this publication to a trademark, proprietary product, or company name is intended for explicit description only and does not imply approval or recommendation to the exclusion of others that may be suitable.

Library of Congress Catalog Card Number: 93-79367
International Standard Book Number: 0-89054-155-8

Printed in the United States of America on acid-free paper

The American Phytopathological Society
3340 Pilot Knob Road
St. Paul, Minnesota 55121-2097, USA

CONTENTS

TABLES

ACKNOWLEDGMENTS

Dr. David Farr provided computer print-outs of data, from the book by Farr et al. *Fungi on Plants and Plant Products in the United States*, in a format which greatly facilitated our abstracting data from that volume. Dr. R.J. Bandoni shared his ideas on a taxonomic scheme for the Heterobasidiomycetes. Constructive reviews were provided by Drs. S.J. Hughes and S.A. Redhead. In the later stages of preparation Ms. Mavis Cheng assisted in the insertion of revisions and standardizing the citation of journal titles. Ms. Mary Ann Beaudin prepared the rough index to the species names.

INTRODUCTION

The lignicolous, corticioid fungi are ecologically and taxonomically heterogeneous. Corticioid refers to a general resemblance that the typically effuse, thin basidiomes (fruitbodies) have to bark. Ecologically most species fruit on dead bark and wood, and they are accepted as saprophytic. However, many have not been critically evaluated. About 40 species decay the wood of living trees; a few dozen are pathogens of agricultural crops and turf grasses; a few are mycorrhizal. Taxonomically these fungi are Basidiomycota of the class Hymenomycetes. The 1163 species treated below are distributed among 21 orders and 54 families (Table 1).

A few genera, e.g., *Auricularia, Climacodon, Creolophus, Hericium, Tremella* and the cyphelloid genera, are treated that do not have corticioid basidiomata, but are taxonomically or ecologically related to this group, thus it is logical that they be included.

Part of our daily routine has been the identification of specimens either collected for research purposes or submitted by scientists and extension personnel, especially in the fields of agriculture and forestry. The literature on these fungi is widely scattered, and it was time consuming and often frustrating having to consult a variety of often obscure books and scientific journals in order to identify specimens. Frequently the best study of a group was that covering another country or region, e.g., *Hymenomycetes de France* by Bourdot and Galzin (50), and *The Corticiaceae of North Europe* by Eriksson and Ryvarden (172-174), Eriksson et al. (169-171) and Hjortstam et al. (294). Older manuals, e.g, *The Hydnaceae of Iowa* by Miller and Boyle (463) and *Revision of the north central Tremellales* by Martin (443), from the United States are useful for identifying specimens, but the nomenclature is outdated and the user must search elsewhere for a currently acceptable name; furthermore many species have been described since these older manuals were published.

Initially we planned to present an annotated key to the species, similar to the one published by Jülich and Stalpers (326). After some months of data compilation we realized that the data were much more voluminous than we had estimated. Finally, we decided to publish our conclusions on systematics, distribution and ecology. The following text also includes the culture characters, in the form of species codes (487), and cites references containing more detailed descriptions and illustrations of the culture characters.

The identification of specimens has been made easier by the publication of several extensive keys, e.g., Julich and Stalpers (326) *The resupinate non-poroid Aphyllophorales of the temperate northern hemisphere*, Hjortstam et al. (294) *The Corticiaceae of North Europe*, and the keys to genera in Ainsworth et al., *The Fungi* (6) by McNabb, McNabb and Talbot, Talbot, and Harrison. But none of these deals specifically with the mycota of North America.

The only monographic treatment of this group for North America is that of E. A. Burt from 1914 to 1926 (78-93). This extensive and detailed work, with descriptions and keys to species, is still useful. A very helpful feature of Burt's publications is the detailed citation of the collections he studied. Thus revisionists can easily find the collections he studied, and Burt's species circumscriptions and geographic distributions can be reevaluated. Burt's monograph is a firm foundation on which to build and expand. Considerable progress has been made since 1926. The concepts for many families, genera and species have been drastically revised, many new species have been described, the taxonomic value of many morphological characters has been reevaluated, some misapplied names have been corrected, and new taxonomic characters have been introduced.

The revision or modernization of Burt's work has been piecemeal by numerous mycologists. Typically, a genus or part of a genus has been monographed, e.g., Burt's *Hypochnus* was revised by Larsen (358, 361, 362) under the names *Tomentella, Pseudotomentella* and *Tomentellastrum*. Other generic revisions are *Aleurodiscus* by Lemke (375, 376), *Coniophora* by Ginns (224, 234), *Merulius* by Ginns (226), and *Thelephora* by Corner (136). In addition, sections of some of the larger genera recognized by Burt have been revised, e.g., part of

TABLE 1. ALPHABETICAL LISTING OF THE ORDERS, FAMILIES, GENERA AND NUMBERS OF SPECIES OF THE LIGNICOLOUS, CORTICIOID BASIDIOMYCOTA IN CANADA AND THE UNITED STATES.

Order / Family / Genus	Species
Agaricales	
Crepidotaceae	
Pellidiscus	2
Phaeosolenia	4
Cyphellopsidaceae	
Calathella	4
Cyphellopsis	3
Merismodes	2
Woldmaria	1
Lachnellaceae	
Flagelloscypha	5
Halocyphina	1
Lachnella	4
Nochascypha	1
Physalacriaceae	
Physalacria	5
Tricholomataceae	
Calyptella	4
Cellypha	3
Stigmatolemma	2
Atheliales	
Atheliaceae	
Amylocorticium	6
Athelia	20
Athelopsis	3
Byssocorticium	3
Byssoporia	1
Caerulicium	1
Cyphellostereum	1
Fibulomyces	4
Hypochnopsis	1
Leptosporomyces	4
Luellia	1
Piloderma	5
Plicatura	2
Tylospora	2
Atractiellales	
Ecchynaceae	
Phleogena	1
Auriculariales	
Auriculariaceae	
Auricularia	4
Mylittopsis	1
Exidiaceae	
Basidiodendron	8
Bourdotia	2
Ductifera	3
Efibulobasidium	1
Eichleriella	5
Exidia	19
Exidiopsis	13
Heterochaete	4
Auriculariales (cont'd)	
Hyaloriaceae	
Heterochaetella	3
Myxarium	3
Protodontia	3
Pseudohydnum	1
Stypella	1
Tremiscus	1
Sebacinaceae	
Sebacina	8
Tremelloscypha	1
Boletales	
Coniophoraceae	
Coniophora	9
Hypochniciellum	3
Leucogyrophana	8
Pseudomerulius	2
Serpula	3
Botryobasidiales	
Botryobasidiaceae	
Botryobasidium	16
Botryohypochnus	2
Candelabrochaete	3
Hypochnella	1
Suillosporium	1
Cantharellales	
?[a]**Clavulinaceae**	
Clavulicium	3
Pterulaceae	
Deflexula	1
?Sparassidaceae	
Sparassis	2
Dacrymycetales	
Dacrymycetaceae	
Calocera	3
Cerinomyces	5
Dacrymyces	23
Dacryonaema	1
Dacryopinax	3
Ditiola	1
Femsjonia	1
Guepiniopsis	1
Heterotextus	2
Gomphales	
Beenakiaceae	
Kavinia	2
Ramaricium	4
Hericiales	
Aleurodiscaceae	
Acanthophysium	16
Aleurobotrys	1
Aleurocystidiellum	2
Aleurodiscus	12
Amyloathelia	1
Dendrothele	17
Globulicium	1
Auriscalpiaceae	
Auriscalpium	1
Gloeodontia	2
Gloiodon	2
Clavicoronaceae	
Clavicorona	6
Echinodontiaceae	
Echinodontium	2
Laurilia	2
Epitheleaceae	
Epithele	1
Jaapia	2
Gloeocystidiellaceae	
Boidinia	2
Gloeocystidiellum	14
Laxitextum	2
Scytinostromella	4
Vesiculomyces	1
Hericiaceae	
Creolophus	1
Dentipellis	3
Hericium	4
Mucronella	4
Steccherinum	1
Lachnocladiaceae	
Dichostereum	5
Scytinostroma	10
Vararia	11
Licrostromataceae	
Licrostroma	1
Punctulariaceae	
Punctularia	2
Vuilleminiaceae	
Corticium	13
Cytidia	5
Dendrocorticium	3
Dentocorticium	1
Laetisaria	3
Limonomyces	2
Pulcherricium	1
Vuilleminia	1
Hymenochaetales	
Hymenochaetaceae	
Asterodon	1
Asterostroma	6

Taxon	Species
Hymenochaetaceae (cont'd)	
Hydnochaete	2
Hymenochaete	22
Hyphodermatales	
Chaetoporellaceae	
Amphinema	3
Hyphodontia	30
Odonticium	2
Hyphodermataceae	
Amylobasidium	1
Basidioradulum	1
Bulbillomyces	1
Cerocorticium	1
Conohypha	2
Hyphoderma	42
Hyphodermella	1
Hypochnicium	15
Intextomyces	1
Melzericium	1
Metulodontia	1
Radulodon	2
Radulomyces	4
Sarcodontia	1
Thujacorticium	1
Steccherinaceae	
Fibricium	2
Irpex	1
Steccherinum	13
Lindtneriales	
Stephanosporaceae	
Cristinia	3
Lindtneria	7
Meruliales	
Climacodontaceae	
Climacodon	2
Mycorrhaphium	2
Meruliaceae	
Amethicium	2
Chondrostereum	1
Dacryobolus	2
Flavophlebia	1
Mycoacia	4
Phlebia	28
Resinicium	4
Skvortzovia	1
Phanerochaetaceae	
Erythricium	4
Meruliopsis	7
Phanerochaete	28
Phlebiopsis	3
Scopuloides	1
Schizophyllaceae	
Auriculariopsis	2
Schizophyllaceae (cont'd)	
Henningsomyces	2
Rectipilus	5
Schizophyllum	3
Platygloeales	
Cystobasidiaceae	
Cystobasidium	1
Helicobasidium	5
Helicogloea	8
Mycogloea	1
Platygloea	16
Polyporales	
Polyporaceae sensu lato	
Hydnopolyporus	2
'Cyphellaceae'[b]	
Glabrocyphella	6
'Maireina'	9
'Phaeoglabrotricha'	1
'Solenia'	5
Stromatocyphella	2
Stromatoscyphaceae	
Porotheleum	4
Sistotremales	
Sistotremaceae	
Brevicellicium	3
Coronicium	2
Galzinia	5
Paullicorticium	4
Phlebiella	13
Repetobasidium	6
Sistotrema	21
Sistotremastrum	2
Sphaerobasidium	2
Trechispora	14
Tubulicrinaceae	
Granulocystis	1
Litschauerella	1
Subulicium	2
Subulicystidium	1
Tubulicium	2
Tubulicrinis	20
Xenasma	4
Xenosperma	2
Stereales	
Chaetodermataceae	
Chaetodermella	1
Cylindrobasidiaceae	
Butlerelfia	1
Ceraceomyces	7
Cylindrobasidium	3
Pirex	1
Stereales (cont'd)	
Mycoboniaceae	
Mycobonia	1
Peniophoraceae	
Dendrophora	3
Peniophora	36
Podoscyphaceae	
Cotylidia	4
Cymatoderma	2
Podoscypha	4
Stereopsis	3
Stereaceae	
Amylostereum	3
Boreostereum	1
Crustoderma	10
Crustomyces	2
Cystostereum	2
Lopharia	3
Stereum	21
Veluticeps	5
Xylobolus	2
Thelephorales	
Thelephoraceae	
Lazulinospora	2
Pseudotomentella	14
Thelephora	17
Tomentella	54
Tomentellastrum	4
Tomentellina	1
Tomentellopsis	5
Tremellales	
Tremellaceae	
Fibulobasidium	1
Sirobasidium	2
Tremella	27
Xenolachne	1
Tulasnellales	
Ceratobasidiaceae	
Ceratobasidium	11
Oliveonia	3
Thanatephorus	3
Uthatobasidium	2
Waitea	1
Ypsilonidium	1
Tulasnellaceae	
Heteroacanthella	1
Tulasnella	24
Incertae sedis	
Incertae sedis	
Entomocorticium	1
Total species	1163

[a] A question mark before a family name indicates uncertainty over the inclusion of the family in that order.

[b] Untenable names are in single quotes. They are used because they contain taxa not yet assigned to acceptable families or genera.

Corticium, section *Athele*, by Liberta (384, 385), and part of *Peniophora* by Burdsall (64), as *Phanerochaete*.

Burt described 218 new species and the reevaluation of them has been a significant step in our understanding of the mycota. Particularly important, in this respect, are the papers by Rogers and Jackson (567) treating many of Burt's species, by Ginns (218-220, 222-224) treating Burt's species of *Coniophora* and *Merulius*, by Liberta (392, 393) treating Burt's species of *Corticium* and *Peniophora*, and by Larsen (357) treating Burt's species of *Hypochnus*.

Floristic studies have added substantially to our knowledge of the distribution and ecology of these fungi, for example, *Stereum and allied genera in the upper Mississippi Valley* by Lentz (379), *Basidiomycetes that decay aspen in North America* by Lindsey and Gilbertson (406), *The genus Peniophora in New York State and adjacent regions* by Slysh (584), *Fungi that decay ponderosa pine* by Gilbertson (197), *The non-stipitate stereoid fungi in the northeastern United States and adjacent Canada* by Chamuris (104) and *Revision of the north central Tremellales* by Martin (443).

The data presented in this work form a base-line for biological surveys, biogeography mapping and biodiversity studies. Although it took considerable effort to compile the scattered literature, the primary effort and major contribution is the evaluation of the data in that literature, and their reorganization into a database reflecting current taxonomic schemes. Not a small part of this evaluation was the problem of determining the proper scientific name for each species, i.e., if a fungus had been given more than one binomial, which was acceptable and which names were to be listed as synonyms. Also, if a fungus had been placed in several genera, which placement would put that species with those most closely related to it. The result is twenty-eight new names and new combinations are proposed (Table 2).

The ecological associations, geographical distribution and number of species of these fungi present in North America are poorly understood. The existing data are widely scattered and needed to be collated to determine geographic patterns, to demonstrate the ecological diversity of the group, and to provide an accurate estimate of the number of species in Canada and the United States.

TABLE 2. NEW COMBINATIONS AND NEW NAMES PROPOSED IN THIS VOLUME

Acanthophysium abietis (Jackson & Lemke) comb. nov.
Acanthophysium farlowii (Burt) comb. nov.
Acanthophysium piceinum (E.M. Lyon & Lemke) comb. nov.
Brevicellicium permodicum (Jackson) comb. nov.
Calathella eruciformis var. *microspora* (W.B. Cooke) comb. nov.
Ceratobasidium ochroleuca (F. Noack) comb. nov.
Corticium griseo-effusum (Larsen & Gilbn.) comb. nov.
Crustomyces pini-canadensis subsp. *subabruptus* (Bourd. & Galzin) comb. nov.
Dacrymyces pennsylvanica (Overh.) comb. nov.
Dendrothele candida var. *sphaerospora* (Coker) comb. nov.
Dichostereum boreale (Pouzar) comb. nov.
Hyphodontia fimbriaeformis (Berk. & Curtis) comb. nov.
Hyphodontia granulosa (Pers.:Fr.) comb. nov.
Hyphodontia latitans (Bourd. & Galzin) comb. nov.
Hyphodontia macrescens (Banker) comb. nov.
Leptosporomyces montanum (Jülich) comb. nov.
Meruliopsis sulphureus (Burt) comb. nov.
Myxarium atratum (Peck) comb. nov.
Pellidiscus pezizoidea (Ellis & Ev.) comb. nov.
Phlebiella insperata (Jackson) comb. nov.
Phlebiella sulphurea (Pers.:Fr.) comb. nov.
Phlebiella tulasnelloideum (Höhnel & Litsch.) comb. nov.
Radulomyces cremoricolor (Berk. & Curtis) comb. nov.
Resinicium praeteritum (Jackson & Dearden) comb. nov.
Scytinostromella olivaceo-album (Bourd. & Galzin) comb. nov.
Serpula lacrimans var. *shastensis* (L. Harmsen) comb. nov.
Tremella diaporthicola nom. nov.
Tubulicrinis corticioides (Overh.) comb. nov.

NUMBERS OF FUNGI

Mycologists have been conservative in their estimates of the numbers of fungi in various taxonomic, ecological, geographic, and other areas, see Hawksworth (271). For example, the *Dictionary of the Fungi* (272) estimated that in the World there were 19 genera and 100 species in the Auriculariales, whereas we tally 18 genera and 81 species just in Canada and the United States. Similarly the Corticiaceae sensu lato in the World was estimated to be composed of 80 genera and over 500 species, but in Canada and the United States we treat about 130 genera and just over 700 species. In 1979, Burdsall (62) stated that, using the books *Fungi that decay ponderosa pine* (197) and *Basidiomycetes that decay aspen in North America* (406), "it will be possible now to identify a large percentage of decay fungi on whatever woody substrate one encounters." These combined floras treat 250 species of lignicolous corticioid Basidiomycota. In 1980, Gilbertson (198), in an overview of the wood-rotting Basidiomycota in North America, tallied approximately 600 species of corticioid fungi in that mycota. Our species count of 1163 (Table 1), is a 94% increase over estimates made only 12 years ago. Furthermore, considering the numerous new species described by Eriksson, Hjortstam, Ryvarden, Hallenberg and colleagues from North Europe, we believe the total mycota in this group in Canada and the United States to be about 1500 species.

TAXONOMIC SCHEME

It is impossible to draw valid conclusions about the ecology and distribution of a group of species if the taxonomy of the group is not sound. And it is important to be aware of the relationships of species and genera to each other, i.e., a taxonomic system. Initially we were inclined to accept the taxonomic outline in the *Dictionary of the Fungi* (272), but it became obvious that, for this group of fungi, that outline is outdated. Our taxonomic outline (Table 1) of the included genera, families and orders of the Hymenomycetes illustrates their heterogeneity.

It is hoped that our taxonomic scheme will focus discussion on the relationships between the suprageneric categories in this group of fungi. Because the principal focus of our work has been at the species level, we have not dealt with the circumscription of families and orders but have accepted the definitions in Bandoni (20) and Jülich (321).

In preparing Table 1 the taxonomic schemes for the subfamilies of the Corticiaceae sensu lato by Parmasto (522, 523) have been helpful. Elements from the outline of the families and orders of Basidiomycota by Jülich (321) have been incorporated. Keller's (336) discussion of spore wall ultrastructure and taxonomy at the suprageneric level guided some of our decisions. Bandoni's (20) revision of the Auriculariales and Tremellales has been largely adopted. Nevertheless, the circumscription of many genera and families is unsettled, primarily because such a small part of the World mycota is known; hence, the taxonomic scheme presented is accordingly a provisional arrangement.

ECOLOGY

If rare or endangered species of fungi are to be protected, the habitats that they occupy must be well understood. Most of these fungi are capable of breaking down cellulose and lignin and have been accepted as being saprophytic simply because they fruit on dead wood, i.e., fallen branches, logs, etc. However, in many cases the association of a basidiome with dead wood is the only evidence there is for saprophytism!

The ecological roles of these species are more diverse than saprophytism. Some are parasitic, causing a variety of diseases from root rot to leaf blights. Rhizoctonia root rot is caused by *Thanatephorus cucumeris* (best known under its anamorph name, *Rhizoctonia solani*) (524). This fungus attacks a wide variety of agricultural crop plants. Violet root rot is caused by *Helicobasidium brebissonii* (anamorph *Rhizoctonia crocorum*) (183). Species in several corticioid genera form *Rhizoctonia* anamorphs, e.g., *Athelia, Botryobasidium, Ceratobasidium, Koleroga, Thanatephorus, Uthatobasidium,* and *Waitea* (330). Several species of *Ceratobasidium*, best known by their *Rhizoctonia* anamorphs, are also pathogens of agricultural crops. *Athelia rolfsii* (anamorph *Sclerotium rolfsii*) causes southern blight of several crops in the southern United States (467, 585). *Chondrostereum purpureum* causes the notorious silver leaf disease of fruit trees (274, 559). Red thread disease of sports turfs and grasses is caused by *Laetisaria fuciformis* and a similar pathogen *Limonomyces roseipellis* causes pink patch disease of turf grasses (592). *Butlerelfia eustacei* causes fisheye rot, a post harvest disease of stored apples (640).

About 40 species in this group have been isolated from heart rot or root rot of living trees. The most economically important are *Echinodontium tinctorium, Stereum sanguinolentum* and

Scytinostroma galactinum (274). The species that decay wood cause either a white rot or a brown rot. In general white rot fungi degrade cellulose and hemicelluloses at approximately the same rates, whereas the lignin is usually degraded at a faster rate (276). The white rotted wood typically becomes soft and fibrous. Brown rotted wood is typically brown, dry, crumbly and is often cracked into small cubes. Brown rotted wood is primarily composed of modified lignin, the cellulose having been degraded (276). The infamous 'dry rot' of buildings is a brown rot that is caused by several fungi. Two of the better known 'dry rot' species are treated below, *Coniophora puteana* and *Serpula lacrimans*.

Several genera, particularly *Aleurodiscus, Corticium,* and *Dendrothele*, occupy xeric habitats, typically the bark of living trees. These fungi in most cases do not penetrate to the cambium and do not cause injury to the tree. Some species are long term occupants of living bark, whereas others, like *Aleurodiscus amorphus*, seem to be primary colonizers of recently killed, suppressed branches in the lower crowns of live trees. Other species, for example *Athelia epiphylla* (370) and some species of *Leptosporomyces* (196), appear to be psychrophilic.

Some species of *Amphinema, Byssocorticium, Byssoporia, Piloderma, Thelephora* and *Tylospora* are ectomycorrhizal, the association having been confirmed in the laboratory (140, 187, 348, 602). These species typically produce delicate, cottony basidiomes in the litter layer or superficially attached to the underside of fallen branches and logs. We suspect that many more species in the Atheliaceae and Thelephoraceae are mycorrhizal.

The ecological role of many species is unknown, as evidenced by the lack of data in the Ecology field for many of the species treated below. Until their role is better understood, we can neither exploit their potential nor evaluate their role in forest ecosystems.

Several species are excellent examples of how fungi can be used by man. *Phlebiopsis gigantea*, which in nature fruits on old conifer logs, has been used on a commercial scale as a biological control agent to prevent *Heterobasidion annosum* (Fr.) Bref. from entering freshly cut conifer stumps and spreading into live trees causing annosus root rot (559). The antibiotic Merulidial (543), produced by *Phlebia tremellosus*, suggests that other useful compounds will be found when this group is better studied. *Phanerochaete chrysosporium* has been extensively used in studies: 1) on the enzymatic activity in the decay of wood, especially in lignin degradation (341, 342); 2) on the biopulping of wood, i.e., the use of fungi in primary pulp manufacture (343); and 3) on the bioremediation of contaminated water and soil (351, 352).

METHODS

Most of the data in this study were derived from published studies. These original sources are cited to allow readers to evaluate the basis for the synonymy, trace specimens from unusual localities or substrates, etc. There are three terms, 'fide', 'sensu,' and 'ex', used in taxonomic writing which may not be familiar to some readers. The three are found primarily in the sections titled 'Basionym' and 'Synonyms'. The word 'fide' means 'according to'. 'Sensu' means 'in the sense of' or 'following the concept of', and is used after an epithet to indicate that the author, whose name follows the word 'sensu', applied the name to a fungus which is considered not to be the same fungus as represented by the type specimen of that name. 'Ex', when seen in the author citation of an epithet, means that name was first validly published by the author(s) whose name appears after the word 'ex' but was attributed to the person(s) whose name(s) precede the 'ex'. The 'ex' citations lengthen the name unduly and are of interest only to the taxonomist delving into the geneology of the name. For the author citations including the 'ex' connotation, we follow the *International code of botanical nomenclature* (250) article 46.3 and shorten the author citation. For example, the name *Lachnella albolivida* was published in a paper authored by W. B. Cooke, who cited the author as "Ellis" and the name can be cited as either *Lachnella albolivida* Ellis ex W.B. Cooke or *L. albolivida* W. B. Cooke. We usually use the shortened form giving only the validating author, in this case W. B. Cooke.

In a few instances we cite herbaria where specimens are permanently stored. The herbaria abbreviations appear as capital letters and follow the listing in Holmgren et al. (299).

FORMAT

The genera are arranged alphabetically and the epithets are in alphabetical order within each genus. In addition, both a genus/species index and an index to species names are provided.

The data were compiled in a standard format in a database structure formulated so one record was equivalent to one species. Each record has the following headings (or fields):

1. Genus name, species epithet, subspecific epithet, author(s) name(s) and year in which the genus-epithet combination was published.
2. Reference to place of publication of the genus-epithet.
3. Basionym.
4. Synonyms.
5. Anamorph Name.
6. Distribution.
7. Hosts.
8. Ecology.
9. Culture Characters.
10. References.
11. Comments.

When the manuscript was complete, the fields named Basionym, Synonyms, Anamorph Name, and Comments that lacked data were removed to save space. Fields named Hosts, Ecology and Culture Characters have been included, even though empty, to emphasize our lack of knowledge.

FAMILY NAMES

Appearing at the beginning of each genus, the family name provides a link to the current taxonomic scheme at the suprageneric level (see Table 1). Genera that are considered to be related are grouped under the same family name. The family name often will be more familiar to readers than the generic name. For example, the family name Dacrymycetaceae conveys taxonomic information on the type of basidia and basidiospores, and the size, form and texture of the basidiomes, whereas the generic name *Dacryonaema*, a genus of the Dacrymycetaceae, by itself conveys little to most of us. In fact, that generic name standing alone prompts questions, such as, to what is it related and what are its features?

The taxonomy of this group of fungi is in a dynamic state. Accordingly, it must be emphasized that there is no consensus on the circumscription of many of the families. The most familiar family name associated with corticioid fungi, Corticiaceae, will not be found in this work. It is a later synonym of the name Vuilleminiaceae and, therefore, has been replaced by the name Vuilleminiaceae.

GENUS, EPITHET AND AUTHOR(S) NAMES

It is not unusual to see in the literature variation in the authors cited for a particular genus-epithet combination. There are a variety of reasons for this, but perhaps the principal reason is changes to the *International code of botanical nomenclature* (250). Authors names are often abbreviated following the list in Farr et al. (183), except for those names in Table 3.

CITATION OF PLACE OF PUBLICATION

In most cases the data have been confirmed by consulting the original publication.

BASIONYM

This is the binomial and the place the name was originally published. It is the name on which a new combination or replacement name, i.e., *nom. nov.*, is based. In most cases the data in the basionym have been confirmed by consulting the original publication.

TABLE 3. ABBREVIATIONS FOR AUTHORS OF FUNGAL NAMES.[a]

Alb. - J.B.A. Albertini	Hol.-Jech. - V. Holubova-Jechova	Oniki - Masaomi Oniki
Araki - Takao Araki	Holway - E.W.D. Holway	Parker-Rhodes - A.R. Parker-Rhodes
Br. - C.E. Broome	Jackson - H.S. Jackson	Parm. - E. Parmasto
Bonorden - H.F. Bonorden	Jech. - V. Holubova-Jechova	Petersen - R.H. Petersen
Boudier - J.L.E. Boudier	Jung - H.S. Jung	Post - H.A. von Post
Boyle - J. Boyle	Karsten - P.A. Karsten	Preuss - C.G.T. Preuss
Brinkm. - W. Brinkmann	Kennedy - L.L. Kennedy	Rauschert - S. Rauschert
Britz. - M. Britzelmayr	Lamarck - J.B.A. Lamarck	Rav. - H. Ravenel
Canfield - E.R. Canfield	Larsen - M.J. Larsen	Ryv. - L. Ryvarden
Curtis - M.A. Curtis	Larsson - K.-H. Larsson	Saksena - H.K. Saksena
Dearness - J. Dearness	Lemke - P.A. Lemke	Sauter - A.E. Sauter
Desm. - J.B.H.J. Desmazieres	Letellier - J.B.L. Letellier	Schrader - H.A. Schrader
Ev. - B.M. Everhart	Libert - M.-A. Libert	Schröter - J. Schröter
Eriksson - John Eriksson	Lloyd - C.G. Lloyd	Schum. - H.C.F. Schumacher
Farlow - W.G. Farlow	Lundell - S. Lundell	Schw. - L.D. Schweinitz
Freeman - G.W. Freeman	McGuire - J.M. McGuire	Solheim - Halvor Solheim
G.H. Cunn. - G.H. Cunningham	Meyer - J.A. Meyer	Strid - Å. Strid
Gilbn. - R.L. Gilbertson	Möller - F.A.G. Möller	Swartz - O.P. Swartz
Hallenb. - N. Hallenberg	Nannf. - J.A. Nannfeldt	Tellería - M.T. Tellería
Halsted - B.D. Halsted	Novobr. - T.I. Novobranova	Trotter - A. Trotter
Harrison - K.A. Harrison	Oberw. - F. Oberwinkler	Welden - A.L. Welden
Hjort. - K. Hjortstam	Ogoshi - Akira Ogoshi	Wells - K. Wells
Höhnel - F.X.R. Höhnel	Olive - L.S. Olive	Wu - S.H. Wu

[a] Only names either abbreviated differently from or not in Farr et al. (183).

SYNONYMS

Two types of synonyms are recognized. Those based upon the same type specimen (obligate synonyms) are typically the transfers of a species epithet from one genus to another; those synonyms based upon different type specimens (facultative synonyms) are so considered when a taxonomist concludes that the types for different epithets are different specimens of the same fungus. Because the facultative synonyms recognized by one author may not be accepted by other authors, a reference is given to the paper where the synonymy was proposed so users can evaluate the basis for the synonymy. With some exceptions, all names listed as synonyms were seen in the literature that contains reports from Canada and the United States. Many additional synonyms exist, but if they have not appeared in the literature treating Canadian or United States specimens they, in most cases, have not been included. In addition, because Burt (78-93) often had broad species concepts, many of the synonyms he listed have not been repeated unless they have been confirmed by other authors. Facultative synonyms are cited only where we have been able to confirm that type specimens have been studied; we have not cited papers that lack explanation of lists of synonyms they include. Occasionally the listing of a synonym is by indirect reference, e.g., *Thelephora floridana* Ellis & Ev. is

TABLE 4. ABBREVIATIONS FOR THE NAMES OF DISTRICTS, PROVINCES, AND STATES OF CANADA AND THE UNITED STATES.

Canada

AB - Alberta
AB-NT - Either area but unspecified
BC - British Columbia
MB - Manitoba
NB - New Brunswick
NF - Newfoundland
NFL - NF Labrador District
NS - Nova Scotia
NT - Northwest Territories
NTF - NT Franklin District
NTK - NT Keewatin District
NTM - NT Mackenzie District
ON - Ontario
PE - Prince Edward Island
PQ - Quebec
SK - Saskatchewan
YT - Yukon Territory

United States

AK - Alaska
AL - Alabama
AR - Arkansas
AZ - Arizona
CA - California
CO - Colorado
CT - Connecticut
DC - District of Columbia
DE - Delaware
FL - Florida
GA - Georgia
IA - Iowa
ID - Idaho
IL - Illinois
IN - Indiana
KS - Kansas
KY - Kentucky
LA - Louisiana
MA - Massachusetts
MD - Maryland
ME - Maine
MI - Michigan
MN - Minnesota
MO - Missouri
MS - Mississippi
MT - Montana
NC - North Carolina
NC-TN - Either area but unspecified, usually Great Smoky Mountains Park
ND - North Dakota
NE - Nebraska
NEW ENGLAND - Comprised of CT, MA, ME, NH, RI, and VT
NH - New Hampshire
NJ - New Jersey
NM - New Mexico
NV - Nevada
NY - New York
OH - Ohio
OK - Oklahoma
OR - Oregon
PA - Pennsylvania
RI - Rhode Island
SC - South Carolina
SD - South Dakota
TN - Tennessee
TX - Texas
UT - Utah
VA - Virginia
VT - Vermont
WA - Washington
WI - Wisconsin
WV - West Virginia
WY - Wyoming

TABLE 5. CULTURE CHARACTERS: DEFINITIONS FOR THE NUMERICAL CODE SYMBOLS.[a]

Code Symbol	Definition
Production of Extracellular Oxidase	
1	Neither laccase nor tyrosinase detected.
2	Both laccase and tyrosinase present.
2a	Only laccase present.
2b	Only tyrosinase present.
Septation of Hyphae	
3	Thin-walled hyphae with a clamp connection at each septum.
3c	A clamp connection present at each true septum.
3i	Clamp connections variable.
3r	Clamp connections rare.
4	Thin-walled hyphae simple-septate in the advancing zone, these hyphae, usually broad, giving off branches which are narrower and with clamp connections, the older part of the mat being composed of such hyphae with clamp connections.
5	Thin-walled hyphae mainly simple-septate but with occasional single or multiple (verticillate) clamp connections, these occurring most frequently on the hyphae of the advancing zone.
6	Thin-walled hyphae consistently simple-septate.
Special Structures Formed by Differentiation of Hyphae	
7	Hyphae remaining thin-walled and undifferentiated.
8	Hyphae differentiated to form fiber hyphae.
8d	Fiber hyphae dextrinoid, e.g., *Scytinostroma*.
9	Hyphae differentiated to form hyphae with clamp connections and with irregularly thickened walls or with scattered thick-walled, refractive areas on walls.
10	Hyphae differentiated to form cuticular cells, closely packed together to form a pseudoparenchyma.
11	Hyphae differentiated through formation of numerous short branches, hooked or recurved, or thick-walled nodules, interlocked to form a plectenchyma.
12	Hyphae differentiated to form numerous contorted, incrusted hyphal tips.
13	Cystidia present on vegetative mycelium.
14	Cystidia present, restricted to the hymenium of fruiting areas.
15	Gloeocystidia present on vegetative mycelium or in hymenium of fruiting areas.
15a	Gloeocystidia sulfo-aldehyde positive.
15b	Gloeocystidia sulfo-aldehyde negative.
15p	Gloeocystidia with schizopapillae.
16	Hyphae aggregated to form conspicuous strands or rhizomorphs.
17	Setae on aerial mycelium or in fruiting areas.
18	Setal hyphae on aerial mycelium.
19	Terminal cells moniliform.
20	Hyphae with minute projections on walls, e.g., *Schizophyllum commune*.
21	Hyphae with resinous masses clinging to walls, e.g., *Stereum sanguinolentum*.
22	Bulbils or knots of hyphae, e.g., *Peniophora pithya*.
23	Sclerotia, e.g., *Leucogyrophana mollusca* and *Sistotrema raduloides*.
24	Hyphae with walls slightly thickened and lumen empty or at least non-staining and with rigid, more or less right-angled branches.
25	Hyphae with walls somewhat thickened and lumen non-staining.
25d	Dichophyses dextrinoid, e.g. *Dichostereum* and *Vararia*.
26	Noteworthy swellings on hyphae.
27	Not in use.
28	Capitulate spines on vegetative hyphae.
29	Asterohyphidia dextrinoid, e.g. *Asterostroma*.
30	Stephanocysts, e.g., *Hyphoderma praetermissum*.
31	Not in use.
Occurrence of Conidia, Chlamydospores and Oidia	
32	Conidia, chlamydospores, and oidia lacking.
33	Conidia present.
34	Chlamydospores present.
35	Oidia present.
Color of Hyphae and Mycelial Mats	
36	Hyphae hyaline and mats white or pale in color.
37	Hyphae yellow or brown when mounted in potassium hydroxide solution and mats yellow or brown, at least in part.

Color Changes in Agar Induced by Growth of Fungus (Reverse side)

38 Reverse unchanged in color.
39 Reverse brown, at least in part or yellow-green in *Leucogyrophana*.
40 Reverse bleached, at least in part.

Rate of Growth Across an 85 mm Petri dish

41 Plates covered in 1 week.
42 Plates covered in 2 weeks.
43 Plates covered in 3 weeks.
44 Plates covered in 4 weeks.
45 Plates covered in 5 weeks.
46 Plates covered in 6 weeks.
47 Plates not covered in 6 weeks.

Production of Basidiomes in Culture

48 Basidiomes regularly produced before the end of 6 weeks.
49 Not in use.

Odor of Cultures

50 Odor fragrant, including the sweet or fruity odors.
51 Odor earthy or musty.
52 Odor suggesting an antiseptic.
53 Odor noteworthy but not as described above.

Host Relationships

54 Associated with decay in broad-leaved trees.
55 Associated with decay in coniferous trees.
56 Occurring in other habits, for example on soil, etc.

Infertility Phenomena

57 Homothallic, completing its life history from single basidiospore.
58 Heterothallic, but mating system unknown.
59 Heterothallic and bipolar (unifactorial).
59a Amphithallic and bipolar.
60 Heterothallic and tetrapolar (bifactorial).
60a Amphithallic and tetrapolar.

Nuclear Behavior

61 Normal: the uninucleate spore germinates into a primary mycelium composed of uninucleate cells; the secondary mycelium is regularly binucleate.
62 Subnormal: after the germination of the binucleate spore, a brief plurinucleate state precedes the appearance of mycelium which rapidly becomes uninucleate; the secondary mycelium is binucleate.
63 Heterocytic: some binucleate or exceptionally uninucleate spores germinate into primary mycelia which remain, at least within the growing cells, pluri- to multinucleate; the secondary mycelium is regularly binucleate.
64 Primary mycelium unknown, secondary mycelium binucleate.
65 Astatocoenocytic: primary mycelium multinucleate, secondary mycelium binucleate or multinucleate depending on the conditions of aeration in the cultures.
66 Holocoenocytic: primary and secondary mycelium multinucleate.
67 Holomonocaryotic: both monospore and polyspore mycelium uninucleate.
68 Holodicaryotic: the binucleate spore germinates into a binucleate mycelium.

---

[a] The definitions are essentially those found in Nobles (487) and include the additions and revisions proposed by Boidin (37), Boidin and Lanquetin (41), Chamuris (101a) and Lanquetin (354).
Within the numerical code, the parentheses indicate variability, e.g., (48) means that basidiomes were formed in some but not all isolates.

listed as a synonym of *Tomentellastrum badium* and we give the reference for this as "Larsen (362)". However, Larsen did not mention *T. badium*, but stated that *Tomentella atroviolacea* Litsch. was a synonym of *Thelephora floridana*. Stalpers (587) linked *floridana* with *badium* when he placed *atroviolacea* in synonymy under *badium*; hence *floridana* = *atroviolacea* = *badium*.

ANAMORPH NAMES

A few fungi in this group produce an 'asexual state,' such as conidia or sclerotia, which function to propagate the species or are a survival state. The name of this morph or state is included if the connection between the basidial state and the anamorph is well established. The best examples are in the genera *Botryobasidium* and *Ceratobasidium*, where nearly all species have an anamorph. *Rhizoctonia* is a well known anamorph name and the teleomorphs (sexual states) of the Rhizoctonias are corticioid Hymenomycetes.

DISTRIBUTION

The range includes all of Canada and the United States, excluding the Hawaiian Islands. Ranges are given by the most commonly cited political subdivisions, i.e., Provinces, States and Districts (Table 4). Localities within States and Provinces are rarely included because of the time and space required to integrate them. This may be too imprecise for some users. However, the sources for the data are cited so localities within States, Provinces and Districts can be traced.

Reporting a species distribution only to the level of Province or State has drawbacks. It is useful to know that a species occurs in Rhode Island (one of the smallest political subdivisions recognized) but less so for Ontario (one of the largest areas) because there is such a great variety of vegetation types in Ontario, i.e., the Carolinian forests, the northern hardwood forest with all its variations, and the boreal forest with its variations. Simply stating 'Ontario' gives no indication whether the fungus has a preference for a particular vegetation zone. Latitude and longitude data are relatively precise and are essential for many computer mapping programs, but they were rarely seen in the sources and are not included.

The bases for the distributional data (and hosts) are the papers cited under the fields Basionym, Synonyms and References.

HOSTS

The introductory phrase for most paragraphs on Hosts could be 'The basidiomes are produced upon,' however to save space we simply list the host plants. Most species fruit on wood, but there are exceptions. Species such as *Sistotrema confluens*, which produce basidiomes arising from leaf litter, are included to provide a complete treatment of that genus. The genus *Tremella* is included, although its species are mycoparasitic, because the basidiomes often occur on wood.

Because of the variety of ways the hosts have been cited in the references, it was necessary to refine and standardize the raw data. For example, the terms hardwood, angiosperm, frondose, and deciduous all refer to the same group of plants and, to be consistent, we cite them all as angiosperm. Scientific names are used where possible and they follow Farr et al. (183). Unambiguous common names have been translated to their scientific name. In instances where only data of a general nature were available, they have been included.

ECOLOGY

The ecological data in the references varied both in quantity and precision. Unfortunately, for most species they were sparse and of a general nature. It was not unusual to find ecological and host data combined in a simple comment, such as 'on wood.' Such comments are not included because, with few exceptions, all the fungi herein occur on wood. Another commonly seen description of a habitat is 'on branches.' This is more helpful, but were the branches rotted on the ground or live on the tree? 'On trunk' is helpful but was the trunk living or dead, upright or fallen, and was the basidiome on the wood or bark, and was it on the top or, as typical, on the underside? Some species are specific to restricted niches, which, if known, can aid both collectors seeking specific species and identifiers in naming unnamed collections.

Ambiguous data are cited as they appeared in the original so the entries are sometimes uneven. We decided not to rewrite the ecological data for each species to make them more readable because we were sometimes uncertain over which interpretation was intended for statements we found ambiguous. To emphasize the fact that different comments in this section are from different authors, a semicolon is used to separate data and comments from the various papers, books, etc. listed in the references for that particular species.

CULTURE CHARACTERS

The culture characters are the features produced when the fungus is grown on agar media in the laboratory. When growing these fungi *in vitro*, most studies follow the protocol of Nobles (487). If known, the culture characters are abbreviated as a species code. The species code is a series of numbers in which each number represents a feature described in Nobles (487) or in subsequent supplements (37, 41, 101a, 354). The definitions for the code numbers are given in Table 5.

The principal work on this group of fungi in North America is that of Nakasone (474), where the culture characters of 277 species are described and illustrated. Stalpers (588) included nearly 200 corticioid Hymenomycetes in an annotated key, and many of these occur in North America. Stalpers developed a species code using nearly 100 characters, and we have translated his codes to the Nobles format only for those species not described by Nobles system. For some species, especially when no data were available for North America, the species code is based upon study of cultures from other localities.

REFERENCES

The numbers in this section correspond to the research papers and books cited in the bibliography, which is also titled References. They direct the user to the original reports. No attempt has been made to list all references on each species in the North American literature. Emphasis was on the taxonomic literature because most of the data on synonymy, distribution and ecology were derived from this. There is a large body of literature on plant pathology, forest pathology and wood products pathology. Access to it can be gained through the large compilations by Conners (116), Farr et al. (183), Ginns (242) and Hepting (274). These secondary sources were also used to expand our sections on distribution and host data.

THE GENERA AND SPECIES

ACANTHOPHYSIUM Cunn. 1963 Aleurodiscaceae

The genus *Acanthophysium* has been segregated from *Aleurodiscus*. Its characteristic features are smooth amyloid basidiospores, sulfocystidia and, in general, the presence of acanthophyses. Three North American species, currently placed in *Aleurodiscus*, have these features and are transferred, below, to *Acanthophysium*: *A. abietis*, *A. farlowii*, and *A. piceinum*.

Acanthophysium abietis (Jackson & Lemke) Ginns & Lefebvre, *comb. nov.*

Basionym *Aleurodiscus abietis* Jackson & Lemke, Canad. J. Bot. 42:225, 1964.

Distribution Canada: BC, NS, ON, PQ.
United States: ID, NY, OR, VT.

Hosts *Abies* sp., *A. balsamea, A. grandis, Tsuga mertensiana.*

Ecology Underside of dead branches and twigs on live trees.

Culture Characters Not known.

References 242, 249, 375.

See comment at beginning of genus treatment.

Acanthophysium bertii (Lloyd) Boidin 1986
Bull. Soc. Mycol. France 101:340.

Basionym *Aleurodiscus bertii* Lloyd, Mycol. Writ. 7 (Note 72):1288, 1924, proposed as a *nom. nov.* to replace *Aleurodiscus cremeus* Burt, Ann. Missouri Bot. Gard. 5:199, 1918, a later homonym of *Aleurodiscus cremeus* Pat. 1915.

Distribution United States: AZ, NM.

Hosts *Cercocarpus betuloides, C. breviflorus, Garrya wrightii, Juniperus monosperma, Quercus emoryi, Q. gambelii, Q. hypoleucoides, Q. reticulata.*

Ecology Decorticated wood; causes a white rot.

Culture Characters
2.3c.8.(15).32.36.38.46-47.53.54 (Ref. 474).

References 216, 375, 474.

Acanthophysium canadense (Skolko) Parm. 1967
Izv. Akad. Nauk Estonsk. SSR, Ser. Biol. 16:378.

Basionym *Aleurodiscus canadensis* Skolko, Canad. J. Res., C, 22:258, 1944.

Distribution Canada: ON, PQ.
United States: MI, MN, NM.

Hosts *Abies balsamea, Betula papyrifera, Picea* spp., *P. glauca, Prunus pensylvanica, Quercus borealis, Thuja occidentalis, Tsuga canadensis*, gymnosperm.

Ecology Lower surface of dead logs.

Culture Characters
2.3c.8.(15).32.36.38.47.51.(53).55.60a (Ref. 474).

References 116, 214, 242, 375, 474.

The records (242) on angiosperms should be reconfirmed. The reports from Arizona (197, 204, 216), are not included above because they were based on collection RLG 9318 which Lindsey (401) suggested was a specimen of *Acanthophysium mesaverdense*.

Acanthophysium cerussatum (Bres.) Boidin 1986
Bull. Soc. Mycol. France 101:340.

Basionym *Corticium cerussatum* Bres., Fungi Tridentini 2:37, 1892.

Synonym *Aleurodiscus cerussatus* (Bres.) Höhnel & Litsch. 1907.

Distribution Canada: AB, AB-NT, BC, MB, ON.
United States: AZ, CO, FL, ID, MN, NM, OR, SD, WA.

Hosts *Abies balsamea, A. concolor, A. lasiocarpa, Acer* sp., *Amelanchier* sp., *A. alnifolia, Cercocarpus breviflorus, Fraxinus velutina, Juniperus monosperma, J. osteosperma, Physocarpus* sp., *P. malvaceus, Pinus banksiana, Populus* sp., *P. tremuloides, Prunus serotina* var. *virens, Pseudotsuga menziesii, Quercus* sp., *Q. arizonica, Q. gambelii, Salix* sp., *Serenoa serrulata, Thuja plicata, Ulmus* sp., *U. americana*, angiosperm, gymnosperm.

Ecology Dead branches; slash; associated with a white rot.

Culture Characters
(1).2.3c.8.32.36.38.44-45.(51).54.55.59 (Ref. 474);
2a.3c.(24-25).32.36.38.(43-45).54.59.63 (Ref. 48).

References 116, 183, 211, 216, 242, 375, 398, 402, 474.

Acanthophysium dendroideum (Ginns) Boidin 1986
Bull. Soc. Mycol. France 101:340.

Basionym *Aleurodiscus dendroideus* Ginns, Canad. Field-Naturalist 96:131, 1982.

Distribution Canada: AB.

Host *Picea glauca.*

Ecology Emergent from under bark scales on branch, perhaps a dead branch on a live tree which is the preferred habitat for some similar species.

Culture Characters Not known.

Known only from the type collection.

Acanthophysium diffissum (Sacc.) Parm. 1967
Izv. Akad. Nauk Estonsk. SSR, Ser. Biol. 16:378.
Basionym *Peniophora diffissa* Sacc., Bull. Soc. Roy. Bot. Belgique 28:79, 1889.
Synonym *Aleurodiscus diffissus* (Sacc.) Burt 1931.
Distribution United States: AZ, CA, CO, OR.
Hosts *Arbutus menziesii, Arctostaphylos nevadensis, A. patula, Ceanothus greggii, Juglans major, Juniperus* sp., *Lycium andersonii, Quercus* sp.
Ecology Dead twigs and branches.
Culture Characters Not known.
References 120, 130, 183, 216, 375.

Acanthophysium farlowii (Burt) Ginns & Lefebvre, *comb. nov.*
Basionym *Aleurodiscus farlowii* Burt, Ann. Missouri Bot. Gard. 5:182, 1918.
Distribution Canada: BC, ON, PQ.
United States: ID, ME, NH, NY, PA.
Hosts *Abies* sp., *A. grandis, Pseudotsuga menziesii, Tsuga canadensis.*
Ecology Recently dead twigs on live trees.
Culture Characters Not known.
References 116, 183, 242, 375.

See comment at beginning of genus treatment.

Acanthophysium fennicum (Laurila) Parm. 1967
Izv. Akad. Nauk Estonsk. SSR, Ser. Biol. 16:378.
Basionym *Aleurodiscus fennicus* Laurila, Ann. Bot. Soc. Zool.-Bot. Fenn. "Vanamo" 10:11, 1939.
Distribution Canada: PQ.
United States: AZ, NH, NY, VT.
Hosts *Picea engelmannii, P. rubens.*
Ecology Branches and twigs.
Culture Characters Not known.
References 183, 216, 242, 375.

Acanthophysium fruticetorum (W.B. Cooke) Boidin 1986
Bull. Soc. Mycol. France 101:339.
Basionym *Aleurodiscus fruticetorum* W.B. Cooke, Mycologia 35:281, 1943.
Distribution United States: CA, CO.
Hosts *Arctostaphylos patula, Artemisia tridentata, Ceanothus velutinus.*
Ecology Twigs and stems; associated with a white rot.
Culture Characters Not known.
References 375, 402.

Acanthophysium lapponicum (Litsch.) Boidin 1986
Bull. Soc. Mycol. France 101:340.
Basionym *Aleurodiscus lapponicus* Litsch., Ann. Mycol. 42:11, 1944.
Distribution Canada: MB. United States: MO.
Hosts *Maclura pomifera, Populus* sp.
Ecology Not known.
Culture Characters Not known.
References 242, 375, 406.

Acanthophysium lividocoeruleum (Karsten) Boidin 1986
Bull. Soc. Mycol. France 101:340.
Basionym *Corticium lividocoeruleum* Karsten, Not. Sällsk. Fauna Fl. Fenn. Förh. 9, N.S. 6:370, 1868.
Synonyms *Aleurodiscus lividocoeruleus* (Karsten) Lemke 1964, *Gloeocystidiellum lividocoeruleum* (Karsten) Donk 1956.
Distribution Canada: AB, AB-NT, BC, MB, NF, NS, ON, PQ, YT. United States: AK, AZ, CO, FL, GA, IA, ID, IN, MD, MI, MN, MS, MT, NC, NH, NM, NY, OR, UT, VT, WA, WI.
Hosts *Abies balsamea, A. concolor, A. grandis, A. lasiocarpa, Acer glabrum, Cupressus arizonica, Juniperus deppeana, J. horizontalis, Larix* sp., *L. occidentalis, Picea* sp., *P. engelmannii, P. glauca, P. mariana, Pinus* sp., *P. aristata* var. *longaeva, P. contorta, P. ponderosa, Populus* sp., *P. tremuloides, Pseudotsuga menziesii, Thuja occidentalis, T. plicata, Tsuga heterophylla*, angiosperm, gymnosperm.
Ecology Primarily on dry, decorticated wood; gymnosperm test flooring; utility poles; associated with a white rot.
Culture Characters
(1).2.3.7.15.32.36.38.44-46.53.55.59 (Ref. 179);
(1).2.3c.(8).(15).26.32.36.39.44-45.(47).50.53.55.59 (Ref. 474);
2a.4.15.32.36.(39).44.55.59.65 (Ref. 48).
References 93, 116, 177, 179, 183, 208, 214, 216, 242, 249, 334, 335, 375, 391, 402, 407, 447, 474, 567 under *Corticium abeuns.*

Acanthophysium macrocystidiatum (Lemke) Boidin 1986
Bull. Soc. Mycol. France 101:339.
Basionym *Aleurodiscus macrocystidiatus* Lemke, Canad. J. Bot. 42:255, 1964.
Distribution Canada: BC. United States: CA.
Hosts *Arbutus menziesii, Arctostaphylos* sp.
Ecology Lower surface of branches which have recently died.
Culture Characters Not known.
References 242, 375.

Acanthophysium mesaverdense (Lindsey) Boidin 1990
Cryptogamie Mycol. 11:180.
Basionym *Aleurodiscus mesaverdensis* Lindsey, Mycotaxon 30:433, 1987.
Distribution United States: AZ, CO.
Hosts *Pinus edulis, P. ponderosa.*
Ecology Bark on underside of live branches and trunks, especially common in drip lines, associated with a white rot.
Culture Characters
2.3.8.25.32.36.(37).38.47.50.55 (Ref. 401).
References 401, 402.

Acanthophysium piceinum (E.M. Lyon & Lemke) Ginns & Lefebvre, *comb. nov.*
Basionym *Aleurodiscus piceinus* E.M. Lyon & Lemke, Canad. J. Bot. 42:264, 1964.

Distribution Canada: ON, PQ. United States: NH, VT.
Hosts *Picea* sp., *P. rubens*.
Ecology Lower surface of dead twigs; stems of standing trees.
Culture Characters Not known.
References 242, 375.

See comment at beginning of genus treatment.

Acanthophysium succineum (Bres.) Boidin 1986
Bull. Soc. Mycol. France 101:340.
Basionym *Aleurodiscus succineus* Bres., Mycologia 17:71, 1925.
Distribution United States: CA, OR.
Hosts *Arbutus menziesii*.
Ecology Dead, decorticated wood.
Culture Characters Not known.
Reference 375.

Acanthophysium weirii (Burt) Nakasone 1990
Mycologia Memoir 15:21.
Basionym *Aleurodiscus weirii* Burt, Ann. Missouri Bot. Gard. 5:203, 1918.
Distribution Canada: BC.
United States: AK, AZ, ID, MI, MT, OR.
Hosts *Abies concolor, A. grandis, Larix* sp., *L. laricina, L. occidentalis, Picea* sp., *P. sitchensis, Pseudotsuga menziesii, Thuja plicata, Tsuga* sp., *T. heterophylla*.
Ecology Bark.
Culture Characters
(1).2.3c.(15).32.36.38.46-47.(50).55 (Ref. 474).
References 97, 183, 216, 242, 375, 474.

ALEUROBOTRYS Boidin 1986 Aleurodiscaceae

Aleurobotrys botryosus (Burt) Boidin, Lanquetin & Gilles 1986
Bull. Soc. Mycol. France 101:355.
Basionym *Aleurodiscus botryosus* Burt, Ann. Missouri Bot. Gard. 5:198, 1918.
Distribution Canada: ON, PQ. United States: AL, CA, CT, FL, GA, LA, MA, MD, NC, PA, SC, TN.
Hosts *Acer spicatum, Juniperus virginiana, Lonicera japonica, Myrica cerifera, Ostrya virginiana, Quercus nigra, Rubus* sp., *R. argutus, R. occidentalis, R. spectabilis, Symphoricarpos* sp., *Syringa* sp., *Thuja occidentalis, Vitis* sp.
Ecology Dead stems; bark of live stems; associated with a white rot.
Culture Characters
2.6.15.26.32.36.40.45-46.(50).(53).54.55.(57) (Ref. 474);
2a.3r.7.32.36.38.44.(54-55).57.66 (Ref. 48).
References 108, 183, 201, 242, 375, 474.

ALEUROCYSTIDIELLUM Lemke 1964 Aleurodiscaceae

Aleurocystidiellum disciformis (DC.:Fr.) Boidin, P. Terra & Lanquetin 1968
Bull. Soc. Mycol. France 84:63.
Basionym *Thelephora disciformis* DC., Flora France 5:31, 1815.
Synonym *Aleurodiscus disciformis* (DC.) Pat. 1894.
Distribution United States: AZ, CA, OR, WA.
Hosts *Quercus* sp., *Q. arizonica, Q. chrysolepis, Q. oblongifolia*.
Ecology Typically on bark.
Culture Characters
2.3c.(8).32.36.38.47.54.60.63 (Ref. 48).
References 133, 216, 375.

Aleurocystidiellum subcruentatum (Berk. & Curtis) Lemke 1964
Canad. J. Bot. 42:277.
Basionym *Stereum subcruentatum* Berk. & Curtis, Proc. Amer. Acad. Arts. 4:123, 1858.
Synonym *Aleurodiscus subcruentatus* (Berk. & Curtis) Burt 1920.
Distribution Canada: BC, NS, ON, PQ.
United States: AK, CA, NH, NY, OR, WA.
Hosts *Abies grandis, Picea* sp., *P. glauca, P. rubens, P. sitchensis, Pseudotsuga menziesii*.
Ecology Bark of live trees; sometimes on bark lesions on suppressed trees.
Culture Characters
2.3c.(8).32.36.38.47.55.60.63 (Ref. 48).
References 93, 183, 242, 249, 274, 375, 391, 481.

ALEURODISCUS Schröter 1888 Aleurodiscaceae

Aleurodiscus amorphus (Pers.:Fr.) Schröter 1888
In F. Cohn, Kryptogamen-Flora Schlesien 3 (1):429.
Basionym *Peziza amorpha* Pers., Synop. Meth. Fung., 657, 1801.
Synonym *Nodularia balsamicola* Peck 1872, fide Ginns (235) and Lemke (375).
Distribution Canada: AB, AB-NT, BC, MB, NB, NF, NS, ON, PE, PQ, SK, YT. United States: AZ, CA, CO, CT, ID, MA, ME, MI, MN, MO, MT, NC-TN, NH, NY, OR, TN, VT, WA, WI, WV.
Hosts *Abies* sp., *A. alba, A. amabilis, A. balsamea, A. concolor, A. fraseri, A. grandis, A. lasiocarpa, A. lasiocarpa* var. *arizonica, A. magnifica, A. procera, Larix* sp., *L. laricina, L. occidentalis, Picea* sp., *P. glauca, P. engelmannii, P. mariana, P. pungens, P. rubens, P. sitchensis, Pinus muricata, P. strobus, Pseudotsuga menziesii, Thuja plicata, Tsuga canadensis, T. mertensiana*, gymnosperm.
Ecology Apparently an early invader of dead or nearly dead branches in the lower crown, primarily on *Abies* spp.; basidiomes produced on bark of recently dead

limbs and stems; parasitic, causing cankers on branches of suppressed trees, main stems and often killing suppressed saplings (274); associated with a white rot.
Culture Characters
(1).2a.6.26.32.36.38.47.55 (Ref. 474); 2a.6.26.32.36.39.47.(53).55.(66) (Ref. 48).
References 120, 183, 216, 235, 242, 249, 327, 332, 375, 391, 402, 447, 474, 517, 532.

From the Pacific Coast to the eastern foothills of the Rocky Mountains, the very similar, *Aleurodiscus grantii* is more common (235).

Aleurodiscus aurantius (Pers.:Fr.) Schröter 1888
In F. Cohn, Kryptogamen-Flora Schlesien 3 (1):429.
Basionym *Corticium aurantium* Pers., Tent. Disp. Meth. Fung., 111, 1794.
Distribution Canada: BC, ON. United States: CA.
Hosts *Arctostaphylos columbiana* var. *tracyi, Lithocarpus densiflora, Rhododendron occidentale, Ribes* sp., *R. sanguineum, Rubus ursinus, Tsuga* sp., *Umbellularia californica, Vaccinium* sp.
Ecology Not known.
Culture Characters
2a.6.15.32.36.(39).47.54.57.66 (Ref. 48).
References 242, 375.

Aleurodiscus croceus Pat. 1893
Bull. Soc. Mycol. France 9:133.
Distribution United States: AZ.
Host *Quercus reticulata.*
Ecology Not known.
Culture Characters Not known.
Reference 216.

Aleurodiscus grantii Lloyd 1920
Mycol. Writ. 6 (Note 62):927.
Distribution Canada: AB, BC, YT.
United States: CA, ID, MT, OR, WA.
Hosts *Abies amabilis, A. concolor, A. grandis, A. lasiocarpa, A. magnifica, A. pinsapo, Picea* sp., *P. sitchensis, Pinus muricata, Pseudotsuga menziesii, Thuja plicata, Tsuga heterophylla, T. mertensiana.*
Ecology Dead branches in live trees; trunks of dead saplings; fallen stems.
Culture Characters Not known.
References 130, 183, 235.

Previously included under *Aleurodiscus amorphus*, but Ginns (235) found that *A. grantii* differed in several respects.

Aleurodiscus laurentianus Jackson & Lemke 1964
Canad. J. Bot. 42:251.
Distribution Canada: PQ. United States: NY.
Hosts *Abies balsamea, Picea mariana.*
Ecology Dead twigs on live trees.
Culture Characters Not known.
References 242, 375.

Aleurodiscus mirabilis (Berk. & Curtis) Höhnel 1909
Sitzungsber. Kaiserl. Akad. Wiss., Math.-Naturwiss. Kl., Abt. 1, 118:818.
Basionym *Psilopeziza mirabilis* Berk. & Curtis, J. Linn. Soc., Bot. 10:364, 1868.
Synonym *Aleurodiscus apiculatus* Burt 1918, fide Lemke (375).
Distribution United States: GA, ID, MT, NC, NC-TN.
Hosts *Betula alleghaniensis, Larix occidentalis, Rhododendron* sp.
Ecology Bark; dead branches.
Culture Characters
2.3c.26.31g.32.36.38.47.(50).(53).54.(57) (Ref. 474); 2a.3c.32.36.38.47.54.57.62 (Ref. 46).
References 110, 183, 327, 375, 474, 517.

The reports (183) from Idaho and Montana were on western larch, quite different from the eastern angiosperm hosts.

Aleurodiscus oakesii (Berk. & Curtis) Höhnel & Litsch. 1907
Sitzungsber. Kaiserl. Akad. Wiss., Math.-Naturwiss. Kl., Abt. 1, 116:802.
Basionym *Corticium oakesii* Berk. & Curtis, Grevillea 1:166, 1873.
Distribution Canada: NB, NS, ON, PQ. United States: AL, AZ, GA, IA, MA, MI, MN, MO, MS, NC, NE, NH, NJ, NY, NC, OH, PA, TN, VA, VT, WI, WV.
Hosts *Acer rubrum, A. saccharum, Betula pumila, Carpinus caroliniana, Carya* sp., *Fraxinus* sp., *Ostrya* sp., *O. virginiana, Quercus* sp., *Q. alba, Q. arizonica, Q. hypoleucoides, Q. macrocarpa, Q. reticulata, Q. rubra, Q. stellata, Q. toumeyi, Salix* sp., *S. nigra, Tilia americana, Ulmus* sp., *U. americana.*
Ecology Bark of live trees.
Culture Characters
2.6.(8).(26).(27).32.36.38.46-47.(53).54.(57) (Ref. 474);
2.6.8.32.36.38.45.54.(57).(66?) (Ref. 48).
References 183, 209, 216, 242, 249, 375, 474, 516.

Aleurodiscus occidentalis Ginns 1990
Mycologia 82:753.
Distribution Canada: BC. United States: ID, WA.
Host *Thuja plicata.*
Ecology Typically on bark on the lower side of dead branches (10-20 mm diam) and twigs in the lower crown of live trees, occasionally collected on dead branches on ground or stems of small dead trees.
Culture Characters Not known.

Known only from the 11 collections cited in the original description.

Aleurodiscus penicillatus Burt 1918
Ann. Missouri Bot. Gard. 5:20.
Distribution Canada: BC, ON, PQ.
United States: CA, ID, NH, NY, NC, OR, VT, WA.

Hosts *Abies grandis, Picea* sp., *P. glauca, P. rubens, P. sitchensis, Pinus* spp., *P. contorta, P. monticola, P. ponderosa, P. sylvestris, Pseudotsuga menziesii, Thuja plicata, Tsuga* spp., *T. canadensis, T. heterophylla, T. mertensiana*.
Ecology Dead branches and twigs.
Culture Characters
2a.3c.9.26.32.36.(38).47.55.58.63 (Ref. 48).
References 61, 242, 375.

Aleurodiscus spiniger D.P. Rogers & Lemke 1964
Canad. J. Bot. 42:265.
Distribution Canada: BC. United States: AZ, ID, OR.
Hosts *Larix occidentalis, Pseudotsuga menziesii, Tsuga heterophylla*.
Ecology Bark; underside of branches.
Culture Characters Not known.
References 204, 242, 375.

Aleurodiscus thujae Ginns 1990
Mycologia 82:754.
Distribution Canada: ON, PQ. United States: MI.
Host *Thuja occidentalis*.
Ecology Typically on bark on the lower side of dead branches (10-17 mm diam) in the lower crown of live trees, also on dead sapling (30 mm diam).
Culture Characters Not known.

Known only from the 10 collections cited in the original description.

Aleurodiscus utahensis Lindsey & Gilbn. 1983
Mycotaxon 18:544.
Distribution United States: UT.
Host *Pinus aristata* var. *longaeva*.
Ecology Not known.
Culture Characters Not known.

Known only from the type collection.

AMETHICIUM Hjort. 1983 — Meruliaceae

Amethicium chrysocreas (Berk. & Curtis) Wu 1990
Acta Bot. Fenn. 142:18.
Basionym *Corticium chrysocreas* Berk. & Curtis, Grevillea 1:178, 1873.
Synonym *Phlebia chrysocrea* (Berk. & Curtis) Burdsall 1975.
Distribution United States: AL, FL, GA, LA, MD, MO, MS, SC, NC, PA, SC, TN, VA, WI.
Hosts *Betula* sp., *Castanea dentata, Fagus grandifolia, Gleditsia triacanthos, Liquidambar styraciflua, Liriodendron tulipifera, Prunus* sp., *P. serotina, Quercus* sp., *Q. alba, Q. coccinea, Q. lyrata, Q. nigra, Q. palustris, Q. phellos, Q. prinus, Q. rubra, Q. rubra* var. *borealis, Q. velutina, Robinia pseudoacacia, Sassafras albidum*, angiosperm.
Ecology Logs; associated with a white rot.
Culture Characters
2.4.(8).14.21.31e.(34).36.(38).40.42.48.(50).53.54.60 (Ref. 474);
2.4.7.34.36.(37).38.(40).42.(48).50.54.60 (composed from the description in 415).
References 93, 183, 415, 474.

Amethicium leoninum (Burdsall & Nakasone) Wu 1990
Acta Bot. Fenn. 142:16.
Basionym *Hyphoderma leoninum* Burdsall & Nakasone, Mycotaxon 17:256, 1983.
Distribution United States: NC, OH, TN.
Hosts *Acer* sp., angiosperm.
Ecology Bark of rotted log; associated with a white rot.
Culture Characters
1.2.3.15.24.32.36.38.45-46.(53).54 (Ref. 75);
(1).2.3c.13.24.32.36.38.45.46.(50).54 (Ref. 474).
References 75, 474.

AMPHINEMA Karsten 1892 — Chaetoporellaceae

Amphinema arachispora Burdsall & Nakasone 1981
Mycologia 73:455.
Distribution United States: GA.
Host *Cornus florida*.
Ecology Not known.
Culture Characters Not known.

Known only from the type collection.

Amphinema byssoides (Pers.:Fr.) Eriksson 1958
Symb. Bot. Upsal. 16 (1):112.
Basionym *Thelephora byssoides* Pers., Synop. Meth. Fung., 577, 1801.
Synonym *Peniophora byssoides* (Pers.) Bres. 1898.
Distribution Canada: AB, BC, MB, NTM, NS, ON, PQ, YT.
United States: AK, AZ, CA, CO, IA, IL, LA, MA, MN, NM, NY, OH, OR, TX, UT, VT, WA.
Hosts *Abies balsamea, A. concolor, A. lasiocarpa, Acer glabrum, Betula* sp., *B. papyrifera, Picea* sp., *P. engelmannii, P. glauca, P. pungens, Pinus aristata* var. *longaeva, P. banksiana, P. contorta* var. *contorta, P. engelmannii, P. ponderosa, P. sylvestris, Populus* sp., *P. balsamifera, P. tremuloides, P. trichocarpa, Pseudotsuga menziesii, Quercus gambelii, Selaginella apoda, Thuja plicata, Tsuga* sp., *T. heterophylla*, angiosperm, gymnosperm.
Ecology Mycorrhizal with, at least, *Picea glauca* (140); basidiomes produced on rotting wood; litter and soil; damp soil and roots; also reported to be associated with a white rot.
Culture Characters
2.3.16.32.36.38.45-46.55 (Ref. 588).
References 116, 127, 130, 183, 208, 214, 216, 242, 249, 267, 390, 391, 399, 402, 406, 407, 464, 584, 615.

Amphinema tomentellum (Bres.) M. Christiansen 1960
Dansk Bot. Ark. 19:229.
Basionym *Kneiffia tomentella* Bres., Ann. Mycol. 1:103, 1903.
Distribution United States: AZ, NM.
Hosts *Quercus gambelii, Picea engelmannii, Populus* sp.
Ecology Not known.
Culture Characters Not known.
References 57, 406.

AMYLOATHELIA Hjort. & Ryv. 1979 Aleurodiscaceae

Amyloathelia amylaceus (Bourd. & Galzin) Hjort. & Ryv. 1979
Mycotaxon 10:202.
Basionym *Corticium amylaceum* Bourd. & Galzin, Bull. Soc. Mycol. France 27:259, 1911.
Synonyms *Aleurodiscus amylaceus* (Bourd. & Galzin) D.P. Rogers & Jackson 1943, *Corticium ermineum* Burt 1926, *Corticium sociatum* Burt 1926. Both fide Lemke (375).
Distribution Canada: BC, MB, ON, PQ.
United States: ID, MI, MT, OR, VT.
Hosts *Picea* sp., *Thuja occidentalis, T. plicata.*
Ecology Bark; decorticated, very rotten wood of logs.
Culture Characters Not known.
References 93, 116, 296, 334, 375.

AMYLOBASIDIUM Ginns 1988 Hyphodermataceae

Amylobasidium tsugae Ginns 1988
Mycologia 80:63.
Distribution Canada: BC. United States: CA, OR.
Host *Tsuga mertensiana.*
Ecology Decorticated wood or in bark crevices of logs and snags; all collections are from the cordillera.
Culture Characters Not known.

Known from the original description and collected in 1991 by Ginns and Burdsall in the area of Whistler, British Columbia.

AMYLOCORTICIUM Pouzar 1959 Atheliaceae

Amylocorticium canadense (Burt) Eriksson & Weresub 1974
Fungi Canadenses No. 45, National Mycological Herbarium, Agriculture Canada, Ottawa.
Basionym *Corticium canadense* Burt, Ann. Missouri Bot. Gard. 13:290, 1926.
Distribution Canada: ON, PQ.
United States: CT, MA, NH, NY.
Hosts *Pinus* sp., *P. strobus, Tsuga canadensis.*
Ecology Decaying wood of logs; associated with a brown cubical rot of dead gymnosperm.
Culture Characters Not known.
References 93, 213, 242, 277, 393, 636.

Amylocorticium cebennense (Bourd.) Pouzar 1959
Ceská Mykol. 13:11.
Basionym *Corticium cebennense* Bourd., Rev. Sci. Bourbonnais Centr. France 23:7, 1910.
Distribution Canada: BC, NS, ON.
United States: AR, CA, ID, MA.
Hosts *Abies balsamea, Arbutus menziesii, Larix occidentalis, Pinus* spp., *P. monticola, P. ponderosa, Thuja plicata*, gymnosperm.
Ecology Associated with a brown cubical rot of dead wood.
Culture Characters Not known.
References 213, 242, 249, 637.

Amylocorticium cumminsii Gilbn. & Lindsey 1989
Mem. New York Bot. Gard. 49:142.
Distribution United States: CO.
Host *Pinus contorta.*
Ecology Floor beams in a house; associated with a brown cubical rot.
Culture Characters Not known.

Known only from the type collection.

Amylocorticium suaveolens Parm. 1968
Consp. Syst. Cort., 197.
Distribution United States: OR.
Host *Pseudotsuga menziesii.*
Ecology Associated with a brown cubical rot.
Culture Characters Not known.
Reference 213.

Amylocorticium subincarnatum (Peck) Pouzar 1959
Ceská Mykol. 13:11.
Basionym *Corticium subincarnatum* Peck, Annual Rep. New York State Mus. 42:124, 1889.
Synonym *Peniophora subincarnata* (Peck) Litsch. 1939.
Distribution Canada: AB, BC, MB, ON, PQ.
United States: AR, AZ, ID, MN, MT, NY, WA.
Hosts *Acer macrophyllum, Larix occidentalis, Picea* spp., *P. engelmannii, P. glauca, P. sitchensis, Pinus* sp., *P. contorta, P. monticola, Pseudotsuga menziesii*, gymnosperm.
Ecology Log; associated with a brown cubical rot.
Culture Characters Not known.
References 92 as *Peniophora subsulphurea*, fide Rogers and Jackson (567), 116, 183, 213, 214, 216, 242, 447, 584, 638.

Amylocorticium subsulphureum (Karsten) Pouzar 1959
Ceská Mykol. 13:11.
Basionym *Corticium subsulphureum* Karsten, Meddeland. Soc. Fauna Fl. Fenn. 6:12, 1881.
Synonym *Peniophora subsulphurea* (Karsten) Höhnel & Litsch. 1906.
Distribution Canada: AB, BC, MB, NS, PQ. United States: AR, AZ, ID, MD, MN, MT, NM, NY.
Hosts *Abies* sp., *Larix occidentalis, Picea* sp., *P. engelmannii, Pinus* sp., *P. ponderosa, P. strobus, P. virginiana*, gymnosperm.

Ecology Decaying decorticated wood; log; associated with a brown cubical rot.
Culture Characters Not known.
References 92, 183, 197, 213, 242, 249, 447, 584.

In Burt (92) only the Idaho collection (Weir 14) is *P. subsulphurea*, fide Rogers and Jackson (567).

AMYLOSTEREUM Boidin 1958 Stereaceae

Amylostereum areolatum (Fr.) Boidin 1958
Rev. Mycol. (Paris) 23:345.
Basionym *Thelephora areolatum* Fr., Elenchus 1:190, 1828.
Distribution North America, specific localities not given.
Hosts Not known for North America.
Ecology Not known for North America.
Culture Characters
2.3c.13.15ap.35.36.39.43-44.55.60.61 (Ref. 42).
Reference 311.

Similar to *A. chailletii* and previously confused with it (311). Since *A. areolatum* is known (42, 311) in Europe to be a symbiont with woodwasps (*Sirex* spp.), perhaps some of the North American reports of *A. chailletii* were, in fact, *A. areolatum*.

Amylostereum chailletii (Pers.:Fr.) Boidin 1958
Rev. Mycol. (Paris) 23:345.
Basionym *Thelephora chailletii* Pers., Mycol. Eur. 1:125, 1822.
Synonyms *Stereum chailletii* (Pers.) Fr. 1838, *Peniophora atkinsoni* Ellis & Ev. 1894, fide Lentz (379).
Distribution Canada: AB, AB-NT, BC, MB, NB, NF, NS, ON, PQ. United States: AK, AZ, CA, CO, CT, IA, ID, ME, MI, MT, NC, NH, NJ, NM, NY, OR, PA, TN, VA, VT, WA, WI.
Hosts *Abies* spp., *A. alba, A. amabilis, A. balsamea, A. concolor, A. fraseri, A. grandis, A. lasiocarpa, Acer macrophyllum, Chamaecyparis thyoides, Cupressus* sp., *Juniperus virginiana, Larix* sp., *L. decidua, L. europa, L. laricina, L. occidentalis, Picea* sp., *P. engelmannii, P. glauca, P. mariana, P. rubens, P. sitchensis, Pinus* spp., *P. contorta, P. monticola, P. sylvestris, Pseudotsuga* sp., *P. menziesii, Sequoia sempervirens, Thuja* sp., *T. occidentalis, T. plicata, Tsuga* spp., *T. canadensis, T. heterophylla*, gymnosperm.
Ecology Isolated from root rot in balsam fir saplings; common sapwood rotting fungus in dead balsam fir; transmitted and inoculated into dead trees by woodwasps; basidiomes produced on moss-covered bark and wood; bark; butt; stump; tree base; up-rooted bottom; cut wood; logs; associated with a white rot.
Culture Characters
2.3c.(11).(13).(15p).21.(22).32.37.39.44-45.(50).55.60 (Ref. 474);
2.3c.13.15ap.32.36.39.43-44.55.60.61 (Ref. 42);
2.3c.8.(13).32.37.39.(43).44.(50).55.60.61 (Ref. 104);
2.3.11.13.36.38.44-46.55 (Ref. 29).
References 31, 89, 104, 116, 183, 216, 242, 249, 327, 379, 402, 448, 474, 519, 594, 595, 645.

Amylostereum laevigatum (Fr.) Boidin 1958
Rev. Mycol. (Paris) 23:345.
Basionym *Thelephora laevigata* Fr., Elenchus 1:224, 1828.
Synonym *Peniophora laevigata* (Fr.) Massee 1889.
Distribution Canada: ON. United States: ME, NY.
Hosts *Juniperus* sp., *Thuja occidentalis*, gymnosperm.
Ecology Bark.
Culture Characters
2.3.22.34.37.39.43.55.60 (Ref. 263);
2.3c.15ap.32.36.38.43.55.60.61 (Ref. 42).
References 92, 104, 183, 584.

ASTERODON Pat. 1894 Hymenochaetaceae

Asterodon ferruginosus Pat. 1894
Bull. Soc. Mycol. France 10:130.
Synonyms *Asterostroma ochrostroma* Burt 1924, fide Rogers and Jackson (567), *Hydnochaete setigera* Peck 1897, fide Gilbertson (189), and Rogers and Jackson (567), *Asterodon setigera* (Peck) Peck 1901.
Distribution Canada: BC, NB, ON, PQ, YT. United States: AK, CT, ID, ME, MI, MT, NH, NY, WA.
Hosts *Abies balsamea, A. lasiocarpa, Betula* sp., *B. alleghaniensis, Picea* sp., *P. engelmannii, P. glauca, P. rubens, Pinus monticola, P. strobus, Pseudotsuga menziesii, Thuja plicata, Tsuga* sp., *T. heterophylla*, gymnosperm.
Ecology Isolated from root rot in saplings of balsam fir; basidiomes produced on bark and decorticated wood; decaying bark; associated with a white rot.
Culture Characters Not known.
References 91, 183, 189, 242, 334, 448, 511, 532, 645.

ASTEROSTROMA Massee 1889 Hymenochaetaceae

Asterostroma andinum Pat. 1893
Bull. Soc. Mycol. France 9:133.
Synonyms *Asterostroma bicolor* Ellis & Ev. 1894, *Asterostroma spiniferum* Burt 1924, *Asterostroma gracile* Burt 1924. All fide Rogers and Jackson (567).
Distribution Canada: BC, NS. United States: AL, DE, FL, IL, KY, LA, MD, MI, MN, NM, NY, SC.
Hosts *Acer* sp., *A. saccharum, Picea* sp., *Thuja plicata*, angiosperm, gymnosperm.
Ecology Rotten wood; associated with a white rot.
Culture Characters
2.6.13.15p.16.29.32.36.38.47.54.55 (Ref. 474).
References 91, 208, 242, 249, 448, 474, 619.

Asterostroma cervicolor (Berk. & Curtis) Massee 1889
J. Linn. Soc., Bot 25:155.
Basionym *Corticium cervicolor* Berk. & Curtis, Grevillea 1:179, 1873.
Distribution Canada: NS, PQ.
United States: AL, AZ, CA, DC, FL, GA, IA, ID, LA, MA, MD, NC, NH, NY, OH, PA, SC, VA, WA
Hosts *Abies balsamea, A. concolor, Liquidambar styraciflua, Picea* sp., *P. glauca, Pinus palustris, P. resinosa, Planera* sp., *Quercus* sp., *Q. arizonica, Q. falcata, Q. nigra*, angiosperm.
Ecology Decaying wood; earth; outside of a flower pot; associated with a white rot.
Culture Characters
2.6.13.15p.(16).29.31c.32.36.38.(39).46-47.54.55 (Ref. 474).
References 91, 108, 183, 204, 216, 242, 249, 448, 474, 619.

Asterostroma laxum Bres. 1920
In Bourdot and Galzin, Bull. Soc. Mycol. France 36:46.
Distribution United States: ID, MD, NC.
Hosts *Abies grandis, Fagus* sp., *Quercus* sp.
Ecology Associated with a white rot.
Culture Characters
2.6.13.29.31c.32.36.38.47.54 (Ref. 474).
References 183, 208, 474.

Asterostroma medium Bres. 1920
Ann. Mycol. 18:49.
Distribution United States: NY.
Hosts Not known.
Ecology Not known.
Culture Characters Not known.
Reference 183.

Based upon a specimen in BPI, which may be misidentified. Hallenberg (258) listed this name as a synonym of *A. cervicolor.*

Asterostroma muscicola (Berk. & Curtis) Massee 1889
J. Linn. Soc., Bot. 25:155.
Basionym *Hymenochaete muscicola* Berk. & Curtis, J. Linn. Soc., Bot. 10: 334, 1868.
Distribution United States: AL, AR, AZ, FL, LA, WV.
Hosts *Fouquieria splendens, Juniperus virginiana, Pseudotsuga menziesii, Quercus* sp., angiosperm.
Ecology Bark of live trees; rotting wood; dead branches of trees covered with moss; dead standing tree; associated with a white rot.
Culture Characters
2.6.17.26.32.37.38.47.50.54 (Ref. 477);
2.6.13.17.26.29.32.36.38.47.54 (Ref. 474).
References 91, 202, 216, 474, 477, 619.

Nakasone (474) inadvertently omitted number 17 for setae from the species code. We have inserted it above.

Asterostroma ochroleucum Bres. 1913
In C. Torrend, Brotéria, Sér. Bot. 11:82.
Distribution United States: NM, TN.
Hosts *Abies concolor, Cornus florida.*
Ecology Not known.
Culture Characters
2.6.(13).(15p).29.32.36.39.46.54 (Ref. 474).
References 208, 474.

Hallenberg (258) listed this name as a synonym of *Asterostroma cervicolor.*

ATHELIA Pers. 1822 — Atheliaceae

Athelia alnicola (Bourd. & Galzin) Jülich 1972
Beih. Willdenowia 7:47.
Basionym *Corticium centrifugum* subsp. *alnicola* Bourd. & Galzin, Hym. France, 198, 1928.
Distribution United States: MA.
Hosts Not known.
Ecology Not known.
Culture Characters Not known.
Reference 312.

Athelia alutacea Jülich 1972
Beih. Willdenowia 7:51.
Distribution United States: NJ.
Hosts Gymnosperm.
Ecology Rotten wood and bark.
Culture Characters Not known.
Reference 312.

Athelia arachnoidea var. ***arachnoidea*** (Berk.) Jülich 1972
Beih. Willdenowia 7:53.
Basionym *Corticium arachnoideum* Berk., Ann. Mag. Nat. Hist., Ser. I, 13:345, 1844.
Synonym *Corticium bisporum* (Schröter) Höhnel & Litsch. 1906, fide Jülich (312).
Distribution Canada: MB, NF, ON, PQ.
United States: AZ, CA, GA, ID, IL, LA, MA, MD, MS, MT, NC, NJ, NY, OR, PA, SC, VT, WA.
Hosts *Dasylirion wheeleri, Heteromeles* sp., *Pinus contorta* var. *latifolia, P. ponderosa, Populus* sp., *Prosopis juliflora, Rosa nutkana, Salix* sp., angiosperm, gymnosperm, herbs.
Ecology Wood and bark of logs; fallen branches; decaying wood; dead leaves and twigs; humus.
Culture Characters
(2).5.24.25.(26).32.36.(39).43-45.(54).(55).(56) (Ref. 588).
References 93, 96, 108, 116, 133, 216, 312, 465.

Athelia arachnoidea var. ***leptospora*** Jülich 1972
Beih. Willdenowia 7:60.
Distribution Canada: ON.
Hosts Angiosperm.
Ecology Fallen leaves.
Culture Characters Not known.
Reference 312.

Athelia bombacina Pers. 1822
Mycol. Eur. 1:85.
Distribution Canada: MB, ON, PQ.
United States: CA, CO, WI.
Hosts *Malus pumili, Picea engelmannii, Pinus edulis, Populus tremuloides, Quercus gambelii*, angiosperm, gymnosperm.
Ecology Apple leaf litter; rotting wood; associated with a white rot.
Culture Characters
1.3.7.32.36.38.42.48.50.56.60 (Ref. 653).
References 130, 183, 312, 384, 402, 653.

The fungus from apple leaf litter (653) is not distinct from *A. bombacina*, fide Burdsall (in litt. 1989). It inhibited formation of pseudothecia of *Venturia inaequalis* (Cooke) Winter and perhaps can be exploited as a biocontrol agent (652).

Athelia coprophila (Wakef.) Jülich 1972
Beih. Willdenowia 7:66.
Basionym *Corticium coprophilum* Wakef., Trans. Brit. Mycol. Soc. 6:480, 1916.
Distribution Canada: ON.
United States: AZ, CO, MA, MN.
Hosts *Acer* sp., *Populus tremuloides, Prosopis juliflora, Quercus gambelii.*
Ecology Associated with a white rot.
Culture Characters Not known.
References 207, 214, 312, 399, 402.

Athelia cystidiolophora Parm. 1967
Izv. Akad. Nauk Estonsk. SSR, Ser. Biol. 16:380.
Distribution Canada: BC.
Hosts Not known.
Ecology Not known.
Culture Characters Not known.
Reference 312.

Athelia decipiens (Höhnel & Litsch.) Eriksson 1958
Symb. Bot. Upsal. 16 (1):86.
Basionym *Corticium decipiens* Höhnel & Litsch., Sitzungsber. Kaiserl. Akad. Wiss., Math.-Naturwiss. Kl., Abt. 1, 117:1116, 1908.
Synonym *Corticium consimile* Bres., fide L.K. Weresub (in litt., note with type Weir 16808 at BPI).
Distribution Canada: AB, AB-NT, BC, MB, NS, ON, PQ. United States: AZ, CA, CO, ID, IL, ME, NC-TN, NH, NJ, NM, NY, PA, UT, WA.
Hosts *Abies concolor, A. fraseri, A. lasiocarpa* var. *arizonica, Alnus oblongifolia, Artemisia tridentata, Carnegiea gigantea, Fouquieria splendens, Larix occidentalis, Picea* sp., *P. engelmannii, P. rubens, Pinus banksiana, P. contorta* var. *contorta, P. ponderosa, Thuja occidentalis*, angiosperm, gymnosperm.
Ecology Perhaps psychrophilic; log; stump; associated with a white rot.
Culture Characters
1.5.7.16.21.32.36.40.41-42.50.54.55 (Ref. 477).
References 53, 183, 196, 208, 216, 242, 249, 312, 327, 402, 477.

The type of *Corticium consimile* is very similar to *C. decipiens* as determined by Litschauer, fide Rogers and Jackson (567).

Athelia epiphylla Pers.:Fr. 1822
Mycol. Eur. 1:84.
Distribution Canada: AB, BC, NS, NTM, ON, PQ.
United States: AZ, CA, CO, ID, IL, MA, MT, NY, OH, UT.
Hosts *Abies lasiocarpa, Alnus oblongifolia, Betula papyrifera, Cupressus arizonica, Cyanophyceae, Eucalyptus* sp., *Picea glauca, Pinus banksiana, P. ponderosa, Populus* sp., *P. tremuloides, Quercus* sp., *Q. hypoleucoides.*
Ecology Colonizing recently felled needles and twigs under snowbanks, perhaps psychrophilic, and a primary nutrient recycling agent; symbiotic with filamentous Cyanophyceae (algae).
Culture Characters
(2).5.23.24.25.(26).32.36.(39).43-45.(54).(55).(56) (Ref. 588).
References 128, 183, 196, 216, 242, 312, 320, 370, 406.

Athelia fibulata M. Christiansen 1960
Dansk Bot. Ark. 19:148.
Distribution Canada: MB, ON, PQ.
United States: ME, NH.
Host *Populus* sp.
Ecology Not known.
Culture Characters Not known.
References 312, 406.

Athelia laxa (Burt) Jülich 1972
Beih. Willdenowia 7:90.
Basionym *Peniophora laxa* Burt, Ann. Missouri Bot. Gard. 12:224, 1925.
Distribution Canada: BC. United States: CT, PA, VA.
Hosts *Tsuga* sp., *T. canadensis*, gymnnosperm.
Ecology Bark with wood underneath wholly decayed.
Culture Characters Not known.
References 92, 312, 584.

Athelia maculare (Lair) Ginns 1992
Mycotaxon 44:204.
Basionym *Corticium maculare* Lair, J. Elisha Mitchell Sci. Soc. 62:216, 1946.
Distribution United States: NC.
Hosts *Quercus alba, Q. stellata.*
Ecology Bark of live trees, causing smooth patch disease.
Culture Characters Not known.
References 245, 350.

Athelia microspora (Karsten) Gilbn. sensu Gilbn. 1974
Fungi That Decay Ponderosa Pine, 42.
Basionym *Grandinia microspora* Karsten, Bidrag Kännedom Finlands Natur Folk 48:365, 1889.

Distribution Canada: AB.
United States: AZ, CO, NM.
Hosts *Picea* sp., *P. pungens, Pinus ponderosa.*
Ecology Associated with a white rot.
Culture Characters Not known.
References 197, 216, 402, 447.

The above references refer to a fungus with small, smooth spores and cylindrical hyphae. However, the accepted concept (294, 326, 395) for *G. microspora* is a species with small, warted spores and hyphae with ampulliform swellings. It is now named *Trechispora microspora*. Apparently Gilbertson in proposing the combination *Athelia microspora* cited the wrong basionym for the fungus he described. Rather than *Grandinia microspora*, the name Gilbertson should have cited is *Corticium microsporum* Bourd. & Galzin, see Bourdot and Galzin (50:195) and Jülich (312:147). Nevertheless, *C. microsporum* is a synonym of *Ceraceomyces sublaevis* fide Jülich (312), but Gilbertson's fungus may be a different species.

Athelia munda (Jackson & Dearden) M. Christiansen 1960
Dansk Bot. Ark. 19:155.
Basionym *Peniophora munda* Jackson & Dearden, Mycologia 43:58, 1951.
Synonym *Leptosporomyces mundus* (Jackson & Dearden) Jülich 1972.
Distribution Canada: BC.
United States: AZ, CA, CO, NJ, WY.
Hosts *Abies lasiocarpa, A. magnifica, Picea engelmannii, Tsuga mertensiana*, gymnosperm.
Ecology Bark; rotting stump; associated with a white rot.
Culture Characters Not known.
References 128, 130, 183, 309, 312, 402.

Athelia neuhoffii (Bres.) Donk 1957
Fungus 27:12.
Basionym *Corticium neuhoffii* Bres. in Neuhoff, Z. Pilzk. 2:179, 1923.
Distribution Canada: AB, ON, PQ. United States: AK, AZ, CT, IA, IN, MA, NH, NJ, NY, PA.
Hosts *Abies balsamea, Betula* sp., *Picea* sp., *P. glauca, Populus tremuloides.*
Ecology Not known.
Culture Characters
(2).5.24.25.(26).32.36.(39).43-45.(54).(55).(56) (Ref. 588).
References 183, 216, 242, 312, 391, 447.

Athelia poeltii Jülich 1978
Persoonia 10:149.
Distribution United States: FL.
Hosts Filamentous Cyanophyceae (algae).
Ecology Basidiolichen in *Sabal-Quercus* forest.
Culture Characters Not known.

Known only from the type collection.

Athelia rolfsii (Curzi) Tu & Kimbrough 1978
Bot. Gaz. (Crawfordsville) 139:460.
Basionym *Corticium rolfsii* Curzi, Boll. Staz. Patol. Veg., Roma II (11):306, 1931.
Anamorph *Sclerotium rolfsii* Sacc. 1911.
Distribution United States: AL, AR, AZ, CA, CT, FL, GA, IA, IL, IN, KS, KY, LA, MA, MD, MO, MS, NC, NJ, NY, OH, OK, OR, PA, SC, TN, TX, VA, WA.
Hosts Cosmopolitan, Farr et al. (183) compiled a list of about 225 genera of host plants.
Ecology A soilborne pathogen, commonly found as the *Sclerotium rolfsii* state, which causes southern blight, also called southern Sclerotium blight and Sclerotium root rot, of turfgrasses and herbaceous plants, many of which are important crop plants, e.g., sugar beets and field beans, also southern root rot of seedlings.
Culture Characters
2.3 & 5.16.23.32.36.(48).56 (Ref. 467).
References 183, 274, 540, 585.

We composed the species code from the description in Mordue (467).

Athelia salicum Pers. 1822
Mycol. Eur. 1:84.
Distribution Canada: AB, BC, NS, ON, PQ.
United States: AZ, CA, CT, ID, IL, ME, MT, NC, NH, NJ, NY, OH, WA, WI.
Hosts *Acer circinatum, Betula occidentalis, Quercus* sp.
Ecology Not known.
Culture Characters Not known.
References 183, 312.

Athelia scutellare (Berk. & Curtis) Gilbn. 1974
Fungi That Decay Ponderosa Pine, 42.
Basionym *Corticium scutellare* Berk. & Curtis, Grevillea 2:4, 1873.
Distribution Canada: BC, MB, ON. United States: AZ, AL, DC, FL, GA, IL, IN, KS, KY, LA, MO, MS, NC, NJ, NM, NY, PA, SC, VA.
Hosts *Acer saccharum, Aesculus octandra, Arbutus menziesii, Arundinaria gigantea* subsp. *tecta, Cytisus scoparius, Liriodendron tulipifera, Morus rubra, Pinus ponderosa, Populus* sp., *Quercus* sp., *Rhododendron canescens*, angiosperm.
Ecology Fallen decaying limbs.
Culture Characters Not known.
References 93, 96, 108, 116, 183, 197, 208, 242.

Burt (93) designated the collection from South Carolina, 2473 in the Curtis Herbarium (FH) as lectotype. The North American concept is presumably based upon that specimen. Interestingly, Hjortstam (288) in 1989 designated Ravenel 1584 (K) as lectotype, and noted that three other collections mentioned in the original description (2473, 5626 and 6091) probably represent three different species. No. 1584 is *Phanerochaete sordida*, fide Hjortstam (288). Hjortstam's designation of another lectotype seems unnecessary and if accepted would leave the North American collections of *A. scutellare* without a name.

Athelia tenuispora Jülich 1972
Beih. Willdenowia 7:120.
Distribution Canada: PQ. United States: CT, NY.
Hosts Not known.
Ecology Not known.
Culture Characters Not known.
Reference 312.

Athelia teutoburgensis (Brinkm.) Jülich 1973
Persoonia 7:383.
Basionym *Corticium teutoburgense* Brinkm., Jahres-Ber. Westfäl. Prov.-Vereins. Wiss. 44:38, 1916.
Synonym *Athelia macrospora* (Bourd. & Galzin) M. Christiansen 1960, fide Jülich (313).
Distribution United States: NJ.
Hosts Not known.
Ecology Not known.
Culture Characters Not known.
Reference 312.

ATHELOPSIS Parm. 1968 Atheliaceae

Athelopsis glaucina (Bourd. & Galzin) Parm. 1968
As Oberw. ex Parm., Consp. Syst. Cort., 42.
Basionym *Corticium glaucinum* Bourd. & Galzin, Hym. France, 207, 1928.
Distribution Canada: YT.
Host *Populus tremuloides*.
Ecology Not known.
Culture Characters
1.3.7.32.36?.38.47.54 (Ref. 588).
Reference 230.

Athelopsis lembospora (Bourd.) Oberw. 1972
Persoonia 7:3.
Basionym *Corticium lembosporum* Bourd., Rev. Sci. Bourbonnais Centr. France 23:10, 1910.
Distribution United States: CA, OR.
Hosts Not known.
Ecology Wood and twigs.
Culture Characters Not known.
Reference 385.

Athelopsis subinconspicua (Litsch.) Jülich 1975
Persoonia 8:292.
Basionym *Corticium subinconspicua* Litsch. in Pilát and V. Lindtner, Glasn. Skopsk. Naucn. Drustva 18:178, 1938.
Synonym *Athelopsis hypochnoidea* Jülich 1971, fide Jülich (317).
Distribution Canada: BC, ON, PQ.
Hosts Not known.
Ecology Fallen log.
Culture Characters Not known.
References 255, 312.

AURICULARIA Bull. 1789 Auriculariaceae

Auricularia auricula-judae (Bull.:Fr.) Wettst. 1885
Verh. Zool.-Bot. Ges. Wien 35:554.
Basionym *Tremella auricula-judae* Bull., Herb. France, tab. 427, fig. 2, 1788.
Synonyms *Auricularia auricula* (L.) Underw. 1902, *Auricularia auricularis* (S.F. Gray) G.W. Martin 1943. Both fide Donk (157).
Distribution Canada: AB, AB-NT, BC, MB, NB, NF, NFL, NS, ON, PQ, YT.
United States: AL, AK, AR, AZ, CA, CO, CT, FL, GA, IA, ID, IN, KS, KY, LA, MA, MD, ME, MN, MO, MS, MT, NC, NE, NH, NY, NJ, OH, OR, PA, SC, TN, TX, UT, VA, WA, WV.
Hosts *Abies* spp., *A. amabilis, A. balsamea, A. concolor, A. grandis, A. lasiocarpa, A. lasiocarpa* var. *arizonica, Acer glabrum, A. negundo, Betula lenta, B. pumila, Carya* sp., *C. illinoensis, Celtis* sp., *Chamaecyparis nootkatensis, Gleditsia* sp., *Juglans cinerea, J. major, Liquidambar* sp., *Picea* sp., *P. engelmannii, P. glauca, P. rubens, Pinus* sp., *P. taeda, Pseudotsuga menziesii, Quercus* sp., *Q. alba, Q. coccinea, Q. emoryi, Q. gambelii, Q. hypoleucoides, Q. nigra, Q. reticulata, Thuja occidentalis, T. plicata, Ulmus* sp., angiosperm.
Ecology Corticated and decorticated limbs; slash; corticated and decorticated logs; associated with a white rot.
Culture Characters
2.3c.7.(8).32.36.38.(40).44-45.53.54.(55).(59).(60) (Ref. 474).
References 28, 96, 107, 120, 127, 163, 181, 183, 214, 216, 242, 249, 273, 328, 332, 402, 417, 419, 443, 448, 474, 498, 499, 509, 610.

The mating system is unifactorial or bifactorial, see Wong and Wells (649).

Auricularia cornea Ehrenb.:Fr. 1820
In C.G.D. Nees von Esenbeck, Horae physicae Berolinenses, 91.
Synonyms *Auricularia nigrescens* (Swartz) Farlow 1905, fide Lowy (422), *Auricularia polytricha* (Mont.) Sacc. 1885, fide Wong and Wells (649).
Distribution United States: AL, FL, LA.
Hosts *Liquidambar styraciflua*, angiosperm.
Ecology Not known.
Culture Characters Not known, but unifactorial (162) and bifactorial (544, 649).
References 28, 162, 417, 419.

The author citation and place of publication for *A. cornea* in Lowy (417) is a *lapsus calami*.

Auricularia fuscosuccinea (Mont.) Henn. 1893
Bot. Jahrb. Syst. 17:19.
Basionym *Exidia fuscosuccinea* Mont., Pl. Cell. Cuba 346, 1841.
Distribution United States: LA, MD, TN.
Hosts *Acer negundo*, angiosperm.

Ecology Not known.
Culture Characters Not known, but unifactorial (252).
References 183, 417, 419.

Auricularia mesenterica (Dickson:Fr.) Pers. 1822
Mycol. Eur. 1:97.
Basionym *Helvella mesenterica* Dickson, Plant. Crypt. Britanniae 1:20, 1785.
Distribution United States: FL.
Hosts Not known.
Ecology Not known.
Culture Characters Not known.
Reference 417.

AURICULARIOPSIS Maire 1902 Schizophyllaceae
Synonym is *Cytidiella* Pouzar 1954, fide Stalpers (591).

Auriculariopsis albomellea (Bondartsev) Kotlaba 1988
Ceská Mykol. 42:239.
Basionym *Cytidia albomellea* Bondartsev, Bolegni Rast. 16:96, 1927.
Synonyms *Cytidiella albomellea* (Bondartsev) Parm. 1968, *Cytidiella melzeri* Pouzar 1954, fide Parmasto (522).
Distribution United States: AZ, WI.
Host *Pinus strobus*.
Ecology Fallen branch; may be associated with a brown rot.
Culture Characters
1.3c.26.32.36.38.46.48.53.54.60 (Ref. 474).
References 226, 474.

Auriculariopsis ampla (Lév.) Maire 1902
Bull. Soc. Mycol. France 18:102.
Basionym *Cyphella ampla* Lév., Ann. Sci. Nat. Bot., Sér. III, 9:126, 1848.
Synonym *Cytidia flocculenta* (Fr.) Höhnel & Litsch. 1907 sensu Aucts., fide Eriksson and Ryvarden (173).
Distribution Canada: ON, YT.
United States: AK, MT, WY.
Hosts *Alnus* sp., *A. sinuata, Betula* sp., *Populus* sp., *P. nigra, P. pyramidalis, Prunus* sp., *Salix* sp., *S. alba, S. glauca*.
Ecology Bark of trees and shrubs.
Culture Characters
1.3.20.32.36.38.43.48.54.60 (Ref. 591).
References 91, 97, 116, 117, 242, 464.

AURISCALPIUM S.F. Gray 1821 Auriscalpiaceae

Auriscalpium vulgare S.F. Gray 1821
Natural Arrangement British Plants 1:650.
Distribution Canada: BC, NS, NTM, ON, PQ. United States: AZ, IA, ID, KS, MA, ME, MN, OR, WA.
Hosts *Pinus* sp., *P. banksiana, P. contorta, P. ponderosa, P. rigida, P. sylvestris, Pseudotsuga menziesii, Zea mays*, gymnosperm.
Ecology Decaying conifer cones on the ground; buried corn cob.
Culture Characters
(1).3.15.32.36.38.45.(48).51.56 (Ref. 529); 60 (Ref. 186).
References 115, 183, 242, 249, 463, 610.

BASIDIODENDRON Rick 1938 Exidiaceae

Basidiodendron caesiocinerea (Höhnel & Litsch.) Luck-Allen 1963
Canad. J. Bot. 41:1036.
Basionym *Corticium caesiocinereum* Höhnel & Litsch., Sitzungsber. Kaiserl. Akad. Wiss., Math.-Naturwiss. Kl., Abt. 1, 117:1116, 1908.
Synonyms *Bourdotia caesiocinerea* (Höhnel & Litsch.) Bourd. & Galzin ex Pilát & Lindtner 1938, *Sebacina caesiocinerea* (Höhnel & Litsch.) D.P. Rogers 1935, *Sebacina cinerella* (Bourd. & Galzin) S. Killermann 1928, fide Wells and Raitviir (630).
Distribution Canada: AB-NT, BC, ON, PQ.
United States: AZ, FL, IA, MA, ME, MO, NM, NY, OR, RI.
Hosts *Arbutus menziesii, Fraxinus velutina, Pinus* sp., *P. banksiana, P. ponderosa, P. strobus, Platanus wrightii, Quercus emoryi*, angiosperm, gymnosperm.
Ecology Very rotten wood, usually decorticated.
Culture Characters Not known.
References 197, 216, 242, 425, 443, 451, 562, 624.

Basidiodendron cinerea (Bres.) Luck-Allen 1963
Canad. J. Bot. 41:1043.
Basionym *Sebacina cinerea* Bres., Fungi Tridentini 2:99, 1900.
Synonym *Bourdotia cinerea* (Bres.) Bourd. & Galzin 1928.
Distribution Canada: BC, ON, PQ. United States: AZ, CA, IA, IN, LA, MI, NM, OH, OR, UT.
Hosts *Abies balsamea, Acer saccharum, Fagus grandifolia, Pinus aristata* var. *longaeva, P. contorta, P. ponderosa, Platanus wrightii, Quercus* sp., *Salix* sp., *Thuja occidentalis, Tilia americana*, angiosperm, gymnosperm.
Ecology Dead wood; decorticated wood; corticated and decorticated limb.
Culture Characters
2.3c.15.32.36.40.47.(53).54.55 (Ref. 474).
References 183, 197, 216, 242, 407, 425, 443, 451, 474, 499, 562, 622, 624.

Basidiodendron eyrei (Wakef.) Luck-Allen 1963
Canad. J. Bot. 41:1034.
Basionym *Sebacina eyrei* Wakef., Trans. Brit. Mycol. Soc. 5:126, 1915.
Synonyms *Bourdotia eyrei* (Wakef.) Bourd. & Galzin 1928, *Sebacina deminuta* Bourd. 1922, fide Wells and Raitviir (630), *Basidiodendron deminuta* (Bourd.) Luck-Allen 1963, *Corticium involucrum* Burt 1926, fide Rogers and Jackson (567).

Distribution Canada: AB, AB-NT, BC, ON, PQ. United States: AK, AZ, CA, CO, IA, IN, LA, MA, MN, MS, NH, NJ, NY, OH, OR, UT, VT, WA.
Hosts *Abies* sp., *Acer* sp., *Arbutus menziesii, Betula alleghaniensis, Fagus* sp., *Fraxinus* sp., *Juglans major, Phellinus* sp., *Picea* sp., *P. pungens, Pinus* sp., *P. aristata* var. *longaeva, P. contorta, Populus* sp., *P. fremontii, P. trichocarpa, Prunus americana, Quercus* sp., *Thuja occidentalis, Ulmus* sp., angiosperm, rarely gymnosperm.
Ecology Decaying wood; usually decorticated angiosperm wood; partially decorticated fallen branch; corticated limb; decaying log; associated with a white rot.
Culture Characters
2.3.7.(27).32.36.38.(39).45-46.(53).54.55 (Ref. 474).
References 93, 116, 183, 214, 216, 242, 245, 402, 407, 425, 443, 448, 451, 474, 499, 502, 562, 563, 624, 643.

Basidiodendron fulvum (Massee) Ginns 1982
Opera Bot. 61:54.
Basionym *Coniophora fulva* Massee, J. Linn. Soc., Bot. 25:136, 1889.
Synonyms *Bourdotia grandinioides* Bourd. & Galzin 1928, fide Ginns (234), *Basidiodendron grandinioides* (Bourd. & Galzin) Luck-Allen 1963, *Sebacina grandinioides* (Bourd. & Galzin) D.P. Rogers 1935.
Distribution Canada: BC, ON.
United States: AZ, FL, NJ, OR.
Hosts *Magnolia* sp., *Populus* sp., *P. tremuloides, P. trichocarpa*, angiosperm.
Ecology Decaying wood.
Culture Characters Not known.
References 183, 216, 234, 242, 443, 451, 563, 624.

Basidiodendron nodosa Luck-Allen 1963
Canad. J. Bot. 41:1045.
Distribution Canada: ON, PQ.
Hosts *Abies* sp., *A. balsamea, Pinus* sp., *Thuja occidentalis*.
Ecology Not known.
Culture Characters Not known.
References 242, 425.

Basidiodendron pini (Jackson & G.W. Martin) Luck-Allen 1963
Canad. J. Bot. 41:1049.
Basionym *Sebacina pini* Jackson & G.W. Martin, Mycologia 32:684, 1940.
Synonym *Bourdotia pini* (Jackson & G.W. Martin) Wells 1959.
Distribution Canada: ON. United States: OR.
Hosts *Pinus* sp., *P. strobus, Tsuga* sp., *T. canadensis*.
Ecology Dead wood.
Culture Characters Not known.
References 116, 242, 425, 437, 443, 451, 624.

Basidiodendron rimosa (Jackson & G.W. Martin) Luck-Allen 1963
Canad. J. Bot 41:1051.
Basionym *Sebacina rimosa* Jackson & G.W. Martin, Mycologia 32:684, 1940.
Distribution Canada: ON.
Hosts *Thuja* sp., *T. occidentalis*.
Ecology Not known.
Culture Characters Not known.
References 116, 242, 425, 437, 443, 451.

Basidiodendron subreniformis Luck-Allen 1963
Canad. J. Bot. 41:1047.
Distribution Canada: ON.
Hosts *Abies* sp., *A. balsamea*.
Ecology Not known.
Culture Characters Not known.
References 242, 425.

BASIDIORADULUM Nobles 1967 Hyphodermataceae

Basidioradulum radula (Fr.:Fr.) Nobles 1967
Mycologia 59:192.
Basionym *Hydnum radula* Fr., Obs. Mycol. 2:271, 1818.
Synonyms *Hyphoderma radula* (Fr.) Donk 1957, *Radulum bennettii* Berk. & Curtis 1873, fide Gilbertson (192), *Corticium colliculosum* Berk. & Curtis 1873, fide Nobles (488), *Corticium hydnans* (Schw.) Burt 1926, fide Nobles (488), *Radulum orbiculare* Grev.:Fr. 1827, fide Nobles (488).
Distribution Canada: BC, NB, NF, NS, ON, PE, PQ. United States: CA, DC, FL, GA, IA, ID, IL, KY, LA, MA, MD, MI, MO, MN, MS, NC, NE, NH, NJ, NY, OH, PA, RI, TX, VT, WA, WI, WV.
Hosts *Abies alba, A. balsamea, Acer* sp., *A. saccharum, A. spicatum, Alnus* sp., *A. incana, A. rubra, A. sinuata, Betula* sp., *B. alba, B. papyrifera, Castanea* sp., *Cornus sericea, C. stolonifera, Eucalyptus* sp., *Fagus grandifolia, Liquidambar styraciflua, Liriodendron tulipifera, Picea* sp., *P. mariana, Pinus* sp., *Populus* sp., *Prunus* sp., *P. cerasus, Pseudotsuga* sp., *P. menziesii, Quercus garryana, Salix* sp., *Sorbus decora, Taxus brevifolia, Tsuga canadensis, T. heterophylla*, angiosperm, gymnosperm.
Ecology Twigs; decaying limbs on the ground; associated with a white rot.
Culture Characters
2.3c.19.32.36.38.46-47.54.55.60 (Ref. 474);
2.3.(7).19.(26).32.36.38.47.54.55.60 (Ref. 488).
References 93, 116, 133, 177, 183, 192, 214, 242, 249, 255, 391, 447, 463, 465, 474, 488, 595.

The type of *C. colliculosum* (Sprague 5297 at K) although "extremely similar to ... *B. radula* ... has shorter spores than normal", fide Hjortstam (288).

BOIDINIA Stalpers & Hjort. 1982 Gloeocystidiellaceae

Boidinia furfuraceum (Bres.) Stalpers & Hjort. 1982
Mycotaxon 14:77.

Basionym *Hypochnus furfuraceus* Bres., Fungi Tridentini 2:97, 1900.
Synonyms *Gloeocystidiellum furfuraceum* (Bres.) Donk 1956, *Gloeocystidium furfuraceum* (Bres.) Höhnel & Litsch. 1908.
Distribution Canada: BC, ON, PQ.
United States: MN.
Hosts *Abies balsamea, Alnus rubra, Picea* sp., *Pinus banksiana, Tsuga mertensiana.*
Ecology Not known.
Culture Characters
2.3c.7.(26).32.36.38.47.55 (Ref. 474).
References 242, 303, 391, 447, 474.

Boidinia propinqua (Jackson & Dearden) Hjort. & Ryv. 1988
Acta Mycol. Sinica 7:79.
Basionym *Corticium propinquum* Jackson & Dearden, Canad. J. Res., C 27:155, 1949.
Synonym *Gloeocystidiellum propinquum* (Jackson & Dearden) Parm. 1965.
Distribution Canada: BC.
Hosts *Pseudotsuga menziesii, Thuja plicata.*
Ecology Not known.
Culture Characters Not known.
References 116, 242, 308.

BOREOSTEREUM Parm. 1968 Stereaceae

Boreostereum radiatum (Peck) Parm. 1968
Consp. Syst. Cort., 187.
Basionym *Stereum radiatum* Peck, Bull. Buffalo Soc. Nat. Hist. 1:62, 1873.
Synonym *Stereum radiatum* var. *reflexum* Peck 1896, fide Lentz (379).
Distribution Canada: AB, AB-NT, BC, MB, ON, PQ.
United States: AZ, IA, ID, MA, ME, MI, MN, MT, NH, NY, PA, TN, VA, VT, WA, WI.
Hosts *Larix laricina, Malus sylvestris, Picea* sp., *P. engelmannii, P. glauca, P. mariana, P. rubens, Pinus* sp., *Populus tremuloides, Thuja occidentalis, Tsuga canadensis*, gymnosperm.
Ecology Charred wood and logs; logs; boards; associated with a white rot.
Culture Characters
1.(2).6.(8).9.21.34.37.39.45-47.(53).55.(57) (Ref. 474);
1.6.8.32.37.39.47.55.57.66 (Ref. 104).
References 89, 104, 183, 214, 242, 327, 334, 379, 448, 474, 512, 519.

BOTRYOBASIDIUM Donk 1931 Botyrobasidiaceae

Botryobasidium ansosum (Jackson & D.P. Rogers) Parm. 1968
Consp. Syst. Cort., 51.
Basionym *Pellicularia ansosa* Jackson & D.P. Rogers, Farlowia 1:103, 1943.
Distribution Canada: BC.
United States: AZ, CA, NM, WA.
Hosts *Abies magnifica, Picea engelmannii, P. sitchensis, Pinus ponderosa*, gymnosperm.
Ecology Bark.
Culture Characters Not known.
References 116, 128, 183, 197, 204, 216, 242, 447, 565.

Botryobasidium aureum Parm. 1965
Izv. Akad. Nauk Estonsk. SSR, Ser. Biol. 14:220.
Anamorph *Haplotrichum aureum* (Pers.) Hol.-Jech. 1976.
Distribution United States: NC, NC-TN.
Hosts *Abies fraseri, Betula alleghaniensis, Fagus grandifolia.*
Ecology Trunk; butt of a log.
Culture Characters Not known.
Reference 327.

Botryobasidium botryosum (Bres.) Eriksson 1958
Symb. Bot. Upsal. 16 (1):53.
Basionym *Corticium botryosum* Bres., Ann. Mycol. 1:99, 1903.
Anamorph *Costantinella micheneri* (Berk. & Curtis) S.J. Hughes 1953.
Distribution Canada: AB-NT, ON.
United States: AK, CO, NM.
Hosts *Abies concolor, Picea sitchensis, Pinus contorta, Populus* sp., *Quercus gambelii.*
Ecology Old branches on ground; associated with a white rot.
Culture Characters
1.6.24.25.26.32.36.38.46-47.(54) (Ref. 588).
References 168, 183, 208, 242, 398, 402.

Botryobasidium candicans Eriksson 1958
Svensk Bot. Tidskr. 52:6.
Anamorph *Haplotrichum capitatum* (Link) Link 1824 (= *Oidium candicans* (Sacc.) Linder, fide Holubova-Jechova, Ceská Mykol. 23:210, 1969).
Distribution United States: AZ, CO, IA, NY, TN, VA.
Hosts *Fagus grandifolia, Picea rubens, Pinus ponderosa, Populus deltoides, Prunus serotina?, Quercus gambellii.*
Ecology Log.
Culture Characters Not known.
References 166, 216, 327, 399.

Botryobasidium conspersum Eriksson 1958
Symb. Bot. Upsal. 16 (1):133.
Anamorph *Haplotrichum conspersum* (Link:Fr.) Hol.-Jech. 1976.
Distribution United States: NC.
Host *Picea rubens.*
Ecology Log.
Culture Characters Not known.
Reference 327.

Botryobasidium croceum Lentz 1967
Mycopathol. Mycol. Appl. 32:6.
Anamorph *Allescheriella crocea* (Mont.) S.J. Hughes 1950.
Distribution Canada: SK. United States: MS.
Host *Acer negundo.*
Ecology Dead tree stump.
Culture Characters Not known.
References 242, 380.

Botryobasidium danicum Eriksson & Hjort. 1969
Friesia 9:11.
Distribution Canada: BC. United States: PA.
Hosts *Abies grandis, Acer circinatum, Alnus* sp., *A. rubra, Betula papyrifera, Cornus pubescens, Pseudotsuga menziesii, Rubus spectabilis, Thuja plicata, Tsuga heterophylla, T. mertensiana*, angiosperm, gymnosperm.
Ecology Bark; branches lying on the ground; stems; logs.
Culture Characters Not known.
References 168, 172.

Botryobasidium intertextum (Schw.) Jülich & Stalpers 1980
Verh. Kon. Ned. Akad. Wetensch. Afd. Natuurk., Tweed Sect. 74:56.
Basionym *Sporotrichum intertextum* Schw., Trans. Amer. Philos. Soc., N.S. 4:271, 1832.
Synonyms *Pellicularia angustispora* Boidin 1951, fide Jülich and Stalpers (326), *Botryobasidium angustisporum* (Boidin) Eriksson 1958.
Distribution United States: OR, PA.
Host *Pseudotsuga menziesii.*
Ecology Not known.
Culture Characters Not known.
References 167, 576.

Botryobasidium medium Eriksson 1958
Symb. Bot. Upsal. 16 (1):54.
Anamorph *Haplotrichum medium* (Hol.-Jech.) Hol.-Jech. 1976.
Distribution Canada: Specific localities not given. United States: AZ, NC, NM, VA.
Hosts *Abies fraseri, Picea rubens, Pinus ponderosa.*
Ecology Bark; trunk; log.
Culture Characters Not known.
References 172, 197, 216, 327.

Botryobasidium obtusisporum Eriksson 1958
Symb. Bot. Upsal. 16 (1):57.
Distribution United States: NC, TN.
Hosts *Abies fraseri, Picea rubens, Rhododendron* sp.
Ecology Moss-covered wood; bark; branch; trunk; log; exposed root.
Culture Characters Not known.
Reference 327.

Botryobasidium pruinatum (Bres.) Eriksson 1958
Svensk Bot. Tidskr. 52:9.
Basionym *Corticium pruinatum* Bres., Ann. Mycol. 1:99, 1903.
Synonyms *Pellicularia pruinata* (Bres.) D.P. Rogers ex Linder 1942, *Corticium botryoideum* Overh. 1934, fide Rogers (565), *Pellicularia coronata* (Schröter) Donk sensu D.P. Rogers (563), fide Rogers (565), *Botryobasidium coronatum* (Schröter) sensu D.P. Rogers, *Botryobasidium laeve* (Eriksson) Parm. 1965, fide Maekawa (430), *Botryobasidium pruinatum* var. *laeve* Eriksson 1958.
Distribution Canada: BC, MB, NS, ON, PQ. United States: AK, AL, AZ, CA, CO, IA, MA, MD, MN, MO, NC, NM, NY, OH, OR, PA, RI, VA, WA.
Hosts *Abies balsamea, A. magnifica, Acer* sp., *A. macrophyllum, A. negundo, A. rubrum, A. saccharum, Alnus* sp., *A. rugosa, Betula* sp., *B. alleghaniensis, Carpinus caroliniana, Carya* sp., *Fagus* sp., *F. grandifolia, Fomes pomaceus, Hamamelis* sp., *Kalmia* sp., *Liquidambar styraciflua, Picea* sp., *P. engelmannii, P. glauca, P. sitchensis, Pinus ponderosa, P. resinosa, P. strobus, Populus* sp., *P. grandidentata, P. tremuloides, Prunus* sp., *P. serotina, Pseudotsuga mucronata, Quercus alba, Q. borealis, Q. macrocarpa, Salix* sp., *Tsuga canadensis, Ulmus* sp., angiosperm, gymnosperm.
Ecology Twig; bark; decorticated branches and logs; ground; associated with a white rot.
Culture Characters
1.6.24.25.26.32.36.(39).43-44.(54).(55) (Ref. 588).
References 116, 128, 166, 183, 197, 204, 214, 216, 242, 249, 327, 391, 402, 406, 447, 517, 563, 565.

Botryobasidium robustius Pouzar & Jech. 1967
Ceská Mykol. 21:69.
Anamorph *Haplotrichum rubiginosum* (Fr.) Hol.-Jech. 1976.
Distribution United States: CA.
Hosts Not known.
Ecology Rotten wood.
Culture Characters
1.6.24.25.26.32.36.(39).46-47.(54).(55) (Ref. 588).

Known only from the specimens cited in the original description.

Botryobasidium simile Pouzar & Hol.-Jech. 1969
Ceská Mykol. 23:99.
Anamorph *Haplotrichum simile* (Berk.) Hol.-Jech. 1976 (= *Oidium aureo-fulvum* (Cooke & Ellis) Linder 1942, = *O. biforme* Linder 1942 and *O. simile* Berk. 1845, fide Pouzar and Holubova-Jechova (loc. cit.) and Hughes (300)).
Distribution Canada: ON.
United States: AL, CT, GA, KN, LA, MA, MI, MO, NC, NJ, NY, OH, PA, SC, TN.
Hosts Not known.
Ecology Not known.

Culture Characters Not known.
References 300, 397.

The above records are based upon Hughes' (300) and Linder's (397) citations of the anamorph. The teleomorph has not been found in North America.

Botryobasidium subalbidum Ginns 1988
Mycologia 80:65.
Distribution United States: NY.
Host *Irpex lacteus*.
Ecology Covering the hymenial surface of an old basidiome on fallen branch from an angiosperm tree.
Culture Characters Not known.

Known only from the type specimen.

Botryobasidium subcoronatum (Höhnel & Litsch.) Donk 1931
Meded. Ned. Mycol. Ver. 18-21:177.
Basionym *Corticium subcoronatum* Höhnel & Litsch., Sitzungsber. Kaiserl. Akad. Wiss., Math.-Naturwiss. Kl., Abt. 1, 116:822, 1907.
Synonym *Pellicularia subcoronata* (Höhnel & Litsch.) D.P. Rogers 1943.
Distribution Canada: AB-NT, BC, MB, NS, NTM, ON, PQ. United States: AZ ,CA, CO, CT, ID, LA, MA, MD, MI, MN, MO, MT, NC, NH, NJ, NM, NY, OR, PA, RI, TN, VT, WA, WI.
Hosts *Abies balsamea, A. concolor, A. fraseri, A. magnifica, Acer* sp., *A. saccharum, Alnus oblongifolia, Arbutus arizonica, Betula* sp., *B. alleghaniensis, Castanea dentata, Fagus grandifolia, Picea* spp., *P. engelmannii, P. glauca, P. rubens, Pinus* sp., *P. ayacahuite, P. contorta, P. leiophylla, P. leiophylla* var. *chihuahuana, P. ponderosa, P. resinosa, P. strobiformis, P. strobus, Populus* sp., *Pseudotsuga menziesii, P. mucronata, Quercus* sp., *Q. kelloggii, Thuja plicata, Tsuga canadensis*, angiosperm, gymnosperm.
Ecology Bark; charred wood; rotting wood; branch; trunk; tree base; stump; exposed root; root bark covered with moss; very rotten log; log; associated with a white rot.
Culture Characters
1.(2).3c.(7).26.32.36.(37).38.(39).47.54.55.58 (Ref. 474).
References 108, 116, 119, 120, 128, 130, 183, 197, 204, 216, 242, 249, 327, 388, 390, 391, 402, 404, 447, 474, 517, 563, 565, 615.

Botryobasidium vagum (Berk. & Curtis) D.P. Rogers 1935
Stud. Nat. Hist. Iowa Univ. 17:17.
Basionym *Corticium vagum* Berk. & Curtis, Grevillea 1:179, 1873.
Synonym *Pellicularia vaga* (Berk. & Curtis) D.P. Rogers ex Linder 1942.
Anamorph *Haplotrichum curtisii* (Berk.) Hol.-Jech. 1976.
Distribution Canada: AB, AB-NT, BC, MB, NB, NS, ON, PQ. United States: AK, AL, AZ, CA, CO, CT, FL, GA, IA, ID, IL, KS, LA, MA, MD, MI, MN, MO, NC, NJ, NM, NY, OH, OR, PA, RI, SC, SD, TN, TX, WA, WI.
Hosts *Abies balsamea, A. fraseri, A. grandis, A. lasiocarpa, A. lasiocarpa* var. *arizonica, A. magnifica, Acer* sp., *A. circinatum, Alnus* sp., *A. rubra, Betula* sp., *B. alba, B. alleghaniensis, Bjerkandera adusta, Castanea dentata, Fagus grandifolia, Hexagonia (Pogonomyces) hydnoides, Kalmia latifolia, Larix occidentalis, Libocedrus decurrens, Ostrya* sp., *Phellinus laevigatus, Picea* sp., *P. canadensis, P. engelmannii, P. glauca, P. mariana, P. rubens, P. sitchensis, Pinus* sp., *P. banksiana, P. cembroides, P. contorta, P. engelmannii, P. monticola, P. ponderosa, P. rigida, P. resinosa, P. strobus, Populus* sp., *P. balsamifera, P. tremuloides, Pseudotsuga menziesii, P. mucronata, Rhododendron* sp., *Quercus* sp., *Sabal palmetto, Salix* sp., *Sambucus caerulea, S. glaucus, Taxodium distichum, Thuja occidentalis, T. plicata, Tilia americana, Tsuga* sp., *T. canadensis, T. heterophylla*, angiosperm, gymnosperm.
Ecology Bark of dead limb; branch; trunk; stump; exposed root; decaying logs; dead fungus; associated with a white rot.
Culture Characters
1.(2).6.(7).26.32.36.37.38.46-47.(53).54.55 (Ref. 474).
References 96, 97, 108, 116, 119, 120, 128, 133, 183, 197, 204, 208, 214, 216, 242, 249, 327, 391, 397, 402, 406, 447, 474, 563, 565, 615.

BOTRYOHYPOCHNUS Donk 1931 — Botryobasidiaceae

Botryohypochnus isabellinus (Fr.) Eriksson 1958
Svensk Bot. Tidskr. 52:2.
Basionym *Hypochnus isabellina* Fr., Obs. Mycol. 2:281, 1818.
Synonyms *Botryobasidium isabellinum* (Fr.) D.P. Rogers 1935, *Pellicularia isabellina* (Fr.) D.P. Rogers 1943.
Distribution Canada: AB-NT, BC, MB, NS, NTM, ON, PQ. United States: AZ, CO, CT, FL, IA, LA, MA, MI, NM, NY, OR, PA, TN, VA, WA, WI.
Hosts *Abies balsamea, A. concolor, Acer* sp., *Betula* sp., *B. alleghaniensis, Liriodendron tulipifera, Picea glauca, Pinus* sp., *P. ponderosa, P. virginiana, Populus* sp., *P. grandidentata, P. tremuloides, P. trichocarpa, Pseudotsuga menziesii, Pteridium latiusculum, Quercus hypoleucoides, Salix* sp., *Tilia americana, Tsuga canadensis*, angiosperm, gymnosperm.
Ecology Bark; soil; associated with a white rot.
Culture Characters
(1).2.6.7.(22).32.37.39.46.53.54.55 (Ref. 474).
References 116, 133, 183, 197, 204, 208, 216, 242, 249, 388, 402, 406, 474, 563, 565.

Botryohypochnus verrucisporus Burdsall & Gilbn. 1982
Mycotaxon 15:334.
Distribution United States: AZ.
Host *Quercus arizonicus.*
Ecology Not known.
Culture Characters Not known.

Known only from the type specimen.

BOURDOTIA (Bres.) Trotter 1925 Exidiaceae
Bourdotia was proposed but not validly published by Bresadola and Torrend in 1913, fide Donk (157).

Bourdotia burtii (Bres.) Wells 1959
Mycologia 51:549.
Basionym *Heterochaete burtii* Bres., Ann. Mycol. 18:51, 1920.
Distribution United States: LA.
Hosts Not known.
Ecology Dead wood, often corticated.
Culture Characters Not known.
Reference 624.

Bourdotia galzinii (Bres.) Trotter 1925
Sylloge Fung. 23:571.
Basionym *Sebacina galzinii* Bres., Ann. Mycol. 6:46, 1908.
Distribution Canada: ON.
United States: AZ, IA, LA, MA, ME.
Hosts *Fraxinus velutina, Populus* sp., *P. tremuloides*, angiosperm.
Ecology Decaying wood.
Culture Characters Not known.
References 216, 406, 443, 451, 624, 630.

BREVICELLICIUM Larsson & Hjort. 1978 Sistotremaceae

Brevicellicium exile (Jackson) Larsson & Hjort. 1978
Mycotaxon 7:118.
Basionym *Corticium exilis* Jackson, Canad. J. Res., C, 28:721, 1950.
Distribution Canada: ON, PQ.
Hosts *Picea* sp., *Pinus* sp., *Thuja occidentalis, Tsuga* sp.
Ecology Not known.
Culture Characters Not known.
References 242, 292, 391.

Very similar to *Trechispora cohaerens*, fide Jülich and Stalpers (326).

Brevicellicium olivascens (Bres.) Larsson & Hjort. 1978
Mycotaxon 7:119.
Basionym *Odontia olivascens* Bres., Fungi Tridentini 2:36, 1892.
Synonyms *Hydnum granulosum* var. *mutabile* Pers. 1825, fide Hjortstam and Larsson (292), and Liberta (395), *Odontia mutabilis* (Pers.) Bres. 1911, *Grandinia mutabilis* (Pers.) Bourd. & Galzin 1914, *Trechispora mutabilis* (Pers.) Liberta 1966, *Trechispora sphaerocystis* Burdsall & Gilbn. 1982, fide K.H. Larsson (in litt. 1992).
Distribution Canada: ON. United States: AZ, IA, IL.
Hosts *Juglans major*, angiosperm, gymnosperm.
Ecology Not known.
Culture Characters
2.3.21.26.32.36.38.(47) (Ref. 588).
References 395, 463.

Reported (395) as *Trechispora mutabilis* but at the level of epithets the earlier *olivascens*, proposed in 1892, has precedence over *mutabilis*, proposed in 1911.

Brevicellicium permodicum (Jackson) Ginns & Lefebvre, *comb. nov.*
Basionym *Corticium permodicum* Jackson, Canad. J. Res., C, 28:721, 1950.
Distribution Canada: ON.
Hosts Angiosperm.
Ecology Not known.
Culture Characters Not known.

Known only from the type and two paratype specimens, all from Lake Timagami. We believe the fungus is congeneric with *B. exile*, the type species of the genus. A paratype 11962 (DAOM 30763) has small, broad basidia (15 x 9 μm), subglobose spores (6.5 x 6.0 μm), and relatively broad, isodiametric subbasidial cells (6 x 7 μm). All are features characteristic of *Brevicellicium.* The simple-septate hyphae (2 μm diameter) distinguish *B. permodicum* from the other species.

BULBILLOMYCES Jülich 1974 Hyphodermataceae

Bulbillomyces farinosus (Bres.) Jülich 1974
Persoonia 8:69.
Basionym *Kneiffia farinosa* Bres., Ann. Mycol. 1:105, 1903.
Synonyms *Peniophora farinosa* (Bres.) Höhnel & Litsch. 1908, *Peniophora candida* (Pers.:Fr.) Lyman 1907, fide Burt (92) and Weresub (635).
Anamorph *Aegerita candida* Pers.:Fr. 1821.
Distribution Canada: ON.
United States: MA, MO, NH, NY, UT, VA.
Hosts *Acer* sp., *Alnus* sp., *Fagus grandifolia, Pinus aristata* var. *longaeva, Populus* sp., *Ulmus* sp., angiosperm, gymnosperm.
Ecology Wood or bark; ground; mostly in very wet places.
Culture Characters
1.(2).3c.23.32.36.40.43-44.(53).54.55.60 (Ref. 474).
References 92, 183, 407, 474, 584, 635.

BUTLERELFIA Weresub & Illman 1980
Cylindrobasidiaceae

Butlerelfia eustacei Weresub & Illman 1980
Canad. J. Bot. 58:145.

Distribution Canada: BC.
United States: ID, IL, NY, OR, VA, WA.
Hosts *Malus* sp., *M. sylvestris*.
Ecology Isolated from fisheye rot of apples in cold storage.
Culture Characters
1.(2a?).3c.7.(26).27.36.38.42-43.48.(50).56.60.61 (Ref. 640).
References 183, 242, 640.

BYSSOCORTICIUM **Singer 1944** Atheliaceae

Byssocorticium atrovirens (Fr.) Singer 1944
As Bondartsev & Singer ex Singer, Mycologia 36:69.
Basionym *Thelephora atrovirens* Fr., Elenchus 1:202, 1828.
Synonyms *Corticium atrovirens* (Fr.) Fr. 1838, *Sporotrichum aeruginosum* Schw. 1832, fide Stalpers (589).
Distribution Canada: NB, NS, ON, PQ.
United States: CA, CT, IA, IL, IN, MA, MI, NC, NH, NJ, NY, OR, PA, SC, VA, VT.
Hosts *Acer* sp., *Betula* sp., *Quercus* sp., gymnosperm.
Ecology Mycorrhizal with some amentiferous trees (602); basidiomes produced on decaying bark; rotting wood; fallen branches; soil.
Culture Characters Not known.
References 93, 128, 130, 242, 249, 312, 576.

Byssocorticium californicum Jülich 1973
Beih. Willdenowia 7:141.
Distribution United States: CA.
Host *Pinus muricata*.
Ecology Bark of log.
Culture Characters Not known.

Known only from the type specimen.

Byssocorticium pulchrum (Lundell) M. Christiansen 1960
Dansk Bot. Ark. 19:158.
Basionym *Corticium pulchrum* Lundell in Lundell and Nannfeldt, Fungi Exs. Suec. No. 1035, Uppsala, 1941.
Distribution Canada: ON. United States: NH, VT.
Hosts Not known.
Ecology Not known.
Culture Characters Not known.
Reference 312.

BYSSOPORIA **Larsen & Zak 1978** Atheliaceae

Byssoporia terrestre (DC.:Fr.) Larsen & Zak 1978
Canad. J. Bot. 56:1123.
Basionym *Boletus terrestre* DC., Flora France 5:39, 1815.
Synonym *Poria terrestris* (DC.) Sacc. 1888.
Distribution United States: AZ, CA, CO, ID, MT, NC, NJ, OR, PA.
Hosts *Picea engelmannii*, *Pinus contorta*, *Populus* sp., *Quercus* sp., *Tsuga heterophylla*, angiosperm, gymnosperm.
Ecology Mycorrhizal, at least, with *Tsuga heterophylla* (348); basidiomes produced on well-rotted wood; log.
Culture Characters Not known.
References 348, 402, 416.

It is uncertain how these records pertain to the varieties, below. However, Larsen and Zak (372) cite numerous J.L. Lowe collections as var. *parksii* and they may have been the basis for the above reports by Lowe (416).

Byssoporia terrestre var. ***aurantiaca*** Larsen & Zak 1978
Canad. J. Bot. 56:1124.
Distribution United States: OR.
Host *Pseudotsuga menziesii*.
Ecology Mycorrhizal; basidiomes produced on brown rotted decayed wood of stumps, logs and debris in second-growth stands.
Culture Characters Not known, but see data in Larsen and Zak (372), and Zak (655).
Reference 372.

Byssoporia terrestre var. ***lilacinorosea*** Larsen & Zak 1978
Canad. J. Bot. 56:1125.
Distribution United States: OR.
Host *Pseudotsuga menziesii*.
Ecology Mycorrhizal; basidiomes produced on brown rotted wood of old stumps, logs and debris in second-growth stands.
Culture Characters Not known, but see data in Larsen and Zak (372), and Zak and Larsen(657).
Reference 372.

Byssoporia terrestre var. ***parksii*** (Murrill) Larsen & Zak 1978
Canad. J. Bot. 56:1126.
Basionym *Poria parksii* Murrill, Mycologia 13:175, 1921.
Distribution Canada: AB. United States: AZ, CA, CO, ID, MT, NC, NJ, NY, OR, PA, VA.
Hosts *Lithocarpus densiflora*, *Picea* sp., *P. engelmannii*, *Pinus* sp., *P. contorta*, *P. ponderosa*, *Populus* sp, *Pseudotsuga menziesii*, *Quercus* sp., *Tsuga heterophylla*, gymnosperm.
Ecology Mycorrhizal at least with Douglas-fir; basidiomes produced on brown rotted wood of old stumps, logs and debris in second-growth stands.
Culture Characters Not known, but see data in Larsen and Zak (372), and Zak and Larsen (657).
Reference 372.

Byssoporia terrestre var. ***sartoryi*** (Bourd. & L. Maire) Larsen & Zak 1978
Canad. J. Bot. 56:1127.
Basionym *Poria sartoryi* Bourd. & L. Maire, Assoc. Fr. Avanc. Sci. 619, 1921.
Distribution United States: OR, WA.

Host *Pseudotsuga menziesii.*
Ecology Mycorrhizal; basidiomes produced on brown rotted wood of old stumps, logs and debris in second-growth stands.
Culture Characters Not known, but see data in Larsen and Zak (372), and Zak (655).
Reference 372.

Byssoporia terrestre var. ***sublutea*** Larsen & Zak 1978
Canad. J. Bot. 56:1128.
Distribution United States: OR.
Host *Pseudotsuga menziesii.*
Ecology Mycorrhizal; basidiomes produced on mineral soil and rotted wood under Douglas-fir and western hemlock.
Culture Characters Not known.
Reference 372.

CAERULICIUM Jülich 1981 Atheliaceae

Caerulicium neomexicanum (Gilbn. & Budington) Jülich 1985
Biblio. Mycol. 85:395.
Basionym *Byssocorticium neomexicanum* Gilbn. & Budington, Mycologia 62:673, 1970.
Distribution United States: AZ, CO, NM.
Hosts *Pinus ponderosa*, gymnosperm.
Ecology Associated with a white rot.
Culture Characters Not known.
References 197, 205, 208, 402.

CALATHELLA D. Reid 1964 Cyphellopsidaceae

Calathella albolivida (W.B. Cooke) Agerer 1983
Mitt. Bot. Staatssamml. München 19:189.
Basionym *Lachnella albolivida* Ellis ex W.B. Cooke, Beih. Sydowia 4:70, 1962.
Distribution United States: CO.
Hosts *Betula* sp., *Salix* sp.
Ecology Bark.
Culture Characters Not known.

Known only from the type specimens.

Calathella dichroa (W.B. Cooke) Agerer 1983
Mitt. Bot. Staatssamml. München 19:195.
Basionym *Lachnella dichroa* W.B. Cooke, Beih. Sydowia 4:71, 1962.
Distribution Canada: MB.
Host *Populus* sp.
Ecology Not known.
Culture Characters Not known.

Known only from the type specimen.

Calathella ellisii Agerer 1983
Mitt. Bot. Staatssamml. München 19:198.
Basionym Proposed as a *nom. nov.* to replace *Peziza campanula* Ellis, Bull. Torrey Bot. Club 8:73, 1881, a later homonym of *Peziza campanula* Nees 1816 which is now *Calyptella campanula.*
Synonyms *Flagelloscypha coloradensis* W.B. Cooke 1962, *Lachnella oregonensis* W.B. Cooke 1962. Both fide Agerer (5).
Distribution United States: CO, OR, UT.
Hosts *Populus* sp., *Salix* sp.
Ecology Branches.
Culture Characters Not known.
Reference 5.

Calathella eruciformis var. ***eruciformis*** (Batsch:Fr.) D. Reid 1964
Persoonia 3:123.
Basionym *Peziza eruciformis* Batsch, Elenchus Fungorum Latine Germanice, 125, 1783.
Synonyms *Cyphella eruciformis* (Batsch) Fr. 1822, *Lachnella eruciformis* var. *eruciformis* (Batsch) W.B. Cooke 1962, *Stromatocyphella lataensis* W.B. Cooke 1961, fide Reid (554).
Distribution Canada: AB, BC, MB, ON.
United States: CA, ID, MI, UT.
Hosts *Populus* sp., *P. tremuloides, P. trichocarpa, Salix* sp.
Ecology Bark; dead wood; fallen branches; twigs; decorticated logs; herbage.
Culture Characters Not known.
References 125, 130, 183, 545, 554.

Calathella eruciformis var. ***microspora*** (W.B. Cooke) Ginns & Lefebvre, *comb. nov.*
Basionym *Lachnella eruciformis* var. *microspora* W.B. Cooke, Beih. Sydowia 4:73, 1962.
Distribution Canada: ON. United States: CA.
Host *Populus* sp.
Ecology Not known.
Culture Characters Not known.
References 125, 128.

The basidiospores are only 3-4.5 x 2-3 μm (125). The new combination is proposed to keep the variety in the same genus with the species.

CALOCERA (Fr.) Fr. 1825 Dacrymycetaceae
Synonyms are *Calopposis* Lloyd 1925, and *Dacryomitra* Tul. 1872, fide McNabb (456).

Calocera cornea (Batsch:Fr.) Fr. 1827
Stirpes Agri Femsionensis 5:67.
Basionym *Clavaria cornea* Batsch, Elenchus Fungorum Latine Germanice, 139, 1783.
Synonyms *Calocera pilipes* Schw. 1832, *Calocera cornea* var. *minima* Coker 1920, *Calocera vermicularis* Lloyd 1923, *Calopposis nodulosa* Lloyd 1925. All fide McNabb (456).
Distribution Canada: AB, BC, MB, NS, ON, PQ.
United States: AZ, GA, ID, IN, LA, MA, MI, MN, NC, NH, NY, OR, SC, WA.

Hosts *Acer* sp., *A. macrophyllum, Alnus rubra, Betula* sp., *B. papyrifera, Fagus* sp., *F. grandifolia, Gleditsia* sp., *Juniperus virginiana, Liriodendron* sp., *Pinus* sp., *P. ayacahuite, P. contorta, P. ponderosa, P. strobiformis, P. strobus, Populus* sp., *P. tremuloides, P. trichocarpa, Pseudotsuga menziesii, Quercus* sp., *Tsuga canadensis*, angiosperm, less frequently gymnosperm.
Ecology Decorticated branches; causes a brown pocket rot and brown basal rot.
Culture Characters Not known.
References 51, 52, 107, 116, 119, 183, 197, 216, 242, 249, 335, 339, 406, 443, 456, 498, 499, 503, 511, 532, 577, 611.

Calocera glossoides (Pers.:Fr.) Fr. 1827
Stirpes Agri Femsonensis 5:67.
Basionym *Clavaria glossoides* Pers., Comment. fung. clav., 68, 1797.
Synonym *Dacryomitra glossoides* Bref. 1888, fide McNabb (456).
Distribution United States: NC, SC.
Hosts *Castanea dentata*, angiosperm.
Ecology Decorticated, rotted logs.
Culture Characters Not known.
References 111, 503.

Calocera viscosa (Pers.:Fr.) Fr. 1827
Stirpes Agri Femsonensis 5:67.
Basionym *Clavaria viscosa* Pers., Neues Mag. Bot. 1:117, 1794.
Distribution Canada: BC, NF, NS, ON, PQ.
United States: ID, MI, MN, NC, NY, TN, WA, WI.
Hosts *Abies balsamea, Picea mariana, Pinus* sp., gymnosperm.
Ecology Old stump; saprophytic on ground; causes a white rot.
Culture Characters Not known.
References 51, 107, 116, 119, 242, 249, 332 as *C. viscera*, 333, 443, 503, 532, 577.

CALYPTELLA Quél. 1886 — Tricholomataceae

Calyptella campanula* var. *campanula (Nees:Fr.) W.B. Cooke 1962
Beih. Sydowia 4:32.
Basionym *Peziza campanula* Nees, System Pilze Schwämme, 71, 1816.
Synonyms *Cyphella sulphurea* (Batsch) Fr. 1874, fide W.B. Cooke (125), *Calyptella campanula* var. *myceliosa* W.B. Cooke 1962, fide Redhead and Traquair (550).
Distribution Canada: ON.
United States: MI, NH, NY, PA.
Hosts *Arctium* sp., *Beta vulgaris, Impatiens* sp., *Scrophularia* sp., *Solanum tuberosum, Trifolium repens, Urtica dioica*.
Ecology Causes a root rot of tomatoes in greenhouses in England (105); in North America on live stems of herbs in damp places; dead stems near the ground.
Culture Characters Not known.
References 80, 125, 550.

Calyptella capula (Holmsk.:Fr.) Quél. 1888
Flore Mycol. France, 25.
Basionym *Peziza capula* Holmsk., Acta Nov. Havn. 1:286, 1799.
Synonym *Cyphella capula* (Holmsk.) Fr. 1838.
Distribution Canada: AB, BC, MB, NF, ON.
United States: FL, ID, IL, LA, MA, MD, NC, NJ, NY, OH, SC, VA, WA.
Hosts *Carex* sp., *Cirsium arvense, Eupatorium* sp., *Foeniculum* sp., *Lactuca serriola* var. *integrata, Lathyrus* sp., *Medicago sativa, Pelargonium zonale, Pteretis pensylvanica, Symphytum officinale,* herbs.
Ecology Dead wood; dead stems; herbs and grass litter; litter and duff; in a flower pot.
Culture Characters Not known.
References 80, 108, 116, 119, 125, 127, 131, 242, 545, 550.

Calyptella griseopallida (Weinm.) Parker-Rhodes 1954
Trans. Brit. Mycol. Soc. 37:332.
Basionym *Cyphella griseopallida* Weinm., Hym. Ross., 522, 1836.
Synonym *Cellypha griseopallida* (Weinm.) W.B. Cooke 1961
Distribution Canada: MB.
United States: IA, NH, NY, OH.
Hosts Not known.
Ecology Bark; dead leaves and twigs on the ground; pulpy rotted wood.
Culture Characters Not known.
References 80, 125, 378, 550.

Calyptella laeta (Fr.) W.B. Cooke 1962
Beih. Sydowia 4:40.
Basionym *Cyphella laeta* Fr., Epicrisis 568, 1838.
Distribution United States: NY.
Hosts Herbs.
Ecology Dead stems of large plants lying on the ground.
Culture Characters Not known.
Reference 80.

Cooke's (125) report from Ontario was redetermined (550) as *Calyptella campanula*.

CANDELABROCHAETE Boidin 1970 — Botryobasidiaceae

Candelabrochaete langloisii (Pat.) Boidin 1970
Cahiers Maboké 8:24.
Basionym *Hypochnus langloisii* Pat., Bull. Soc. Mycol. France 24:3, 1908.
Synonyms *Pellicularia langloisii* (Pat.) D.P. Rogers 1943, *Botryobasidium langloisii* (Pat.) Gilbn. & Budington 1970, *Phanerochaete insolita* Burdsall & Nakasone 1981, fide Burdsall (63).

Distribution United States: AZ, FL, LA, MS, NM.
Hosts *Carya* sp., *Liquidambar styraciflua, Pinus ponderosa*, angiosperm.
Ecology Decorticated wood; rotten wood; associated with a white rot.
Culture Characters
1.3r.26.32.36.38.43-44.54 (Ref. 474).
References 63, 64, 74, 197, 204, 216, 474, 565.

Candelabrochaete magnahypha (Burt) Burdsall 1984 Mycotaxon 19:391.
Basionym *Peniophora magnahypha* Burt, Ann. Missouri Bot. Gard. 12:238, 1926.
Distribution United States: FL.
Hosts Angiosperm.
Ecology Decaying wood; apparently associated with a white rot.
Culture Characters Not known.
References 63, 92.

Known only from the type specimen. Apparently mistaken to be a synonym of *C. langloisii* by Rogers and Jackson (567).

Candelabrochaete septocystidia (Burt) Burdsall 1984 Mycotaxon 19:392.
Basionym *Peniophora septocystidia* Burt, Ann. Missouri Bot. Gard. 12:260, 1926.
Distribution United States: MI, MN, NC, NJ, NY, TN, WI.
Hosts *Acer* sp., *A. saccharum, Betula alleghaniensis, Liriodendron tulipifera, Pinus* sp., *Populus* sp., *P. balsamifera*, angiosperm, gymnosperm.
Ecology Moss-covered bark; branch; logs; associated with a white rot.
Culture Characters
1.(2).6.7.(26).32.36.38.46-47.54.55 (Ref. 474).
References 61, 63, 214, 327, 474.

CELLYPHA Donk 1959 Tricholomataceae

Cellypha goldbachii (Weinm.) Donk 1959 Persoonia 1:85.
Basionym *Cyphella goldbachii* Weinm., Hym. Ross., 522, 1836.
Synonym *Cyphella caricina* Peck 1880, fide W.B. Cooke (125) and Donk (155).
Distribution Canada: ON. United States: MI, NY, PA.
Hosts *Carex* sp., *Impatiens* sp., *Juncus* sp., *Typha latifolia, Picea* sp., *Rubus* sp., herbs, woody plants, grasses, sedges.
Ecology Dead twigs; stems; leaves; culms.
Culture Characters Not known.
References 80, 125, 183, 517, 546.

Peziza cuticulosa Dickson:Fr. if the same fungus is an earlier name, see Donk (155).

Cellypha musaecola (Berk. & Curtis) W.B. Cooke 1961 Beih. Sydowia 4:55.
Basionym *Cyphella musaecola* Berk. & Curtis, J. Linn. Soc., Bot. 10:337, 1868.
Synonym *Cyphella fumosa* Cooke 1891, fide W.B. Cooke (125).
Distribution United States: SC.
Host *Gladiolus* sp.
Ecology Rotting leaves.
Culture Characters Not known.
Reference 80.

Cellypha subgelatinosa (Berk. & Curtis) W.B. Cooke 1961 Beih. Sydowia 4:58.
Basionym *Cyphella subgelatinosa* Berk. & Curtis, Grevillea 2:5, 1873.
Distribution United States: SC.
Host *Alnus serrulata.*
Ecology Not known.
Culture Characters Not known.
Reference 80.

CERACEOMYCES Jülich 1972 Cylindrobasidiaceae

Ceraceomyces borealis (Romell) Eriksson & Ryv. 1973 Cort. N. Europe 2:205.
Basionym *Merulius borealis* Romell, Ark. Bot. 11(3):27, 1911.
Synonym *Athelia borealis* (Romell) Parm. 1967, *Merulius gyrosus* Burt 1917, fide Ginns (226).
Distribution Canada: AB, BC, ON, PQ. United States: AL, CO, ID, ME, MI, MN, NC, NY, TN, WA.
Hosts *Abies* sp., *A. amabilis, Betula* sp., *B. papyrifera, Picea* sp., *P. glauca, P. mariana, Pinus* sp., *P. ponderosa, Populus* sp., *P. tremuloides, Quercus* sp., angiosperm, particularly aspen, gymnosperm.
Ecology Saprophytic, associated with a brown rot (226) or a white rot (406).
Culture Characters
2.3c.21.32.36.(38).40.46-47.(53).54.55 (Ref. 474);
2.3c.16.21.34.36.38.46.55 (Ref. 474, isolate JLL 8014);
(1).2.3.22.26.34.36.38.47.(48).54.55 (Ref. 226).
References 214, 226, 242, 402, 406, 447, 474.

Ceraceomyces serpens (Tode:Fr.) Ginns 1976 Canad. J. Bot. 54:147.
Basionym *Merulius serpens* Tode, Abh. Hallischen Naturf. Ges. 1:355, 1783.
Synonyms *Byssomerulius serpens* (Tode) Parm. 1967, *Ceraceomerulius serpens* (Tode) Eriksson & Ryv. 1973, *Merulius porinoides* Fr. 1821, fide Ginns (226), *Merulius ceracellus* Berk. & Curtis 1872, fide Ginns (226), *Merulius farlowii* Burt 1917, fide Ginns (226).
Distribution Canada: AB, BC, MB, NS, NTM, ON, PQ. United States: AK, AZ, CA, CO, FL, ID, MI, MN, MT, NC, NJ, NM, NY, SC, WA, WI.
Hosts *Abies* sp., *A. balsamea, A. concolor, A. magnifica, Acer* sp., *A. glabrum, A. saccharum, Alnus rubra,*

Betula sp., *B. alleghaniensis, Fagus* sp., *F. grandifolia, Larix occidentalis, Magnolia* sp., *Ostrya virginiana, Picea* sp, *P. engelmannii, P. glauca, P. mariana, Pinus* sp., *P. banksiana, P. cembra, P. ponderosa, P. resinosa, P. strobus, Platanus wrightii, Populus* sp., *P. tremuloides, P. trichocarpa, Quercus gambelii, Q. hypoleucoides, Thuja plicata*, angiosperm, gymnosperm.

Ecology Saprophytic, associated with a white rot.

Culture Characters
2.3c.(26).31c.32.36.38.(44).45-46.(50).(53).54.55.(59).(60) (Ref. 474);
1.(2).3.7.(26).32.36.40.43-47.(50).54.55.59 (Ref. 226).

References 116, 120, 128, 183, 197, 204, 208, 216, 226, 242, 249, 255, 391, 402, 406, 447, 474.

Ceraceomyces subapiculatus (Bres.) Ginns 1992 Mycotaxon 44:210.

Basionym *Corticium subapiculatum* Bres., Mycologia 17:69, 1925.

Synonym *Peniophora subapiculata* (Bres.) Burt 1926.

Distribution United States: ID, MT.

Hosts *Pinus monticola, P. ponderosa.*

Ecology Underside of charred, decorticated but not obviously decayed log; decorticated wood.

Culture Characters Not known.

References 92, 245.

Several of the collections cited by Burt are not cited here because they were misidentified (245, 567).

Ceraceomyces sublaevis (Bres.) Jülich 1972 Beih. Willdenowia 7:147.

Basionym *Corticium sublaeve* Bres., Ann. Mycol. 1:95, 1903.

Synonym *Peniophora sublaevis* (Bres.) Höhnel & Litsch. 1908.

Distribution Canada: BC, ON, PE, PQ. United States: CO, CT, NC, NH, NY, OH, PA, RI, TN, WA.

Hosts *Abies balsamea, Picea engelmannii, P. rubens, Rhododendron* sp., angiosperm, gymnosperm.

Ecology Branch; log; associated with a white rot.

Culture Characters
2.3.(24).(25).32.36.(39).46-47.55 (Ref. 588).

References 242, 312, 327, 402, 584.

Ceraceomyces sulphurinus (Karsten) Eriksson & Ryv. 1978 Cort. N. Europe 5:895.

Basionym *Tomentella sulphurina* Karsten, Bidrag Kännedom Finlands Natur Folk 48:420, 1889.

Synonym *Peniophora sulphurina* (Karsten) Höhnel & Litsch. 1908, *Phanerochaete sulphurina* (Karsten) Budington & Gilbn. 1973.

Distribution Canada: AB, BC, ON. United States: AL, AZ, DC, ID, KY, NH, NJ, NM, NY, OR, PA, SD.

Hosts *Picea* sp., *Pinus ponderosa*, gymnosperm.

Ecology Bark.

Culture Characters Not known.

References 92, 169, 197, 204, 208, 216, 242, 447, 584.

Ceraceomyces tessulatus (Cooke) Jülich 1972 Beih. Willdenowia 7:154.

Basionym *Corticium tessulatum* Cooke, Grevillea 6:132, 1878.

Synonyms *Athelia tessulata* (Cooke) Donk 1957, *Corticium apiculatum* Bres. 1925, fide Jülich (312), *Corticium illaqueatum* Bourd. & Galzin 1911, fide Jülich (312).

Distribution Canada: AB, AB-NT, BC, NS, ON, PQ. United States: AL, AZ, ID, LA, MD, ME, MO, NH, NM, NY, OR, RI, SC, VT, WA.

Hosts *Abies lasiocarpa, Alnus tenuifolia, Castanea* sp., *Picea* sp., *P. glauca, Pinus banksiana, P. ponderosa, Pseudotsuga menziesii*, angiosperm.

Ecology Decaying branches; bark of decaying logs; bark on ground.

Culture Characters Not known.

References 53, 93, 197, 204, 216, 242, 312, 447.

Most of the specimens that Burt (93) referred to *C. apiculatum* are not cited here because they were misidentified (567).

Ceraceomyces violascens (Fr.:Fr.) Jülich 1972 Beih. Willdenowia 7:162.

Basionym *Himantia violascens* Fr., Obs. Mycol. 1:211, 1815.

Distribution Canada: ON. United States: NY.

Hosts Not known.

Ecology Not known.

Culture Characters Not known.

Reference 312.

CERATOBASIDIUM D.P. Rogers 1935 Ceratobasidiaceae

Ceratobasidium anceps (Bres. & Syd.) Jackson 1949 Canad. J. Res., C, 27:243.

Basionym *Tulasnella anceps* Bres. & Syd., Ann. Mycol. 8:490, 1910.

Anamorph *Sclerotium deciduum* J.J. Davis, fide Jackson (304).

Distribution Canada: ON, PQ. United States: ME, NH, NY, OR, WA, WI.

Hosts Farr et al. (183) list 32 genera of host plants. It includes trees, herbs, and ferns. The references below add the genera *Acer, Achillea, Apocynum, Athyrium, Carex, Chrysanthemum, Corylus, Cystopteris, Dennstaedtia, Dicentra, Dryopteris, Epilobium, Habenaria, Hackelia, Hieracium, Lactuca, Linnaea, Lonicera, Lycopus, Maianthemum, Petasites, Plantago, Prenanthes, Prunella, Pteretis, Pyrola, Pyrus, Ribes, Streptopus, Taraxacum, Tiarella, Trillium, Viburnum, and Viola* to the list.

Ecology Pathogenic on live petioles and leaves, causing brown necrotic patches; basidiomes produced on the underside of leaves in advance of the necrotic tissue.

Culture Characters
2.6.(22).(24).(25).26.32.36.38.43-45.55 (Ref. 588).

References 116, 183, 242, 304, 330, 443.

Ceratobasidium calosporum D.P. Rogers 1935
Stud. Nat. Hist. Iowa Univ. 17:5.
Distribution United States: IA.
Host *Ulmus* sp.
Ecology Bark of dead branch; dead wood.
Culture Characters Not known.
References 443, 563.

Ceratobasidium cornigerum (Bourd.) D.P. Rogers 1935
Stud. Nat. Hist. Iowa Univ. 17:5.
Basionym *Corticium cornigerum* Bourd., Rev. Sci. Bourbonnais Centr. France 35:4, 1922.
Anamorph *Ceratorhiza goodyearea-repentis* (Costantin & Dufour) R.T. Moore 1987 (≡ *Rhizoctonia goodyearea-repentis* Costantin & Dufour 1920).
Distribution Canada: AB, ON. United States: CO, IA, IL, NEW ENGLAND, NY, OH, OR.
Hosts *Coeloglossum, Platanthera obtusata, Quercus* sp., *Q. gambelii, Rhus glabra, Salix* sp., *Ulmus* sp., angiosperm.
Ecology Bark and wood; decayed wood; mycorrhizal with exotic (106) and native orchids (138a).
Culture Characters
(2).6.(19).(24).(25).32.36.(39).42-44.(54).(55) (Ref. 588).
References 137a, 183, 387, 400, 402, 443, 563.

Ceratobasidium gramineum (Ikata & T. Matsuura) Oniki, Ogoshi & Araki 1986
Trans. Mycol. Soc. Japan 27:156.
Basionym *Corticium gramineum* Ikata & T. Matsuura in Matsuura, Byochugai Zasshi 17:453, 1910 (original not seen).
Synonym *Ceratobasidium cereale* D. Murray & L.L. Burpee 1984, fide Oniki et al. (loc. cit.).
Anamorph *Ceratorhiza cerealis* (Van der Hoeven) R.T. Moore 1987 (≡ *Rhizoctonia cerealis* Van der Hoeven 1977).
Distribution Canada: AB.
United States: AR, NC, NY, OH, PA.
Hosts *Agrostis* spp., *A. palustris, A. stolonifera, Cynodon dactylon, Festuca arundinacea, F. pratensis, Goodyera repens, Poa annua, P. pratensis, Lolium perenne, Triticum aestivum, Zoysia japonica.*
Ecology Isolated from foliage, causes sharp eyespot and pre-emergence damping-off of cereals; in virulence tests caused post-emergence damping-off in a variety of cereals; yellow patch of turf grasses; mycorrhizal with *Goodyera*.
Culture Characters Not known.
References 77, 139, 183, 330, 408, 449, 531, 574, 585.

This fungus is widespread, geographically, and has been reported (330) to infect species of Chenopodiaceae, Gramineae, Leguminosae, Liliaceae, Malvaceae, and Solanaceae. Thus the host range appears to be much broader than the epithets imply.

Ceratobasidium obscurum D.P. Rogers 1935
Stud. Nat. Hist. Iowa Univ. 17:6.
Distribution Canada: AB. United States: IA, MA.
Hosts *Amerorchis rotundifolia, Ulmus* sp.
Ecology Endophyte in orchid roots; lower side of rotted log; dead wood.
Culture Characters Not known.
References 138, 443, 563.

Ceratobasidium ochroleuca (F. Noack) Ginns & Lefebvre, *comb. nov.*
Basionym *Hypochnus ochroleuca* F. Noack in Saccardo, Sylloge Fung. 16:197, 1902.
Synonyms *Corticium stevensii* Burt 1918, proposed as a *nom. nov.* for *H. ochroleuca* because there already was a *Corticium ochroleuca* Bres. 1892, *Pellicularia koleroga* Cooke sensu D.P. Rogers 1943, fide Talbot (599).
Distribution United States: AL, FL, GA, IN, LA, MS, NC, OK, SC, TX, WA.
Hosts Farr et al. (183) list 48 genera of host plants, which include a wide variety of orchard and ornamental crops.
Ecology Thread blight of twigs and leaves, causing leaves to dry and fall.
Culture Characters Not known.
References 84, 93, 96, 108, 183, 274, 565.

The discussion of the epithets and species concepts by Burt (84) is pertinent. Twenty-five years later, Rogers (565) updated the circumscription, synonymy, and the preferred name. And in 1965 Talbot's (599) comment (under *C. stevensii*) put the earlier comments in perspective. Talbot considered this fungus to be a species of *Ceratobasidium*. We agree and propose the transfer. Neither *Corticium* nor *Hypochnus*, genera where this species has been placed, is currently suitable.

Ceratobasidium oryzae-sativae P.S. Gunnel & R.K. Webster 1987
Mycologia 79:732.
Anamorph *Ceratorhiza oryzae-sativae* (Sawada) R.T. Moore 1989 (≡ *Sclerotium oryzae-sativae* Sawada, Trans. Nat. Hist. Soc. Formosa 9:138, 1919 or Descriptive Catalogue of the Formosa Fungi, Dept. Agric. Gov. Res. Inst. Formosa, Japan 2:171, 1922 (neither seen by us), *Rhizoctonia oryzae-sativae* (Sawada) Mordue 1974).
Distribution United States: CA.
Host *Oryza sativa.*
Ecology Parasitic on live plants, causes aggregate sheath spot.
Culture Characters Described by Mordue (1974, CMI Descript. 409) but not translatable to a useful species code.

Although widespread in Asia, this fungus in North America is restricted to California, where it was first detected in the late 1960s.

Ceratobasidium ramicola Tu, D.A. Roberts & Kimbrough 1969
Mycologia 61:781-782.
Anamorph *Ceratorhiza ramicola* (G.F. Weber & D.A. Roberts) R.T. Moore 1987 (≡ *Rhizoctonia ramicola* G.F. Weber & D.A. Roberts 1951).
Distribution United States: FL.
Hosts *Buxus microphylla, Carissa* sp., *Cinnamomum camphora, Elaeagnus* sp., *E. pungens, Erythrina* sp., *Euonymus japonica, Feijoa sellowiana, Ilex cornuta, I. crenata, I. vomitoria, Lagerstroemia indica, Ligustrum lucidum, L. sinense, Myrica cerifera, Pittosporum* sp., *P. tobira, Rhaphiolepis indica, R. umbellata, Rhododendron* sp.
Ecology Parasitic on twigs, petioles and leaf blades; causes thread blight of foliage.
Culture Characters Not known.
References 183, 274, 330, 605.

Nearly all the host records are based upon the presence of the anamorph.

***Ceratobasidium* species A**
Anamorph *Ceratorhiza fragariae* (S.S. Husain & W.E. McKeen) R.T. Moore 1987 (≡ *Rhizoctonia fragariae* S.S. Husain & W.E. McKeen 1963).
Distribution Canada: ON. United States: CA, WV.
Host *Fragaria* sp.
Ecology Parasitic, causes black root rot and winter killing diseases.
Culture Characters Not known.
References 183, 330.

Presumably a species of *Ceratobasidium*, but the teleomorph has not been found.

***Ceratobasidium* species B**
Anamorph *Sclerotium hydrophilum* Sacc. 1892, in W. Rothert, Bot. Zeitung (Berlin) 50:322.
Distribution Canada: MB.
United States: AR, CA, ID, LA, MN, NY, WI.
Hosts *Alisma* sp., *Echinochloa colonum, E. crusgalli, Glyceria fluitans, Nuphar luteum* var. *variegatum, Nymphaea odorata, Oryza sativa, Sagittaria* sp., *S. sagittifolia, Trapa natans, Typha latifolia, Zizania aquatica, Zizaniopsis miliacea.*
Ecology Wet habitats; causes stem rot of *Zizania aquatica*; leaf sheath rot of *Echinochloa*; sheath rot and stem rot of *Oryza sativa*; petioles; dead culms; culms and leaves.
Culture Characters Not known, but growth on PDA was 35-79 mm diam in 4 days, with sclerotia beginning to form in 4-5 days and mycelium not pigmented (541).
References 183, 541.

Similarities in the dolipore septum led Punter et al. (541) to conclude that this fungus was "probably near *Ceratobasidium cornigerum*".

***Ceratobasidium* species C**
Anamorph *Rhizoctonia endophytica* var. *endophytica* Saksena & Vaartaja 1960, Canad. J. Bot. 38: 936-937.
Distribution Canada: SK.
Hosts *Pinus banksiana, P. sylvestris, Picea glauca, P. pungens*, angiosperms.
Ecology Apparently an endophyte in roots, also isolated from damping-off in seedlings.
Culture Characters Not known.

Binucleate and anastomoses with mycelium of a *Ceratobasidium* (600).

CERINOMYCES Martin 1949 Dacrymycetaceae
Synonym is *Ceracea* Cragin in part, fide Martin (443).

Cerinomyces canadensis (Jackson & G.W. Martin) G.W. Martin 1949
Mycologia 41:85.
Basionym *Ceracea canadensis* Jackson & G.W. Martin, Mycologia 32: 693, 1940.
Distribution Canada: ON.
Hosts Gymnosperm.
Ecology Not known.
Culture Characters Not known.
References 437, 443, 454.

Cerinomyces ceraceus Ginns 1982
Canad. J. Bot. 60:519.
Distribution United States: MS.
Host *Magnolia grandiflora.*
Ecology Bark of old, rotted, 3 cm diam. branch, causes a uniform brown rot.
Culture Characters
1.6.13.33.36.38.47.54 (Ref. 231).
References 231, 577.

Cerinomyces crustulinus (Bourd. & Galzin) G.W. Martin 1949
Mycologia 41:85.
Basionym *Ceracea crustulina* Bourd. & Galzin, Bull. Soc. Mycol. France 39:266, 1924.
Distribution Canada: BC. United States: LA, MI.
Hosts Angiosperm, gymnosperm.
Ecology Dead wood.
Culture Characters Not known.
References 172, 419, 454, 502, 577.

Two reports (52, 440) were subsequently (443) referred to *Cerinomyces pallidus*.

Cerinomyces lagerheimii (Pat.) McNabb 1964
New Zealand J. Bot. 2:421.
Basionym *Ceracea lagerheimii* Pat., Bull. Soc. Mycol. France 9:141, 1893.

Distribution United States: FL, SC.
Hosts Angiosperm, gymnosperm.
Ecology Dead wood.
Culture Characters Not known.
Reference 454.

Cerinomyces pallidus G.W. Martin 1949
Mycologia 41:83.
Synonym *Ceracea crustulina* sensu Brasfield 1940, fide McNabb, 1964 (454).
Distribution Canada: ON.
United States: CO, IA, IL, WI.
Hosts *Malus* sp., *Quercus* sp., *Tsuga canadensis*, angiosperm, gymnosperm.
Ecology Decayed wood.
Culture Characters Not known.
References 52, 388, 390, 394, 440, 441, 443, 454.

CEROCORTICIUM Henn. 1899 Hyphodermataceae

Cerocorticium molle (Berk. & Curtis) Jülich 1975
Persoonia 8:219.
Basionym *Corticium molle* Berk. & Curtis, J. Linn. Soc., Bot. London 10:336, 1868.
Synonyms *Corticium armeniacum* Sacc. 1888, *Corticium ceraceum* Berk. & Rav. ex Massee 1890. Both fide Burt (93).
Distribution United States: AL, LA, NC, NJ, OH, VA, SC.
Hosts Angiosperms.
Ecology Decaying trunks.
Culture Characters Not known.
References 93, 316, 616.

CHAETODERMELLA Rauschert 1988 Chaetodermataceae
Proposed as a *nom. nov.* to replace *Chaetoderma* Parm. 1968, which is a later homonym of *Chaetoderma* Kützing 1843.

Chaetodermella luna (D.P. Rogers & Jackson) Rauschert 1988
Haussknechtia 4:52.
Basionym *Peniophora luna* Romell ex D.P. Rogers & Jackson, Farlowia 1:320, 1943.
Synonyms *Chaetoderma luna* (Romell) Parm. 1968, *Peniophora odorata* sensu Burt 1926, fide Rogers and Jackson (567).
Distribution Canada: AB, AB-NT, BC, YT. United States: AK, AZ, CO, ID, MT, NM, UT, WA, WY.
Hosts *Abies* sp., *A. amabilis*, *A. lasiocarpa*, *A. lasiocarpa* var. *lasiocarpa*, *Larix* sp., *L. occidentalis*, *Picea* sp., *P. engelmannii*, *P. glauca*, *P. mariana*, *P. sitchensis*, *Pinus* sp., *P. aristata* var. *longaeva*, *P. albicaulis*, *P. contorta*, *P. monticola*, *P. ponderosa*, *Populus tremuloides*, *Pseudotsuga* sp., *P. menziesii*, *Thuja* sp., *Tsuga heterophylla*, gymnosperm.
Ecology Causes decay in mature and overmature *Pinus contorta*; a minor trunk rot in *Abies lasiocarpa* and *Picea engelmannii*; associated with a brown cubical rot.
Culture Characters
1.(2).3c.9.32.36.38.47.(50).(53).55.58 (Ref. 474);
1.(2).3.7.32.36.39.45-47.50.55 (Ref. 452).
References 92, 116, 119, 133, 144, 183, 197, 199, 204, 208, 216, 242, 255, 274, 402, 407, 452, 474.

CHONDROSTEREUM Pouzar 1959 Meruliaceae

Chondrostereum purpureum (Pers.:Fr.) Pouzar 1959
Ceská Mykol. 13:18.
Basionym *Thelephora purpurea* Pers., Disp. Meth. Fun., 30, 1797.
Synonyms *Stereum purpureum* (Pers.) Fr. 1838, *Stereum micheneri* Berk. & Curtis 1873, fide Lentz (379), *Corticium nyssae* Berk. & Curtis 1873, fide Lentz (379) and Hjortstam (288), *Stereum rugosiusculum* Berk. & Curtis 1872, fide Lentz (379).
Distribution Canada: AB, AB-NT, BC, MB, NB, NF, NFL, NS, ON, PE, PQ, SK, YT. United States: AK, AL, CA, CT, DE, FL, GA, IA, ID, IN, KS, MA, MD, ME, MI, MN, MO, MT, ND, NJ, NH, NY, OH, OK, OR, PA, VA, VT, WA, WI, WV, WY.
Hosts *Abies* sp., *A. amabilis*, *A. balsamea*, *A. grandis*, *A. lasiocarpa*, *Acer* sp., *A. macrophyllum*, *A. negundo*, *A. nigrum*, *A. rubrum*, *A. rubrum* var. *rubrum*, *A. saccharinum*, *A. saccharum*, *A. spicatum*, *Aesculus hippocastanum*, *Alnus* sp., *A. rubra*, *A. rugosa*, *A. sinuata*, *A. tenuifolia*, *Amelanchier canadensis*, *Arbutus* sp., *A. menziesii*, *Betula* sp., *B. alba*, *B. alleghaniensis*, *B. occidentalis*, *B. papyrifera*, *B. pendula*, *B. populifolia*, *B. pumila*, *Carpinus caroliniana*, *Carya* sp., *C. cordiformis*, *Castanea* sp., *Celtis* sp., *C. laevigata*, *Cornus florida*, *Crataegus* sp., *Cydonia oblonga*, *Fagus* sp., *F. grandifolia*, *Larix laricina*, *Liriodendron tulipifera*, *Malus* sp., *x M. adstringens*, *M. baccata*, *M. pumila*, *M. sylvestris*, *Nyssa* sp., *N. sylvatica*, *Ostrya virginiana*, *Picea glauca*, *Platanus occidentalis*, *Populus* sp., *P. balsamifera*, *P. deltoides*, *P. grandidenta*, *P. nigra*, *P. nigra* var. *italica*, *P. tremuloides*, *P. trichocarpa*, *Prunus* sp., *P. armeniaca*, *P. avium*, *P. cerasus*, *P. domestica*, *P. emarginata*, *P. laurocerasus*, *P. padus*, *P. persica*, *P. pumila*, *Pseudotsuga menziesii*, *Pyrus communis*, *Quercus* sp., *Q. garryana*, *Q. stellata*, *Salix* sp., *S. babylonica*, *S. nigra*, *Sorbus americana*, *Syringa vulgaris*, *Thuja plicata*, *Tsuga heterophylla*, *Ulmus* sp., *U. americana*, *Vitis* sp., *V. labrusca*, angiosperm, gymnosperm.
Ecology Causes silver leaf disease, the fungus produces toxins which produce the foliar symptoms and ultimately kill branches or the entire tree; causes a minor trunk rot in *Betula alleghaniensis*; basidiomes produced on old sticks; dead stems; dead stumps; logs; pulpwood; associated with a white rot.

Culture Characters
(1).2.3i.8.13.21.32.36.38.(39).42.(50).54.(55).60 (Ref. 474);
2.3.(12).15.(25).32.36.38.42.54.(55).60 (Ref. 575);
2.3c or 3i.15b.32.36.38.41-42.48.54.(55).60.61 (Ref. 104).
References 89, 96, 97, 104, 116, 127, 132, 133, 178, 180, 181, 183, 242, 249, 328, 335, 340, 379, 391, 406, 448, 474, 519, 595, 603.

CLAVICORONA Doty 1947 — Clavicoronaceae

Clavicorona cristata (Kauffm.) Doty 1947
Lloydia 10:40.
Basionym *Craterellus cristatus* Kauffm., Pap. Michigan Acad. Sci. 11:172, 1929
Distribution Canada: PQ. United States: OR, WA.
Hosts *Pseudotsuga menziesii*, gymnosperm.
Ecology Decayed logs.
Culture Characters Not known.
Reference 148.

Clavicorona divaricata Leathers & A.H. Sm. 1967
Mycologia 59:458.
Distribution United States: AZ, ID.
Hosts *Abies* sp., *Populus* sp., *P. deltoides, P. tremuloides*, gymnosperm.
Ecology Rotting logs.
Culture Characters Not known.
References 148, 183.

Clavicorona piperata (Kauffm.) Leathers & A.H. Sm. 1967
Mycologia 59:461.
Basionym *Clavaria piperata* Kauffm., Pap. Michigan Acad. Sci. 8:146, 1928.
Synonym *Clavicorona avellanea* Leathers & A.H. Sm. 1967, fide Dodd (148).
Distribution Canada: BC. United States: ID, WA.
Hosts *Pinus monticola, Tsuga* sp., gymnosperm.
Ecology Rotting logs.
Culture Characters Not known.
References 148, 183.

Clavicorona pyxidata (Pers.:Fr.) Doty 1947
Lloydia 10:43.
Basionym *Clavaria pyxidata* Pers., Neues Mag. Bot. 1:117, 1794.
Synonym *Clavicorona coronata* (Schw.) Doty 1947, fide Dodd (148).
Distribution Canada: MB, NS, ON, PQ.
United States: AL, CO, CT, DE, IA, IL, IN, MA, MD, ME, MI, MN, MO, NC, NH, NJ, NM, NY, OH, OR, PA, SC, TN, UT, VA, VT, WA, WI, WY.
Hosts *Acer* sp., *A. saccharinum, A. saccharum, Alnus* sp., *Betula* sp., *B. alleghaniensis, Carya* sp., *Cornus florida, Fraxinus americana, Liquidambar* sp., *L. styraciflua, Populus* sp., *P. tremuloides, Quercus* sp., *Q. alba, Rhododendron* sp., *Salix* sp., *Tsuga* sp., *T. canadensis, Ulmus* sp., angiosperm, gymnosperm.
Ecology Slash; logs; associated with a white rot.
Culture Characters
2.3.11.15.26.34.36.44-45.(48).53.54.55.60 (Ref. 148).
References 148, 161, 183, 214, 242, 249, 402, 406, 610.

Clavicorona taxophila (Thom) Doty 1947
Lloydia 10:39.
Basionym *Craterellus taxophilus* Thom, Bot. Gaz. (Crawfordsville) 37:215, 1904.
Distribution Canada: BC, ON, PQ.
United States: CA, ID, MI, NY, OR.
Hosts *Pinus* sp., *Pseudotsuga menziesii, Taxus canadensis, Thuja plicata*, gymnosperm.
Ecology Fallen needles and twigs on ground under gymnosperms, probably lignicolous (148).
Culture Characters Not known.
References 148, 161.

Clavicorona turgida (Lév.) Corner 1957
Darwiniana 11:195.
Basionym *Clavaria turgida* Lév., Ann. Sci. Nat. Bot., Sér. III, 5:155, 1846.
Distribution United States: NC.
Hosts Not known.
Ecology Rotten wood.
Culture Characters Not known.
Reference 148.

CLAVULICIUM Boidin 1957 — Clavulinaceae

Clavulicium delectabile (Jackson) Hjort. 1973
Svensk Bot. Tidskr. 67:106.
Basionym *Corticium delectabile* Jackson, Canad. J. Res., C, 26:145, 1948.
Distribution Canada: ON.
Hosts Not known.
Ecology Decayed; decorticated wood; humus; ground.
Culture Characters Not known.
Reference 303.

Jülich (317) reduced *Corticium delectabile* to synonymy under *Membranomyces spurius* (Bourd.) Jülich, but Eriksson and Ryvarden (170) preferred to keep these two names separate and placed them in *Clavulicium*.

Clavulicium macounii (Burt) Eriksson & Boidin 1968
In Parmasto, Consp. Syst. Cort., 165.
Basionym *Corticium macounii* Burt, Ann. Missouri Bot. Gard. 13:256, 1926.
Synonym *Corticium vinososcabens* Burt 1926, fide Rogers and Jackson (567), and Hjortstam (281).
Distribution Canada: BC, PQ.
United States: NH, NY, VT, WI.
Hosts *Abies lasiocarpa, A. "rubra", Picea glauca, Pinus* sp., *Tsuga canadensis*.
Ecology Decaying wood.
Culture Characters Not known.
References 93, 116, 242, 281, 393.

Liberta (393) and Pouzar (538) seem to have been mistaken in associating this name with a fungus which has four sterigmate basidia (281).

Clavulicium venosum (Berk. & Rav.) Ginns 1992 Mycotaxon 44:213.
Basionym *Corticium venosum* Berk. & Rav., Grevillea 1:177, 1873
Distribution United States: SC.
Hosts Not known.
Ecology Well-rotted wood of logs.
Culture Characters Not known.
References 93, 245.

Known only from the type specimen.

CLIMACODON Karsten 1881 Climacodontaceae
Synonym is *Donkia* Pilát 1936, fide Maas Geesteranus (427).

Climacodon pulcherrimus (Berk. & Curtis) Nikolaeva 1961 [Cryptog. Plants USSR] 6, Fungi (2):194.
Basionym *Hydnum pulcherrimum* Berk. & Curtis, Hooker's J. Bot. Kew Gard. Misc. 1:235, 1849.
Synonyms *Donkia pulcherrima* (Berk. & Curtis) Pilát 1936, *Steccherinum pulcherrimum* (Berk. & Curtis) Banker 1906, *Hydnum kauffmanii* Peck 1907, fide Lloyd (411), *Hydnum australe* Lloyd 1919, fide Lloyd (411).
Distribution Canada: MB, ON. United States: AL, AR, AZ, DC, DE, FL, GA, IA, IN, KY, LA, MA, MD, MI, MS, NC, NJ, NM, OH, PA, SC, TX, WI.
Hosts *Abies* sp., *A. concolor, Acer negundo, Betula nigra, B. papyrifera, Carya* sp., *Fagus* sp., *Liquidambar* sp., *L. styraciflua, Pinus* sp., *P. caribaea, P. leiophylla* var. *chihuahuana, P. ponderosa, Platanus wrightii, Populus* sp., *Quercus* sp., *Q. alba, Q. laurifolia, Q. lyrata, Q. macrocarpa, Q. nigra, Q. phellos, Q. virginiana*, angiosperm.
Ecology Dead wood; logs; associated with a white rot.
Culture Characters
(1).2.5.21.34.36.40.41-42.48.50.53.54.(57) (Ref. 474).
References 25, 109, 116, 183, 197, 204, 208, 216, 333, 428, 463, 474, 593.

The Ontario report is based upon three collections in DAOM (100981, 182520, and 182846).

Climacodon septentrionalis (Fr.) Karsten 1881 Rev. Mycol. (Toulouse) 3:20.
Basionym *Hydnum septentrionale* Fr., Syst. Mycol. 1:414, 1821.
Synonym *Steccherinum septentrionale* (Fr.) Banker 1906.
Distribution Canada: AB, AB-NT, MB, NS, ON, PQ. United States: AL, IA, IN, MA, MD, MI, MN, MS, NJ, NY, OH, PA, TN, TX.
Hosts *Acer* sp., *A. negundo, A. rubrum, A. rubrum* var. *rubrum, A. saccharinum, A. saccharum, Betula alleghaniensis, B. papyrifera, Carya* sp., *Fagus* sp., *F. grandifolia, Liquidambar styraciflua, Liriodendron tulipifera, Malus sylvestris, Nyssa* sp., *Populus* sp., *P. tremuloides, Quercus coccinea, Q. nigra, Q. phellos, Tilia americana, Ulmus* sp., angiosperm.
Ecology Causes a heart rot in species of *Acer* and *Fagus*; a minor trunk rot in *Carya* species; basidiomes typically produced on live trunks of maple; logs; associated with a white rot.
Culture Characters
2.6.8.13.36.38.(39).44-45.50.54 (Ref. 474).
References 25, 116, 183, 242, 249, 274, 333, 406, 463, 474.

CONIOPHORA DC. 1815 Coniophoraceae

Coniophora arida var. ***arida*** (Fr.) Karsten 1868 Not. Sällsk. Fauna Fl. Fenn. Förh. 9:370.
Basionym *Thelephora arida* Fr., Elenchus 1:197, 1828.
Synonym *Coniophora cookei* Massee 1889, fide Ginns (234).
Distribution Canada: AB-NT, BC, MB, NB, NS, NT, NT, ON, PE, PQ, YT. United States: AL, AZ, CA, CO, CT, FL, GA, IA, ID, IL, IN, KS, KY, LA, MA, MD, ME, MO, MT, NC, NC-TN, NJ, NM, NY, OH, OR, PA, RI, SC, VT, WA, WI.
Hosts *Abies* sp., *A. concolor, A. fraseri, A. grandis, A. lasiocarpa, Acer* sp., *A. saccharum, Alnus rubra, Betula alleghaniensis, B. populifolia, Castanea* sp., *Chamaecyparis thyoides, Chilopsis linearis, Fagus* sp., *Juglans major, Juniperus monosperma, Picea* sp., *P. engelmannii, P. rubens, Pinus* sp., *P. attenuata, P. ayacahuite, P. banksiana, P. cembroides, P. contorta, P. echinata, P. edulis, P. ponderosa, P. resinosa, P. rigida, P. strobiformis, P. strobus, P. virginiana, Platanus wrightii, Populus* sp., *Populus tremuloides, Prunus serotina, Pseudotsuga menziesii, Pyrus* sp., *Quercus* sp., *Rhododendron maximum, Taxodium distichum, Thuja plicata, Tsuga* sp., angiosperm, gymnosperm.
Ecology Rarely associated with decay behind wounds in live spruce; basidiomes produced on dead, rotting wood and bark of branches; slash; trunks; logs; stumps; discarded lumber, burlap matting; associated with a brown rot.
Culture Characters
1.5.(7).(11).16.32.36.(38).(39).43.(44).50.53.54.55.56.57 (Ref. 234).
References 96, 116, 119, 183, 197, 199, 204, 208, 216, 234, 242, 249, 327, 391, 402, 406, 448.

Coniophora arida var. ***suffocata*** (Peck) Ginns 1982 Opera Bot. 61:24.
Basionym *Corticium suffocatum* Peck, Annual Rep. New York State Mus. 30:48, 1878.
Synonyms *Coniophora suffocata* (Peck) Massee 1889, *Coniophora betulae* Karsten 1896, fide Ginns (234), *Hypochnus flavobrunneus* Dearness & Bisby 1929, fide Ginns (234).

Distribution As in var. *arida*.
Hosts As in var. *arida*.
Ecology As in var. *arida*.
Culture Characters As in var. *arida*.
References 116, 133, 199, 208, 234, 242, 249, 402.

Coniophora eremophila Lindsey & Gilbn. 1975
Mycotaxon 2:86.
Distribution United States: AZ.
Hosts *Carnegiea gigantea, Chilopsis* sp., *C. linearis, Juglans* sp., *J. major, Juniperus* sp., *J. monosperma, Olneya* sp., *O. tesota, Prosopis* sp., *P. juliflora, Saguaro* sp., *Sambucus* sp., *S. mexicana*.
Ecology Saprophytic on dead desert angiosperms and juniper; associated with a brown rot.
Culture Characters Not known.
References 199, 207, 211, 234, 403.

Coniophora fusispora (Cooke & Ellis) Sacc. 1888
Sylloge Fung. 6:650.
Basionym *Corticium fusisporum* Cooke & Ellis, Grevillea 8:11, 1879.
Distribution Canada: BC, NS, ON.
United States: AL, FL, MA, NH, NJ, NY, OR, WA.
Hosts *Juniperus virginiana, Picea* sp., *P. excelsa, Pinus* spp., *P. banksiana, P. niger, P. rigida, P. strobus, P. sylvestris, Platanus* sp., *Pseudotsuga menziesii, Quercus* sp., *Tsuga canadensis*, gymnosperm.
Ecology Occasionally associated with butt rot in live conifers; basidiomes produced near the base of live trees; dead fallen trunks; sound or well-rotted wood and bark; leaves; needles and debris on the ground; associated with a brown rot.
Culture Characters
1.(2).5.7.11.16.32.36.38.(39).45.53.54.55.56 (Ref. 234).
References 116, 183, 199, 202, 234, 242, 249, 448.

Coniophora hanoiensis Pat. 1907
Bull. Soc. Mycol. France 23:76.
Distribution United States: NJ.
Host *Nyssa* sp.
Ecology Rotting wood.
Culture Characters Not known.
Reference 234.

Coniophora marmorata Desm. 1823
Cat. pls. ommises bot. belgique et Fl. Nord Fr., 18.
Distribution United States: NY.
Host *Picea* sp.
Ecology Loose cement floor in basement; spruce floorboard; under wall-paper in house; wall of hot house; decaying trunks in hot house; exterior of flower pot.
Culture Characters
1.5.8.32.36-37.38.(44).45.55.56 (Ref. 234).
Reference 234.

Coniophora olivacea (Fr.:Fr.) Karsten 1882
Bidrag Kännedom Finlands Natur Folk 37:162.
Basionym *Hypochnus olivaceus* Fr., Obs. Mycol. 2:282, 1818.
Synonyms *Coniophorella olivacea* (Fr.) Karsten 1889, *Thelephora sistotremoides* Schw. 1822, *Hymenochaete ellisii* Berk. & Cooke 1876, *Corticium brunneoleum* Berk. & Curtis 1873, *Corticium leucothrix* Berk. & Curtis 1873, *Rhinotrichum corticioides* Cooke 1883, *Zygodesmus indigoferus* Ellis & Ev. 1885. All fide Ginns (234).
Distribution Canada: AB, BC, MB, NS, ON, PQ, SK, YT. United States: AL, AZ, CA, CO, GA, IA, ID, LA, MA, MD, ME, MN, MO, MT, NC, NH, NJ, NM, NY, OH, OR, PA, SC, VT, WA, WV.
Hosts *Abies* sp., *A. balsamea, A. grandis, A. lasiocarpa, A. magnifica, Alnus rubra, Betula alleghaniensis, Fagus sylvatica, Picea* sp., *P. engelmannii, P. glauca, P. mariana, Pinus* sp., *P. banksiana, P. contorta, P. echinata, P. leiophylla* var. *chihuahuana, P. monticola, P. ponderosa, P. strobus, P. sylvestris, Platanus wrightii, Populus* sp., *P. tremuloides, Pseudotsuga menziesii, Quercus* sp., *Q. alba, Q. gambelii, Rhamnus purshiana, Taxodium distichum, Thuja occidentalis, T. plicata, Tsuga canadensis, T. mertensiana*, angiosperm, gymnosperm.
Ecology Causing decay in live spruce; occasionally on boards and timbers in buildings; slash; logs; associated with a brown cubical rot.
Culture Characters
1.5.(7).(11).(16).(21).(22).(26).(32).(34).(35).(36). (37).(38).(39).(42).43-44.(47).(50).54.55.56.57 (Ref. 234).
References 116, 119, 120, 128, 130, 133, 183, 197, 199, 204, 208, 214, 216, 234, 242, 249, 390, 402, 406, 448.

Coniophora prasinoides (Bourd. & Galzin) Bourd. & Galzin 1928
Hym. France, 361.
Basionym *Coniophora olivascens* Bourd. & Galzin subsp. *prasinoides* Bourd. & Galzin, Bull. Soc. Mycol. France 39:115, 1923.
Distribution United States: MD, MO.
Hosts *Pinus* sp. (southern yellow pine), *Salix* sp., *Vitis vinifera*.
Ecology Board in a basement; used flooring stored in the open; tarpaulin.
Culture Characters
1.5.7.16.32.36.38.(40).44-45.(48).55.56 (Ref. 234 and Ginns, in litt.).
Reference 234.

Coniophora puteana (Schum.:Fr.) Karsten 1868
Not. Sällsk. Fauna Fl. Fenn. Förh. 9:370.
Basionym *Thelephora puteana* Schum., Enumerario Plant. Saellandiae 2:397, 1803.
Synonyms *Corticium kalmiae* Peck 1893, *Coniophora kalmiae* (Peck) Burt 1917, *Coniophora puteana* var. *rimosa* Peck 1892, *Coniophora puteana* var.

tuberculosa Peck 1892. All fide Ginns (234).
Distribution Canada: AB, AB-NT, BC, MB, NB, NF, NS, ON, PE, PQ, SK, YT.
United States: AK, AR, AZ, CA, CO, DC, IA, ID, IL, MA, ME, MI, MO, MT, NH, NJ, NM, NY, OH, OR, PA, TN, VA, VT, WA, WV.
Hosts *Abies* sp., *A. balsamea, A. concolor, A. grandis, A. lasiocarpa, Acer* sp., *A. saccharum, Alnus* sp., *A. incana, A. oblongifolia, A. rubra, A. tenuifolia, Betula* sp., *B. alleghaniensis, Carya* sp., *Castanea dentata, Cornus nuttallii, Crataegus douglasii, Cupressus macrocarpa, Fagus grandifolia, Fraxinus* sp., *F. oregona, Larix laricina, L. occidentalis, Liriodendron tulipifera, Malus sylvestris, Picea* sp., *P. engelmannii, P. glauca, P. mariana, P. rubens, Pinus* sp., *P. banksiana, P. contorta, P. engelmannii, P. monticola, P. muricata, P. ponderosa, P. resinosa, P. strobus, Platanus occidentalis, Populus* sp., *P. tremuloides, Prunus* sp., *P. serotina, Pseudotsuga menziesii, Quercus* sp., *Q. garryana, Q. rubra* var. *borealis, Salix* sp., *Thuja occidentalis, T. plicata, Tsuga canadensis, T. heterophylla, Ulmus americana*, angiosperm, gymnosperm.
Ecology Causing heart rot, less commonly a root rot, in live gymnosperm trees; causes dry rot; destroys boards and timbers in buildings; basidiomes produced on logs; dead stems; causes a brown cubical rot.
Culture Characters
1.(2).5.7.(16).(21).(26).(32).(35).(36).37.39.(42).43-44.(47).53.54.55.56.57 (Ref. 234 and Ginns, in litt.).
References 30, 116, 133, 181, 183, 197, 199, 204, 208, 216, 234, 242, 249, 274, 340, 391, 402, 406, 414, 448, 521, 595, 645.

Coniophora submembranacea (Berk. & Br.) Sacc. 1888 Sylloge Fung. 6:649.
Basionym *Thelephora submembranacea* Berk. & Br., J. Linn. Soc., Bot. 14:64, 1875.
Synonym *Coniophora inflata* Burt 1917, fide Ginns (234).
Distribution United States: AZ.
Hosts *Carnegiea gigantea, Olneya tesota, Prosopis juliflora*.
Ecology Dead, fallen stems and branches.
Culture Characters Not known.
Reference 216.

CONOHYPHA Jülich 1975 — Hyphodermataceae

Conohypha albocremea (Höhnel & Litsch.) Jülich 1975 Persoonia 8:304.
Basionym *Corticium albocremeum* Höhnel & Litsch., Wiesner Festschrift, Wien, 61, 1908.
Synonym *Hyphoderma albocremeum* (Höhnel & Litsch.) Eriksson & Strid 1975.
Distribution United States: LA.
Host *Juniperus phoenicea*.
Ecology Not known.
Culture Characters Not known.
Reference 202.

Conohypha terricola (Burt) Jülich 1976 Persoonia 8:442.
Basionym *Peniophora terricola* Burt, Ann. Missouri Bot. Gard. 12:237, 1925.
Synonym *Hyphoderma terricola* (Burt) K.J. Martin & Gilbn. 1977.
Distribution Canada: AB, ON.
United States: LA, NM, NY.
Hosts *Picea glauca, Populus* sp., *P. tremuloides, Tsuga canadensis*.
Ecology Not known but basidiomes sometimes on the ground.
Culture Characters Not known.
References 92, 183, 208, 242, 406, 584.

CORONICIUM Eriksson & Ryv. 1975 — Sistotremaceae

Coronicium alboglaucum (Bourd. & Galzin) Jülich 1975 Persoonia 8:299.
Basionym *Corticium alboglaucum* Bourd. & Galzin, Bull. Soc. Mycol. France 27: 251, 1911.
Synonym *Xenasma alboglaucum* (Bourd. & Galzin) Liberta 1960.
Distribution Canada: ON. United States: MI, OH.
Hosts *Acer* sp., *A. saccharum, Betula alleghaniensis, Tilia* sp.
Ecology Not known.
Culture Characters
(1).2.3c.13.(31d).32.36.38.44-45.54.58 (Ref. 474).
References 242, 384, 474, 492.

Coronicium proximum (Jackson) Jülich 1975 Persoonia 8:300.
Basionym *Corticium proximum* Jackson, Canad. J. Res., C, 28:722, 1950.
Distribution Canada: ON. United States: VT.
Hosts Angiosperm.
Ecology Not known.
Culture Characters Not known.
References 307, 317.

CORTICIUM Pers. 1794 — Vuilleminiaceae

Synonym is *Laeticorticium* Donk 1956, fide Jülich (322).

Corticium appalachiensis (Burdsall & Larsen) Larsen 1990 In Nakasone, Mycologia Memoir 15:58.
Basionym *Laeticorticium appalachiensis* Burdsall & Larsen, J. Elisha Mitchell Sci. Soc. 91:243, 1976.
Distribution United States: MS, TN.
Hosts *Nyssa sylvatica, Vitis* sp., angiosperm.
Ecology Not known.
Culture Characters
2.3c.7.32.36.38.43.(51).54 (Ref. 474).
References 60, 368, 474.

Corticium boreoroseum Boidin & Lanquetin 1983
Bull. Soc. Mycol. France 99:275.
Basionym Proposed as a *nom. nov.* for *Laeticorticium lundellii* Eriksson, Symb. Bot. Upsal. 16 (1):74, 1958, because there already was a *Corticium lundellii* Bourd. in Eriksson 1949.
Synonyms *Corticium lundellii* (Eriksson) Jülich 1982, *Dendrocorticium lundellii* (Eriksson) Larsen & Gilbn. 1974.
Distribution Canada: AB, BC, YT.
Hosts *Picea glauca, Pinus contorta, Populus balsamifera, Salix* sp.
Ecology Not known.
Culture Characters Not known.
References 174, 368.

Corticium canfieldiae (Larsen & Gilbn.) Boidin & Lanquetin 1983
Bull. Soc. Mycol. France 99:275.
Basionym *Laeticorticium canfieldiae* Larsen & Gilbn., Canad. J. Bot. 52:687 1974.
Distribution United States: AZ.
Host *Juglans major.*
Ecology Associated with a white rot.
Culture Characters
(1).2.3c.27.32.36.38.43-44.54.(59) (Ref. 474).
References 367, 368, 474.

Known only from the type collection.

Corticium cremeoalbidum (Larsen & Nakasone) Larsen 1990
In Nakasone, Mycologia Memoir 15:59.
Basionym *Laeticorticium cremeoalbidum* Larsen & Nakasone, Mycologia 76:528, 1984.
Distribution United States: FL.
Host *Vitis* sp.
Ecology Associated with a white rot.
Culture Characters
2.3c.26.27.32.36.40.43.(53).54 (Ref. 474);
2.3.7.24.27.32.36.40.43.54 (Ref. 371).
References 371, 474.

Corticium durangense (Larsen & Gilbn.) Boidin & Lanquetin 1983
Bull. Soc. Mycol. France 99:275.
Basionym *Laeticorticium durangense* Larsen & Gilbn., Canad. J. Bot. 52:688 1974.
Distribution United States: TX.
Host *Pinus* sp.
Ecology Cone epiphylls.
Culture Characters Not known.
Reference 368.

Corticium efibulatum (Larsen & Nakasone) Larsen 1990
In Nakasone, Mycologia Memoir 15:60.
Basionym *Laeticorticium efibulatum* Larsen & Nakasone, Mycologia 76:530, 1984.
Distribution United States: MS.
Host *Vaccinium* sp.
Ecology Bark.
Culture Characters
2.6.8.32.36.38.(40).43.(53).54 (Ref. 474);
2.6.7.32.36.40.43.54 (Ref. 371).
References 371, 474.

Corticium floridense (Larsen & Nakasone) Larsen 1990
In Nakasone, Mycologia Memoir 15:61.
Basionym *Laeticorticium floridense* Larsen & Nakasone, Mycologia 76:531, 1984.
Distribution United States: FL.
Hosts *Myrica* sp., angiosperm.
Ecology Not known.
Culture Characters
2.3c.7.32.36.38.(40).42-43.54 (Ref. 474);
2.3.21.32.36.38.(40).42-43.54 (Ref. 371).
References 371, 474.

Corticium griseo-effusum (Larsen & Gilbn.) Ginns & Lefebvre, *comb. nov.*
Basionym *Laeticorticium griseo-effusum* Larsen & Gilbn., Norw. J. Bot. 24:105, 1977.
Distribution United States: AZ.
Host *Pseudotsuga menziesii.*
Ecology Branches.
Culture Characters Not known.

Known only from the type specimen. Fungi previously placed in *Laeticorticium* are now treated under *Corticium sensu stricto*. The new combination is proposed to place *L. griseo-effusum* with morphologically related species.

Corticium lombardiae (Larsen & Gilbn.) Boidin & Lanquetin 1983
Bull. Soc. Mycol. France 99:275.
Basionym *Laeticorticium lombardiae* Larsen & Gilbn., Mycologia 70:206, 1978.
Distribution United States: WI.
Host *Populus* sp.
Ecology Fallen branches
Culture Characters Not known.
References 369, 406.

Corticium minnsiae (Jackson) Boidin & Lanquetin 1983
Bull. Soc. Mycol. France 99:275.
Basionym *Aleurodiscus minnsiae* Jackson, Canad. J. Res., C, 28:67, 1950.
Synonym *Laeticorticium minnsiae* (Jackson) Donk 1956.
Distribution Canada: BC, ON, PQ. United States: CA, ME, MI, NC, NH, NY, OR, PA, TN, VT.
Hosts *Abies lasiocarpa, ?Picea* sp., *Pinus contorta, P. ponderosa, Pseudotsuga menziesii, Tsuga canadensis, T. heterophylla.*
Ecology Branches; twigs.
Culture Characters Only a partial species code is known, 3c.55.59 (Ref. 48).
References 116, 197, 305, 368.

Corticium mississippiense (Larsen & Gilbn.) Larsen 1990
In Nakasone, Mycologia Memoir 15:62.
Basionym *Laeticorticium mississippiense* Lentz ex Larsen & Gilbn., Canad. J. Bot. 52:688, 1974.
Distribution United States: AZ, FL, MS.
Hosts *Cornus* sp., *Juglans major, Morus* sp., *Quercus* sp., *Q. emoryi, Sabal palmetto*, angiosperm.
Ecology Wood on the ground; associated with a white rot.
Culture Characters
2.3c.7.(27).32.36.38.(40).43-44.(51).54 (Ref. 474).
References 367, 368, 474.

Corticium pini (Jackson) Boidin & Lanquetin 1983
Bull. Soc. Mycol. France 99:275.
Basionym *Aleurodiscus pini* Jackson, Canad. J. Res., C, 28:74, 1950.
Synonym *Laeticorticium pini* (Jackson) Donk 1956.
Distribution Canada: ON, PQ.
United States: CT, MA, MI, NH, NY, PA.
Hosts *Pinus* sp., *P. strobus*.
Ecology Not known.
Culture Characters Only a partial species code is known, 3c.55.59 (Ref. 48).
References 116, 305, 368.

Corticium roseum Pers.:Fr. 1794
Neues Mag. Bot. 1:111.
Synonyms *Aleurodiscus roseus* (Pers.) Höhnel & Litsch. 1906, *Laeticorticium roseum* (Pers.) Donk 1956.
Distribution Canada: AB-NT, BC, MB, NS, ON, NB. United States: AL, AZ, CA, CO, DC, GA, IA, ID, IL, IN, MA, MD, ME, MI, MN, MO, MT, NC, NJ, NM, NY, OH, OR, PA, VT, WA, WI.
Hosts *Abies magnifica, Acer rubrum, Alnus* sp., *A. incana, A. oblongifolia, Arctostaphylos patula, Carya* sp., *Juglans major, Liriodendron tulipifera, Nyssa sylvatica, Populus* sp., *P. tremuloides, P. trichocarpa, Quercus kelloggii, Rhus glabra, Salix* sp., *S. sericea, Tsuga mertensiana, Ulmus* sp., angiosperm.
Ecology Limbs; bark and wood of branches and logs; associated with a white rot.
Culture Characters
(1).3c.12.32.36.(39).44.53.54.63 (Ref. 48).
References 93, 96, 103, 108, 116, 120, 128, 130, 133, 183, 204, 208, 216, 249, 334, 368, 402 , 406.

COTYLIDIA Karsten 1881 — Podoscyphaceae

Cotylidia aurantiaca (Pers.) Welden 1958
Lloydia 21:40.
Basionym *Thelephora aurantiaca* Pers. in Gaudichaud, Voy. Uranié Bot., 176, 1821.
Distribution United States: OH.
Hosts Not known.
Ecology Not known.
Culture Characters Not known.
Reference 613.

Cotylidia diaphana (Schw.) Lentz 1955
U.S. Dept. Agric. Monograph 24:12.
Basionym *Thelephora diaphana* Schw. in Berk. & Curtis, J. Acad. Nat. Sci. Philadelphia, Ser. II, 2:278, 1854.
Synonyms *Stereum diaphanum* (Schw.) Cooke in Sacc. 1888, *Thelephora sullivantii* Mont. 1856, fide Lentz (379), *Thelephora willeyi* G.W. Clinton in Peck 1874, fide Lentz (379).
Distribution Canada: PQ. United States: AL, CA, IA, IL, IN, MI, MN, MO, NY, OH, PA, UT, WA, WV.
Hosts Not known.
Ecology Ground; amongst humus and twigs; moist woods of angiosperm or gymnosperm species; lawns under trees.
Culture Characters Not known.
References 89, 379, 519, 555, 610 , 612.

Cotylidia pannosa (Sowerby:Fr.) D. Reid 1965
Beih. Nova Hedwigia 18:81.
Basionym *Helvella pannosa* Sowerby, Coloured figures of English fungi or mushrooms, pl. 155 & caption, 1799.
Synonym *Craterella pallida* Pers., 1797, fide Reid (555), *Thelephora pallida* (Pers.) Pers. 1801, *Stereum pallidum* (Pers.) Cooke 1909.
Distribution United States: CT, IA, NC, VT, WA, WV.
Hosts Not known.
Ecology Ground in woods.
Culture Characters Not known.
References 89, 555.

Cotylidia undulata (Fr.) Karsten 1881
Rev. Mycol. (Toulouse) 3 (9):22.
Basionym *Thelephora undulata* Fr., Elenchus 1:164, 1828.
Synonyms *Thelephora exigua* Peck 1902, *Stereum exiguum* (Peck) Burt 1920, *Stereum tenerrimum* Berk. & Rav. 1873. All fide Lentz (379) and Reid (555).
Distribution United States: IA, MA, MI, MN, NC, NY, SC, WI.
Hosts Not known.
Ecology Ground, usually among mosses.
Culture Characters Not known.
References 89, 379, 555.

We follow Reid (555) in attributing the epithet to Fries and citing the place of publication of the basionym as the Elenchus.

CREOLOPHUS Karsten 1879 — Hericiaceae

Creolophus cirrhatus (Pers.:Fr.) Karsten 1879
Meddeland. Soc. Fauna Fl. Fenn. 5:42.

Basionym *Hydnum cirrhatum* Pers., Syn. Meth. Fung., 558, 1801.
Distribution Canada: AB. United States: CO, NM.
Host *Picea glauca*.
Ecology Log; stump; ground; possibly from buried wood.
Culture Characters Not known.
Reference 269.

The Alberta report is based upon a collection in DAOM (No. 183815).

CRISTINIA Parm. 1968 Stephanosporaceae

Cristinia helvetica (Pers.) Parm. 1968
Consp. Syst. Cort., 48.
Basionym *Hydnum helveticum* Pers., Mycol. Eur. 2:184, 1825.
Synonym *Grandinia helvetica* (Pers.) Fr. 1874.
Distribution Canada: AB-NT, BC, ON, PQ.
United States: IA, WI.
Hosts *Acer* sp., *A. saccharum, Fagus* sp., *Pinus contorta, Populus* sp., *P. trichocarpa, Quercus* sp., *Q. alba, Tilia* sp., angiosperm.
Ecology Associated with a white rot.
Culture Characters
1.3c.13.26.32.36.38.43-44.(50).(53).54.55.58 (Ref. 474).
References 116, 242, 255, 406, 463, 474.

Cristinia mucida (Bourd. & Galzin) Eriksson & Ryv. 1975
Cort. N. Europe 3:311.
Basionym *Radulum mucidum* Bourd. & Galzin, Bull. Soc. Mycol. France 30:247, 1914.
Distribution Canada: BC, PQ.
Hosts *Alnus* sp., *Populus* sp., *Ulmus* sp., angiosperm.
Ecology Not known.
Culture Characters Not known.
References 255, 406.

Cristinia sonorae Nakasone & Gilbn. 1978
Mycologia 70:271.
Distribution United States: AZ.
Host *Fouquieria splendens*.
Ecology Associated with a white rot.
Culture Characters
2.3.7.13.32.36.38.47.(48).50.54 (Ref. 477);
2.3c.13.26.32.36.38.(40).46-47.(48).54.59 (Ref. 474).
References 474, 477.

CRUSTODERMA Parm. 1968 Stereaceae

Crustoderma carolinense Nakasone 1984
Mycologia 76:41.
Distribution United States: NC.
Host *Castanea* sp.
Ecology Dead wood; associated with a brown rot.
Culture Characters Not known.

Known only from the type specimen.

Crustoderma corneum (Bourd. & Galzin) Nakasone 1984
Mycologia 76:45.
Basionym *Peniophora gigantea* subsp. *cornea* Bourd. & Galzin, Hym. France, 318, 1928.
Synonym *Peniophora cornea* (Bourd. & Galzin) Eriksson 1950, *Phlebia cornea* (Bourd. & Galzin) Parm. 1967.
Distribution Canada: ON.
United States: AZ, MT, NM.
Hosts *Pinus contorta, P. engelmannii, P. ponderosa, Quercus hypoleucoides*, gymnosperm.
Ecology Associated with a brown rot.
Culture Characters
1.3.7.9.22.(27).32.36.38.(46).47.53.55 (Ref. 471);
1.(2).3c.9.27.32.36.(38).(39).46-47.50.53.55 (Ref. 474).
References 170, 197, 216, 471, 474.

Crustoderma dryinum (Berk. & Curtis) Parm. 1968
Consp. Syst. Cort., 88.
Basionym *Corticium dryinum* Berk. & Curtis, Grevillea 1:179, 1873.
Synonyms *Peniophora dryina* (Berk. & Curtis) D.P. Rogers & Jackson 1943, *Coniophora laeticolor* (Karsten) Karsten 1889, fide Rogers and Jackson (567), *Peniophora tabacina* Burt 1926, fide Rogers and Jackson (567), *Peniophora weiri* Bres. 1925, fide Nakasone (471).
Distribution Canada: BC, ON, PQ.
United States: AK, AL, AZ, CO, ID, MD, MO, MS, NM, OR, VA, WA, WI.
Hosts *Abies* sp., *A. grandis, Acer saccharum, Picea* sp., *P. engelmannii, P. glauca, P. sitchensis, Pinus monticola, P. ponderosa, Populus* sp., *P. tremuloides, Pseudotsuga menziesii, Quercus* sp., *Tsuga heterophylla*, angiosperm, gymnosperm.
Ecology Decaying wood and bark of logs; dead trees and limbs; mine timbers; associated with a brown rot.
Culture Characters
1.3i.6.7.34.36.38.43.50.55.58 (Ref. 246).
References 53, 92, 97, 116, 177, 183, 197, 199, 204, 208, 216, 242, 246, 308, 335, 391, 402, 406, 447, 471, 584.

Crustoderma flavescens Nakasone & Gilbn. 1982
Mycologia 74:601.
Distribution United States: MD, MO, NC, WI.
Hosts *Castanea dentata, Quercus* sp., *Sequoia* sp., angiosperm.
Ecology Isolated from rot in redwood cooling towers; associated with brown rot.

Culture Characters
1.3.9.21.34.36.43.53.54.55 (Ref. 478);
1.4.9.21.27.34.36.38.(39).43.(50).53.54.55 (Ref. 474).
References 474, 478.

Crustoderma longicystidia (Litsch.) Nakasone 1984
Mycologia 76:47.
Basionym *Peniophora longicystidia* Litsch., Oesterr. Bot. Z. 72(2):131, 1928.
Distribution Canada: BC.
Hosts *Abies lasiocarpa, Tsuga heterophylla.*
Ecology Isolated from unseasoned lumber.
Culture Characters
1.3i.(4).9.34.36.38.44.(48).55.59 (Ref. 246).
Reference 246.

Crustoderma marianum Nakasone 1984
Mycologia 76:43.
Distribution United States: MD.
Hosts *Pinus* sp., *Robinia pseudoacacia*, angiosperm.
Ecology Post; associated with a brown rot.
Culture Characters
1.4.7.(9).34.36.38.43-44.53.54.55 (Ref. 471);
1.4.9.27.34.36.39.43-44.(50).(53).54.55 (Ref. 474).
References 471, 474.

Crustoderma nakasoneae Gilbn. & Blackwell 1988
Mycotaxon 33:376.
Distribution United States: LA.
Hosts *Pinus* sp., *P. taeda.*
Ecology Fallen tree; fence post; associated with a brown rot.
Culture Characters Not known.

Known only from the type specimen.

Crustoderma opuntiae Nakasone & Gilbn. 1982
Mycologia 74:603.
Distribution United States: AZ.
Host *Opuntia fulgida.*
Ecology Dead jumping cholla; associated with a brown rot.
Culture Characters
1.3.9.21.32.36.38.43.53.55.59 (Ref. 478);
1.3c.9.26.32.36.38.44.53.54.59 (Ref. 474).
References 474, 478.

Crustoderma resinosum (Jackson & Dearden) Gilbn. 1981
Mycotaxon 12:377.
Basionym *Peniophora resinosa* Jackson & Dearden, Canad. J. Res., C, 27:147, 1949.
Synonym *Hyphoderma resinosum* (Jackson & Dearden) K.J. Martin & Gilbn. 1977.
Distribution Canada: AB, BC.
United States: MT, OR, WA.
Hosts *Picea glauca, P. sitchensis, Pinus contorta, Pseudotsuga menziesii*, gymnosperm.
Ecology Dead wood; associated with a brown rot.
Culture Characters
1.5.9.21.34.36.38.43.53.55 (Ref. 478);
1.4.9.13.21.27.34.36.38.(39).43.53.55 (Ref. 474).
References 116, 199, 242, 308, 447, 474, 478.

Crustoderma testatum (Jackson & Dearden) Nakasone 1985
Mycotaxon 22:416.
Basionym *Corticium testatum* Jackson & Dearden, Canad. J. Res., C, 27:151, 1949.
Distribution Canada: BC. United States: OR.
Hosts *Pseudotsuga menziesii*, gymnosperm.
Ecology Associated with a brown rot.
Culture Characters
1.3.9.34.36.38.47.53 (composed from Ref. 472);
1.3c.9.21.34.36.38.47.(50).(51).55 (Ref. 474).
References 308, 472, 474.

CRUSTOMYCES Jülich 1978 — Stereaceae

Crustomyces expallens (Bres.) Hjort. 1987
Windahlia 17:56.
Basionym *Corticium expallens* Bres., Ann. Mycol. 6:43, 1908.
Synonyms *Laeticorticium expallens* (Bres.) Eriksson & Hjort. 1981, *Phlebia expallens* (Bres.) Parm. 1967.
Distribution Canada: AB, BC, ON.
Hosts *Populus balsamifera, P. trichocarpa*, angiosperm.
Ecology Not known.
Culture Characters
2a.3c.15b.32.36.38.47.54.58.61 (Ref. 257).
References 116, 170, 257.

Crustomyces pini-canadensis subsp. ***pini-canadensis*** (Schw.) Jülich 1978
Persoonia 10:140.
Basionym *Radulum pini-canadensis* Schw., Trans. Amer. Philos. Soc., N.S., 4:164, 1832.
Synonyms *Corticium pini-canadensis* (Schw.) D.P. Rogers & Jackson 1943, *Cystostereum pini-canadensis* (Schw.) Parm. 1968, *Peniophora piceina* Overh. 1930, fide Rogers and Jackson (567) and Chamuris (101), *Cystostereum piceina* (Overh.) Lindsey & Gilbn. 1978.
Distribution Canada: BC, MB, NF, ON, PE, PQ.
United States: CA, MI, MN, MO, MT, NC, NH, NY, OR, PA, WA, WI, WV.
Hosts *Abies balsamea, Acer* sp., *A. saccharinum, A. macrophyllum, Alnus* sp., *Betula* sp., *Larix occidentalis, Liriodendron tulipifera, Picea* sp., *P. abies, P. rubens, Pinus* sp., *P. contorta, P. resinosa, Populus* sp., *P. balsamifera, P. tremuloides, Pseudotsuga menziesii, Quercus* sp., *Q. alba, Sequoia sempervirens, Thuja occidentalis, Tsuga canadensis*, angiosperm, gymnosperm.
Ecology Mine timbers; associated with a white rot.
Culture Characters
1.2.3.15.26.32.36.38.43-46.(48).(50).(51).54.55.60 (Ref. 179);

(1).2.3c.15.32.36.38.43-46.(48).(50).(53).54.55.60 (Ref. 474);
2.3.15b.(19).(26).32.36.38.44.(45).(48).53.54.55.58.61 (Ref. 101).

References 101, 116, 179, 210, 214, 242, 406, 448, 474, 515.

Crustomyces pini-canadensis subsp. ***subabruptus*** (Bourd. & Galzin) Ginns & Lefebvre, *comb. nov.*

Basionym *Odontia subabrupta* Bourd. & Galzin, Hym. France, 439, 1928.

Synonyms *Crustomyces subabruptus* (Bourd. & Galzin) Jülich 1978, *Cystostereum subabruptum* (Bourd. & Galzin) Eriksson & Ryv. 1975, *Cystostereum pini-canadensis* subsp. *subabruptus* (Bourd. & Galzin) Chamuris 1986.

Distribution Canada: BC. United States: CA, WA.

Hosts *Acer macrophyllum, Alnus* sp., *Lithocarpus densiflora, Sequoia sempervirens, Salix* sp., angiosperm.

Ecology Not known.

Culture Characters
2.3.15b.(19).(26).32.36.38.44.(45).(48). 53.54.55.58.61 (Ref. 101).

References 101, 116, 173, 242.

Because we accept this species as a member of the genus *Crustomyces*, it is necessary to transfer the variety *subabruptus* to *Crustomyces*.

CYLINDROBASIDIUM Jülich 1974 — Cylindrobasidiaceae

Cylindrobasidium corrugum (Burt) Ginns 1982 Opera Bot. 61:54.

Basionym *Coniophora corrugis* Burt, Ann. Missouri Bot. Gard. 13:310, 1926.

Synonym *Corticium corruge* (Burt) D.P. Rogers & Jackson 1943.

Distribution Canada: AB, AB-NT, BC. United States: AK, AZ, CA, CO, ID, NM, OR, UT, WA, WY.

Hosts *Abies* sp., *A. concolor, A. grandis, A. lasiocarpa, A. magnifica, Acer glabrum, Juniperus* sp., *J. scopulorum, Picea* sp., *P. engelmannii, P. glauca, P. pungens, Pinus* sp., *P. aristata* var. *longaeva, P. contorta, P. flexilis, P. murrayana, P. ponderosa, Populus* sp., *P. tremuloides, Pseudotsuga menziesii, Ribes* sp., *R. lacustre, Sambucus racemosus, Thuja plicata, Tsuga* sp., *T. heterophylla, T. mertensiana*, gymnosperm.

Ecology Perhaps psychrophilic; causes a decay of dead sapwood in *Picea engelmannii*; basidiomes produced on live foliage; branches near the ground; logs; stones and litter after melting of snow pack.

Culture Characters Not known.

References 93, 116, 120, 128, 130, 133, 183, 196, 197, 208, 216, 242, 243, 274, 402, 406, 407, 448, 509.

Cylindrobasidium laeve (Pers.:Fr.) Chamuris 1984 Mycotaxon 20:587.

Basionym *Corticium laeve* Pers., Neues Mag. Bot. 1:110, 1794.

Synonyms *Corticium evolvens* (Fr.) Fr. 1838, fide Chamuris (99) and Jülich (315), *Cylindrobasidium evolvens* (Fr.) Jülich 1974.

Distribution Canada: AB, AB-NT, BC, MB, NB, NF, NS, ON, PE, PQ. United States: AK, AZ, CO, DC, GA, MI, MO, NH, NY, OR, VT, WA.

Hosts *Abies* sp., *A. balsamea, A. lasiocarpa, A. lasiocarpa* var. *lasiocarpa, Acer* sp., *A. macrophyllum, A. saccharum, Aesculus hippocastanum, Alnus* sp., *A. rubra, Betula* sp., *B. alleghaniensis, Cytisus scoparius, Fagus* sp., *Liquidambar styraciflua, Malus pumila, Oemleria cerasiformis, Osmaronia cerasiformis, Picea* sp., *P. glauca, P. mariana, Pinus contorta, P. contorta* var. *contorta, P. strobus, Populus* sp., *P. tremuloides, P. trichocarpa, Salix* sp., *S. scouleriana, Tsuga heterophylla, Ulmus americana*, angiosperm, gymnosperm.

Ecology Bark; leaves; fallen decaying limbs; logs; associated with a white rot.

Culture Characters
(1).2.3c.7.(26).32.36.38.(39).43.(53).54.55.60 (Ref. 474);
1.(2).3.26.32.36.38.39.42.54.55.60 (Ref. 583).

References 93, 96, 97, 116, 242, 249, 255, 315, 391, 402, 406, 474, 521, 580, 603.

Chamuris (99) pointed out the priority of *laeve*, published in 1794, over the more commonly seen *evolvens*, published in 1815.

Cylindrobasidium torrendii (Bres.) Hjort. 1983 Mycotaxon 17: 571.

Basionym *Peniophora torrendii* Bres. in Torrend, Brotéria, Sér. Bot. 11(2):77, 1913.

Synonyms *Peniophora albula* Atk. & Burt, 1925, fide Hjortstam (280), *Corticium albulum* (Atk. & Burt) D.P. Rogers & Jackson 1943, *Cylindrobasidium albulum* (Atk. & Burt) Eriksson & Hjort. 1976, *Peniophora burkei* Burt 1926, fide Rogers and Jackson (567).

Distribution Canada: ON. United States: AL, DC, FL, GA, IA, IL, IN, MA, MD, ME, MI, MS, NH, NJ, NY, PA, TN, VA, WA, WI, WV.

Hosts *Acer macrophyllum, Fagus* sp., *Fraxinus oregona, Juglans nigra, Liriodendron tulipiferae, Pinus ponderosa, Populus* sp., *P. trichocarpa, Prunus* sp., *Quercus durandii, Rhus typhina, Salix* sp., *S. nigra, Sassafras albidum*, angiosperm.

Ecology Bark of fallen decaying branches; rough angiosperm bark; wood products (stored willow pulpwood; beechwood chips; yellow poplar and sassafras mine timbers; lumber); associated with a white rot.

Culture Characters
1.2.3.7.14.(26).32.36.38.42-45.(48).53.54.60 (Ref. 179);
(1).2.3c.7.(14).(26).32.36.38.(39).42-43.(48).(50).(53).54.60 (Ref. 474).
References 92, 96, 133, 178, 179, 333, 474, 511.

CYMATODERMA Jungh. 1840 Podoscyphaceae
Synonym is *Cladoderris* Pers. ex Berk. 1842, fide Reid (552).

Cymatoderma caperatum (Berk. & Mont.) D. Reid 1956
Kew Bull. 10:635.
Basionym *Thelephora caperata* Berk. & Mont., Ann. Sci. Nat. Bot., Sér. III, 11:241, 1849.
Synonym *Stereum caperatum* (Berk. & Mont.) Massee 1890
Distribution United States: FL, GA, LA.
Hosts *Ostrya virginiana, Quercus* sp., angiosperm.
Ecology Fallen branches; dead stumps; trunks; roots.
Culture Characters
2.3c.8.32.36.(38).40.45.53.54.60 (Ref. 474).
References 89, 474, 555, 617.

Cymatoderma fuscum (Cooke) D. Reid 1959
Kew Bull. 13:526.
Basionym *Cladoderris fusca* Cooke, Grevillea 10:123, 1882.
Synonyms *Cladoderris floridana* Lloyd 1915, fide Reid (552), *Cymatoderma fuscum* var. *floridanum* (Lloyd) D. Reid 1959.
Distribution United States: FL, LA.
Hosts Angiosperm.
Ecology Dead wood; logs.
Culture Characters Not known.
References 91, 555, 617.

CYPHELLOPSIS Donk 1931 Cyphellopsidaceae

Cyphellopsis anomala (Pers.:Fr.) Donk 1931
Medd. Nedl. Myc. Ver. 18-20:128.
Basionym *Peziza anomala* Pers., Obs. Mycol. 1:29, 1796.
Synonym *Solenia anomala* (Pers.) Fuckel 1871.
Distribution Canada: AB, BC, MB, NF, NS, NT, ON, PQ. United States: AK, AL, AZ, CA, CO, CT, DE, GA, IA, ID, IL, IN, KS, LA, MA, MD, ME, MI, MN, MO, MT, NC, ND, NE, NH, NJ, NM, NY, OH, OR, PA, SC, TN, VA, VT, WA, WI, WV.
Hosts *Abies balsamea, A. concolor, Acer* sp., *A. macrophyllum, A. rubrum* var. *rubrum, Alnus* spp., *A. crispa* var. *mollis, A. rubra, A. rugosa, A. sinuata, A. tenuifolia, Betula* sp., *B. alleghaniensis, B. papyrifera, B. populifolia, Corylus* sp., *C. californica, C. cornuta, Fagus grandifolia, Fraxinus nigra, Juglans major, Lindera* sp., *Liriodendron tulipifera, Populus* sp., *P. tremuloides, P. trichocarpa, Prunus* sp., *P. emarginata, Pyrus malus, Quercus* sp., *Q. reticulata, Robinia pseudoacacia, Salix* spp., *S. alaxensis, S. lasiandra, S. scouleriana, Ulmus* sp., *U. americana, Umbellularia californica*, angiosperm.
Ecology Old wood; dead wood; broken bark; twigs and branches; decorticated twig; slash; logs; underside of old logs; associated with a white rot.
Culture Characters Not known.
References 91, 96, 97, 103, 116, 120, 125, 128, 129, 130, 133, 183, 204, 208, 214, 216, 242, 249, 332, 334, 335, 344, 402, 406, 545.

Cyphellopsis confusa (Bres.) D. Reid 1964
Persoonia 3:110.
Basionym *Solenia confusa* Bres., Ann. Mycol. 1:84, 1903.
Distribution Canada: BC. United States: KY, MI.
Hosts *Alnus* sp., *Populus* sp., angiosperm.
Ecology Decorticated wood.
Culture Characters Not known.
References 242, 333, 545.

Cyphellopsis subglobispora D. Reid 1961
Kew Bull. 15:265.
Distribution Canada: BC.
Hosts Angiosperm.
Ecology Not known.
Culture Characters Not known.
Reference 545.

CYPHELLOSTEREUM D. Reid 1965 Atheliaceae

Cyphellostereum laeve (Fr.) D. Reid 1965
Beih. Nova Hedw. 18:337.
Basionym *Cantharellus laevis* Fr., Syst. Mycol. 1:324, 1821.
Distribution Canada: BC, ON. United States: CT, ID, MT, NH, NY, OR, VT, WA.
Hosts *Atrichum selwynii, Ditrichum ?ambigum, Mnium glabrescens, Pogonatum contortum, P. macounii, ?Polytrichum commune, P. formosum, ?P. juniperum*, mosses.
Culture Characters Not known.
References 173, 545, 547, 548, 555.

CYSTOBASDIUM (Lagerh.) Neuhoff 1924 Cystobasidiaceae

Cystobasidium proliferans Olive 1952
Mycologia 44:564.
Distribution United States: NJ.
Host *Pinus* sp. (yellow pine).
Ecology Insect excreta on inner surface of bark of dead standing trees.
Culture Characters Not known.

Known only from the type specimen.

CYSTOSTEREUM Pouzar 1959 Stereaceae

Cystostereum australe Nakasone 1983
Mycotaxon 17:270.
Distribution United States: FL, GA.
Hosts *Carya* sp., *Ulmus* sp., angiosperm.
Ecology Associated with a white rot.
Culture Characters
2.3.(8).15.34.36.38.47.54 (Ref. 470);
2.3c.(8).15p.34.36.38.(39).47.54 (Ref. 474).
References 470, 474.

Cystostereum murraii (Berk. & Curtis) Pouzar 1959
Ceská Mykol. 13:18.
Basionym *Thelephora murraii* Berk. & Curtis, J. Linn. Soc., Bot. 10:329, 1869.
Synonyms *Stereum murraii* (Berk. & Curtis) Burt 1920, *Stereum tuberculosum* Fr. 1874, fide Lentz (379), *Stereum pulverulentum* Peck 1900, fide Burt (89) and Lentz (379), *Corticium effusum* Overh. 1930, fide Lentz (379).
Distribution Canada: BC, NB, NF, NS, ON, PQ. United States: CT, FL, IA, ID, IL, LA, MA, ME, MI, MN, NH, NY, OR, PA, RI, TN, VA, VT, WI, WV.
Hosts *Acer* sp., *A. negundo, A. pennsylvanicum, A. rubrum, A. saccharum, A. saccharinum, A. spicatum, Betula* sp., *B. alba, B. alleghaniensis, B. lenta, B. nigra, B. occidentalis, B. papyrifera, B. populifolia, B. pumila, Carpinus caroliniana, Fagus grandifolia, Ficus aurea, Fraxinus* sp., *F. nigra, Malus* sp., *M. malus, M. sylvestris, Ostrya* sp., *O. murrayi, O. virginiana, Picea* sp., *P. rubens, Pinus* sp., *Populus* sp., *P. grandidentata, P. tremuloides, Quercus* sp., *Salix* sp., *Tilia americana, Tsuga* sp., *Ulmus americana*, angiosperm, gymnosperm.
Ecology Causes a minor trunk rot in *Betula alleghaniensis*; sometimes causes cankers on species of *Acer* and *Betula*, less commonly on *Fagus* and *Ostrya*, the associated decay can be extensive; basidiomes produced on rotting logs and limbs; stumps; associated with a white rot.
Culture Characters
2.3c.8.15p.32.36.38.47.51.54.(55).60 (Ref. 474);
2.3.8.15.32.36.38.47.53.54.58 (Ref. 487);
2.3c.8.15b.32.36.38.47.53.54.60.61 (Ref. 104).
References 89, 104, 116, 143, 173, 183, 214, 242, 245, 249, 255, 274, 332, 333, 373, 379, 391, 406, 448, 470, 474, 484, 511, 515, 519, 597.

CYTIDIA Quél. 1888 Vuilleminiaceae

Cytidia lanata W.B. Cooke 1951
Mycologia 43:205.
Distribution United States: ID.
Host *Betula* sp.
Ecology Bark.
Culture Characters Not known.
Known only from the type specimen.

Cytidia patelliformis (Burt) Welden 1958
Mycologia 50:304.
Basionym *Stereum patelliforme* Burt, Ann. Missouri Bot. Gard. 7:182-183, 1920.
Distribution United States: CA, NM, WA.
Hosts *Acer* sp., *A. macrophyllum, Corylus cornuta, C. californica, Crataegus douglasii, Fraxinus latifolia, F. oregona, Purshia tridentata, Quercus* sp., *Rhus diversiloba, Rosa nutkana*, angiosperm.
Ecology Fallen branches.
Culture Characters Not known.
References 89, 119, 133, 183, 614.

Cytidia pezizoidea Pat. 1900
Essai taxonomique, 54.
Basionym Proposed as a *nom. nov.* for *Corticium pezizoideum* Pat., J. Bot. (Morot) 5:314, 1891, which is a later homonym of *Corticium pezizoideum* Ellis & Ev. 1888.
Synonym *Cytidia tremellosa* Lloyd 1912, fide W.B. Cooke (117), and Martin (438).
Distribution United States: LA.
Hosts Angiosperm.
Ecology Bark of decaying limbs; rotting wood.
Culture Characters Not known.
References 91, 117.

Cytidia salicina (Fr.) Burt 1924
Ann. Missouri Bot. Gard. 11:10.
Basionym *Thelephora salicina* Fr., Syst. Mycol. 1:442, 1821.
Distribution Canada: BC, MB, NB, NF, NS, ON, PQ, YT. United States: AK, AL, CO, CT, ID, IN, LA, MA, ME, MI, MN, MT, NC, ND, NH, NM, NY, OH, OR, PA, VA, VT, WA, WI, WV, WY.
Hosts *Alnus* spp., *A. incana, Betula papyrifera, Populus* spp., *P. tremula, P. tremuloides, P. trichocarpa, Prunus serotina, Salix* sp., *S. alaxensis, S. aurita, S. capraea, S. cordata, S. discolor, S. fragilis, S. fluviatilis, S. grandifolia, S. lasiandra, S. longipes, S. nigra, S. pentandra, S. purpurea, S. scouleriana*, angiosperm.
Ecology Limbs.
Culture Characters
(1).2.3c.26.27.(31d).32.36.38.45.53.54.60 (Ref. 474).
References 91, 97, 103, 117, 132, 133, 173, 181, 242, 249, 344, 402, 464, 474, 533.

Cytidia stereoides W.B. Cooke 1951
Mycologia 43:206.
Distribution United States: CA.
Hosts *Ceanothus velutinus, Cercocarpus ledifolius, Purshia tridentata, Quercus kelloggii.*
Ecology Dead branches of shrubs.
Culture Characters Not known.
References 117, 120, 128.

DACRYMYCES Nees:Fr. 1816 Dacrymycetaceae
Synonym is *Arrhytidia* Berk. & Curtis 1849, fide McNabb (461)

Dacrymyces abietinus (Pers.) Schröter
var. ***triseptata*** Olive 1948
Mycologia 40:598.
Distribution United States: LA.
Host *Pinus* sp.
Ecology Dead limbs.
Culture Characters Not known.

Known only from the type specimen. Probably a variant of *D. variisporus*, see McNabb (461).

Dacrymyces aquaticus Bandoni & G.C. Hughes 1984
Mycologia 76:63.
Distribution Canada: BC.
Hosts Gymnosperm.
Ecology Sodden wood; wood floating in pond.
Culture Characters Not known.

Known only from the type specimen.

Dacrymyces capitatus Schw. 1832
Trans. Amer. Philos. Soc., N.S. 4:186.
Synonyms *Dacrymyces involutus* Schw. 1832, *Arrhytidia involuta* (Schw.) Coker 1928, *Dacrymyces stipitatus* (Bourd. & Galzin) Neuhoff 1936, *Dacryomitra nuda* (Berk. & Br.) Pat. 1900. All fide McNabb (461).
Distribution Canada: AB, AB-NT, BC, ON, PQ.
United States: AZ, GA, IA, ID, KS, LA, MA, MI, MS, NC, NJ, NY, OR, PA, WA.
Hosts *Acer glabrum, Alnus* sp., *A. rugosa, A. sinuata, Arbutus menziesii, Gleditsia triacanthos, Larix* sp., *Picea* sp., *Pinus* sp., *P. contorta, Platanus occidentalis, Pseudotsuga menziesii, Quercus* sp., *Taxodium* sp., *Tsuga heterophylla*, angiosperm, gymnosperm.
Ecology Dead corticated and decorticated wood; branches; causes a uniform brown rot or a brown pocket rot.
Culture Characters Not known.
References 51, 52, 107, 111, 116, 132, 183, 242, 443, 461, 498, 499, 558, 577.

Some reports under the epithet *involutus* may be *Darcymyces corticioides*, see McNabb (461:474).

Dacrymyces chrysocomus (Bull.:Fr.) Tul. 1853
Ann. Sci. Nat. Bot., Sér. III, 19:211.
Basionym *Peziza chrysocomus* Bull., Histoire Champignons France, 254, 1791.
Synonym *Guepiniopsis chrysocomus* (Bull.) Brasfield 1938.
Distribution Canada: BC. United States: GA, WA.
Hosts *Abies amabilis, Pinus* sp., *Tsuga heterophylla, T. mertensiana.*
Ecology Corticated branches.
Culture Characters Not known.
References 133, 242, 498.

The records in Brasfield (51) are not cited here because the fungus is "most probably" (461) an undescribed *Heterotextus* (458).

Dacrymyces chrysospermus Berk. & Curtis 1873
Grevillea 2:20.
Synonyms *Tremella palmata* Schw. 1832, fide Burt (90) and McNabb (461), *Dacrymyces palmatus* (Schw.) Bres. in Höhnel 1904, *Dacrymyces palmatus* (Schw.) Burt 1921 is a superfluous transfer.
Distribution Canada: AB, AB-NT, BC, NB, NF, NS, ON, PQ, YT.
United States: AL, AZ, CA, CT, ID, IN, LA, MA, MI, MN, NH, NM, NY, OR, PA, SC, VT, WA, WI.
Hosts *Abies balsamea, A. concolor, A. grandis, Acer spicatum, Alnus* sp., *Picea* sp., *P. engelmannii, P. glauca, P. mariana, Pinus* sp., *P. engelmannii, P. ponderosa, P. strobiformis, Pseudotsuga menziesii, Tsuga* sp., *T. canadensis*, gymnosperm.
Ecology Saprophytic on stumps; logs and brush; stub; causes a uniform brown rot or a brown pocket rot.
Culture Characters Not known.
References 51, 90, 116, 119, 120, 132, 133, 181, 197, 214, 216, 242, 249, 335, 443, 448, 461, 499, 554, 577, 611.

Tremella palmata Schw. 1832 cannot be used for this fungus because it is a later homonym of *T. palmata* Schum. 1803. The oldest epithet for this fungus is *D. chrysospermus*.

Dacrymyces cokeri McNabb 1973
New Zealand J. Bot. 11:475.
Basionym Proposed as a *nom. nov.* for *Dacrymyces pallidus* Coker, J. Elisha Mitchell Sci. Soc. 35:171, 1920, which is a later homonym of *Dacrymyces pallidus* Lloyd 1919.
Distribution United States: NC.
Hosts *Pinus* sp., gymnosperm.
Ecology Decorticated and corticated fallen branches.
Culture Characters Not known.
References 107, 461.

Dacrymyces corticioides Ellis & Ev. 1885
J. Mycol. 1:149.
Distribution United States: NC, NH, NJ, NY.
Hosts Gymnosperm.
Ecology Sticks.
Culture Characters Not known.
References 332, 461.

Previously misidentified as *D. involutus*, fide McNabb (461).

Dacrymyces dacryomitriformis McNabb 1973
New Zealand J. Bot. 11:461.
Synonym *Dacryomitra dubia* Lloyd 1917, nom. nud., fide McNabb (461).

Distribution United States: KY, MI, NC, NY, OH, PA, WV.
Hosts *Pinus* sp., angiosperm.
Ecology Log.
Culture Characters Not known.
References 107, 461, 593.

Dacrymyces dictyosporus G.W. Martin 1958
Mycologia 50:939.
Distribution United States: AZ, NM.
Hosts *Pinus engelmannii, P. ponderosa.*
Ecology Causing a uniform brown rot.
Culture Characters Not known.
References 197, 216, 577.

Dacrymyces ellisii Coker 1920
J. Elisha Mitchell Sci. Soc. 35:167.
Synonym *Dacrymyces harperi* Bres. 1920, fide Kobayasi *et al.* (344).
Distribution Canada: BC, NS, ON, PQ.
United States: AK, AZ, GA, ID, IL, LA, MA, NC, NE, TN, VT, WI.
Hosts *Acer* sp., *Alnus* sp., *A. crispa, Betula* sp., *Ligustrum* sp. *Picea* sp., *Prunus persica, Quercus* sp., *Ulmus* sp., angiosperm.
Ecology Bark of dead limb; fallen limb; corticated or decorticated angiosperm wood; dead privet stems.
Culture Characters Not known.
References 51, 90, 107, 116, 119, 216, 249, 344, 443, 448, 498, 499.

It is uncertain, to us, which epithet, *ellisii* or *harperi*, was published first, and we follow Kobayasi's synonymy. Records may be a mixture of *D. capitatus* and *D. stillatus*, see McNabb (461).

Dacrymyces enatus (Berk. & Curtis) Massee 1891
J. Mycol. 6:182.
Basionym *Tremella enata* Berk. & Curtis, Grevillea 2:20, 1873.
Synonyms *Arrhytidia enata* (Berk. & Curtis) Coker 1928, *Dacrymyces gangliformis* Brasfield 1940, fide McNabb (461).
Distribution United States: GA, IA, MA, NJ, NY, NC, SC.
Hosts *Alnus* sp., *A. incana, Pinus strobus, Quercus* sp.
Ecology Typically dead branches of alder on the ground.
Culture Characters Not known.
References 51, 52, 111, 444, 461.

Dacrymyces fuscominus Coker 1920
J. Elisha Mitchell Sci. Soc. 35:171.
Distribution United States: IA, OH, LA, NC.
Hosts *Quercus* sp., angiosperm.
Ecology Not known.
Culture Characters Not known.
References 51, 107, 419, 443.

Dacrymyces hyalina Quél. 1888
Fl. Mycol., 17.
Distribution United States: NC, NH, PA.
Hosts *Pinus strobus, ?Tsuga canadensis.*
Ecology Log.
Culture Characters Not known.
References 511, 512.

Perhaps specimens of *D. tortus*, see Kennedy (337).

Dacrymyces marginatus McNabb 1973
New Zealand J. Bot. 11:498.
Distribution United States: SC.
Hosts Gymnosperm.
Ecology Not known.
Culture Characters Not known.

Known only from the type specimen.

Dacrymyces minor Peck 1878
Annual Rep. New York State Mus. 30:49.
Synonym *Dacrymyces deliquescens* var. *minor* (Peck) Kennedy 1958
Distribution Canada: NS, ON, PQ.
United States: AZ, CO, GA, IA, LA, NC, NY, VT.
Hosts *Abies* sp., *A. balsamea, Baccharis sarothroides, Cercidium microphyllum, Cornus* sp., *Corylus* sp., *Fraxinus* sp., *Lagerstroemia* sp., *Ligustrum chinense, Maclura pomifera, Malus* sp., *M. sylvestris, Ostrya virginiana, Parkinsonia microphylla, Picea* spp., *Pinus* sp., *P. strobus, Platanus* sp., *Populus* sp., *Prosopis juliflora, Quercus* sp., *Q. emoryi, Q. gambelii, Q. rubra, Salix* sp., *Ulmus* sp., angiosperm.
Ecology Decorticated wood; associated with a brown rot.
Culture Characters Not known.
References 51, 90 as *D. subochraceus* sensu Burt 1921, see McNabb (461), 107, 116, 183, 207, 216, 242, 249, 398, 402, 443, 498, 499, 503, 533, 577.

Dacrymyces minutus (Olive) McNabb 1973
New Zealand J. Bot. 11:497.
Basionym *Guepiniopsis minuta* Olive, Bull. Torrey Bot. Club 81:334, 1954.
Distribution Canada: BC. United States: NC, SC, WA.
Hosts *Pinus contorta, Tsuga* sp.
Ecology Decorticated twigs.
Culture Characters Not known.
References 242, 461, 504.

Dacrymyces nigrescens Lowy 1954
Bull. Torrey Bot. Club 81:300.
Distribution United States: LA, NC.
Hosts Angiosperm.
Ecology Dead wood.
Culture Characters Not known.
References 418, 461.

Dacrymyces ovisporus Bref. 1888
Unters. Gesammtgeb. Myk. 7:158.
Distribution Canada: BC.
Hosts Not known.
Ecology Not known.
Culture Characters Not known.
Reference 17.

Dacrymyces pedunculatus (Berk. & Curtis) Coker 1920
J. Elisha Mitchell Sci. Soc. 35:166.
Basionym *Exidia pedunculata* Berk. & Curtis, Grevillea 2:19, 1873.
Synonym *Dacryomitra pedunculata* (Berk. & Curtis) Burt 1921.
Distribution United States: NC, SC.
Hosts *Pinus* sp., *P. taeda.*
Ecology Fallen, corticated branches.
Culture Characters Not known.
References 90, 107, 461.

Dacrymyces pennsylvanica (Overh.) Ginns & Lefebvre, *comb. nov.*
Basionym *Guepinia pennsylvanicus* Overh., Mycologia 32:261, 1940.
Distribution United States: PA.
Host *Betula* sp.
Ecology Bark of dead standing tree.
Culture Characters Not known.

Known only from the type specimen. In general appearance, much like *D. tortus* but differs in the shorter or more roughened cortical cells and the broader spores (520). Clearly this fungus is congeneric with *D. tortus*. Thus the new combination in the genus *Dacrymyces* is proposed.

Dacrymyces stillatus Nees:Fr. 1816
System Pilze Schwämme, 89.
Synonyms *Dacrymyces abietinus* (Pers.) Schröter 1888, fide McNabb (461), *Dacrymyces deliquescens* (Bull.) sensu Aucts., fide McNabb (461).
Distribution Canada: AB-NT, NF, NS, NTM, ON.
United States: AK, AL, AZ, CA, CO, GA, IA, ID, LA, MO, NC, NJ, NY, OH, OR, SC, TX, VT, WA, WI.
Hosts *Abies balsamea, A. grandis, A. magnifica, Alnus rubra, Arctostaphylos patula, Calocedrus decurrens, Cercocarpus ledifolius, Chamaecyparis nootkatensis, Eucalyptus* sp., *Fagus* sp., *F. grandifolia, Larix occidentalis, Libocedrus decurrens, Picea engelmannii, P. glauca, Pinus* sp., *P. attenuata, P. banksiana, P. contorta, P. contorta, P. ponderosa, P. resinosa, P. strobus, P. sylvestris, P. taeda, Populus* sp., *Prosopis juliflora, Purshia tridentata, Quercus chrysolepis, Q. coccinea, Q. kelloggii, Tsuga mertensiana,* gymnosperm, less commonly angiosperm.
Ecology Logs; old railroad ties; causes a brown rot.
Culture Characters Not known.
References 51, 90, 116, 119, 120, 127, 183, 197, 216, 242, 249, 273, 402, 406, 443, 465, 498, 499, 577.

Dacrymyces tortus (Willd.) Fr. 1828
Elenchus 2:36.
Basionym *Tremella torta* Willd., Neues Mag. Bot. 4:18, 1788.
Synonym *Dacrymyces punctiformis* Neuhoff 1934, fide McNabb (461).
Distribution Canada: AB-NT, ON, PQ.
United States: AZ, CA, ID, LA, OR, MA, MT, NC, NJ, NM, NV, WA.
Hosts *Picea engelmannii, Pinus* sp., *P. contorta* var. *contorta, P. ponderosa, P. strobus, Quercus* sp., gymnosperm.
Ecology Rotting wood; stump; saprophytic on wood on the ground.
Culture Characters Not known.
References 51, 52, 119, 130, 197, 216, 242, 443, 448, 497, 533, 558.

Dacrymyces variisporus McNabb 1973
New Zealand J. Bot. 11:461.
Distribution United States: MI, SC, WA.
Hosts *Pinus ponderosa*, angiosperm, gymnosperm.
Ecology Not known.
Culture Characters Not known.
References 183, 461, 558.

DACRYOBOLUS Fr. 1849 Meruliaceae

Dacryobolus karstenii (Bres.) Parm. 1968
As Oberw. ex Parm., Consp. Syst. Cort., 98.
Basionym *Stereum karstenii* Bres., Atti Imp. Regia Accad. Rovereto, Ser. III, 3:108, 1897.
Synonym *Peniophora crassa* Burt 1901, proposed as a *nom. nov.* because there already was a *Peniophora karstenii* Massee 1889, *Peniophora verticillata* Burt 1926, fide Weresub (635).
Distribution Canada: BC, MB, NS, ON, PQ.
United States: AK, AL, AZ, CA, CO, ID, MA, ME, MT, NC, ND, NH, NJ, NM, NY, OR, PA, SC, TN, VT, WA.
Hosts *Abies* sp., *A. grandis, Chamaecyparis nootkatensis, Corylus californica, C. cornuta, Picea* sp., *P. engelmannii, P. glauca, P. sitchensis, Pinus* sp., *P. contorta, P. leiophylla* var. *chihuahuana, P. monticola, P. ponderosa, Populus* sp., *Pseudotsuga* sp., *P. menziesii, Quercus gambellii, Thuja* sp., *T. plicata, Tsuga* sp., *T. heterophylla*, gymnosperm.
Ecology Decaying logs; associated with a brown cubical rot.
Culture Characters Not known.
References 92, 93, 97, 116, 120, 128, 133, 183, 197, 199, 208, 216, 242, 249, 491, 584, 634, 635.

Dacryobolus sudans (Alb. & Schw.:Fr.) Fr. 1849
Summa veg. Scand. 2:404.
Basionym *Hydnum sudans* Alb. & Schw., Conspectus Fungorum, 272, 1805.
Synonym *Odontia sudans* (Alb. & Schw.) Bres. 1897.

Distribution Canada: BC, ON.
United States: AZ, CO, FL, IA, ID, MD, MI, MN, MS, MT, NC, NM, NY, TN, WI, WY, UT.
Hosts *Cornus florida, Juniperus monosperma, J. virginiana, Larix occidentalis, Malus sylvestris, Picea glauca, Pinus aristata var. longaeva, P. edulis, P. ponderosa, Platanus wrightii, Populus* sp., *Quercus hypoleucoides, Q. reticulata, Rhododendron* sp., *Tsuga heterophylla*, angiosperm, gymnosperm.
Ecology Dead branches; associated with a brown rot.
Culture Characters
1.3c.9.(14).21.27.32.36.38.44-45.(48).50.53.54.55.60 (Ref. 474).
References 116, 183, 197, 199, 202, 204, 208, 211, 216, 242, 406, 407, 447, 463, 474.

DACRYONAEMA Nannf. 1947 — Dacrymycetaceae

Dacryonaema rufum (Fr.:Fr.) Nannf. 1947
Svensk Bot. Tidskr. 41:336.
Basionym *Sphaeronaema rufum* Fr., Kongl. Vetensk. Acad. Handl. 39:357, 1818.
Distribution Canada: BC.
Host Gymnosperm, probably *Pinus* sp.
Ecology Not known.
Culture Characters Not known.
Reference 55.

DACRYOPINAX G.W. Martin 1948 — Dacrymycetaceae

Dacryopinax elegans (Berk. & Curtis) G.W. Martin 1948
Lloydia 11:116.
Basionym *Guepinia elegans* Berk. & Curtis, Hooker's J. Bot. Kew Gard. Misc. 1:239, 1849.
Synonym *Guepinia biformis* Peck 1900, fide Martin (443).
Distribution Canada: specific localities not given (443).
United States: AL, GA, IA, IL, LA, NC, NY, OH, SC, TN.
Hosts *Acer negundo, Ulmus* sp., angiosperm.
Ecology Decaying wood; trunks; logs.
Culture Characters Not known.
References 51, 107, 419, 443, 457, 498.

Dacryopinax petaliformis (Berk. & Curtis) McNabb 1965
New Zealand J. Bot. 3:70.
Basionym *Guepinia petaliformis* Berk. & Curtis, Grevillea 2:5, 1873.
Distribution United States: AL.
Hosts Not known.
Ecology Dead wood.
Culture Characters Not known.

Known only from the type specimen.

Dacryopinax spathularia (Schw.) G.W. Martin 1948
Lloydia 11:116.
Basionym *Merulius spathularia* Schw., Schriften Naturf. Ges. Leipzig 1:97, 1822.
Synonyms *Guepinia spathularia* (Schw.) Fr. 1828, *Dacryopsis ceracea* Coker 1920, fide McNabb (457), *Dacryomitra ceracea* (Coker) Brasfield 1938.
Distribution United States: AR, AZ, GA, IA, ID, KS, LA, MA, MD, MI, MN, MO, NC, NJ, NY, OH, PA, VA, WV.
Hosts *Abies grandis, Acer* sp., *A. negundo, A. rubrum, Castanea dentata, Citrus* sp., *Fagus* sp., *Lagerstroemia indica, Liquidambar* sp., *Liriodendron tulipifera, Malus pumila, M. sylvestris, Picea* sp., *Pinus* sp., *P. elliottii, P. palustris, Platanus occidentalis, P. wrightii, Prunus serotina, Quercus* sp., *Q. emoryi, Q. prinus, Q. velutina, Sequoia sempervirens, Taxodium* sp., *Tsuga canadensis*, angiosperm, gymnosperm, cedar.
Ecology Corticated wood; old fallen trunks; causes a uniform brown rot.
Culture Characters Not known.
References 51, 52, 96, 107, 127, 183, 216, 419, 443, 444, 457, 498, 499, 577, 611.

DEFLEXULA Corner 1950 — Pterulaceae

Deflexula ulmi (Peck) Corner 1950
Ann. Bot. Mem. 1:400.
Basionym *Mucronella ulmi* Peck, Annual Rep. New York State Mus. 54:154, 1901.
Distribution United States: IA, NY, PA.
Hosts *Fraxinus* sp., *Quercus* sp., *Salix* sp., *Ulmus* sp., *U. americana.*
Ecology Bark of live and dead trees.
Culture Characters Not known.
References 189, 463, 510.

DENDROCORTICIUM Larsen & Gilbn. 1974 — Vuilleminiaceae

Dendrocorticium piceinum Larsen & Gilbn. 1977
As Lemke ex Larsen & Gilbn., Norw. J. Bot. 24:113.
Distribution Canada: ON. United States: SC.
Hosts *Picea* sp., *Taxodium distichum.*
Ecology Not known.
Culture Characters Not known.

Known only from the type specimens.

Dendrocorticium roseocarneum (Schw.) Larsen & Gilbn. 1977
Norw. J. Bot. 24:115.
Basionym *Thelephora roseocarnea* Schw., Schriften Naturf. Ges. Leipzig 1:107, 1822.
Synonyms *Stereum roseocarneum* (Schw.) Fr. 1851, *Laxitextum roseocarneum* (Schw.) Lentz 1955, *Laeticorticium roseocarneum* (Schw.) Boidin 1958, *Corticium lilacino-fuscum* Berk. & Curtis 1873, fide Burt (89), Lentz (379), and Larsen and Gilbertson (368), *Stereum lilacino-fuscum* (Berk. & Curtis) Lloyd 1919, *Corticium subrepandum* Berk. & Cooke

1878, fide Burt (89), Lentz (379), and Hjortstam (288).

Distribution Canada: NB, NF, NS, ON. United States: AZ, CT, DC, GA, IA, IN, MA, MD, MI NC, NH, NJ, NY, OH, PA, TX, VA, VT, WI, WV.

Hosts *Acer* sp., *A. negundo, A. nigrum, A. rubrum, A. rubrum* var. *rubrum, A. saccharinum, A. saccharum, Alnus incana, A. rugosa, Crataegus* sp., *Fagus* sp., *F. grandifolia, Liriodendron tulipifera, Malus domestica, Nyssa sylvatica, Prunus pensylvanica, Quercus* sp., *Q. arizonica, Q. emoryi, Salix babylonica*, angiosperm.

Ecology Bark and wood of limb; fallen limbs; dead branches.

Culture Characters
2.3c.12.21.32.36.38.44.(53).54.60 (Ref. 474);
2a.3c.12.21.32.36.38.44.54.60.63 (Ref. 48).

References 89, 96, 108, 116, 127, 183, 216, 249, 368, 379, 474, 511, 519.

Dendrocorticium violaceum Larsen & Gilbn. 1977
As Jackson ex Larsen & Gilbn., Norw. J. Bot. 24:116.

Distribution Canada: NS, ON, PQ.

Hosts *Acer saccharum, Alnus* sp., *Betula alleghaniensis, Fraxinus* sp., *F. nigra, Picea* sp., *Populus* sp., *Prunus* sp., angiosperm.

Ecology Not known.

Culture Characters Not known.

References 249, 368.

DENDROPHORA (Parm.) Chamuris 1987 Peniophoraceae

Dendrophora albobadia (Schw.:Fr.) Chamuris 1987
Mycotaxon 28:544.

Basionym *Thelephora albobadia* Schw., Schriften Naturf. Ges. Leipzig 1:108, 1822.

Synonyms *Stereum albobadium* (Schw.) Fr. 1838, *Peniophora albobadia* (Schw.) Boidin 1961, *Thelephora albomarginata* Schw. in Berk. 1847, fide Burt (89) and Lentz (379), *Peniophora albomarginata* (Schw.) Massee 1889, *Stereum bizonatum* Berk. & Curtis 1873, fide Burt (89) and Lentz (379), *Stereum heterosporum* Burt 1920, fide Welden (620).

Distribution United States: AL, AZ, CA, DC, FL, GA, IA, ID, IL, IN, KS, LA, MD, MO, MS, NC, NJ, NM, NY, OH, OK, PA, SC, TN, TX, VA, WV.

Hosts *Acer glabrum, A. negundo, Albizia julibrissin, Alnus* sp., *A. oblongifolia, Baccharis halimifolia, B. sarothroides, Brassica oleracea, Calycanthus floridus, Carpinus* sp., *C. caroliniana, Carya* sp., *Castanea* sp., *Catalpa* sp., *Celtis occidentalis, C. reticulata, Cercis canadensis, Condalia mexicana, Diospyros* sp., *Forsythia* sp., *Fortunella* sp., *Fouquieria splendens, Fraxinus velutina, Haplopappus laricifolius, Hibiscus syriacus, Juglans nigra, Juniperus virginiana, Kolkwitzia amabilis, Ligustrum amurense, L. vulgare, Liriodendron tulipifera, Lycium* sp., *Maclura* sp., *M. pomifera, Magnolia* sp., *M. acuminata, Malus sylvestris, Ostrya* sp., *Philadelphus* sp., *Pinus sylvestris, Platanus* sp., *P. occidentalis, P. wrightii, Prosopis juliflora, Prunus* sp., *P. persica, P. serotina, Ptelea angustifolia, Quercus* sp., *Q. alba, Q. arizonica, Q. gambelii, Q. nigra, Q. virginiana, Rhus vernix, Sambucus canadensis, Simmondsia chinensis, Spiraea* sp., *Syringa vulgaris, Umbellularia californica, Vaccinium* sp., *Viburnum* sp., *Vitis arizonica, V. rotundifolia*, angiosperm.

Ecology Dead twigs; fallen branches; crevices of the bark of dead limbs and logs; decorticated log; associated with a white rot.

Culture Characters
2.3c.21.26.32.37.(39).(40).42-43.(50).54.(55).60 (Ref. 474);
2.3.21.24.26.32.37.39.42-43.54.60 (Ref. 477);
2.3c.8.32.36.39.(40).42.43.53.54.60.61 (Ref. 104).

References 89, 96, 102, 104, 108, 127, 183, 201, 207, 208, 216, 379, 474, 477, 519.

Burt's (89) records of *Stereum heterosporum* from Arizona, California and Oregon are not cited here, because most have been redetermined as *Peniophora malençonii* subsp. *americana* (see 102).

Dendrophora erumpens (Burt) Chamuris 1987
Mycotaxon 28:544.

Basionym *Stereum erumpens* Burt, Ann. Missouri Bot. Gard. 7:209, 1920.

Synonym *Peniophora erumpens* (Burt) Boidin 1959.

Distribution United States: AL, AR, DC, GA, IA, IL, IN, MD, MT, ND, NJ, NY, OR, PA, RI, SD, VA, WA.

Hosts *Acer rubrum, Alnus* sp., *A. rubra, A. tenuifolia, Betula papyrifera, Carpinus caroliniana, Castanea* sp., *C. dentata, Catalpa bignonioides, Malus* sp., *M. sylvestris, Prunus* sp., *P. avium, Salix* sp., *S. scouleriana, Sassafras albidum, Vitis rotundifolia*, angiosperm.

Ecology Dead limbs.

Culture Characters Not known.

References 89, 104, 183, 379.

Dendrophora versiformis (Berk. & Curtis) Chamuris 1987
Mycotaxon 28:544.

Basionym *Stereum versiforme* Berk. & Curtis, Grevillea 1:164, 1873.

Synonyms *Peniophora versiformis* (Berk. & Curtis) Bourd. & Galzin 1928, *Peniophora ellisii* Massee 1889, fide Lentz (379).

Distribution Canada: MB, ON, PQ.
United States: AL, AR, AZ, CT, DC, DE, GA, IA, IL, IN, MA, MD, MI, MO, MS, NH, NJ, NY, PA, RI, SC, TN, VA, VT, WV.

Hosts *Acer* sp., *A. negundo, A. nigrum, A. rubrum, A. rubrum* var. *rubrum, A. saccharinum, A. saccharum, Alnus* sp., *A. pumila, Amelanchier spicata, Betula* sp., *B. alleghaniensis, B. lenta, Carpinus caroliniana, Carya* sp., *Castanea* sp., *C. crenata, C. dentata, C. sativa, Cornus florida, Crataegus spathulata,*

Diospyros virginiana, Fraxinus sp., *Hamamelis virginiana, Liquidambar styraciflua, Liriodendron tulipifera, Malus* sp., *M. pumila, M. sylvestris, Myrica cerifera, Populus balsamifera, P. grandidentata, Prunus* sp., *P. persica, P. serotina, Quercus* sp., *Q. alba, Q. ilicifolia, Q. nigra, Rhus* sp., *R. typhina, Salix* sp., *Ulmus americana*, angiosperm.
Ecology Bark of dead limbs; dead branches and twigs; charred chestnut wood; associated with a white rot.
Culture Characters
2.3c.7.(22).(26).(27).32.36.(37).(38).39.(40).42-43.(50).(53).54.60 (Ref. 474);
2.3c.8.32.36.(37).39.42.53.54.60.61 (Ref. 104).
References 89, 96, 104, 116, 164, 183, 204, 216, 379, 474, 519.

DENDROTHELE Höhnel & Litsch. 1907 Aleurodiscaceae
Synonym is *Aleurocorticium* Lemke 1964, fide Lemke (377).

Dendrothele acerina (Pers.:Fr.) Lemke 1965
Persoonia 3:366.
Basionym *Corticium acerinum* Pers., Obs. Mycol. 1:37, 1796.
Synonyms *Aleurocorticium acerinum* (Pers.) Lemke 1965, *Aleurodiscus acerinus* (Pers.) Höhnel & Litsch. 1907.
Distribution Canada: MB, ON.
United States: AZ, FL, IA, IL, LA, MA, MD, MI, MO, NC, NH, NY, PA, TX, VT, WI.
Hosts *Acer* sp., *A. saccharum, Carya* sp., *Castanea dentata, Fraxinus* sp., *F. americanus, Mortonia scabrella, Morus rubra, Quercus* sp. *Q. alba, Q. macrocarpa, Q. virginiana, Tilia americana, Ulmus* sp.
Ecology Bark of live trees.
Culture Characters
(2).(3c).7.32.36.(39).47.54.58.(61) (Ref. 48).
References 108, 116, 183, 216, 242, 376, 511, 516.

Dendrothele alliacea (Quél.) Lemke 1965
Persoonia 3:366.
Basionym *Corticium alliacea* Quél., Assoc. Franç. Avancem. Sci. (Rouen) 12:505, 1884.
Synonyms *Aleurocorticium alliaceum* (Quél.) Lemke 1964, *Aleurodiscus acerinus* var. *alliaceus* (Quél.) Bourd. & Galzin 1928.
Distribution Canada: MB, ON, PQ.
United States: AR, IA, ME, NY, VT.
Hosts *Acer* sp., *A. saccharum, Carya tomentosa, Fagus grandifolia, Fraxinus* sp., *Ostrya virginiana, Quercus alba, Q. macrocarpa, Tilia americana, Ulmus* sp.
Ecology Bark of live trees.
Culture Characters
2.3?.6?.22.32.36.(39).47.54.58.(61?) (Ref. 48).
References 116, 242, 376.

Dendrothele bispora Burdsall & Nakasone 1983
Mycotaxon 17:253.
Distribution United States: FL.
Host *Taxodium distichum.*
Ecology Not known.
Culture Characters Not known.

Known only from the type specimen.

Dendrothele candida var. ***candida*** (Schw.:Fr.) Lemke 1965
Persoonia 3:366.
Basionym *Thelephora candida* Schw., Schriften Naturf. Ges. Leipzig 1:110, 1822.
Synonyms *Aleurodiscus candidus* (Schw.) Burt 1918, *Aleurocorticium candidum* (Schw.) Lemke 1964, *Aleurodiscus crassus* Lloyd 1920, fide Lemke (376).
Distribution Canada: BC, ON. United States: CA, CT, FL, GA, IL, IN, LA, MA, MD, MO, NC, NH, OH, OR, PA, SC, TN, TX, VA, WV.
Hosts *Acer* sp., *A. macrophyllum, A. rubrum, A. saccharum, Carya* sp., *C. illinoensis, Castanea* sp., *Fraxinus* sp., *F. americana, Lithocarpus densiflora, Quercus* sp., *Q. alba, Q. chrysolepis, Q. falcata, Q. garryana, Q. nigra, Q. obtusiloba, Q. prinus, Q. rubra, Q. stellata, Q. velutina, Ulmus alata.*
Ecology Bark of live trees, especially oaks.
Culture Characters
2.6.7.(26).32.36.38.47.54 (Ref. 474);
2a.6.7.32.36.(39).47.54 (Ref. 48).
References 108, 116, 183, 242, 376, 474, 516.

Dendrothele candida var. ***sphaerospora*** (Coker) Ginns & Lefebvre, *comb. nov.*
Basionym *Aleurodiscus candidus* var. *sphaerosporus* Coker, J. Elisha Mitchell Sci. Soc. 36:155, 1921.
Distribution United States: NC.
Host *Ulmus alata.*
Ecology Not known.
Culture Characters Not known.

Known only from the type specimen. The new combination is proposed to keep this variety in the same genus as the species.

Dendrothele dryina (Pers.) Lemke 1965
Persoonia 3:366.
Basionym *Thelephora dryina* Pers., Mycol. Eur. 1:152, 1822.
Synonym *Aleurocorticium dryinum* (Pers.) Lemke 1964, *Aleurodiscus acerinus* var. *dryinus* (Pers.) Bourd. & Galzin 1913.
Distribution Canada: ON, PQ.
United States: AZ, NH.
Hosts *Acer* sp., *A. saccharum, Arbutus arizonica.*
Ecology Bark of live trees.
Culture Characters
2a.3c.(8).(9).(12).32.36.(39).47.54.64 (Ref. 48).
References 116, 216, 242, 376.

Dendrothele griseocana (Bres.) Bourd. & Galzin 1913 Bull. Soc. Mycol. France 28:354.

Basionym *Corticium griseocanum* Bres., Fungi Tridentini 2:58, 1892.

Synonym *Aleurocorticium griseocanum* (Bres.) Lemke 1965, *Aleurodiscus griseocanus* (Bres.) Höhnel & Litsch. 1908.

Distribution Canada: MB, ON, PQ. United States: AZ, FL, IA, LA, MI, MO, MS, NEW ENGLAND, NY, OH, PA.

Hosts *Acer* sp., *A. campestre, A. rubrum, Carya texana, Crataegus* sp., *Fraxinus* sp., *F. americana, Juniperus deppeana, J. virginiana, Ligustrum* sp., *Liriodendron tulipifera, Lonicera tatarica, Malus pumila, Ostrya virginiana, Pinus strobus, Platanus occidentalis, Populus balsamifera, Quercus* sp., *Q. alba, Q. arizonica, Q. macrocarpa, Q. toumeyi, Rhus toxicodendron, Salix* sp., *Syringa vulgaris, Thuja occidentalis, Tilia americana, Ulmus* sp., *U. americana, Vitis* sp., *V. riparia.*

Ecology Bark of live and dead branches.

Culture Characters 2a.6.26.32.36.38.47.54.58.61 (Ref. 48).

References 116, 183, 201, 216, 242, 376, 563.

Perhaps the epithet is misapplied in North America, because North American specimens have clamp connections and basidia with four spores, whereas the descriptions of European specimens report simple septa and two-spored basidia.

Dendrothele incrustans (Lemke) Lemke 1965 Persoonia 3:366.

Basionym *Aleurocorticium incrustans* Lemke, Canad. J. Bot. 42:739, 1964.

Distribution Canada: BC. United States: AZ, CA, LA.

Hosts *Abies grandis, Acer macrophyllum, Arbutus menziesii, Ceanothus thyrsiflorus, Holodiscus discolor, Juniperus deppeana, J. pachyphloea, J. virginiana, Picea sitchensis, Pseudotsuga menziesii, Ribes sanguineum, Spiraea douglasii, Taxus brevifolia, Thuja plicata, Tsuga heterophylla.*

Ecology Dead branches; associated with a white rot.

Culture Characters Not known.

References 183, 202, 211, 242, 376.

Dendrothele itihummensis Gilbn. & Blackwell 1985 Mycotaxon 24:331.

Distribution United States: AL, FL, LA, MS.

Hosts *Juniperus virginiana, J. phoenicea.*

Ecology Dead branches; associated with a white rot.

Culture Characters Not known.

References 201, 202.

Dendrothele macrodens (Coker) Lemke 1965 Persoonia 3:366.

Basionym *Aleurodiscus macrodens* Coker, J. Elisha Mitchell Sci. Soc. 36:155, 1921.

Synonym *Aleurocorticium macrodens* (Coker) Lemke 1965.

Distribution United States: CT, GA, LA, MA, MI, NC, NH, NY, OH, OR, PA, WV.

Hosts *Fraxinus* sp., *F. americana, F. nigra, Nyssa sylvatica, Quercus kelloggii, Salix* sp.

Ecology Bark of live trees.

Culture Characters Not known.

References 93, 376.

Dendrothele macrospora (Bres.) Lemke 1965 Persoonia 3:366.

Basionym *Corticium acerinum* var. *macrosporum* Bres., Ann. Mycol. 1:96, 1903.

Distribution Canada: AB-NT.

Host *Populus* sp.

Ecology Not known.

Culture Characters 2a.3c.(9).(12).(26).32.36.38.44-45.54.60.61. (Ref. 48).

Reference 242.

Dendrothele maculata (Jackson & Lemke) Lemke 1965 Persoonia 3:367.

Basionym *Aleurocorticium maculatum* Jackson & Lemke, Canad. J. Bot. 42:742, 1964.

Distribution Canada: ON, PQ. United States: CT, ME, VA.

Hosts *Acer* sp., *Corylus* sp., *C. cornuta, C. rostrata, Prunus* sp., *P. serotina, P. virginiana, Rhododendron maximum.*

Ecology Not known.

Culture Characters 2.6.26.(27).32.36.38.44.54 (Ref. 474).

References 242, 376, 474.

Dendrothele microspora (Jackson & Lemke) Lemke 1965 Persoonia 3:367.

Basionym *Aleurocorticium microsporum* Jackson & Lemke, Canad. J. Bot. 42:745, 1964.

Distribution Canada: ON, PQ. United States: AL, CT, FL, IA, IL, LA, MA, MD, MI, MO, NC, NH, NJ, NY, PA, RI, TN, VA, VT, WI.

Hosts *Acer* sp., *A. negundo, A. platanoides, A. rubrum, A. rubrum* var. *rubrum, A. saccharum, Betula alleghaniensis, Carya* sp., *C. glabra, C. ovata, Fraxinus* sp., *Liquidambar styraciflua, Liriodendron tulipifera, Nyssa sylvatica, Ostrya virginiana, Platanus occidentalis, Populus balsamifera, Quercus* sp., *Q. alba, Salix* sp., *Taxodium distichum, Tilia americana, Ulmus* sp., *U. americana.*

Ecology Not known.

Culture Characters 2.3c.7.32.36.38.47.(50).54.60 (Ref. 474).

References 183, 242, 376, 474.

Dendrothele nivosa (Höhnel & Litsch.) Lemke 1965 Persoonia 3:367.

Basionym *Aleurodiscus nivosus* Berk. & Curtis ex Höhnel & Litsch., Sitzungsber. Kaiserl. Akad. Wiss., Math.-Naturwiss. Kl., Abt.1, 116:808, 1907.

Synonym *Aleurocorticium nivosum* (Höhnel & Litsch.) Lemke 1965.

Distribution Canada: ON. United States: AL, AR, CT, FL, GA, IN, KS, LA, MA, MO, MS, NC, NJ, NY, OH, OK, OR, PA, SC, TN, TX.
Hosts *Chamaecyparis* sp., *C. thyoides*, especially *Juniperus* spp., *J. phoenica*, *J. scopulorum*, *J. silicicola*, *J. virginiana*, *Thuja occidentalis*.
Ecology Bark on trunks and branches of living trees.
Culture Characters
(1).2.3c.26.31g.32.36.39.47.(50).(51).55 (Ref. 474).
References 183, 201, 202, 242, 376, 474, 516.

Dendrothele pachysterigmata (Jackson & Lemke) Lemke 1965
Persoonia 3:367.
Basionym *Aleurocorticium pachysterigmatum* Jackson & Lemke, Canad. J. Bot. 42:750, 1964.
Distribution Canada: ON.
United States: LA, MN, NY.
Hosts *Carya ovata*, *Juniperus virginiana*, *Thuja occidentalis*.
Ecology Bark and decorticated wood of dead branches, associated with a white rot.
Culture Characters Not known.
References 201, 214, 242, 376.

Dendrothele seriata (Berk. & Curtis) Lemke 1965
Persoonia 3:367.
Basionym *Stereum seriatum* Berk. & Curtis, J. Linn. Soc., Bot. 10:332, 1868.
Synonym *Aleurocorticium seriatum* (Berk. & Curtis) Lemke 1964.
Distribution United States: FL, LA.
Hosts *Fraxinus* sp., *F. pennsylvanica*.
Ecology Not known.
Culture Characters Not known.
References 183, 376.

Dendrothele strumosa (Fr.) Lemke 1965
Persoonia 3:367.
Basionym *Stereum strumosum* Fr., Nov. Symb. Mycol., 95, 1851.
Synonyms *Aleurocorticium strumosum* (Fr.) Lemke 1965, *Aleurodiscus strumosus* (Fr.) Burt 1918.
Distribution United States: FL, LA, SC.
Hosts *Carya* sp., *Fraxinus pennsylvanica*, *F. tomentosa*, *Liquidambar styraciflua*, *Quercus* sp.
Ecology Not known.
Culture Characters Not known.
References 183, 376.

Dendrothele subfusispora Burdsall & Nakasone 1983
Mycotaxon 17:255.
Distribution United States: FL.
Host *Ostrya virginiana*.
Ecology Not known.
Culture Characters Not known.

Known only from the type specimen.

DENTIPELLIS Donk 1962 Hericiaceae

Dentipellis dissita (Berk. & Cooke) Maas G. 1974
Persoonia 7:551.
Basionym *Hydnum dissitum* Berk. & Cooke, J. Linn. Soc., Bot. 15:387, 1876.
Synonym *Hydnum macrodon* sensu L.W. Miller 1933 and Aucts., fide Ginns (240), *Mycoacia macrodon* sensu L.W. Miller.
Distribution Canada: ON, PQ. United States: DC, FL, IA, MD, MI, NC, NY, OH, WI.
Hosts *Acer* sp., *A. rubrum*, *Fagus grandifolia*, *Ostrya virginiana*, *Populus tremuloides*, *Quercus* sp., *Q. borealis*, *Q. rubra*, *?Tilia americana*, angiosperm.
Ecology Rotted, decorticated log; stump; solid log; associated with a white rot.
Culture Characters
2a.3.(15).(21).34.36.38.44.(48).54.59 (Ref. 240);
2.3c.15(p).21.34.36.38.(39).44.53.54.59 (Ref. 474).
References 240, 444, 474.

The Alaska record (97) of *Mycoacia macrodon* may have been based on a specimen of *D. fragilis*, because *D. dissita* appears to be restricted to eastern North America and *D. fragilis* is known in North America only from the northwest.

Dentipellis fragilis (Pers.:Fr.) Donk 1962
Persoonia 2:233.
Basionym *Hydnum fragile* Pers., Synop. Meth. Fung. 561, 1801.
Distribution United States: AK, WA.
Hosts *Populus balsamifera*, *P. trichocarpa*.
Ecology Rotten wood; associated with a white rot.
Culture Characters Not known.
References 183, 240.

Dentipellis leptodon (Mont.) Maas G. 1974
Persoonia 7:558.
Basionym *Hydnum leptodon* Mont., Ann. Sci. Nat. Bot., Sér. II, 20:366, 1843.
Synonyms *Hydnum separans* Peck 1897, fide Ginns (240), *Dentipellis separans* (Peck) Donk 1962, *Odontia separans* (Peck) C.A. Brown 1935, *Hydnum ohioense* Berk. 1845, fide Ginns, see below.
Distribution Canada: ON. United States: NY.
Hosts *Acer saccharum*, *Betula alleghaniensis*, *?Fraxinus* sp., *Sambucus glauca*.
Ecology Bark; dead wood; associated with a white rot.
Culture Characters Not known.
References 56, 240.

Maas Geesteranus (428) referred the name *Hydnum ohioensis* to *?Gyrodontium*. We have examined the type specimen at Kew (K), labelled "41 *Hydnum* sp. *Hydnum fernandesianum* Mont. var. *ohioense* Berk. ... March (rare). Herb. Hooker 1867". It consists of two pieces glued to paper, no writing on the paper, each piece is 2-3 x 2-3 cm. Macroscopically the subiculum and margin are pallid, the spines brownish red, acicular, 3-4 x 0.1

mm. The trama of the spines has many slightly refractive gloeocystidia with contents homogeneous, curving into the hymenium and the apex often moniliform. Basidiospores broadly ellipsoid, 4.2 x 3.0-3.6 µm, perhaps weakly amyloid. Although we consider the name *Hydnum ohioense* to be a synonym of *Dentipellis leptodon*, the California collection (now BPI 259730) identified by E. A. Burt (465) as *H. ohioense* is not *D. leptodon*. In North America both published accounts and herbarium specimens labelled "*separans*" were based upon specimens of *D. dissita*, thus only the recent records in Ginns (240), where the confusion was recognized and corrected, are cited.

DENTOCORTICIUM (Parm.) Larsen & Gilbn. 1974 Vuilleminiaceae

Dentocorticium sulphurellum (Peck) Larsen & Gilbn. 1974 Norw. J. Bot. 21:226.
Basionym *Hydnum sulphurellum* Peck, Annual Rep. New York State Mus. 31:38, 1879.
Synonym *Laeticorticium sulphurellum* (Peck) Gilbn. 1962.
Distribution Canada: ON, PQ. United States: MN, NC, NY.
Hosts *Acer* sp., *A. spicatum*, *Liriodendron tulipifera*, *Salix* sp., *Tilia americana*, angiosperm.
Ecology Not known.
Culture Characters Not known.
References 214, 242, 368.

DICHOSTEREUM Pilát 1926 Lachnocladiaceae

Dichostereum boreale (Pouzar) Ginns & Lefebvre, *comb. nov.*
Basionym *Vararia borealis* Pouzar, Ceská Mykol. 36:72, 1982.
Synonyms *Vararia granulosa* (Pers.:Fr.) Laurila 1939 sensu Aucts., fide Pouzar (537), *Dichostereum granulosum* (Pers.:Fr.) Boidin & Lanquetin 1977 sensu Aucts.
Distribution Canada: AB, BC, ON. United States: AZ, CO, ID, MI, MN, MT, NH, NM, OR, SD, WA.
Hosts *Abies* sp., *A. amabilis*, *A. concolor*, *A. grandis*, *A. lasiocarpa*, *Larix occidentalis*, *Picea* sp., *P. engelmannii*, *P. glauca*, *Pinus albicaulis*, *P. contorta*, *P. monticola*, *P. ponderosa*, *P. resinosa*, *P. strobus*, *Pseudotsuga menziesii*, *Tsuga canadensis*, *T. heterophylla*, gymnosperm.
Ecology Logs; associated with a white rot.
Culture Characters
2.3.7.33.34.36.38.44-46.55.59 (Ref. 487);
2.3c.15p.33.34.36.38.(39).44-45.(50).55.60 (Ref. 474);
2a.3c.15.25d.33.34.36.39.44.55.59.62 (Ref. 355).
References 116, 119, 193, 197, 204, 208, 214, 216, 242, 332, 355, 402, 448, 450, 463, 474.

Traditionally this fungus has been known by the epithet *granulosa*. In 1982 Pouzar (537) concluded that the epithet *granulosa* applied to a distinctly different fungus, and he proposed the name *V. borealis* for the fungus that had been misnamed as *granulosa*. This fungus (as *granulosa*) has been placed in *Dichostereum* (40). And we agree that the fungus is better placed in that genus than in *Vararia*. Thus the new epithet *borealis* is combined in *Dichostereum*. The mating system has been reported as bipolar as well as tetrapolar, which suggests misinterpretation of data, misidentified specimens, or a species complex. The report (183) on *Crataegus brevispina* is suspect, because all other reports were on gymnosperms.

Dichostereum dura (Bourd. & Galzin) Pilát 1926 Ann. Mycol. 24:223.
Basionym *Asterostromella dura* Bourd. & Galzin, Bull. Soc. Mycol. France 36:74-75, 1920.
Distribution United States: GA, LA.
Hosts Angiosperm.
Ecology Leaves.
Culture Characters
2.3c.15.25d.33.34.36.(38).(39).43.54.60.63 (Ref. 355).
Reference 355.

The specimens were listed as "aff. *dura*" by Lanquetin (355). She found haploid cultures in this group were incompatible with *D. effuscatum*, *D. pallescens*, *D. ramulosum* (Boidin & Lanquetin) Boidin & Lanquetin and *D. rhodosporum* (Wakef.) Boidin & Lanquetin. Thus she rejected Rogers and Jackson's (567) listing of *dura* as a synonym of *pallescens*. Due to a lack of cultures she was not able to determine whether the two North American collections were compatible with European strains.

Dichostereum effuscatum (Cooke & Ellis) Boidin & Lanquetin 1977 Mycotaxon 6:284.
Basionym *Corticium effuscatum* Cooke & Ellis, Grevillea 9:103, 1881.
Synonym *Vararia effuscata* (Cooke & Ellis) D.P. Rogers & Jackson 1943.
Distribution Canada: BC, NB, NF, NS, ON, PE, PQ. United States: CT, DC, GA, IA, ID, IL, KY, LA, MA, MD, ME, MI, MN, MO, NH, NJ, NY, OH, PA, TX, VA, WA, WI.
Hosts *Acer* sp., *A. saccharinum*, *A. saccharum*, *A. rubrum*, *Alnus* sp., *A. incana*, *Betula alleghaniensis*, *B. nigra*, *Carpinus* sp., *Carya ovata*, *Fraxinus nigra*, *Juglans cinerea*, *Populus* sp., *Quercus* sp., *Q. alba*, *Salix* sp., *Tsuga canadensis*, *Ulmus* sp., angiosperm.
Ecology Branch on ground; decaying wood and bark of logs; rotten log.
Culture Characters
2.3c.15.25d.33.34.36.(39).43.48.54.60.63 (Ref. 355);
2.3c.15p.33.34.36.39.42-43.(48).54.60 (Ref. 474).
References 93, 116, 183, 193, 214, 242, 249, 353, 355, 474, 483, 618.

Dichostereum pallescens (Schw.) Boidin & Lanquetin 1977
Mycotaxon 6:284.
Basionym *Thelephora pallescens* Schw., Trans. Amer. Philos. Soc., N.S. 4:167, 1832.
Synonyms *Vararia pallescens* (Schw.) D.P. Rogers & Jackson 1943, *Thelephora insinuans* Schw. 1832, *Corticium thelephoroides* Ellis & Ev. 1885. All fide Rogers and Jackson (567).
Distribution Canada: BC, ON. United States: AL, AZ, CA, DC, FL, GA, IL, LA, MA, MD, ME, MO, NC, NH, NY, OR, PA, RI, TN, TX, VA, VT, WA, WI.
Hosts *Acer saccharum, Alnus* sp., *A. oblongifolia, Betula alleghaniensis, B. papyrifera, Castanea dentata, Pinus echinata, P. palustris, Pseudotsuga menziesii, Quercus* sp., *Q. borealis, Q. rubra, Rubus spectabilis, Thuja plicata, Tsuga heterophylla, Ulmus americana*, angiosperm.
Ecology Logs; old logs; stumps; dead roots; fallen wood; associated with a white rot.
Culture Characters
2.3c.8v.15p.(32).(33).(34).37.38.(39).44.54.60 (Ref. 474);
2a.3c.15.25d.33.(34).37.39.44.54.60.63 (Ref. 355).
References 116, 183, 193, 216, 242, 353, 355, 474, 616, 618.

Dichostereum sordulentum (Cooke & Massee) Boidin & Lanquetin 1981
Bull. Soc. Mycol. France 96:384.
Basionym *Corticium sordulentum* Cooke & Massee, Grevillea 16:69, 1888.
Distribution United States: AR, FL, GA, IL, LA, MD, MO, NC.
Hosts *Acer rubrum, Carpinus caroliniana, Liquidambar styraciflua, Quercus nigra, Q. prinus, Salix* sp., *Ulmus* sp., angiosperm.
Ecology Associated with a white rot.
Culture Characters
2.3c.15p.33.34.36.37.39.42-43.(53).54.60 (Ref. 474);
2a.3c.15.25d.33.34.36.(38).(39).43.54.60.63 (Ref. 355).
References 40, 474.

DITIOLA Fr. 1822 — Dacrymycetaceae

Ditiola radicata var. ***radicata*** (Alb. & Schw.):Fr. 1822
Syst. Mycol. 2:170.
Basionym *Helotium radicatum* Alb. & Schw., Conspectus Fungorum, 348, 1805.
Distribution Canada: AB, AB-NT, BC, NS, ON. United States: AZ, CA, GA, IA, MA, MD, NC, NH, NY, OH, VT.
Hosts *Abies balsamea, Betula* spp., *Pinus* sp., *P. contorta, Quercus* sp., *Q. coccinea*, cedar, angiosperm, gymnosperm.
Ecology Fallen, corticated limbs; usually decorticated wood; associated with a brown rot.
Culture Characters
1.2.6.22.33.34.36.?38.?47 (composed from the description in 265).
References 107, 116, 127, 183, 216, 242, 265, 338.

Some of the above reports may have been based upon specimens of var. *gyrocephala*, because Kennedy (338) considered *Tremella stipitata* to be a synonym of var. *radicata*, whereas McNabb (460) placed the epithet under var. *gyrocephala*.

Ditiola radicata var. ***gyrocephala*** (Berk. & Curtis) Kennedy 1964
Mycologia 56:302.
Basionym *Coryne gyrocephala* Berk. & Curtis, Hooker's J. Bot. Kew Gard. Misc. 1:239, 1849.
Synonyms *Dacryomitra brunnea* G.W. Martin 1934, fide McNabb (460), *Tremella stipitata* Peck 1875, fide McNabb (460), *Dacryomitra stipitata* (Peck) Burt 1921.
Distribution Canada: NS, ON.
United States: LA, MA, ME, MN, MS, NJ, NY, SC.
Hosts *Acer* sp., *Magnolia grandiflora*, angiosperm, gymnosperm.
Ecology Decaying wood; decorticated wood.
Culture Characters Not known.
References 51, 52, 90, 249, 338, 434, 443, 460, 499.

DUCTIFERA Lloyd 1917 — Exidiaceae

Synonym is *Gloeotromera* Ervin 1956, fide Wells (623).

Ductifera micropera (Kalchbr. & Cooke) Wells 1958
Mycologia 50:413.
Basionym *Tremella micropera* Kalchbr. & Cooke, Grevillea 9:18, 1880.
Synonyms *Seismosarca tomentosa* Olive 1947, fide Wells (623), *Exidia tomentosa* (Olive) Olive 1954.
Distribution United States: GA.
Hosts Angiosperm.
Ecology Fallen corticated limbs.
Culture Characters Not known.
References 498, 623.

Ductifera pululahuana (Pat.) Wells 1958
Mycologia 50:412.
Basionym *Tremella pululahuana* Pat., Bull. Soc. Mycol. France 9:138, 1893.
Synonyms *Gloeotromera pululahuana* (Pat.) Ervin 1935, *Sebacina pululahuana* (Pat.) D.P. Rogers 1935.
Distribution United States: IA, IN, LA, MO, OH, TN.
Hosts Angiosperm.
Ecology Dead wood.
Culture Characters Not known.
References 563, 622, 623.

Ductifera sucina (Möller) Wells 1958
Mycologia 50:413.

Basionym *Exidia sucina* Möller, Protobas. 169, 1895.
Synonyms *Gloeotromera sucina* (Möller) Wells 1957, *Exidia cystidiata* Olive 1954, fide Wells (623), *Sebacina lactescens* Burt 1926, fide Wells (623).
Distribution United States: NC, NM.
Hosts *Cornus florida, Quercus* sp.
Ecology Decorticated limb; dead wood.
Culture Characters Not known.
References 504, 622, 623.

ECHINODONTIUM Ellis & Ev. 1900 Echinodontiaceae

Echinodontium ballouii (Banker) H. Gross 1964
Mycopath. Mycol. Appl. 24:5.
Basionym *Steccherinum ballouii* Banker, Bull. Torrey Bot. Club 36:341, 1909.
Distribution United States: NJ.
Host *Chamaecyparis thyoides.*
Ecology Not known.
Culture Characters Not known.

Known only from the type specimen.

Echinodontium tinctorium (Ellis & Ev.) Ellis & Ev. 1900
Bull. Torrey Bot. Club 27:49.
Basionym *Fomes tinctorius* Ellis & Ev., Bull. Torrey Bot. Club 22:362, 1895.
Distribution Canada: AB, AB-NT, BC. United States: AK, AZ, CA, CO, ID, MT, NM, OR, UT, WA.
Hosts *Abies* sp., *A. amabilis, A. concolor, A. grandis, A. lasiocarpa, A. lasiocarpa* var. *arizonica, A. magnifica, A. procera, Larix occidentalis, Picea engelmannii, P. glauca, P. sitchensis, Pseudotsuga menziesii, Thuja plicata, Tsuga* sp., *T. heterophylla, T. mertensiana,* gymnosperm.
Ecology Heart rot of live trees, although known as brown stringy rot, it is a white rot.
Culture Characters
2.3c.8.13.(15).34.36.37.39.47.(53).55 (Ref. 474);
2.3.8.13.34.37.39.47.50.53.55 (Ref. 487);
60 (Ref. 647).
References 130, 183, 217, 242, 251, 274, 474, 647.

EFIBULOBASIDIUM Wells 1975 Exidiaceae

Efibulobasidium rolleyi (Olive) Wells 1975
Mycologia 67:152.
Basionym *Exidia rolleyi* Olive, Bull. Torrey Bot. Club 85:95, 1958.
Distribution United States: IA.
Hosts *Quercus* sp., *Ulmus* sp.
Ecology Oak leaves in moist chamber; corticated elm wood.
Culture Characters Not known.
Reference 627.

EICHLERIELLA Bres. 1903 Exidiaceae

Eichleriella alliciens (Berk. & Cooke) Burt 1915
Ann. Missouri Bot. Gard. 2:746.
Basionym *Stereum alliciens* Berk. & Cooke, J. Linn. Soc., Bot. 15:389, 1876.
Synonym *Exidiopsis alliciens* (Berk. & Cooke) Wells 1961.
Distribution United States: FL, MS.
Hosts *Ilex decidua, Quercus* sp., angiosperm.
Ecology Dead wood; associated with a white rot.
Culture Characters
(1).2.3c.26.32.36.38.40.46-47.54 (Ref. 474).
References 474, 625.

Eichleriella deglubens (Berk. & Br.) D. Reid 1970
Trans. Brit. Mycol. Soc. 55:436.
Basionym *Radulum deglubens* Berk. & Br., Ann. Mag. Nat. Hist., Ser. IV, 15:32, 1875.
Synonyms *Radulum spinulosum* Berk. & Curtis, 1873 sensu Burt 1915 sensu N. Amer. Aucts., fide Reid (556), *Eichleriella spinulosa* (Berk. & Curtis) Burt sensu N. Amer. Aucts.
Distribution Canada: AB, BC, MB, NTM, ON.
United States: AK, ID, OR.
Hosts *Populus balsamifera, P. tremuloides, P. trichocarpa, Salix* sp., angiosperm, especially on *Populus* species.
Ecology Associated with a white rot.
Culture Characters
2.3c.31g.32.36.38.(40).47.53.54 (Ref. 474).
References 81, 183, 242, 340, 406, 443, 474, 625.

Eichleriella leveilliana (Berk. & Curtis) Burt 1915
Ann. Missouri Bot. Gard. 2:744.
Basionym *Corticium leveillianum* Berk. & Curtis, Hooker's J. Bot. Kew Gard. Misc. 1:238, 1849.
Distribution Canada: NS, ON.
United States: AL, FL, GA, LA, NC, SC, TX.
Hosts *Alnus incana, A. rugosa, Crataegus* sp., *Liquidambar* sp., *Populus* sp.
Ecology Dead limbs; grape arbor.
Culture Characters Not known.
References 81, 107, 183, 242, 249, 625.

Perhaps restricted to the southern United States, thus reports from Nova Scotia and Ontario (242, 249) should be confirmed. Burt's (81) specimen from New York was redetermined as *Corticium roseum*, fide G. Chamuris (in litt. May 1987).

Eichleriella pulvinata Coker 1928
J. Elisha Mitchell Sci. Soc. 43:236.
Distribution United States: NC.
Host *Carya* sp.
Ecology Fallen twigs.
Culture Characters Not known.

References 111, 625.

Probably an *Exidia* rather than *Exidiopsis*, fide Wells (625).

Eichleriella schrenkii Burt 1915
Ann. Missouri Bot. Gard. 2:744.
Synonym *Exidiopsis leucophaea* (Bres.) Wells 1961 sensu N. Amer. Aucts., fide Wells and Raitviir (632).
Distribution United States: AZ, TX.
Hosts *Acacia greggii, Celtis reticulata, Prosopis* sp., *P. juliflora*.
Ecology Dead wood, often corticated; associated with a white rot.
Culture Characters
2.3c.26.32.36.40.47.54 (Ref. 474).
References 81, 207, 216, 474, 625.

ENTOMOCORTICIUM H.S. Whitney, Bandoni & Oberw. 1987 Incertae sedis

Entomocorticium dendroctoni H.S. Whitney 1987
Canad. J. Bot. 65:96.
Distribution Canada: BC. United States: MT.
Hosts *Pinus contorta* var. *latifolia, P. ponderosa*.
Ecology Growing in pupal chambers and on the walls of larval galleries of *Dendroctonus ponderosae* Hopkins on *Pinus* spp. Disseminated by the beetle as the spores are borne symmetrically and lack an active release mechanism.
Culture Characters Not known.

Known only from the type specimen.

EPITHELE Pat. 1900 Epitheleaceae

Epithele sulphurea Burt 1919
Ann. Missouri Bot. Gard. 6:265.
Distribution United States: FL.
Host *Sabal palmetto*.
Ecology Not known.
Culture Characters Not known.

Known only from the type specimen.

ERYTHRICIUM Eriksson & Hjort. 1970 Phanerochaetaceae

Erythricium chaparralum Burdsall & Gilbn. 1982
Mycotaxon 15:335.
Distribution United States: AZ.
Host *Yucca* sp.
Ecology Dead flower stalks.
Culture Characters Not known.

Known only from the type specimen.

Erythricium laetum (Karsten) Eriksson & Hjort. 1970
Svensk Bot. Tidskr. 64:166.
Basionym *Hyphoderma laetum* Karsten, Rev. Mycol. (Toulouse) 11:206, 1889.
Synonym *Corticium laetum* (Karsten) Bres. 1903.
Distribution United States: MI, MT, ND, NY.
Hosts *Alnus* sp., *A. tenuifolia, Betula* sp., *B. populifolia, Populus* sp., mosses.
Ecology Not known for the woody plants but on live mosses.
Culture Characters Not known.
References 93, 183, 406, 511.

Erythricium salmonicolor (Berk. & Br.) Burdsall 1985
Mycologia Memoir 10:151.
Basionym *Corticium salmonicolor* Berk. & Br., J. Linn. Soc., Bot. 14:71, 1875.
Synonym *Phanerochaete salmonicolor* (Berk. & Br.) Jülich 1975.
Anamorph *Necator decretus* Massee 1898.
Distribution United States: FL, LA.
Hosts *Cercis canadensis, Ficus* sp. (edible fig), *F. carica, Malus* sp., *Pyrus* sp., economic woody plants.
Ecology Parasitic, causes pink disease of small branches and shoots.
Culture Characters
2.6.24.25.(26).32.36.39.43-47.54 (Ref. 588).
References 93, 274, 317.

Erythricium species A
Distribution United States: AZ.
Hosts *Populus tremuloides, Pseudotsuga menziesii*.
Ecology Gymnosperm needles and twigs on the ground; branches under snow; perhaps psychrophilic.
Culture Characters Not known.
References As *Corticium lepidum* (Romell) Bourd. & Galzin 1928 in 196, 204, 216.

These Arizona reports were based upon misidentified specimens of a species of *Erythricium*, close to *E. laetum*, fide Eriksson *et al.* (169:934).

EXIDIA Fr. 1822 Exidiaceae

Exidia alba (Lloyd) Burt 1921
Ann. Missouri Bot. Gard. 8:366.
Basionym *Exidiopsis alba* Lloyd, Mycol. Writ. 4(Letter 44):8, 1913.
Synonym ?*Exidia villosa* Neuhoff 1935 sensu Gourley (249), see Bodman (35).
Distribution Canada: MB, NS, PQ.
United States: AL, FL, IA, IL, KY, LA, MI, MN, MO, NY, OH, SC, WI, WV.
Hosts *Betula alleghaniensis*, angiosperm.
Ecology Dead wood; branches; log.
Culture Characters Not known.
References 90, 249, 419, 443, 503, 532, 593, 610.

Exidia beardsleei Lloyd 1920
Mycol. Writ. 6:984.
Basionym Proposed as a *nom. nov.* for *Exidia uva-passa* Lloyd, Mycol. Writ. 6:898, 1919, which is a later homonym of *E. uva-passa* Lloyd 1918.
Distribution United States: NC, NJ, NY.
Hosts *Quercus* sp., *Robinia pseudoacacia*.
Ecology Corticated, dead branches; decorticated wood.
Culture Characters Not known.
References 107, 410, 593.

Exidia candida Lloyd 1916
Mycol. Writ. 5:620.
Distribution Canada: AB-NT, BC, NT, NTM. United States: ID, NY, PA, WA.
Hosts *Acer macrophyllum, Alnus rubra, Betula* sp., *B. grandulosa, Corylus* sp., *C. californica, C. cornuta*.
Ecology Not known.
Culture Characters Not known.
References 90, 133, 183, 242, 410, 593.

Exidia cartilaginea Lundell & Neuhoff 1935
In Neuhoff, Pilze Mittleur. 2a:19.
Distribution Canada: ON.
Hosts Not known.
Ecology Not known.
Culture Characters Not known.
Reference 421.

Exidia cokeri Olive 1958
J. Elisha Mitchell Sci. Soc. 74:41.
Distribution United States: GA, NC.
Hosts *Carya* sp., *Fagus*? sp., *Quercus* sp., angiosperm.
Ecology Bark of fallen limbs.
Culture Characters Not known.
References 495, 498.

Detailed description and illustrations in Olive (495). The name was proposed in 1944 but not validly published with a Latin description until 1958.

Exidia compacta Lowy 1955
Lloydia 18:149.
Distribution United States: LA.
Host *Quercus virginiana*.
Ecology Dead branches.
Culture Characters Not known.

Known only from the type specimen.

Exidia crenata (Schw.) Fr. 1823
Syst. Mycol. 2:226.
Basionym *Tremella crenata* Schw., Schriften Naturf. Ges. Leipzig 1:80, 1822.
Distribution Canada: BC.
Host *Alnus sinuata*.
Ecology Not known.
Culture Characters Not known.
Reference 242.

Exidia gelatinosa (Bull.) Duby 1803
Bot. Gallicum 2:731.
Basionym *Peziza gelatinosa* Bull., Histoire Champignons France, 239, tab. 460, 1791.
Distribution United States: NC.
Hosts *Betula nigra, Prunus serotina, Quercus* sp., *Q. alba, Vitis rotundifolia*.
Ecology Fallen limbs.
Culture Characters Not known.
References 96, 107.

Perhaps *E. recisa* is another name for the same fungus.

Exidia glandulosa (Bull.) Fr. 1822
Syst. Mycol. 2:224.
Basionym *Tremella glandulosa* Bull., Histoire Champignons France, 220, 1791.
Synonym *Exidia spiculosa* (Pers.) Sommerf. 1826, fide Donk (157).
Distribution Canada: BC, MB, NB, NF, NS, ON, PE, PQ, YT. United States: AK, AZ, CA, CO, FL, GA, ID, LA, MD, MN, MI, MS, NC, NH, NY, OH, OR, WA, WI.
Hosts *Abies balsamea, Acer saccharum, Alnus* sp., *A. crispa* var. *mollis, A. rugosa, A. rugosa* var. *americana, A. sinuata, A. tenuifolia, Betula* sp., *B. alleghaniensis, B. lenta, B. papyrifera, Carya* sp., *C. illinoensis, Fagus* sp., *F. grandifolia, Holodiscus discolor, Myrica* sp., *Persea* sp., *Pinus* sp., *Populus* sp., *P. tremuloides, P. trichocarpa, Prunus* sp., *Pseudotsuga menziesii, Quercus* sp., *Q. alba, Q. gambelii, Q. rubra* var. *borealis, Q. velutina, Salix* sp., *S. alaxensis, S. babylonia, S. discolor, S. nigra, Sorbus americana, Tecoma radicans, Ulmus americana, Vitis aestivalis*, angiosperm, rarely gymnosperm.
Ecology Bark; dead branches on live trees; fallen branches; dead corticated branches; limbs; slash; associated with a white rot.
Culture Characters
2.3c.7.(21).32.36.38.(40).45.(50).(51).54.55.59 (Ref. 474).
References 96, 103, 107, 116, 119, 127, 181, 183, 214, 216, 242, 249, 334, 335, 344, 402, 410, 443, 465, 474, 498, 499, 509, 511, 580, 611, 642.

Apparently a species complex (157). One segregate is *E. plana*, see Donk (157), but there does not appear to be any easy method for segregating that fungus in the above references.

Exidia glandulosa forma ***populi*** Neuhoff 1936
Ark. Bot. 28A:15.
Distribution Canada: AB-NT, BC, NTM, YT. United States: CO.
Hosts *Alnus* sp., *Quercus gambelii, Salix* sp.
Ecology Not known.
Culture Characters Not known.
References 181, 242, 398.

Exidia hispidula Lowy 1957
Mycologia 49:899.
Distribution United States: LA.
Hosts Not known.
Ecology Lower surface of rotting, corticated log.
Culture Characters Not known.

Known only from the type specimen.

Exidia minutissima Coker 1928
J. Elisha Mitchell Sci. Soc. 43:234.
Distribution United States: NC.
Host *Quercus* sp.
Ecology Fallen corticated branch.
Culture Characters Not known.

Known only from the type specimen.

Exidia pinicola (Peck) Coker 1920
J. Elisha Mitchell Sci. Soc. 35:150.
Basionym *Tremella pinicola* Peck, Annual Rep. New York State Mus. 39:44, 1886.
Distribution Canada: YT. United States: NY.
Hosts *Picea glauca, Pinus* sp.
Ecology Not known.
Culture Characters Not known.
References 181, 242.

Until recently known only from the type specimen which was on *Pinus* branches in New York State.

Exidia plana (A. Wigg.) Donk 1966
Persoonia 4:228.
Basionym *Tremella plana* A. Wigg., Prim. Fl. Hols. 95, 1780.
Synonyms ?*Tremella myricae* Berk. & Cooke 1878, *Exidia applanata* Schw. 1832, ?*Exidia spiculata* Schw. 1832. All fide Donk (157).
Distribution United States: FL, GA, NY, PA, VT.
Hosts *Betula* sp., *Myrica* sp., *Persea* sp., *Quercus* sp., *Salix* sp.
Ecology Bark of fallen decaying limbs.
Culture Characters Not known.
References 90, 157, 498.

Exidia recisa (Ditmar):Fr. 1822
Syst. Mycol. 2:223.
Basionym *Tremella recisa* Ditmar in Sturm, Deutschl. Fl. 3:27, 1813.
Distribution Canada: AB, AB-NT, NS, ON, PQ, YT. United States: CA, GA, ID, LA, MD, MN, NC, New England, OH.
Hosts *Acer* sp., *A. spicatum, Adenostoma fasciculatum, Amelanchier alnifolia, Betula nigra, Liquidambar styraciflua, Malus* sp., *Prunus serotina, Quercus* sp., *Q. alba, Q. coccinea, Salix* sp., *S. discolor*, angiosperm.
Ecology Dead limbs.
Culture Characters Not known.
References 116, 119, 127, 181, 183, 214, 227, 242, 249, 443, 465, 498, 499, 642.

Exidia repanda Fr. 1822
Syst. Mycol. 2:225.
Distribution Canada: ON. United States: AL, ID, LA, ME, MN, NJ, WA.
Hosts *Alnus incana, Betula papyrifera, Quercus* sp., *Salix scouleriana, S. sitchensis*, angiosperm.
Ecology Decorticated wood.
Culture Characters Not known.
References 183, 214, 443, 502.

Exidia saccharina (Alb. & Schw.):Fr. 1822
Syst. Mycol. 2:225.
Basionym *Tremella spiculosa* Pers. var. *saccharina* Alb. & Schw., Conspectus Fungorum, 302, 1805, *Tremella saccharina* var. *foliacea* (Bref.) Bres. 1900 sensu N. Amer. Aucts., fide Ginns (below).
Distribution Canada: AB, AB-NT, BC, MB, NS, ON, PQ. United States: AK, AZ, MI, New England, WA, WI.
Hosts *Abies balsamea, Alnus sinuata, Picea* sp., *Pinus banksiana, P. contorta, P. ponderosa, P. resinosa, P. strobus, Pseudotsuga menziesii, Quercus* sp., *Sorbus* sp., gymnosperm.
Ecology Not known.
Culture Characters Not known.
References 116, 119, 183, 216, 242, 249, 443, 532, 642.

The variety *foliacea* was reported by Conners (116). We have seen one (Manitoba, DAOM 205679) of the two specimens cited and it is *E. saccharina*.

Exidia tremelloides Olive 1951
Mycologia 43:682.
Distribution United States: LA.
Hosts *Magnolia* sp., *Quercus* sp., *Rhamnus* sp.
Ecology Bark; rarely on corticated portions of limbs.
Culture Characters Not known.
References 419, 502.

Exidia umbrinella Bres. 1900
Fungi Tridentini 2:98, tab. 209.
Distribution United States: OR.
Hosts *Amelanchier* sp., *Salix* sp.
Ecology Sticks.
Culture Characters Not known.
Reference 335.

Exidia zelleri Lloyd 1920
Mycol. Writ. 6:931.
Distribution United States: CA, OR.
Hosts *Quercus agrifolia, Sambucus glauca*.
Ecology Not known.
Culture Characters Not known.
Reference 593.

EXIDIOPSIS (Brefeld) Möller 1895 Exidiaceae

Exidiopsis calospora Bourd. & Galzin 1924
Bull. Soc. Mycol. France 39:263.
Synonym *Sebacina calospora* (Bourd. & Galzin) Bourd. & Galzin 1928.
Distribution Canada: ON. United States: IA.
Hosts Angiosperm.
Ecology Dead wood.
Culture Characters Not known.
References 443, 451, 625.

Exidiopsis diversa Wells 1987
Mycologia 79:280.
Distribution Canada: BC.
Host *Populus* sp.
Ecology Not known.
Culture Characters Only partial species code known: 60 (Ref. 628).

Known only from the type specimen.

Exidiopsis fugacissima (Bourd. & Galzin) Sacc. & Trotter 1912
Sylloge Fung. 21:452.
Basionym *Sebacina fugacissima* Bourd. & Galzin, Bull. Soc. Mycol. France 25:28, 1909.
Distribution United States: AZ, CA, IA, NC.
Hosts *Pinus engelmannii*, angiosperm.
Ecology Usually well-rotted wood.
Culture Characters Not known.
References 216, 443, 451, 625, 643.

Exidiopsis fuliginea Rick 1906
Brotéria 5:8.
Synonyms *Sebacina adusta* Burt 1915, *Sebacina variseptata* Olive 1948. Both fide Wells (625).
Distribution Canada: BC.
United States: FL, ID, LA, WA.
Hosts *Betula microphylla, Populus* sp., *P. deltoides, P. tremuloides, P. trichocarpa, Trachycarpus fortunei, Tsuga heterophylla*, angiosperm.
Ecology Rotten wood; limb; decorticated trunk; dead bases of leaves and fruiting stalks of windmill palm.
Culture Characters Not known.
References 81, 116, 183, 419, 451, 499, 622, 625.

Exidiopsis glaira (Lloyd) Wells 1957
Lloydia 20:48.
Basionym *Tremella glaira* Lloyd, Mycol. Writ. 5 (Note 60):874, 1919.
Synonym *Sebacina opalea* Bourd. & Galzin 1924, fide Wells (625).
Distribution Canada: ON.
United States: AZ, IA, KY, NC, NY.
Hosts *Pinus leiophylla* var. *chihuahuana*, angiosperm.
Ecology Dead wood, usually decorticated; logs.
Culture Characters Not known.
References 216, 443, 451, 622, 625.

Exidiopsis laccata (Bourd. & Galzin) Luck-Allen 1962
Mycologia 53:340.
Basionym *Sebacina laccata* Bourd. & Galzin, Bull. Soc. Mycol. France 39:262, 1924.
Distribution United States: CA.
Hosts Not known.
Ecology Dead wood.
Culture Characters Not known.
Reference 625.

Exidiopsis macrospora (Ellis & Ev.) Wells 1961
Mycologia 53:351.
Basionym *Corticium macrosporum* Ellis & Ev., Bull. Torrey Bot. Club 27:49, 1900.
Synonyms *Eichleriella macrospora* (Ellis & Ev.) G.W. Martin 1944, *Sebacina macrospora* (Ellis & Ev.) Burt 1915, ?*Sebacina monticola* Burt 1915, fide Wells (625).
Distribution Canada: BC, NS, ON. United States: CO, FL, IA, OH, MO, NH, NY, TN, TX, WI.
Hosts *Abies lasiocarpa, Acer negundo, Alnus rugosa* var. *americana, Picea engelmanni, P. glauca, Pinus strobus, Pseudotsuga menziesii*, angiosperm.
Ecology Dead wood or bark attached to tree; bark of log; limbs.
Culture Characters
2.3c.7.32.37.39.46-47.53.54 (Ref. 474).
References 81, 242, 249, 443, 448, 474, 625.

Perhaps identical with *E. calcea* sensu European specimens, fide Wells and Raitviir (631).

Exidiopsis molybdea (McGuire) Ervin 1957
Mycologia 49:123.
Basionym *Sebacina molybdea* McGuire, Lloydia 4:17, 1941.
Synonym *Sebacina atra* McGuire 1941, fide Wells (625).
Distribution United States: AZ, GA, IA, MA, NC, VT.
Hosts *Diospyros* sp., *Populus* sp., *Prunus* sp., *Quercus* sp., *Q. gambelii*, angiosperm.
Ecology Usually well-rotted wood; rotten wood and bark of old limbs; branch; logs.
Culture Characters Not known.
References 216, 443, 451, 495, 498, 625.

Exidiopsis paniculata Wells & Bandoni 1987
Mycologia 79:283.
Distribution Canada: BC.
Host *Robinia pseudoacacia.*
Ecology Not known.
Culture Characters Not known.

Known only from the type specimen.

Exidiopsis plumbescens (Burt) Wells 1957
Lloydia 20:53.
Basionym *Sebacina plumbescens* Burt, Ann. Missouri Bot. Gard. 3:241, 1916 a *nom. nov.* for *Sebacina*

plumbea Burt, Ann. Missouri Bot. Gard. 2:765, 1915, which is a later homonym of *Sebacina plumbea* Bres. & Torrend 1913.
Synonyms *Exidiopsis grisea* sensu N. Amer. Aucts., e.g. Wells (625), fide Wells and Raitviir (631), *Sebacina umbrina* D.P. Rogers 1935, fide Wells (625).
Distribution Canada: BC, NB, NS, ON.
United States: CA, GA, IA, LA, OR, WA.
Hosts *Acer* sp., *A. circinatum, A. spicatum, Alnus* sp., *A. rubra, Arctostaphylos manzanita, Betula* sp., *Fraxinus* sp., *Liriodendron tulipifera, Lithocarpus densiflora, Populus* sp., *P. trichocarpa, Quercus* sp., *Q. garryana, Q. kelloggii, Q. lobata, Salix* sp., *Umbellularia californica*, angiosperm.
Ecology Fallen branch; dead wood, often corticated.
Culture Characters Only a partial species code is known: 60 (Ref. 633).
References 81, 133, 242, 249, 443, 451, 498, 502, 563, 622, 625, 628, 633.

In their recent treatment of *E. plumbescens*, Wells and Wong (633) dealt only with this fungus from the "West Coast". Presumably the earlier reports, especially Wells (625), from other regions in North America are of the same fungus. Wells and Raitviir (631) pointed out that *Exidiopsis grisea* (Pers.) Bourd. & Maire is restricted to gymnospermous wood and is known only from Europe and the U.S.S.R. Thus, they reject Wells's (625) synonymy of *E. plumbescens* under *E. grisea*. Apparently the North American records of *E. grisea* are to be redisposed as *E. plumbescens* and we have done that here.

Exidiopsis prolifera (D.P. Rogers) Ervin 1957
Mycologia 49:123.
Basionym *Sebacina prolifera* D.P. Rogers, Mycologia 28:350, 1936.
Distribution United States: IA.
Hosts *Ulmus* sp., angiosperm.
Ecology Dead wood, usually decorticated; well-rotted wood.
Culture Characters Not known.
References 443, 451, 564, 625.

Exidiopsis species A
Synonyms *Exidiopsis calcea* (Pers.) Wells 1962 sensu N. Amer. Aucts., fide Wells and Raitviir (631), *Sebacina calcea* (Pers.) Bres. 1892 sensu N. Amer. Aucts.
Distribution Canada: AB, BC, MB, NS, NTM, ON, PE, SK, YT.
United States: AK, AZ, CA, FL, GA, IA, ID, IL, LA, ME, MI, MN, MT, NC, NH, NY, VT, WA.
Hosts *Abies balsamea, A. lasiocarpa, Acer* sp., *A. macrophyllum, Cercocarpus breviflorus, C. montanus* var. *paucidentatus, Fraxinus* sp., *Haplopappus laricifolius, Juniperus deppeana, J. monosperma, J. pachyphloea, J. phoenicea, J. virginiana, Lycium* sp., *Paulownia tomentosa, Philadelphus* sp., *P. lewisii, Picea* sp., *P. engelmannii, P. glauca, P. mariana, Pinus* sp., *P. rigida, Populus* sp., *P. balsamifera, Prosopis juliflora, Pseudotsuga menziesii, Quercus* sp., *Ribes* sp., *Robinia pseudoacacia, Salix* sp., *Sambucus* sp., *S. caerulea, S. glaucus, Thuja occidentalis, Tsuga* sp., *T. canadensis, Ulmus* sp., angiosperm, gymnosperm.
Ecology Bark and wood of dead branches; live limbs; dead, corticated or decorticated wood; associated with a white rot.
Culture Characters
2.3c.7.32.36.40.46-47.50.54.55 (Ref. 474).
References 81, 96, 97, 107, 116, 133, 183, 202, 207, 211, 214, 216, 242, 249, 406, 443, 448, 451, 465, 474, 511, 625.

Wells and Raitviir (631) pointed out that *E. calcea* in Martin (443) and in Wells (625) was incorrectly applied, and the North American fungus was probably an unnamed species. Since North American authors would, most likely, have been following these authors, we list them here.

Exidiopsis sublivida (Pat.) Wells 1962
Mycologia 53:347.
Basionym *Heterochaete sublivida* Pat., Bull. Soc. Mycol. France 24:2, 1908.
Distribution United States: LA.
Hosts Angiosperm.
Ecology Rotten wood; bark of decaying wood.
Culture Characters Not known.
References 34, 90, 625.

FEMSJONIA Fr. 1849 — Dacrymycetaceae

Femsjonia peziziformis (Lév.) Karsten 1876
Bidrag Kännedom Finlands Natur Folk 25:352.
Basionym *Exidia pezizaeformis* Lév., Ann. Sci. Nat. Bot., Sér. III, 9:127, 1848.
Synonyms *Femsjonia luteoalba* Fr., 1849, fide Donk (157), and Eriksson et al. (175), *Guepinia luteoalba* (Fr.) Lloyd 1922, *Ditiola conformis* Karsten 1871, fide Donk (157), and Eriksson et al. (175), *Femsjonia radiculata* (Sowerby) G.W. Martin sensu G.W. Martin 1952, fide Donk (157).
Distribution Canada: BC, NS, ON, PQ.
United States: GA, ME, MI, NH, NY, OH, SC, VT.
Hosts *Abies balsamea, Alnus* sp., *Betula* sp., *B. alleghaniensis, Fagus grandifolia, Liriodendron tulipifera, Tsuga heterophylla*, angiosperm, rarely gymnosperm.
Ecology Corticated limbs; corticated wood; branches.
Culture Characters Not known.
References 51, 90, 116, 242, 249, 443, 498, 503, 520, 532, 593.

FIBRICIUM Eriksson 1958 Steccherinaceae

Fibricium lapponicum Eriksson 1958
Symb. Bot. Upsal. 16 (1):112.
Distribution Canada: BC.
Hosts Not known.
Ecology Not known.
Culture Characters Not known.
Reference 173.

Fibricium rude (Karsten) Jülich 1974
Persoonia 8:81.
Basionym *Corticium rude* Karsten, Bidrag Kännedom Finlands Natur Folk 37:143, 1882.
Synonyms *Fibricium greschikii* (Bres.) Eriksson 1958, fide Jülich (315), *Peniophora greschikii* (Bres.) Bourd. & Galzin 1928, *Peniophora alba* Burt 1926, fide Rogers and Jackson (567), *Peniophora subcremea* Höhnel & Litsch. 1906, fide Rogers and Jackson (567).
Distribution Canada: AB-NT, BC, MB, NS, ON. United States: AK, AZ, CO, ID, MI, MT, NM, NY, WA, WI.
Hosts *Abies balsamea, A. concolor, Alnus oblongifolia, A. rubra, Amelanchier* sp., *A. alnifolia, Juglans major, Juniperus* sp., *Picea* sp., *P. engelmannii, P. sitchensis, Pinus* sp., *P. banksiana, P. ponderosa, Populus* sp., *Pseudotsuga menziesii, Symphoricarpos occidentalis, Thuja plicata*, angiosperm, gymnosperm.
Ecology Bark and wood; twigs; needles; slash; associated with a white rot.
Culture Characters
2.3c.(8).26.32.36.38.45-46.54.55.60 (Ref. 474); 2a.3c.(7).(26).32.36.38.45.55.58.61 (Ref. 262).
References 92, 116, 119, 197, 204, 216, 242, 245, 249, 402, 406, 448, 474, 584.

FIBULOBASIDIUM Bandoni 1979 Tremellaceae

Fibulobasidium inconspicuum Bandoni 1979
Canad. J. Bot. 57:264.
Distribution United States: LA.
Host *Quercus virginiana.*
Ecology Fallen branches.
Culture Characters Not known.

Known only from the type specimen.

FIBULOMYCES Jülich 1972 Atheliaceae

Fibulomyces canadensis Jülich 1972
Beih. Willdenowia 7:178.
Distribution Canada: MB, ON.
Host *Acer* sp.
Ecology Bark.
Culture Characters Not known.

Known only from the type specimens.

Fibulomyces fusoideus Jülich 1972
Beih. Willdenowia 7:178.
Distribution Canada: ON, PQ.
Hosts Not known.
Ecology Not known.
Culture Characters Not known.

Known only from the type specimens.

Fibulomyces mutabilis (Bres.) Jülich 1972
Beih. Willdenowia 7:182.
Basionym *Corticium mutabile* Bres., Fungi Tridentini 2:59, 1892.
Synonym *Athelia mutabilis* (Bres.) Donk 1957.
Distribution Canada: BC, MB, ON. United States: AZ, CO, IA, ID, IL, IN, MI, MN, MO, OR, PA, WV.
Hosts *Abies lasiocarpa, Carya* sp., *Pinus ponderosa, Pseudotsuga menziesii, Quercus gambelii*, angiosperm.
Ecology Slash; associated with a white rot.
Culture Characters Not known.
References 183, 214, 312, 399, 402.

Fibulomyces septentrionalis (Eriksson) Jülich 1972
Beih. Willdenowia 7:187.
Basionym *Athelia septentrionalis* Eriksson, Symb. Bot. Upsal. 16 (1):88, 1958.
Distribution Canada: BC. United States: AZ, MN.
Hosts *Betula papyrifera, Picea engelmannii.*
Ecology Not known.
Culture Characters
2.3c.21.(23).26.32.36.38.46-47.54.55 (Ref. 474).
References 173, 216, 312, 474.

FLAGELLOSCYPHA Donk 1951 Lachnellaceae

Flagelloscypha faginea (Libert) W.B. Cooke 1961
Beih. Sydowia 4:60.
Basionym *Cyphella faginea* Libert, Crypt. Arden., 331, 1837.
Synonym *Cyphella peckii* Sacc. 1888, *nom. nov.* for *Cyphella candida* Peck 1875, which is a later homonym of *Cyphella candida* Jungh. 1838, fide Agerer (4).
Distribution United States: NY.
Host *Osmunda cinnamomea.*
Ecology Dead stems.
Culture Characters Not known.
References 4, 80.

Flagelloscypha langloisii (Burt) Agerer 1975
Sydowia 27:227.
Basionym *Cyphella langloisii* Burt, Ann. Missouri Bot. Gard. 1:368, 1914.
Distribution United States: IA, LA.
Hosts *Arundinaria* sp., unidentified wood.
Ecology Dead stems; decaying wood on the ground.
Culture Characters Not known.
References 2, 80, 378.

Flagelloscypha libertiana (Cooke) Agerer 1979
Persoonia 10:339.
Basionym *Cyphella libertiana* Cooke, Grevillea 8:81, 1880.
Synonym *Lachnella rosae* W.B. Cooke 1962, fide Agerer (3).
Distribution United States: ID, WA.
Host *Rosa spauldingii.*
Ecology Dead canes.
Culture Characters Not known.
Reference 125.

Flagelloscypha minutissima (Burt) Donk 1951
Lilloa 22:312.
Basionym *Cyphella minutissima* Burt, Ann. Missouri Bot. Gard. 1:367, 1914.
Synonyms *Flagelloscypha citrispora* (Pilát) D. Reid 1964, fide Agerer 1975 (2), *Flagelloscypha faginea* sensu W.B. Cooke 1962, fide Reid (554).
Distribution Canada: BC, MB, SK.
United States: IA, IL, LA, ME, MI, NH, NJ, VT.
Hosts *Alnus* sp., *Picea* sp., *Populus* sp., *Poria* sp., *Pteridium aquilinum, ?Pseudotsuga menziesii, Vicia ?gigantea.*
Ecology Rotting sticks; branches.
Culture Characters Not known.
References 2, 80, 116, 125, 131, 378, 545.

Flagelloscypha orthospora (Bourd. & Galzin) R. Bertault & Malençon 1976
Acta Phytotax. Barc. 19:30.
Basionym *Cyphella villosa* var. *orthospora* Bourd. & Galzin, Bull. Soc. Mycol France 26:224, 1910.
Synonym *Lachnella ciliata* (Sauter) W.B. Cooke 1962 sensu W.B. Cooke, fide Agerer (5) and see under Excluded names.
Distribution Canada: ON.
United States: ME, OR, VA.
Hosts Not known.
Ecology Not known.
Culture Characters Not known.
Reference 125.

FLAVOPHLEBIA Larsson & Hjort. 1977 — Meruliaceae

Flavophlebia sulfureo-isabellina (Litsch.) Larsson & Hjort. 1977
Mycotaxon 5:475.
Basionym *Corticium sulfureo-isabellinum* Litsch. in Pilát, Sborn. Nár. Mus. v Praze, Rada B, Prír. Vedy 2B:43, 1940.
Synonym *Phlebia sulfureo-isabellina* (Litsch.) Parm. 1968.
Distribution Canada: BC, NS, ON, PQ.
United States: MN, NH.
Hosts *Abies balsamea, A. lasiocarpa.*
Ecology Not known.
Culture Characters Not known.
References 116, 214, 242, 249, 291, 303.

GALZINIA Bourdot 1922 — Sistotremaceae

Galzinia cymosa D.P. Rogers 1944
Mycologia 36:100.
Distribution United States: MA, NC.
Hosts *Pinus resinosa, P. rigida*, angiosperm.
Ecology Very rotten wood; firm wood of log.
Culture Characters Not known.
References 183, 505, 566.

Galzinia geminispora Olive 1954
Mycologia 46:794.
Distribution United States: NC.
Hosts *Castanea dentata, Populus* sp., angiosperm.
Ecology Very rotten, decorticated limb.
Culture Characters Not known.
References 324, 505.

Galzinia incrustans (Höhnel & Litsch.) Parm. 1965
Izv. Akad. Nauk Estonsk. SSR, Ser. Biol. 14:225.
Basionym *Corticium incrustans* Höhnel & Litsch., Sitzungsber. Kaiserl. Akad. Wiss., Math.-Naturwiss. Kl., Abt. 1, 115:1602, 1906.
Synonym *Corticium roseopallens* Burt, 1907, fide Rogers and Jackson (567).
Distribution Canada: BC, ON, PQ. United States: IA, ID, LA, MA, MD, ME, MN, MO, NY, WI, VT.
Hosts *Acer* sp., *Alnus* sp., *Betula papyrifera, Fagus* sp., *Pinus contorta, Phellinus* sp., *Populus* sp., *P. tremuloides, Quercus* sp., *Salix* sp., *Ulmus americana*, angiosperm.
Ecology Not known.
Culture Characters
1.3c.26.27.33.36.(38).(39).(40).43-44.(48).(50).54.59 (Ref. 474).
References 93, 116, 173, 242, 394, 474, 482, 562, 563.

Galzinia occidentalis D.P. Rogers 1944
Mycologia 36:102.
Distribution United States: AZ, OR.
Hosts *Pinus contorta, P. leiophylla* var. *chihuahuana, Pseudotsuga menziesii, P. mucronata.*
Ecology Not known.
Culture Characters Not known.
References 183, 204, 216, 566.

Galzinia pedicellata Bourd. 1922
Assoc. Franç. Avancem. Sci. (Rouen) 45:578.
Distribution United States: PA, OR.
Hosts *Pinus contorta, P. pungens.*
Ecology Decorticated, partly decayed logs.
Culture Characters Not known.
References 183, 566.

GLABROCYPHELLA W.B. Cooke 1961 — Cyphellaceae

Glabrocyphella ailanthi W.B. Cooke 1961
Beih. Sydowia 4:45.
Distribution United States: NJ.

Host *Ailanthus altissima.*
Ecology Bark.
Culture Characters Not known.

Known only from the type specimen.

Glabrocyphella bananae (Cooke) W.B. Cooke 1961
Beih. Sydowia 4:46.
Basionym *Cyphella bananae* Cooke, Grevillea 6:132, 1878.
Distribution United States: FL.
Host *Musa* sp.
Ecology Dead banana leaf.
Culture Characters Not known.
References 80, 125.

Glabrocyphella dermatoides W.B. Cooke 1961
As Ellis ex W.B. Cooke, Beih. Sydowia 4:47.
Distribution Canada: ON.
Host *Corylus americana.*
Ecology Not known.
Culture Characters Not known.

Known only from the type specimen.

Glabrocyphella ellisiana W.B. Cooke 1961
Beih. Sydowia 4:47.
Distribution United States: NJ.
Hosts *Oenothera* sp., *Populus* sp., *Zea mays.*
Ecology Herbage.
Culture Characters Not known.

Known only from the type specimens.

Glabrocyphella filicicola (Berk. & Curtis) W.B. Cooke 1961
Beih. Sydowia 4:49.
Basionym *Cyphella filicicola* Berk. & Curtis, Grevillea 2:5, 1873.
Distribution United States: NC.
Host Fern.
Ecology Dead plant.
Culture Characters Not known.
Reference 80.

Glabrocyphella ohiensis W.B. Cooke 1961
Beih. Sydowia 4:49.
Distribution United States: OH.
Hosts Not known.
Ecology Rotten wood.
Culture Characters Not known.

Known only from the type specimen.

GLOBULICIUM Hjort. 1973 — Aleurodiscaceae

Globulicium hiemale (Laurila) Hjort. 1973
Svensk Bot. Tidskr. 67:109.
Basionym *Corticium hiemale* Laurila, Ann. Bot. Soc. Zool.-Bot. Fenn. "Vanamo" 10(4):4, 1939.
Synonyms *Aleurodiscus hiemalis* (Laurila) Eriksson 1958, *Corticium probatum* Jackson 1948, fide Jülich and Stalpers (326).
Distribution Canada: MB, ON.
United States: AK, OR.
Hosts *Abies balsamea, Thuja plicata, Picea sitchensis,* gymnosperm.
Ecology Not known.
Culture Characters Not known.
References 116, 242, 303, 376.

GLOEOCYSTIDIELLUM Donk 1931 — Gloeocystidiellaceae

Gloeocystidiellum clavuligerum (Höhnel & Litsch.) Nakasone 1982
Mycotaxon 14:320.
Basionym *Gloeocystidium clavuligerum* Höhnel & Litsch., Sitzungsber. Kaiserl. Akad. Wiss., Math.-Naturwiss. Kl., Abt. 1, 115:1603, 1906.
Synonym *Corticium vesiculosum* Burt 1926, fide Nakasone (469).
Distribution Canada: BC, PQ.
United States: IL, LA, MA, MI, MN, MS, NY, WI.
Hosts *Acer* sp., *A. macrophyllum, A. saccharum, Alnus* sp., *Ostrya virginiana, Populus* sp., *P. balsamifera, P. tremuloides, P. trichocarpa, Pseudotsuga menziesii, Tilia americana, Ulmus americana,* angiosperm.
Ecology Bark and decorticated branches on the ground; logs; associated with a white rot.
Culture Characters
2.3.15.(26).32.36.38.39.46-47.48.53.54 (Ref. 469);
2.3c.(15p).(26).32.36.38.(39).45-47.48.53.54.(55).58 (Ref. 474);
2a.3c.15a.(26).32.36.38.45.54.58 (Ref. 255).
References 245, 255, 469, 474.

Gloeocystidiellum convolvens (Karsten) Donk 1956
Fungus 26:9.
Basionym *Corticium convolvens* Karsten, Bidrag Kännedom Finlands Natur Folk 37:148, 1882.
Synonym *Peniophora convolvens* (Karsten) Höhnel & Litsch. 1906.
Distribution United States: AZ, GA, NC, NM, TN.
Hosts *Populus* sp., *P. tremuloides,* angiosperm.
Ecology Not known.
Culture Characters Not known.
References 204, 208, 216, 406, 584.

Gloeocystidiellum fimbriatum Burdsall, Nakasone & Freeman 1981
Syst. Bot. 6:422.
Distribution United States: FL, LA.
Hosts *Acer negundo, Carpinus caroliniana, Ostrya virginiana, Quercus* sp., *Q. nigra,* angiosperm.
Ecology Associated with a white rot.
Culture Characters
2.6.15.32.36.38.43-44.54 (Ref. 76).
References 76, 474.

Gloeocystidiellum karstenii (Bourd. & Galzin) Donk 1956 Fungus 26:9.

Basionym *Gloeocystidium karstenii* Bourd. & Galzin, Hym. France, 254, 1928.

Synonym *Corticium ravum* Burt 1926, fide Ginns (245).

Distribution Canada: BC, NB, PE, PQ. United States: AL, AK, AZ, LA, MN, SC, WI.

Hosts *Acer rubrum, A. saccharinum, Nyssa* sp., *Populus* sp., *P. balsamifera, P. tremuloides, P. trichocarpa, Pseudotsuga menziesii, Quercus* sp., angiosperm, gymnosperm.

Ecology Isolated uncommonly from decay behind wounds on trembling aspen; basidiomes produced on fallen branches; bark of angiosperm twigs, 15-20 mm diam; logs; associated with a white rot.

Culture Characters 2.5.8.15.32.36.(37).38.43-44.(50).(53).54 (Ref. 474).

References 93, 97, 216, 242, 245, 349, 406, 474.

Several of the collections recorded by Burt (93) as *C. ravum* were misidentified (245) and they are not cited here.

Gloeocystidiellum lactescens (Berk.) Boidin 1951 Compt. Rend. Hebd. Séances Acad. Sci. 233:1668.

Basionym *Thelephora lactescens* Berk. in Smith, Engl. Fl. 5(2):169, 1836.

Synonyms *Corticium lactescens* (Berk.) Berk. 1860, *Megalocystidium lactescens* (Berk.) Jülich 1978, *Corticium epigaeum* Ellis & Ev. 1885, fide Ginns (245).

Distribution Canada: BC, MB, NF, ON, PQ. United States: AZ, CA, LA, MA, ME, MI, NC, NH, NY, OH, OR, TN, VT, WA, WI.

Hosts *Agave palmeri, Alnus oblongifolia, Carya* sp. *Fraxinus americana, Juglans major, Platanus wrightii, Populus* sp., *P. tremuloides, Quercus* sp., *Q. garryana, Ulmus americana*, angiosperm.

Ecology Decaying wood of logs; ground over soil; associated with a white rot.

Culture Characters 2.6.15.32.36.38.(39).(40).45-47.54.58 (Ref. 474).

References 93, 116, 133, 183, 204, 216, 245, 406, 474.

Gloeocystidiellum leucoxanthum (Bres.) Boidin 1951 Compt. Rend. Hebd. Séances Acad. Sci. 233:825.

Basionym *Corticium leucoxanthum* Bres., Fungi Tridentini 2:57, 1898.

Synonym *Megalocystidium leucoxanthum* (Bres.) Jülich 1978.

Distribution Canada: AB, AB-NT, BC, NB, NF, NS, YT. United States: AK, AZ, CA, CO, MN, WI.

Hosts *Acer saccharum, Alnus* sp., *A. rugosa* var. *americana, A. sinuata, Picea engelmannii, Populus* sp., *P. tremuloides, Quercus chrysolepis, Q. hypoleucoides, Q. garryana*, angiosperm, gymnosperm.

Ecology Associated with a white rot.

Culture Characters 2.3c.(8).15.(26).27.32.36.38.47.(53).54.(55).57 (Ref. 474).

References 116, 128, 173, 181, 183, 214, 216, 242, 249, 402, 406, 474.

The records compiled by Lindsey (402) on *Picea engelmannii* and gymnosperms need confirmation because this widespread, relatively common species seems restricted to angiosperms.

Gloeocystidiellum luridum (Bres.) Boidin 1951 Compt. Rend. Hebd. Séances Acad. Sci. 233:1668.

Basionym *Corticium luridum* Bres., Fungi Tridentini 2:59, 1898.

Synonym *Megalocystidium luridum* (Bres.) Jülich 1978.

Distribution Canada: MB. United States: AK, AZ, CA, ID, OH.

Hosts *Abies magnifica, Alnus rubra, A. sinuata, Picea engelmannii, Populus* sp.

Ecology Bark and wood; rotten wood.

Culture Characters Not known.

References 93, 116, 120, 128, 130 as "*lividum*", 183, 216.

We suspect that this epithet has been misapplied and the above reports were based upon specimens of *G. leucoxanthum*.

Gloeocystidiellum marianus Burdsall, Nakasone & Freeman 1981 Syst. Bot. 6:424.

Distribution United States: MD.

Host *Quercus* sp.

Ecology Not known.

Culture Characters Not known.

Known only from the type specimen.

Gloeocystidiellum ochraceum (Fr.) Donk 1956 Fungus 26:9.

Basionym *Thelephora ochraceus* Fr., Obs. Mycol. 1:151, 1815.

Synonym *Corticium ochraceum* (Fr.) Fr. 1838.

Distribution Canada: AB-NT, BC, NS, NTM, PQ. United States: AK, ID, MA, NC, WA.

Hosts *Abies balsamea, Alnus* sp., *Betula alleghaniensis, Picea engelmannii, P. glauca, P. sitchensis, Pinus resinosa, P. strobus, Pseudotsuga menzeisii, Tsuga heterophylla*, gymnosperm.

Ecology Decorticated wood, moss-covered cut wood; butt.

Culture Characters (1).2.6.15.(31g).32.36.38.44-45.(50).(51).(54).55.(57) (Ref. 474).

References 93, 183, 242, 249, 327, 391, 474.

Most of Burt's (93) reports were misidentified, see Ginns (245), and they are not included here. One report (116) of *Corticium versatum* from British Columbia was based upon a specimen of *G. ochraceum* (245). The single specimen from Minnesota in Gilbertson and Lombard (214) was redetermined as *G. karstenii* by Freeman (in litt. 1978).

Gloeocystidiellum porosum (Berk. & Curtis) Donk 1931
Medd. Nedl. Mycol. Ver. 18-20:156.
Basionym *Corticium porosum* Berk. & Curtis in Berk. & Br., Ann. Mag. Nat. Hist., Ser. V, 3:211, 1879.
Synonyms *Corticium pruni* Overh. 1929, fide Rogers and Jackson (567) and Nakasone (469), *Corticium stramineum* Bres. in Brinkm. 1900, fide Rogers and Jackson (567) and Nakasone (469).
Distribution Canada: BC, MB, NB, NS, ON, PQ. United States: AK, AZ, CO, IL, KY, MD, ME, MI, MN, MO, MT, NC, NH, NJ, NM, NY, OR, SC, VT, WA, WI.
Hosts *Abies lasiocarpa* var. *lasiocarpa, Acer* sp., *A. rubrum, A. spicatum, Alnus* sp., *A. rubra, Carya* sp., *Fagus grandifolia, Liquidambar styraciflua, Picea engelmannii, P. pungens, P. rubens, Pinus resinosa, Populus* sp., *P. balsamifera, P. tremuloides, P. trichocarpa, Prunus* sp., *Quercus* sp., *Salix* sp., *S. alaxensis, Tilia americana*, angiosperm, gymnosperm.
Ecology Typically on bark of twigs and branches; sometimes on decorticated wood or decayed wood; associated with a white rot.
Culture Characters
2.3.15.(26).32.36.38.46-47.50.54.55 (Ref. 469);
2.3c.15(p).(26).32.36.38.(39).46-47.50.54.55.60 (Ref. 474);
2a.3c.7.15a.32.36.38.46.54.58 (Ref. 255);
2.3.15.32.36.38.45-47.50.54.60 (Ref. 431).
References 93, 116, 133, 183, 204, 208, 214, 216, 242, 249, 255, 387, 402, 406, 469, 474, 514, 567 under *Corticium abeuns*.

In 1982 Nakasone (469) segregated *G. clavuligerum* from the *G. porosum* complex. Therefore reports of *G. porosum* prior to 1982 may have included some specimens of *G. clavuligerum*.

Gloeocystidiellum subasperisporum (Litsch.) Eriksson & Ryv. 1975
Cort. N. Europe 3:443.
Basionym *Corticium subasperisporum* Litsch., Ann. Mycol. 39:125, 1941.
Synonym *Corticium electum* Jackson 1948, fide Eriksson and Ryvarden (173).
Distribution Canada: ON.
Hosts *Pinus* sp., *Thuja occidentalis*.
Ecology Not known.
Culture Characters Not known.
References 116, 303.

Gloeocystidiellum tropicalis Burdsall, Nakasone & Freeman 1981
Syst. Bot. 6:428.
Distribution United States: FL.
Host *Dipholis salicifolia*.
Ecology Not known.
Culture Characters Not known.

Known only from the type specimen.

Gloeocystidiellum turpe Freeman 1981
Syst. Bot. 6:430.
Distribution United States: FL.
Hosts *Liquidambar styraciflua, Ostrya virginiana*.
Ecology Not known.
Culture Characters Not known.

Known only from the type specimens.

Gloeocystidiellum wakullum Burdsall, Nakasone & Freeman 1981
Syst. Bot. 6:431.
Distribution United States: FL, MS.
Hosts *Taxodium distichum*, angiosperm shrub.
Ecology Not known.
Culture Characters
2.5.15.32.36.38.43-44.(50).54 (Ref. 76);
2.5.8.(15).16.32.36.(37).38.(39).43-45.50.53.54 (Ref. 474).
References 76, 474.

GLOEODONTIA Boidin 1966 — Auriscalpiaceae

Gloeodontia columbiensis Burdsall & Lombard 1976
As Burt ex Burdsall & Lombard, Mem. New York Bot. Gard. 28:17.
Distribution Canada: AB, BC. United States: ID, MN, MT, OR.
Hosts *Acer* sp., *A. glabrum, A. macrophyllum, Alnus* sp., *Pinus* sp., *Populus* sp., *P. tremuloides, P. trichocarpa, Salix* sp., angiosperm.
Ecology Dead wood; associated with a white rot.
Culture Characters
2.3.7.15.32.36.40.45.54.(55).59 (Ref. 70);
2.3c.(8).15ap.32.36.38.(40).45-47.(50).(53).54.55.59 (Ref. 474).
References 70, 406, 474.

Gloeodontia discolor (Berk. & Curtis) Boidin 1966
Cahiers Maboké 4:22.
Basionym *Irpex discolor* Berk. & Curtis, Grevillea 1:145, 1873.
Synonym *Odontia eriozona* Bres. 1925, fide Burdsall and Lombard (70).
Distribution Canada: AB, AB-NT. United States: AL, AZ, FL, GA, LA, MS, SC, TN, TX.
Hosts *Acer* sp., *Bursera simaruba, Liquidambar styraciflua, Populus tremuloides, Quercus* sp., *Q. nigra, Salix gooddingii, Swietenia mahagoni*, angiosperm.
Ecology Underside of logs.

Culture Characters
2.3c.(8).15ap.(16).32.36.(38).39.45-47.(50).54.59 (Ref. 474);
2.3.12.15.22.32.37.39.47.54.59.61 (Ref. 38);
2.3.8.15.32.36.38.39.45.54.59 (Ref. 70).
References 53, 70, 190, 192, 216, 242, 429, 474.

The Canadian reports (242), perhaps based on a single specimen, are significantly beyond the principal geographic range of the species and need confirmation.

GLOIODON Karsten 1879 — Auriscalpiaceae

Gloiodon occidentale Ginns 1988
Mycologia 80:66.
Distribution Canada: BC.
Host *Tsuga heterophylla*.
Ecology Cut end of the butt log and associated stump; hollow in decayed log.
Culture Characters Not known.

Known only from the type specimens.

Gloiodon strigosus (Swartz:Fr.) Karsten 1879
Meddeland. Soc. Fauna Fl. Fenn. 5:42.
Basionym *Hydnum strigosum* Swartz, Kongl. Vetensk. Acad. Handl., 250, 1810.
Distribution Canada: ON, PQ.
United States: AZ, IA, MN, PA.
Hosts *Abies lasiocarpa* var. *arizonica, Fraxinus nigra, Populus grandidentata*.
Ecology Stump; old log.
Culture Characters
1.3.15.26.33.36.38.46.(48).52 (Ref. 529); 60 (Ref. 186).
References 216, 242, 463, 466, 597, 610.

GRANULOCYSTIS Hjort. 1986 — Tubulicrinaceae

Granulocystis flabelliradiata (Eriksson & Hjort.) Hjort. 1986
Mycotaxon 25:277.
Basionym *Phanerochaete flabelliradiata* Eriksson & Hjort., Cort. N. Europe 6:1073, 1981.
Distribution Canada: ON. United States: NH.
Hosts Angiosperm.
Ecology Not known.
Culture Characters Not known.
Reference 285.

In Europe this fungus occurs on decayed wood, especially branches in piles (170).

GUEPINIOPSIS Pat. 1883 — Dacrymycetaceae

Guepiniopsis buccina (Pers.:Fr.) Kennedy 1959
Mycologia 50:888.
Basionym *Peziza buccina* Pers., Synop. Meth. Fung. 2:659, 1801.
Synonyms *Guepinia peziza* Tul. 1853, fide Martin (443), *Guepiniopsis merulinus* (Pers.) Pat. 1887, fide McNabb (458), *Guepiniopsis torta* Pat. 1883, fide McNabb (458).
Distribution Canada: AB, AB-NT, BC, ON.
United States: CA, CO, IA, ID, MA, MI, MN, NC, NY, OH, OR, SC, WA, VT.
Hosts *Abies grandis, A. lasiocarpa, Acer glabrum, Artemisia tridentata, ?Juniperus* sp., *Pinus contorta* var. *contorta, P. contorta* var. *latifolia, P. murrayana, P. ponderosa, P. strobus, Pseudotsuga menziesii, Quercus* sp., *Salix lasiolepsis*, angiosperm, gymnosperm.
Ecology Saprophytic on rotten wood; dead wood; stump.
Culture Characters Not known.
References 51, 52, 119, 130, 183, 242, 443, 458, 465, 503.

HALOCYPHINA J. & E. Kohlmeyer 1965 — Lachnellaceae

Halocyphina villosa J. & E. Kohlmeyer 1965
Nova Hedwigia 9:100.
Distribution United States: FL.
Host Mangrove.
Ecology Marine, on roots in tidal zone.
Culture Characters Not known.
Reference 247.

HELICOBASIDIUM Pat. 1885 — Cystobasidiaceae

Helicobasidium brebissonii (Desm.) Donk 1958
Taxon 7:164.
Basionym *Protonema brebissonii* Desm., Pl. crypt. Nord. Fr. No. 651, 1834.
Synonym *Helicobasidium purpureum* Pat. 1885, fide Donk (157).
Anamorph *Rhizoctonia crocorum* (Pers.:Fr.) DC.
Distribution Canada: BC, PE.
United States: AK, AZ, IA, KS, MA, MI, MO, MT, NC, ND, NE, NY, OH, OK, OR, TX, WA, WI.
Hosts *Acer negundo, Ambrosia trifida, Ampelopsis* sp., *Asparagus officinalis, Carya* sp. (pecan), *Catalpa* sp., *Celtis laevigata, Daucus carota, Fraxinus* sp., *Helianthus annuus, Iresine* sp., *Malvaviscus arboreus, Medicago sativa, Melia azedarach, Morus alba, Parthenocissus quinquefolia, Phytolacca* sp., *Ratibida columnifera, Rhus toxicodendron, Rivina humilis, Salix* sp., *Sambucus* sp., *S. canadensis, Solanum elaeagnifolium, S. tuberosum, Trifolium pratense, Ulmus americana, Verbesina virginica, Viguiera* sp., *Viola* sp.
Ecology Causes violet root rot of a variety of woody plants, especially in the southern half of the United States; often on seedlings; basidiomes produced on dead wood; ground or encrusting live plants.

Culture Characters Not known.
References 183, 242, 274, 443.

Helicobasidium candidum G.W. Martin 1940
Mycologia 32:692.
Distribution Canada: PQ. United States: AZ.
Hosts *Acer* sp., *Quercus hypoleucoides*.
Ecology Not known.
Culture Characters Not known.
References 216, 437.

Helicobasidium corticioides Bandoni 1955
Mycologia 47:918-919.
Distribution Canada: AB.
United States: CO, UT, WY.
Hosts *Abies* sp., *A. lasiocarpa, Picea* sp., *P. engelmannii, Pinus aristata* var. *longaeva, P. contorta, Populus tremuloides, Quercus gambelii*, gymnosperm.
Ecology Pole; log; associated with a brown pocket rot beneath basal wounds in live trees.
Culture Characters
1.6.7.(26).32.36.38.47.55 (Ref. 452).
References 12, 144, 183, 242, 402, 407, 448, 452.

Helicobasidium holospirum Bourd. 1922
Assoc. Franç. Avancem. Sci. (Rouen) 45:576.
Distribution Canada: NS. United States: NY.
Hosts *Dryopteris austriaca* var. *spinulosa, D. intermedia, D. marginalis*.
Ecology Not known.
Culture Characters Not known.
References 183, 242, 249.

Helicobasidium peckii Burt 1921
Ann. Missouri Bot. Gard. 8:395.
Distribution United States: NY.
Host *Picea* sp.
Ecology Bark.
Culture Characters Not known.

Known only from the type specimen.

HELICOGLOEA Pat. 1892 — Cystobasidiaceae

Helicogloea augustispora Olive 1951
Bull. Torrey Bot. Club 78:107.
Distribution United States: NC.
Hosts Angiosperm.
Ecology Very rotten stump in a rhododendron thicket.
Culture Characters Not known.

Known only from the type specimen.

Helicogloea caroliniana (Coker) G. Baker 1936
Ann. Missouri Bot. Gard. 23:92.
Basionym *Saccoblastia ovispora* var. *caroliniana* Coker, J. Elisha Mitchell Sci. Soc. 35:121, 1920.
Synonym *Saccoblastia caroliniana* (Coker) Coker 1928.
Distribution United States: NC.
Host *Quercus* sp.
Ecology In hollow tree.
Culture Characters Not known.
References 8, 107.

Helicogloea contorta G. Baker 1946
Mycologia 38:634.
Distribution United States: IA.
Host *Quercus macrocarpa*.
Ecology Not known.
Culture Characters Not known.
Reference 443.

Helicogloea farinacea (Höhnel) D.P. Rogers 1944
Stud. Nat. Hist. Iowa Univ. 18:66.
Basionym *Helicobasidium farinacea* Höhnel, Sitzungsber. Kaiserl. Akad. Wiss., Math.-Naturwiss. Kl., Abt. 1, 116:84, 1907.
Distribution Canada: MB, ON, PQ.
United States: AZ.
Hosts *Corylus* sp., *Ostrya virginiana, Picea mariana, Populus* sp., *Quercus arizonica, Salix* sp., angiosperm, gymnosperm.
Ecology Not known.
Culture Characters Not known.
References 216, 237, 443.

Helicogloea lagerheimii Pat. 1892
Bull. Soc. Mycol. France 8:121.
Synonym *Saccoblastia sebacea* Bourd. & Galzin 1909, fide Baker (10).
Distribution Canada: BC, MB, ON. United States: AZ, CA, IA, LA, MO, NEW ENGLAND, OH, OR.
Hosts *Acer* sp., *Betula* sp., *Fraxinus velutina, Picea mariana, Quercus* sp., *Q. hypoleucoides, Salix* sp., *Trachycarpus fortunei*, angiosperm.
Ecology Dead limb; old leaf bases; fruiting stalks of the windmill palm.
Culture Characters Not known.
References 8, 10, 216, 242, 419, 443, 499, 562.

Helicogloea longispora G. Baker 1946
Mycologia 38:634.
Synonym *Helicogloea parasitica* Olive 1951, fide Olive (504).
Distribution United States: LA, NC, OR.
Hosts *Exidia glandulosa, Gloeotulasnella* sp., *Sebacina* sp., also reported on *Pseudotsuga menziesii*.
Ecology Mycoparasitic.
Culture Characters Not known.
References 183, 419, 502, 504.

Helicogloea pinicola (Bourd. & Galzin) G. Baker 1936
Ann. Missouri Bot. Gard. 23:89.
Basionym *Saccoblastia pinicola* Bourd. & Galzin, Bull. Soc. Mycol. France 25:16, 1909.
Distribution Canada: MB, ON.
Hosts *Alnus* sp., *Amelanchier* sp., *Corylus rostrata, Populus* sp., *Salix* sp., angiosperm.
Ecology Not known.

Culture Characters Not known.
References 8, 10, 116.

Helicogloea terminalis Olive 1954
Bull. Torrey Bot. Club 81:331.
Distribution United States: NC.
Host *Betula* sp.
Ecology Wood and occasionally on bark of limb.
Culture Characters Not known.

Known only from the type specimens.

HENNINGSOMYCES Kuntze 1898 Schizophyllaceae
Synonym is *Solenia* Pers.:Fr. 1794, non *Solenia* Lour. 1790, see Agerer (1).

Henningsomyces candidus (Pers.:Fr.) Kuntze 1898
Rev. Generum Plantarum 3(2):483.
Basionym *Solenia candida* Pers., Neues Mag. Bot. 1:116, 1794.
Synonyms *Solenia candida* var. *polyporoidea* Peck 1886, fide W.B. Cooke (131), *Solenia polyporoidea* (Peck) Burt 1924.
Distribution Canada: BC, NS, ON, PQ.
United States: AZ, CA, CO, CT, DE, FL, GA, IA, ID, KY, LA, MA, ME, MI, MO, NC, NH, NJ, NM, NY, OH, OR, PA, SC, TN, TX, VA, VT, WA.
Hosts *Abies* sp., *A. balsamea, A. concolor, A. magnifica, Betula* sp., *B. papyrifera, Calocedrus decurrens, Fraxinus velutina, Juniperus virginiana, Libocedrus decurrens, Picea* sp., *Pinus ponderosa, P. strobus, Populus* sp., *P. tremuloides, Prosopis juliflora, Prunus serotina, Pseudotsuga menziesii, Quercus agrifolia, Taxodium* sp., *Tsuga canadensis, Washingtonia filifera*, angiosperm, gymnosperm.
Ecology Bark and wood; decorticated and decaying wood; rotten wood; twigs; dead branches; rotten logs; old plank; associated with a white rot.
Culture Characters Not known.
References 91, 125, 128, 130, 183, 197, 201, 202, 204, 207, 208, 216, 242, 249, 378, 402, 406.

Cooke (125) listed numerous hosts from many countries but those from Canada and the United States cannot be segregated.

Henningsomyces pubera (W.B. Cooke) D. Reid 1964
Persoonia 3:119.
Basionym *Solenia pubera* Rom. ex W.B. Cooke, Beih. Sydowia 4:26, 1961.
Distribution Canada: BC, ON. United States: CT, MI.
Hosts ?*Tsuga* sp., gymnosperm.
Ecology Not known.
Culture Characters Not known.
References 1, 242, 545, 554.

HERICIUM Pers. 1794 Hericiaceae

Hericium abietis (Hubert) Harrison 1964
Canad. J. Bot. 42:1208.
Basionym *Hydnum abietis* Weir ex Hubert, Outline of Forest Pathology, 305, 1931.
Distribution Canada: BC.
United States: AK, CA, CO, ID, MT, OR, WA.
Hosts *Abies* sp., *A. amabilis, A. concolor, A. grandis, A. lasiocarpa, A. procera, Picea engelmannii, P. sitchensis, Pseudotsuga menziesii, Tsuga heterophylla, T. mertensiana*.
Ecology Causes minor decay in live *Abies grandis* and *Tsuga heterophylla*; basidiomes reported on old windfall; old log; associated with a white pocket rot.
Culture Characters
2a.3.15.34.36.(38).(39).47.(48).51.(53).55.60 (Ref. 239).
References 116, 183, 239, 242, 268, 274.

Hericium americanum Ginns 1984
Mycotaxon 20:43.
Synonym *Hericium coralloides* sensu N. Amer. Aucts., fide Ginns (238).
Distribution Canada: AB-NT, BC, NS, ON, PE, PQ.
United States: AZ, CO, GA, IA, ID, ME, MI, MN, MT, NC, NH, NY, OR, PA, TN, VT, WA, WV.
Hosts *Abies concolor, A. grandis, Acer* sp., *A. negundo, A. nigrum, A rubrum* var. *rubrum, A. saccharinum, A. saccharum, Betula* sp., *B. alleghaniensis, B. papyrifera, B. subcordata, Carya* sp., *C. ovata, Fagus* sp., *F. grandifolia, Larix laricina, Picea engelmannii, Platanus* sp., *Populus* sp., *P. balsamifera, P. tremuloides, P. trichocarpa, Pseudotsuga menziesii, Quercus* sp., *Q. garryana, Q. hypoleucoides, Q. rubra, Q. rubra* var. *borealis, Tsuga* sp., *T. canadensis, T. heterophylla, Ulmus thomasii*.
Ecology Causes heart rot in *Acer saccharum* and *Fagus* sp.; basidiomes reported on log; dead trunk; associated with a white rot.
Culture Characters
2a.3.15.34.36.(38).(44).46.(47).(48).(51).(53).54.55.60 (Ref. 239).
References 96, 116, 183, 214, 238, 239, 242, 268, 274, 402, 406, 463, 511, 610.

Hericium coralloides (Scop.:Fr.) Pers. 1794
Neues Mag. Bot. 1:109.
Basionym *Hydnum coralloides* Scop., Flora Carniolica, Ed. 2, 2:472, 1772.
Synonyms *Hericium caput-ursi* (Fr.) Banker 1906, fide Harrison (268), *Hericium laciniatum* (Leers) Banker 1906, fide Harrison (268), *Hericium ramosum* (Bull.) Letellier 1826, fide Ginns (239).
Distribution Canada: AB, BC, NS, NT, ON, PQ, SK.
United States: AK, AZ, CA, IA, ME, MI, MN, NC, NC-TN, NY, OR, VA, VT, WA, WV.

Hosts *Acer* sp., *A. glabrum, A. rubrum* var. *rubrum, A. saccharum, Alnus* sp., *A. tenuifolia, Betula* sp., *B. alleghaniensis, B. papyrifera, Carya* sp., *Fagus* sp., y*F. grandifolia, Fraxinus* sp., *Picea* spp., *P. engelmannii, P. glauca, Populus* sp., *P. balsamifera, P. tremuloides, P. trichocarpa, Quercus agrifolia, Q. hypoleucoides, Salix* sp., *Ulmus* sp., angiosperm.
Ecology Causing decay in live trees; logs; decorticated windfall.
Culture Characters
2a.3.15.32.36.38.43-45-47.(48).(51).53.54.60 (Ref. 239).
References 116, 183, 216, 238, 239, 242, 249, 268, 327, 332, 406, 463, 610.

The reports (242) on *Picea* from western Canada need to be confirmed, because they may have been specimens of *H. abietis*.

Hericium erinaceus subsp. ***erinaceus*** (Bull.:Fr.) Pers. 1825
Mycol. Eur. 2:153.
Basionym *Hydnum erinaceus* Bull., Herb. France, pl. 34, 1780.
Distribution United States: AL, AR, AZ, FL, GA, IA, KS, LA, MD, MI, MN, MS, NC, NJ, NY, OR, PA, VA, WV.
Hosts *Acer* sp., *A. nigrum, A. saccharinum, A. saccharum, Betula alleghaniensis, Carya aquatica, Castanea dentata, Celtis laevigata, Diospyros virginiana, Fagus grandifolia, Fraxinus* sp., *Liquidambar styraciflua, Liriodendron tulipifera, Nyssa* sp., *N. sylvatica, Platanus occidentalis, Pseudotsuga menziesii, Quercus* sp., *Q. alba, Q. falcata, Q. garryana, Q. hypoleucoides, Q. laurifolia, Q. lyrata, Q. nigra, Q. nuttallii, Q. phellos, Q. prinus, Q. rubra, Q. rubra* var. *borealis, Q. velutina, Salix nigra*.
Ecology Causes heart rot or trunk rot in a variety of tree species; the most destructive trunk rot in *Platanus occidentalis*; a principal trunk rot in *Quercus*.
Culture Characters
2a.3.15.(32).(34).36.(38).(39).45-46-47.(48).(51).(53).54.60 (Ref. 239).
References 96, 183, 216, 239, 274, 335, 463.

Conners (116) and Ginns (242) compiled Canadian reports from New Brunswick, Nova Scotia and Ontario, but they may have been misdetermined specimens because neither Ginns (239) nor Harrison (268) were able to confirm its presence in Canada.

Hericium erinaceus subsp. ***erinaceo-abietis*** Burdsall, O.K. Miller & Nishijima 1978
Mycotaxon 7:4.
Distribution United States: VA.
Host *Quercus* sp.
Ecology Live trunk.
Culture Characters Not known.
Reference 238.

HETEROACANTHELLA Oberw. 1990 Tulasnellaceae

Heteroacanthella acanthophysa (Burdsall) Oberw. 1990
Trans. Mycol. Soc. Japan 31:211.
Basionym *Platygloea acanthophysa* Burdsall, Mycotaxon 27:500, 1986.
Distribution United States: WI.
Host *Tsuga canadensis*.
Ecology Bark.
Culture Characters Not known.

Known only from the type specimen.

HETEROCHAETE Pat. 1892 Exidiaceae

Heterochaete andina Pat. & Lagerh. 1892
Bull. Soc. Mycol. France 8:120.
Distribution United States: FL, KY, LA.
Hosts Angiosperm.
Ecology Dead fallen branches; dead wood.
Culture Characters Not known.
References 34, 35, 90, 443.

Heterochaete crassa Bodman 1949
Mycologia 41:531.
Distribution United States: FL.
Hosts Not known.
Ecology Not known.
Culture Characters Not known.
References 34, 35.

Heterochaete shearii (Burt) Burt 1921
Ann. Missouri Bot. Gard. 8:377.
Basionym *Sebacina shearii* Burt, Ann. Missouri Bot. Gard. 2:758, 1915.
Distribution United States: DC, FL, LA.
Hosts *Berberis vulgaris, Juniperus virginiana*, angiosperm.
Ecology Dead branches; associated with a white rot.
Culture Characters Not known.
References 34, 35, 90, 201.

Heterochaete spinulosa (Berk. & Curtis) D. Reid 1970
Trans. Brit. Mycol. Soc. 55:434.
Basionym *Radulum spinulosum* Berk. & Curtis, Grevillea 1:146, 1873.
Distribution United States: AL.
Host *Cephalanthus* sp.
Ecology Not known.
Culture Characters Not known.
References 192, 556.

The type specimen of *R. spinulosum* is from Alabama and it is the only collection cited here. Other specimens cited as *Eichleriella spinulosa*, e.g., Burt (81) and Martin (443), are not thought to be the same as the type and they are listed under *E. deglubens*. See discussion by Reid (556).

HETEROCHAETELLA (Bourd.) Bourd. & Galzin 1928 Hyaloriaceae

Heterochaetella bispora Luck-Allen 1960
Canad. J. Bot. 38:563.
Distribution Canada: ON.
Host _Ulmus_ sp.
Ecology Not known.
Culture Characters Not known.
References 242, 424.

Heterochaetella brachyspora Luck-Allen 1960
Canad. J. Bot. 38:566.
Distribution Canada: ON.
Host _Ulmus_ sp.
Ecology Not known.
Culture Characters Not known.
References 242, 424.

Heterochaetella dubia (Bourd. & Galzin) Bourd. & Galzin 1928
Hym. France, 51.
Basionym _Heterochaete dubia_ Bourd. & Galzin, Bull. Soc. Mycol. France 25:30, 1909.
Synonym _Sebacina dubia_ (Bourd. & Galzin) Bourd. 1922.
Distribution Canada: ON, PQ. United States: IA, LA, MO, OR.
Hosts _Acer_ sp., _Betula_ sp., _Fagus_ sp., _F. grandifolia_, _Pinus strobus_, _Populus_ sp., _Ulmus_ sp., angiosperm, gymnosperm.
Ecology Not known.
Culture Characters Not known.
References 116, 242, 424, 443, 451, 502, 562.

HETEROTEXTUS Lloyd 1922 Dacrymycetaceae

Heterotextus alpinus (Tracy & Earle) G.W. Martin 1932
Mycologia 24:217.
Basionym _Guepinia alpina_ Tracy & Earle in Greene, Plant. Baker. 1:23, 1901.
Synonyms _Guepiniopsis alpina_ (Tracy & Earle) Brasfield 1938, _Ditiola shopei_ Coker 1930, fide McNabb (460), _Guepinia monticola_ Tracy & Earle 1901, fide Brasfield (51), _Heterotextus monticolus_ (Tracy & Earle) Lloyd 1922.
Distribution Canada: BC, ON. United States: AR, AZ, CA, CO, ID, MT, NM, OR, UT, WA, WY.
Hosts _Abies_ sp., _A. concolor_, _A. lasiocarpa_, _A. lasiocarpa_ var. _arizonica_, _A. magnifica_, _Alnus rubra_, _Picea engelmannii_, _Pinus albicaulis_, _P. ponderosa_, _Tsuga heterophylla_, _T. mertensiana_, gymnosperm.
Ecology Basidiomes produced under snow at high elevations; slash; gymnosperm logs; associated with a brown rot.
Culture Characters Not known.
References 51, 113, 120, 121, 130, 133, 183, 197, 216, 242, 402, 443, 448, 459, 509, 554.

Better placed in _Heterotextus_ than in _Guepiniopsis_, fide McNabb (458).

Heterotextus luteus (Bres.) McNabb 1965
New Zealand J. Bot. 3:221.
Basionym _Guepinia lutea_ Bres. in Harriman, Alaska Expedition 5:42, 1904.
Synonym _Guepinia occidentalis_ Lloyd 1916, fide McNabb (459).
Distribution Canada: BC. United States: AK, WA.
Hosts _Pinus_ sp., gymnosperm.
Ecology Decorticated wood.
Culture Characters Not known.
Reference 459.

HYDNOCHAETE Bres. 1896 Hymenochaetaceae

Hydnochaete olivacea (Schw.:Fr.) Banker 1914
Mycologia 6:234.
Basionym _Sistotrema olivaceum_ Schw., Schriften Naturf. Ges. Leipzig 1:101, 1822.
Distribution Canada: NB, NS, ON. United States: AL, AR, CT, FL, GA, ID, IL, IN, KY, LA, MA, MD, ME, MI, MN, MO, MS, NC, NH, NJ, NY, OH, PA, SC, TN, TX, VA, VT, WA, WI, WV.
Hosts _Acer rubrum_, _A. spicatum_, _Alnus crispa_, _A. incana_, _A. rubra_, _A. rugosa_, _Betula_ sp., _B. alleghaniensis_, _B. lenta_, _B. pumila_, _Carpinus caroliniana_, _Fagus grandifolia_, _Populus_ sp., _Prunus serotina_, _Pseudotsuga menziesii_, _Quercus_ sp., _Q. alba_, _Q. nigra_, _Q. rubra_, _Q. rubra_ var. _borealis_, _Q. velutina_, _Salix nigra_, angiosperm.
Ecology Twig; moss- or lichen-covered bark.
Culture Characters Not known.
References 59, 116, 178, 183, 217, 242, 249, 327, 573.

Hydnochaete tabacina (Berk. & Curtis) Ryv. 1982
Mycologia 15:441.
Basionym _Irpex tabacinus_ Berk. & Curtis in Fries, Nova Acta Regiae Soc. Sci. Upsal., Ser. III, 1:106, 1851.
Distribution United States: FL, GA, LA, MI, NC, SC.
Hosts _Quercus_ sp., _Q. alba_.
Ecology Not known.
Culture Characters Not known.
References 217, 573.

HYDNOPOLYPORUS D. Reid 1962 Polyporaceae s.l.

Hydnopolyporus fimbriatus (Fr.) D. Reid 1962
Persoonia 2:151.
Basionym _Polyporus fimbriatus_ Fr., Linnaea 5:520, 1830.
Distribution United States: FL, LA.
Hosts Angiosperm.
Ecology Often from buried wood; associated with a white rot.

Culture Characters
(2).6.24.25.32.336.38.43-44.48.54 (Ref. 588).
References 217, 553.

Hydnopolyporus hartmannii (Mont.) D. Reid 1962
Persoonia 2:151.
Basionym *Thelephora hartmannii* Mont., Ann. Sci. Nat. Bot., Sér. II, 20:366, 1843.
Synonyms *Cotylidia hartmannii* (Mont.) Welden 1958, *Stereum hartmannii* (Mont.) Lloyd 1913.
Distribution United States: NC, SC.
Hosts Not known.
Ecology Decaying wood and bark; dead herbaceous stems.
Culture Characters Not known.
References 89, 553, 613.

HYMENOCHAETE Lév. 1846 Hymenochaetaceae

Hymenochaete anomala Burt 1918
Ann. Missouri Bot. Gard. 5:358.
Distribution United States: LA.
Hosts Not known.
Ecology Decorticated wood.
Culture Characters Not known.
Reference 551.

Hymenochaete arida (Karsten) Sacc. 1891
Sylloge Fung. 9:228.
Basionym *Hymenochaetella arida* Karsten, Fin. Vet.-Soc. Bidr. Nat. Folk 48:428, 1889.
Distribution United States: AZ, CA, CO, MI, NM, NY, SC, VT.
Hosts *Artemisia tridentata, Ceanothus fendleri, C. velutinus, Corylus* sp., *Juniperus deppeana, Ostrya virginiana, Prosopis juliflora, Quercus gambelii, Vaccinium arboreum*, angiosperm.
Ecology Bark of dead branches; dead branch bases; dead fallen trees; associated with a white rot.
Culture Characters Not known.
References 85, 130, 183, 207, 208, 212, 216, 398, 402.

Hymenochaete badio-ferruginea (Mont.) Lév. 1846
Ann. Sci. Nat. Bot., Sér. III, 5:152.
Basionym *Stereum badio-ferrugineum* Mont., Ann. Sci. Nat. Bot., Sér. II, 20:367, 1843.
Distribution Canada: BC, MB, NB, NS. United States: ME, NH, NC, NJ, NS, NY, OR, VT, WA.
Hosts *Abies* sp., *A. balsamea, Acer* spp., *Alnus rugosa, Betula* sp., *B. papyrifera, B. pumila, Fagus* sp., *F. grandifolia, Populus* sp., *Quercus garryana, Taxus brevifolia, Tsuga heterophylla, Vaccinium parvifolium*, angiosperm.
Ecology Erect rotting stumps.
Culture Characters Not known.
References 85, 116, 133, 183, 242, 249.

The features which characterize this species have not been discussed since Burt's (85) description in 1918. He thought it might be a form of *H. tabacina* and several of the references echo that opinion. A critical comparison of the two fungi is needed.

Hymenochaete borealis Burt 1918
Ann. Missouri Bot. Gard. 5:317.
Distribution Canada: ON. United States: NJ, VT.
Hosts *Carya* sp., angiosperm.
Ecology Decorticated wood.
Culture Characters Not known.

Known only from the type specimen.

Hymenochaete cervina Berk. & Curtis 1868
J. Linn. Soc., Bot. 10:334.
Distribution United States: LA.
Hosts Not known.
Ecology Dead limbs; decorticated wood.
Culture Characters Not known.
Reference 85.

Hymenochaete cinnamomea (Pers.:Fr.) Bres. 1897
Atti Imp. Regia Accad. Rovereto, Ser. III, 3:66.
Basionym *Thelephora cinnamomea* Pers., Mycol. Eur. 1:141, 1822.
Distribution Canada: BC, MB, YT.
United States: AZ, CA, CO, IL, KS, NC, NC-TN, ND, NE, NY, OH, UT, VA, WA.
Hosts *Acer macrophyllum, Alnus* sp., *Betula* sp., *B. alleghaniensis, Ceanothus* sp., *Corylus americana, Fagus* sp., *F. grandifolia, Picea* sp., *P. glauca, Pinus aristata* var. *longaeva, Populus* spp., *Quercus* sp., *Q. gambelii, Q. hypoleucoides, Salix* sp., *Symphoricarpos occidentalis, Thuja plicata*, angiosperm, gymnosperm.
Ecology Bark and decaying wood; twig; slash; log; associated with a white rot.
Culture Characters Not known.
References 85, 116, 119, 133, 183, 204, 216, 242, 327, 332, 402, 407, 448.

Hymenochaete corrugata (Fr.) Lév. 1846
Ann. Sci. Nat. Bot., Sér. III, 5:152.
Basionym *Thelephora corrugata* Fr., Obs. Mycol. 1:154, 1815.
Synonym *Hymenochaete agglutinans* Ellis 1874, fide Reeves and Welden (551).
Distribution Canada: MB, NB, NS, ON, PE, PQ.
United States: AK, AL, AZ, CT, FL, GA, ID, KY, LA, MA, MD, ME, MI, MS, MT, NC, NH, NJ, NY, OH, PA, SC, TN, TX, VT, WI, WV.
Hosts *Abies balsamea, A. lasiocarpa* var. *arizonica, Acer* sp., *A. negundo, A. nigrum, A. pensylvanicum, A. rubrum, A. rubrum* var. *rubrum, A. saccharum, A. spicatum, Aesculus* sp., *Alnus* sp., *A. incana, A. rugosa, A. rugosa* var. *americana, A. serrulata, Amelanchier* sp., *A. alnifolia, Betula* sp., *B. alleghaniensis, B. papyrifera, Carpinus* sp., *C. caroliniana, Castanea* sp., *Corylus* sp., *C. cornuta, Fagus* sp., *F. grandifolia, Fraxinus nigra, Lindera (Benzoin)* sp., *Magnolia* sp., *Malus* sp., *Ostrya*

virginiana, Populus sp., *P. grandidentata, P. tremuloides, Quercus* sp., *Q. nigra, Q. rubra, Q. rubra* var. *borealis, Rhododendron* sp., *Salix* sp., *Thuja occidentalis, T. plicata, Tsuga canadensis, Viburnum dentatum*, angiosperm, rarely gymnosperm.
Ecology Lichen-covered bark; twig; live branches where they rub together; corticated branches; dead fallen limbs and trunks; logs.
Culture Characters
2.6.7.(11).32.36.38.39.42-43.50.54.(55) (Ref. 487).
References 85, 96, 97, 108, 116, 132, 183, 204, 216, 242, 249, 327, 333, 511, 513, 551.

Hymenochaete agglutinans is a sterile mycelial pad which develops where one branch or small stem lies against another, at least one of which is live. The pad 'glues' the two together. See also *H. tabacina*.

Hymenochaete corticolor Berk. & Rav. 1873
Grevillea 1:165.
Distribution Canada: NS.
United States: FL, MD, NJ, SC.
Hosts *Fagus grandifolia, Magnolia* sp., *Quercus* sp., *Ulmus* sp., angiosperm.
Ecology Bark of live trees; stems.
Culture Characters Not known.
References 85, 116, 242, 249, 551.

Hymenochaete curtisii (Berk.) Morgan 1888
J. Cincinnati Soc. Nat. Hist. 10:197.
Basionym *Stereum curtisii* Berk., Grevillea 1:164, 1873.
Distribution Canada: MB.
United States: AL, AR, CT, DC, FL, GA, IA, KY, LA, MA, MD, MN, MO, MS, NC, ND, NE, NJ, NY, OH, OK, OR, PA, SC, TN, TX, VA, WA, WI.
Hosts *Acer* sp., *Carpinus caroliniana, Castanea dentata, Quercus* sp., *Q. alba, Q. garryana, Q. macrocarpa, Q. marilandica, Q. nigra, Q. stellata*.
Ecology Bark; twigs; branch; dead branches; rotting limbs.
Culture Characters Not known.
References 85, 96, 108, 116, 133, 183, 327, 332, 333.

Hymenochaete epichlora (Berk. & Curtis) Cooke 1880
Grevillea 8:147.
Basionym *Corticium epichlorum* Berk. & Curtis, Grevillea 1:178, 1873.
Distribution United States: AL, LA.
Hosts *Symplocos* sp., *Vitis* sp., angiosperm.
Ecology Bark of dead plants.
Culture Characters Not known.
Reference 85.

Hymenochaete episphaeria (Schw.:Fr.) Massee 1890
J. Linn. Soc., Bot. 27:111.
Basionym *Thelephora episphaeria* Schw. in Fries, Elenchus 1:225, 1828.
Distribution United States: IL, MA, NY, OH, PA, VT.
Hosts *Acer* sp., *Alnus* sp., *Castanea* sp., angiosperm.
Ecology Lower side of dead limbs.
Culture Characters Not known.
References 85, 183.

Hymenochaete fuliginosa (Pers.) Lév. sensu Burt 1918
Ann. Missouri Bot. Gard. 5:365.
Basionym *Thelephora fuliginosa* Pers., Myc. Eur. 1:145, 1822.
Distribution Canada: BC. United States: CO, KY, MD, MI, MN, NC, NC-TN, OH, VT.
Hosts *Abies fraseri, Alnus incana, Betula* sp., *Picea engelmannii, P. glauca, Pseudotsuga menziesii, Rhododendron* sp., *Taxus brevifolia, Tsuga heterophylla, Thuja occidentalis, T. plicata*, angiosperm.
Ecology Bark; moss-covered wood; decorticated rotting wood; log; associated with a white rot.
Culture Characters Not known.
References 85, 116, 214, 242, 327, 334, 402, 448.

Burt (85) based his concept on specimens identified by Bresadola. Burt stated that two species may have been labelled *H. fuliginosa*, but he does not clearly indicate whether his concept followed the type. Reeves and Welden (551) also indicated that at least two species are involved. Presumably North American authors have followed Burt.

Hymenochaete fulva Burt 1918
Ann. Missouri Bot. Gard. 5:354.
Distribution United States: LA.
Hosts Angiosperm.
Ecology Rotting fallen limbs.
Culture Characters Not known.

The holotype is from Jamacia and two paratypes are listed from Lousiana. Reeves and Welden (551) published a detailed description of the holotype, which differed in some details from the original description.

Hymenochaete leonina Berk. & Curtis 1868
J. Linn. Soc., Bot. 10:334.
Distribution United States: AR, AZ, LA.
Hosts *Fraxinus velutina, Quercus arizonica*, angiosperm.
Ecology Not known.
Culture Characters Not known.
References 85, 216, 554.

Hymenochaete opaca Burt 1918
Ann. Missouri Bot. Gard. 5:364.
Distribution United States: AL, FL, LA.
Hosts Not known.
Ecology Dead wood.
Culture Characters Not known.
Reference 551.

Hymenochaete pinnatifida Burt 1918
Ann. Missouri Bot. Gard. 5:355.
Distribution United States: AL, FL, GA, LA, MS.
Hosts *Alnus rugosa, Elaeagnus umbellata, Pinus taeda, Prunus serotina*, angiosperm.

Ecology Bark of fallen limbs.
Culture Characters Not known.
References 85, 96, 183, 374.

Hymenochaete rigidula Berk. & Curtis 1868
J. Linn. Soc., Bot. 10:334.
Distribution United States: TN.
Hosts Angiosperm.
Ecology Dead wood.
Culture Characters Not known.
Reference 551.

Hymenochaete rubiginosa (Dickson:Fr.) Lév. 1846
Ann. Sci. Nat. Bot., Sér. III, 5:151.
Basionym *Helvella rubiginosa* Dickson, Fasc. Pl. Crypt. Brit. 1:20, 1785.
Distribution Canada: BC, ON.
United States: AL, AR, AZ, CA, CO, CT, FL, IL, IN, KN, KY, LA, MA, MD, ME, MO, NE, NC, NJ, NY, OH, OR, PA, TN, VA, VT, WI, WV.
Hosts *Abies lasiocarpa, Acer saccharum, Castanea dentata, Holodiscus discolor, Juglans major, Myrica cerifera, Prosopis juliflora, Quercus* sp., *Q. agrifolia, Q. alba, Q. bicolor, Q. garryana, Q. macrocarpa, Q. prinus, Q. stellata, Rhamnus crocea* subsp. *insula, Sambucus caerulea, S. glauca*, angiosperm
Ecology Lower side dead limbs; decaying logs and stumps; associated with a white rot.
Culture Characters
2.6.8.11.17.(34).37.39.45-47.(53).54 (Ref. 582).
References 85, 116, 183, 207, 216, 242, 333, 402, 465.

Hymenochaete sallei Berk. & Curtis 1868
J. Linn. Soc., Bot. 10:333.
Distribution United States: FL, NC, SC.
Hosts *Quercus* sp., angiosperm.
Ecology Dead wood; dead twigs; prostrate limbs; base of trees.
Culture Characters Not known.
References 85, 183.

Hymenochaete spreta Peck 1879
Annual Rep. New York State Mus. 30:47.
Distribution Canada: BC, NS, ON, PE.
United States: AL, AZ, CA, DC, FL, GA, ID, IN, KY, MI, MT, NH, NJ, NY, OH, OR, PA, TN, VA, VT, WA, WV.
Hosts *Acer macrophyllum, A. spicatum, Alnus* sp., *Arbutus menziesii, Ceanothus velutinus, Fagus grandifolia, Populus* sp., *Prunus* sp., *Pseudotsuga menziesii, Quercus hypoleucoides, Thuja plicata*, angiosperm, gymnosperm.
Ecology Decaying wood; bark; log.
Culture Characters Not known.
References 85, 116, 120, 133, 183, 216, 242, 249, 327, 334.

Hymenochaete tabacina (Sowerby:Fr.) Lév. 1846
Ann. Sci. Nat. Bot., Sér. III, 5:152.
Basionym *Auricularia tabacina* Sowerby, Brit. Fun., pl. 25, 1797.
Distribution Canada: AB, BC, NB, NF, NFL, NS, NTM, ON, PE, PQ. United States: AK, CA, CT, GA, ID, KY, MD, ME, MI, MT, NC, ND, NH, NJ, NY, OK, OR, TN, WA.
Hosts *Abies* sp., *A. balsamea, A. grandis, A. lasiocarpa, Acer* sp., *A. macrophyllum, A. negundo, A. nigrum, A. pensylvanicum, A. rubrum, A. saccharinum, A. saccharum, A. spicatum, Alnus* sp., *A. fruticosa, A. incana, A. rubra, A. rugosa, A. rugosa* var. *americana, A. sinuata, A. tenuifolia, Amelanchier* sp., *A. alnifolia, Arbutus menziesii, Arctostaphylos patula, Aristolochia* sp., *Betula* sp., *B. alleghaniensis, B. lenta, B. papyrifera, Castanea* sp., *C. dentata, Castanopsis chrysophylla, Ceanothus velutinus, Corylus* sp., *C. cornuta, Fagus* sp., *F. grandifolia, Fraxinus nigra, Holodiscus discolor, Larix laricina, L. occidentalis, Lithocarpus densiflora, Lyonothamnus floribundus, Ostrya virginiana, Paxistima myrsinites, Physocarpus malvaceus, Picea* sp., *P. engelmannii, P. glauca, P. rubens, Pinus contorta, P. ponderosa, Populus* sp., *P. grandidentata, P. tremuloides, Prunus* sp., *P. emarginata, P. pensylvanica, P. serotina, P. virginiana, Pseudotsuga menziesii, Quercus* sp., *Q. alba, Q. garryana, Q. rubra* var. *borealis, Q. stellata, Rosa nutkana, Rubus* sp., *Salix* sp., *S. alaxensis, Sambucus* sp., *Sequoia sempervirens, Sorbus scopulina, Symphoricarpos albus, Taxus brevifolia T. canadensis, Thuja occidentalis, T. plicata, Tsuga* sp., *T. heterophylla, Ulmus* sp., *U. americana, Vaccinium* sp., *V. parvifolium*, angiosperm.
Ecology Causes a minor trunk rot in *Abies balsamea*; a gluing fungus which can kill cambium where a dead branch is cemented to a live one; rots dead sapwood; basidiomes produced on bark; twig; dead limbs; branch; decorticated branches; trunk.
Culture Characters
2.6.11.32.37.38.43.54.55 (Ref. 487).
References 85, 96, 97, 102, 116, 119, 120, 130, 132, 133, 183, 242, 249, 327, 328, 332, 333, 334, 340, 344, 448, 484, 511, 579, 603.

Hymenochaete corrugata also glues branches, twigs, etc. together.

Hymenochaete tenuis Peck 1887
Annual Rep. New York State Mus. 40:57.
Distribution Canada: BC, MB.
United States: AK, AZ, FL, MI, NM, NY, PA, VT.
Hosts *Abies balsamea, A. lasiocarpa, Picea* sp., *P. glauca, Pinus ponderosa, Pseudotsuga menziesii, Sabal* sp., *Thuja* sp., *T. occidentalis, Tsuga* sp., gymnosperm.

Ecology Bark; decorticated wood of fallen limbs.
Culture Characters Not known.
References 85, 116, 183, 197, 216, 242, 448.

HYPHODERMA Wallr. Hyphodermataceae
Synonym is *Atheloderma* Parm. 1968, fide Jülich (315).

Hyphoderma albicans (Pers.) Nakasone 1990
Mycologia Memoir 15:142.
Basionym *Hydnum granulosum* var. *albicans* Pers., Mycol. Eur. 2:184, 1825.
Synonyms *Hyphodontia albicans* (Pers.) Parm. 1968, *Odontia albicans* (Pers.) L.W. Miller & Boyle 1943.
Distribution United States: IA, MN, MS.
Hosts *Populus* sp., *Quercus* sp., *Tilia americana*, angiosperm.
Ecology Much decayed wood; associated with a white rot.
Culture Characters
2.3c.(12).21.26.32.36.38.44.(48).54.55.59 (Ref. 474).
References 214, 383, 406, 463, 474.

Hyphoderma amoenum (Burt) Donk 1957
Fungus 27:14.
Basionym *Peniophora amoena* Burt, Ann. Missouri Bot. Gard. 12:276, 1926.
Synonyms *Corticium pilosum* Burt 1926, fide Rogers and Jackson (567), *Corticium subalbum* Burt 1926, fide Gilbertson et al. (209), *Peniophora subalba* (Burt) D.P. Rogers & Jackson 1943, *Peniophora montana* Burt 1926, fide Liberta (392).
Distribution Canada: BC.
United States: AL, AZ, CO, GA, ID, MO, VT, WY.
Hosts *Agave parryi, Alnus* sp., *Arbutus menziesii, Garrya wrightii, Holodiscus discolor, Picea* sp., *Pinus contorta, Prosopis* sp., *P. juliflora, Prunus serotina* subsp. *virens, Quercus arizonica, Q. garryana, Q. hypoleucoides, Q. oblongifolia, Sageretia wrightii, Tsuga* sp., *Vitis* sp., angiosperm, gymnosperm.
Ecology Badly decayed wood; small dead limbs; bark of fallen limbs; associated with a white rot.
Culture Characters
2.3c.26.32.36.38.46-47.(53).54.55 (Ref. 474).
References 92, 93, 116, 119, 183, 204, 207, 216, 242, 392, 402, 474, 584.

Hyphoderma anasaziense Lindsey 1986
Mycotaxon 27:328.
Distribution United States: CO.
Host *Quercus gambelii*.
Ecology Slash; associated with a white rot.
Culture Characters Not known.

Known only from the type specimen.

Hyphoderma argillaceum (Bres.) Donk 1957
Fungus 27:14.
Basionym *Corticium argillaceum* Bres., Fungi Tridentini 2:63, 1898.
Synonym *Peniophora argillaceum* (Bres.) Bres. 1911, *Peniophora fusca* Burt 1926, fide Liberta (392).
Distribution Canada: ON, PQ. United States: AL, AZ, CO, FL, ID, IL, MI, MN, NM, WI.
Hosts *Abies concolor, Acer* sp., *Betula* sp., *Juniperus deppeana, Picea* sp., *P. engelmannii, Pinus monticola, P. ponderosa, P. strobus, Populus* sp., *P. tremuloides, Poria* sp., *Tsuga canadensis*, angiosperm, gymnosperm.
Ecology Very rotten, decorticated wood; associated with a white rot.
Culture Characters
2.3c.26.32.36.38.47.(53).54.55 (Ref. 474).
References 92, 183, 197, 204, 208, 214, 216, 242, 267, 388, 391, 392, 402, 406, 474, 584.

Hyphoderma assimile (Jackson & Dearden) Donk 1957
Fungus 27:15.
Basionym *Peniophora assimilis* Jackson & Dearden, Mycologia 43:55, 1951.
Distribution Canada: Specific locality not given.
United States: CA.
Host *Purshia tridentata*.
Ecology Not known.
Culture Characters Not known.
References 120, 128, 309.

Not *Crustoderma* as proposed by Nakasone (471), fide Stalpers (590).

Hyphoderma baculorubrense Gilbn. & Blackwell 1984
Mycotaxon 20:89.
Distribution United States: FL, LA, MI, MS, TX.
Hosts *Juniperus phoenicea, J. virginiana, Quercus virginiana, Q. virginiana* var. *fusiformis*.
Ecology Bark; associated with a white rot.
Culture Characters
2.3c.26.30.31d.32.36.38.47.54 (Ref. 474).
References 183, 200, 202, 474.

Hyphoderma budingtonii Lindsey & Gilbn. 1977
Mycotaxon 5:313.
Distribution United States: AZ.
Hosts *Populus* sp., *P. tremuloides*.
Ecology Fallen stem; associated with a white rot.
Culture Characters Not known.
References 183, 405, 406.

Hyphoderma clavigerum (Bres.) Donk 1957
Fungus 27:15.
Basionym *Kneiffia clavigera* Bres., Ann. Mycol. 1:103, 1903.
Synonyms *Peniophora clavigera* (Bres.) Bourd. & Galzin 1913, *Peniophora odontioides* Burt 1926, fide Liberta (392).
Distribution Canada: MB, ON, PQ.
United States: AR, AZ, GA, MN, MS.
Hosts *Abies balsamea, Betula papyrifera, Carnegiea gigantea, Chilopsis linearis, Picea* sp., *Quercus hypoleucoides*, angiosperm.

Ecology Decaying wood.
Culture Characters Not known.
References 92, 214, 216, 242, 383, 391, 392, 403, 447, 584.

Hyphoderma comptum (Jackson) Jülich 1974
Persoonia 8:80.
Basionym *Peniophora compta* Jackson, Canad. J. Res., C, 26:138, 1948.
Distribution Canada: ON.
Hosts *Pinus strobus*, gymnosperm.
Ecology Rotten wood.
Culture Characters Not known.
References 116, 302, 318, 584.

Hyphoderma cremeoalbum (Höhnel & Litsch.) Jülich 1974
Persoonia 8:80.
Basionym *Corticium cremeoalbum* Höhnel & Litsch., Wiesner Festschrift, Wien, 63, 1908.
Distribution United States: CO, NC, VA.
Hosts *Abies fraseri, Picea engelmannii.*
Ecology Trunk; associated with a white rot.
Culture Characters Not known.
References 327, 402.

Hyphoderma definitum (Jackson) Donk 1957
Fungus 27:15.
Basionym *Corticium definitum* Jackson, Canad. J. Res., C, 26:149, 1948.
Distribution Canada: ON. United States: AZ, NM.
Hosts *Pinus engelmanii, P. ponderosa*, gymnosperm.
Ecology Not known.
Culture Characters Not known.
References 197, 204, 216, 303.

Hyphoderma deserticola Gilbn. & Lindsey 1975
Great Basin Naturalist 35:293.
Distribution United States: AZ.
Host *Juniperus deppeana.*
Ecology Not known.
Culture Characters Not known.

Known only from the type specimen.

Hyphoderma deviatum (Lundell) Eriksson & Ryv. 1976
Cort. N. Europe 4:573.
Basionym *Gloeocystidium triste* var. *deviatum* Lundell, Fung. Exs. Suec., No. 1852, 1950.
Distribution Canada: BC.
Host *Pseudotsuga menziesii.*
Ecology Not known.
Culture Characters Not known.

Known in North America from only one collection (174:573).

Hyphoderma echinocystis Eriksson & Strid 1975
Cort. N. Europe 3:471.
Distribution United States: FL, NY.
Hosts *Sabal palmetto*, angiosperm.
Ecology Not known.
Culture Characters
2.3c.26.(30).32.36.38.54.59 (Ref. 474).
References 74, 474.

Hyphoderma fouquieriae Nakasone & Gilbn. 1978
Mycologia 70:272.
Distribution United States: AZ.
Hosts *Fouquieria splendens, Opuntia fulgida.*
Ecology Associated with a white rot.
Culture Characters
2.3.(6).7.16.20.(25).(26)32.(34).37.40.44-46.53.54.59 (Ref. 477);
2.3c.(16).(21).31e.32.36.38.(40).45-46.54.59 (Ref. 474).
References 474, 477.

Hyphoderma guttuliferum (Karsten) Donk 1962
Persoonia 2:222.
Basionym *Gloeocystidium guttuliferum* Karsten, Bidrag Kännedom Finlands Natur Folk 48:430, 1889.
Synonym *Peniophora guttulifera* (Karsten) Sacc. 1891.
Distribution Canada: BC, MB, ON.
United States: AL, AZ, AR, IL, LA, ME, MI, MN, NH, NJ, NY, OH, OR, VT.
Hosts *Acer* sp., *A. negundo, A. saccharum, Alnus oblongifolia, Betula* sp., *B. papyrifera, Platanus wrightii, Populus* sp., *P. grandidentata, P. tremuloides, Quercus* sp., *Q. arizonica, Salix* sp., angiosperm, gymnosperm.
Ecology Associated with a white rot.
Culture Characters
2.3c.16.21.(23).26.27.30.32.36.38.45-47.(53).54.(55).58 (Ref. 474).
References 92, 116, 214, 216, 267, 383, 406, 474, 584.

Hyphoderma heterocystidium (Burt) Donk 1957
Fungus 27:15.
Basionym *Peniophora heterocystidia* Burt, Ann. Missouri Bot. Gard. 12:293, 1926.
Synonym *Peniophora kauffmanii* Burt 1926, fide Rogers and Jackson (567).
Distribution Canada: ON, PQ.
United States: CT, DC, FL, IN, IL, KY, MA, MD, MI, MO, MS, NJ, NY, OH, PA, VT.
Hosts *Acer* sp., *A. rubrum, A. saccharum, Betula* sp., *B. papyrifera, Carpinus* sp., *C. caroliniana, Fagus* sp., *F. grandifolia, Fraxinus* sp., *Robinia pseudoacacia, Juglans cinerea, Magnolia* sp., *Ulmus* sp., angiosperm.
Ecology Fallen limbs.
Culture Characters
2.3c.(14).26.33.(36).37.(38).(39).(40).43.44.(48).(50).54.59 (Ref. 474);
2.3.7.14.15.33.36.38.40.43.44.48.54.59 (Ref. 487).
References 92, 116, 242, 267, 453, 474, 584.

Hyphoderma inusitata (Jackson & Dearden) Ginns 1984
Mycotaxon 21:327.
Basionym *Peniophora inusitata* Jackson & Dearden, Canad. J. Res., C, 27:150, 1949.

Distribution Canada: BC.
Host *Populus trichocarpa*.
Ecology Wood of branch.
Culture Characters Not known.
References 116, 242, 308.

Hyphoderma involutum (Jackson & Dearden) Hjort. & Ryv. 1979.
Mycotaxon 9:505.
Basionym *Peniophora involuta* Jackson & Dearden, Mycologia 43:54, 1951.
Distribution United States: WY.
Hosts Gymnosperm.
Ecology Rotting wood.
Culture Characters Not known.

Known only from the type specimen in North America, but collected several times on *Picea* in Finland and Sweden (295).

Hyphoderma lapponicum (Litsch.) Ryv. 1971
Rep. Kevo Subarctic Res. Sta. 8:149.
Basionym *Gloeocystidium lapponicum* Litsch., Ann. Mycol. 39:133, 1941.
Distribution Canada: MB. United States: MO.
Hosts *Maclura pomifera, Populus* sp., angiosperm.
Ecology Logs and slash; associated with a white rot.
Culture Characters Not known.
References 375, 406.

Hyphoderma leptaleum (Ellis & Ev.) Ginns 1992
Mycotaxon 44:202.
Basionym *Corticium leptaleum* Ellis & Ev. in Millspaugh & Nuttall, Publ. Field Mus. Nat. Hist., Bot. Ser. 1:170, 1896.
Distribution United States: WV.
Host *Magnolia fraseri*.
Ecology Bark with adjacent wood rotted.
Culture Characters Not known.
References 93, 245.

Known only from the type specimen.

Hyphoderma litschaueri (Burt) Eriksson & Strid 1975
Cort. N. Europe 3:481.
Basionym *Corticium litschaueri* Burt, Ann. Missouri Bot. Gard. 13: 259, 1926.
Synonym *Corticium septentrionale* Burt 1926, fide Rogers and Jackson (567).
Distribution Canada: MB, NB.
United States: FL, MN, ND, NY, OR, WI.
Hosts *Acer* sp., *A. saccharinum, A. saccharum, Alnus* sp., *Amelanchier ?alnifolia, Betula papyrifera, Malus* sp., *Pinus resinosa, Quercus macrocarpa, Tilia americana*, angiosperm.
Ecology Bark; decaying, weathered wood; branches; associated with a white rot.
Culture Characters
2.3c.7.(26).32.36.38.46.47.54.(55).59 (Ref. 474).
References 93, 116, 173, 214, 242, 474, 567 under *Corticium abeuns*.

Hyphoderma medioburiense (Burt) Donk 1957
Fungus 27:15.
Basionym *Peniophora medioburiensis* Burt, Ann. Missouri Bot. Gard. 12:328, 1925.
Distribution Canada: BC, ON, PQ.
United States: AZ, CO, IA, MN, MT, VT.
Hosts *Abies lasiocarpa, Carya* sp., *Picea engelmannii, Pinus ponderosa, Populus* sp., *P. tremuloides, Quercus gambelii*, angiosperm, gymnosperm.
Ecology Wood and bark of fallen rotten limb; associated with a white rot.
Culture Characters Not known.
References 92, 183, 214, 216, 400, 402, 406, 584.

Hyphoderma mirabile (Parm.) Jülich 1974
Persoonia 8:80.
Basionym *Atheloderma mirabile* Parm., Consp. Syst. Cort., 200, 1968.
Distribution United States: MD, NC, NY, TN.
Hosts *Acer* sp., *Pinus* sp., *Tsuga* sp., angiosperm, gymnosperm.
Ecology Not known.
Culture Characters
1.3c.31e.32.36.38.44-45.(50).52.(53).54.55 (Ref. 474).
Reference 474.

Hyphoderma mutatum (Peck) Donk 1957
Fungus 27:15.
Basionym *Corticium mutatum* Peck, Annual Rep. New York State Mus. 43:69, 1890.
Synonyms *Peniophora mutata* (Peck) Höhnel & Litsch. 1906, *Peniophora allescheri* (Bres.) Sacc. & Syd. 1902, fide Rogers and Jackson (567).
Distribution Canada: BC, MB, NB, NS, ON, PE, PQ.
United States: AL, CA, CO, FL, GA, ID, IL, IN, KY, MD, ME, MI, MN, MO, MS, NC, ND, NH, NJ, NY, OH, PA, SD, VT, WA, WI.
Hosts *Acer* sp., *A. rubrum, A. saccharum, Aesculus hippocastanum, Aralia spinosa, Carpinus* sp., *Castanea dentata, Corylus californica, C. cornuta, Fagus* sp., *Heteromeles* sp., *Juglans cinerea, Liquidambar styraciflua, Liriodendron tulipifera, Magnolia tripetala, Pinus contorta, Populus* sp., *P. angustifolia, P. grandidentata, P. tremuloides, P. trichocarpa, Quercus* sp., *Tilia* sp., *T. americana, Ulmus americana, Viburnum rufidulum*, angiosperm.
Ecology Bark; bark of dead fallen branches; twigs; slash; underside of decorticated log; decaying logs; associated with a white rot.
Culture Characters
2.3c.7.(13).33.36.(37).(38).40.43.(50).(53).54.59 (Ref. 474);
2.3.7.33.36.40.43.44.54.59 (Ref. 487).
References 92, 96, 108, 116, 133, 177, 183, 242, 249, 255, 267, 333, 383, 402, 406, 453, 465, 474, 480, 511, 584.

Hyphoderma nudicephalum Gilbn. & Blackwell 1988 Mycotaxon 33:378.
Distribution United States: LA, TX.
Hosts *Liquidambar styraciflua, Nyssa sylvatica, Quercus nigra, Q. phellos*.
Ecology Associated with a white rot.
Culture Characters Not known.

Known only from the type specimen.

Hyphoderma obtusiforme Eriksson & Strid 1975 Cort. N. Europe 3:493.
Distribution United States: FL.
Host *Juniperus virginiana*.
Ecology Basidiome in contact with a basidiocarp of *Steccherinum ochraceum*; associated with a white rot.
Culture Characters Not known.
Reference 202.

Hyphoderma pallidum (Bres.) Donk 1957 Fungus 27:15.
Basionym *Corticium pallidum* Bres., Fungi Tridentini 2:59, 1898.
Synonyms *Corticium ochrofarctum* Burt 1926, fide Rogers and Jackson (567), *Corticium tsugae* Burt 1926, fide de Vries (606), *Hyphoderma tsugae* (Burt) Eriksson & Strid 1975.
Distribution Canada: AB-NT, BC, ON, PQ.
United States: AZ, FL, ID, NC, NH, NM, TN.
Hosts *Abies fraseri, Pinus* sp., *P. banksiana, P. engelmannii, P. ponderosa, P. strobus, P. taeda, Picea* sp., *Juniperus pachyphloea, Populus trichocarpa, Rubus spectabilis, Tsuga canadensis*, gymnosperm.
Ecology Bark; trunk; tree base; stump; log; associated with a white rot.
Culture Characters
2.3c.26.(30).32.36.38.47.(53).(54).55.58 (Ref. 474).
References 93, 197, 204, 211, 216, 242, 255, 327, 474.

Spore shape was the only feature Eriksson and Ryvarden (173) found to distinguish *H. pallidum* with allantoid spores from *H. tsugae* with ellipsoid spores. However, de Vries (606) found some European specimens to have spores of both shapes and, as a result, placed *H. tsugae* in synonymy under *H. pallidum*.

Hyphoderma pilosum (Burt) Gilbn. & Budington 1970 J. Arizona Acad. Sci. 6:93.
Basionym *Peniophora pilosa* Burt, Ann. Missouri Bot. Gard. 12:291, 1925.
Distribution Canada: ON.
United States: AL, AZ, FL, MN, NM, NY.
Hosts *Abies balsamea, Alnus oblongifolia, Arbutus arizonica, Pinus* sp., *P. ponderosa, P. strobiformis, Populus* sp., *P. tremuloides, Quercus emoryi, Q. hypoleucoides*, angiosperm, gymnosperm.
Ecology Associated with a white rot.
Culture Characters
2.3c.26.(31d).32.36.(38).(40).44.54.55 (Ref. 474).
References 92, 197, 204, 208, 214, 216, 474, 584, 635.

Perhaps Coniophoraceae, fide Weresub (635).

Hyphoderma populneum (Peck) Donk 1957 Fungus 27:15.
Basionym *Stereum populneum* Peck, Annual Rep. New York State Mus. 47:145, 1894.
Synonym *Peniophora populnea* (Peck) Burt 1926.
Distribution Canada: MB, NF, ON.
United States: AZ, NM, NY.
Hosts *Populus* sp., *P. tremuloides*.
Ecology Bark of prostrate trunks; associated with a white rot.
Culture Characters
2.3c.26.33.36.38.43-44.54.59 (Ref. 474);
2.3.7.33.36.38.40.43.54.59 (Ref. 487).
References 93, 116, 208, 216, 242, 406, 453, 474, 584.

Hyphoderma praetermissum (Karsten) Eriksson & Strid 1975
Cort. N. Europe 3:505.
Basionym *Corticium praetermissum* Karsten, Bidrag Kännedom Finlands Natur Folk 48:423, 1889.
Synonyms *Peniophora albugo* Burt 1926, fide Liberta (392), *Peniophora pertenuis* (Karsten) Burt, fide Rogers and Jackson (567), *Peniophora taxodii* Burt 1926, fide Liberta (392), *Peniophora tenuis* (Pat.) Massee sensu Aucts., *Hyphoderma tenue* (Pat.) Donk sensu Aucts., fide Eriksson and Ryvarden (173).
Distribution Canada: AB, BC, MB, NB, NS, ON, PQ.
United States: AZ, CO, CT, DC, FL, IA, ID, IL, LA, MA, MD, MI, MN, MS, MT, NC, NH, NJ, NM, NY, OR, TX, VA, WA, WI.
Hosts *Abies* sp., *A. balsamea, A. concolor, A. fraseri, A. lasiocarpa, Acer* sp., *A. negundo, A. saccharum, Alnus* sp., *A. oblongifolia, A. rubra, A. tenuifolia, Betula* sp., *B. alleghaniensis, Cornus florida, Juniperus virginiana, Picea* sp., *P. engelmannii, P. glauca, P. pungens, P. rubens, Pinus* sp., *P. ayacahuite, P. banksiana, P. nigra, P. ponderosa, P. resinosa, P. strobiformis, Platanus wrightii, Populus* sp., *P. tremuloides, Prunus* sp., *Pseudotsuga menziesii, Quercus* sp., *Q. arizonica, Q. gambelii, Q. hypoleucoides, Q. nigra, Q. virginiana, Taxodium distichum, Thuja plicata*, angiosperm, gymnosperm.
Ecology Decaying wood and bark; dead branches; dead, fallen trees; log; trail stairway; creosote- or penta-treated southern pine poles; test stakes; associated with a white rot.
Culture Characters
2.3c.13.26.(27).30.32.36.38.44-47.(53).54.55. (57).(58).59 (Ref. 474);
2a.3c.13.(26).32.36.38.45-46.54.55.59 (Ref. 255).
References 58, 92, 116, 119, 133, 179, 183, 197, 201, 202, 204, 208, 214, 216, 242, 249, 255, 267, 327, 383, 390, 391, 392, 398, 402, 406, 447, 474, 584, 654.

Hyphoderma probata (Jackson) Jülich 1974
Persoonia 8:80.
Basionym *Peniophora probata* Jackson, Canad. J. Res., C., 26:134, 1948.
Distribution Canada: ON. United States: WI.
Hosts *Pinus* sp., *P. banksiana, Tsuga canadensis*, gymnosperm.
Ecology Associated with a white rot.
Culture Characters
2.3c.27.31d.32.36.38.47.55 (Ref. 474).
References 116, 302, 474, 584.

Hyphoderma puberum (Fr.) Wallr. 1833
Fl. Crypt. Germany 2:576.
Basionym *Thelephora pubera* Fr., Elenchus 1:215, 1828.
Synonyms *Peniophora pubera* (Fr.) Sacc. 1888, *Phlebia pubera* (Fr.) M. Christiansen 1960, *Peniophora tenuissima* Peck 1912, fide Rogers and Jackson (567), *Peniophora tenella* Burt 1926, fide Liberta (392).
Distribution Canada: AB-NT, BC, MB, NB, NS, NTM, ON, PQ. United States: AL, AZ, CO, DC, FL, ID, IL, KY, LA, MA, MD, MI, MN, MO, MT, NC, NH, NJ, NM, NY, OR, RI, VA, WA, WI.
Hosts *Abies balsamea, A. lasiocarpa, Acer* sp., *A. saccharum, Alnus* sp., *A. oblongifolia, A. rubra, Arbutus menziesii, Betula alleghaniensis, Juglans cinerea, Picea* sp., *P. engelmannii, P. glauca, Pinus* sp., *P. contorta, P. ponderosa, P. strobus, Populus* sp., *P. tremuloides, Quercus* sp., *Q. macrocarpa, Salix* sp., *Thuja plicata*, angiosperm, gymnosperm, southern pine.
Ecology Bark; dead wood; logs and limbs; test stakes; house sills; shingles; creosote- or penta-treated pine poles; associated with a white rot.
Culture Characters
2.3.21.(30).32.36.38.45-47.54.55.59 (Ref. 179);
2.3c.(13).26.30.(31a).32.36.38.44-46.54.55.59 (Ref. 474).
References 92, 116, 179, 183, 197, 208, 214, 216, 242, 249, 255, 267, 327, 335, 391, 392, 402, 406, 447, 474, 584, 654.

Hyphoderma rimosum Burdsall & Nakasone 1983
Mycotaxon 17:259.
Distribution United States: FL, MD, MS.
Hosts *Acer* sp., *Cornus florida, Ilex vomitoria, Lagerstroemia indica, Liquidambar styraciflua, Quercus* sp., *Q. virginiana*, angiosperm.
Ecology Fallen branch; associated with a white rot.
Culture Characters
2.3.7.32.36.38.45-47.54 (Ref. 75);
2.3c.13.16.32.36.38.45-46.50.54 (Ref. 474).
References 75, 474.

Hyphoderma roseocremeum (Bres.) Donk 1957
Fungus 27:15.
Basionym *Corticium roseocremeum* Bres., Ann. Mycol. 3:163, 1905.
Distribution Canada: BC, NS, PQ.
United States: AZ, NC, NC-TN, VA.
Hosts *Abies balsamea, A. fraseri, A. lasiocarpa, Fraxinus velutina, Platanus wrightii, Prunus serotina* var. *virens, Quercus emoryi, Q. hypoleucoides, Q. reticulata.*
Ecology Live tree trunk; trunk; log.
Culture Characters Only a partial species code is known: 60 (Ref. 36).
References 116, 216, 242, 249, 255, 391.

Both host and geographic ranges between Canada and the United States suggest that two species may be involved. Cooke (127) reported an Ohio collection on *Rosa* sp. as "near *roseocremeum*".

Hyphoderma rubropallens (Schw.) Ginns 1992
Mycotaxon 44:208.
Basionym *Thelephora rubropallens* Schw., Trans. Amer. Philos. Soc., N.S., 4:168, 1832.
Synonym *Corticium rubropallens* (Schw.) Massee 1890.
Distribution United States: PA.
Hosts Not known.
Ecology Bark and litter.
Culture Characters Not known.
References 93, 245.

Known only from the type collection. The Alabama report by Burt (93) was based upon a misidentified specimen, fide Ginns (245).

Hyphoderma rude (Bres.) Hjort. & Ryv. 1980
Mycotaxon 10:275.
Basionym *Odontia rudis* Bres., Ann. Mycol. 18:42, 1920.
Distribution United States: FL, GA, MD.
Hosts *Acer* sp., *Castanea* sp., *Prunus* sp., *Quercus* sp., angiosperm.
Ecology Associated with a white rot.
Culture Characters
2.3c.13.(21).26.30.32.36.38.47.(50).54 (Ref. 474).
Reference 474.

Hyphoderma sambuci (Pers.) Jülich 1974
Persoonia 8:80.
Basionym *Thelephora sambuci* Pers., Mycol. Eur. 1:152, 1822.
Synonyms *Hypochnus sambuci* (Pers.) Sacc. 1888, *Peniophora sambuci* (Pers.) Burt 1925, *Hyphodontia sambuci* (Pers.) Eriksson 1958, *Corticium serum* (Pers.) Fr. 1874, fide Burt (92), and Rogers and Jackson (567), *Peniophora thujae* Burt 1926, fide Eriksson and Ryvarden (174).
Distribution Canada: BC, MB, NS, ON, PQ.
United States: AR, AZ, CA, CO, GA, ID, IL, KS, LA, MA, ME, MI, MN, MO, NH, NJ, NM, NY, OH, PA, TX, VA, VT, WA.
Hosts *Abies lasiocarpa, Acer saccharum, Alnus* sp., *Carex* sp., *Chrysanthemum* sp., *Cytisus scoparius, Fouquieria splendens, Fraxinus* sp., *Juglans nigra, Juniperus* sp., *J. ashei, Maclura pomifera, Morus* sp.,

Picea sp., *P. pungens, Pinus ponderosa, P. strobus, P. taeda, Populus* sp., *P. tremuloides, P. trichocarpa, Prosopis juliflora, Quercus emoryi, Robinia neomexicana, Salix* sp., *Sambucus* sp., *S. callicarpa, S. glauca, S. racemosa, S. racemosa* var. *arborescens, S. racemosa* var. *callicarpa, Thuja* sp., *T. plicata, Tsuga* sp., *T. canadensis, Typha* sp., angiosperm, gymnosperm.

Ecology Bark and wood; dead branches; trunks; associated with a white rot.

Culture Characters
2.3.7.22.32.36.38.46-47.50.54 (Ref. 477); 60 (Ref. 36).

References 92, 96, 116, 127, 174, 183, 201, 207, 208, 214, 216, 242, 249, 267, 383, 391, 402, 406, 447, 465, 477, 539, 584.

Hyphoderma setigerum (Fr.) Donk 1957
Fungus 27:15.

Basionym *Thelephora setigera* Fr., Elenchus 1:208, 1828.

Synonyms *Peniophora setigera* (Fr.) Höhnel & Litsch. 1906, *Odontia setigera* (Fr.) L.W. Miller 1934, *Hydnum cristulatum* Fr. 1821, fide Jülich and Stalpers (326), *Odontia cristulata* (Fr.) Fr. 1838, *Hyphoderma cristulatum* (Fr.) Donk 1957, *Corticium berkeleyi* Cooke 1890, fide Rogers and Jackson (567), *Odontia acerina* Peck 1900, fide Rogers and Jackson (567) and Gilbertson (189), *Peniophora aspera* (Pers.) Sacc. 1916, fide Fries loc. cit. 1828.

Distribution Canada: AB-NT, BC, MB, NB, NF, NS, ON, PE. United States: AK, AZ, CA, CO, FL, GA, IA, ID, IL, LA, MD, MI, MN, MS, MT, NC, NM, NY, TN, VA, WA, WI.

Hosts *Abies* sp., *A. balsamea, A. fraseri, A. grandis, A. lasiocarpa, Acer* sp., *A. macrophyllum, A. negundo, A. pensylvanicum, A. saccharum, Alnus* sp., *A. oblongifolia, A. rubra, A. rugosa, A. sinuata, Betula* sp., *B. alleghaniensis, B. papyrifera, Chilopsis linearis, Cornus florida, Corylus californica, C. cornuta, Cytisus scoparius, Fagus* sp., *F. grandifolia, Larix lyallii, Liquidambar* sp., *Liridendron tulipifera, Maclura pomifera, Magnolia* sp., *Persea* sp., *Picea* sp., *P. glauca, P. mariana, P. rubens, Pinus banksiana, P. contorta, P. echinata, P. ponderosa, P. resinosa, P. strobus, Planera* sp., *Platanus occidentalis, P. wrightii, Populus* sp., *P. trichocarpa, Prunus* sp., *P. emarginata, P. serotina* subsp. *virens, Quercus alba, Q. arizonica, Q. gambelii, Q. kelloggii, Q. toumeyi, Rhus divaricatum, R. diversiloba, Ribes divaricatum, Salix alaxensis, Sorbus* sp., *Tilia americana, Tsuga canadensis, T. heterophylla, Ulmus* sp., angiosperm, gymnosperm.

Ecology Bark; rotting wood; twig; trunk; log; associated with a white rot.

Culture Characters
2.3c.7.(26).32.36.38.43-44.(48).(50).(53).54.55.59 (Ref. 474).

References 56, 96, 97, 116, 120, 128, 133, 183, 189, 197, 214, 216, 242, 249, 267, 327, 344, 383, 402, 406, 447, 463, 474, 584, 595.

Corticium berkeleyi sensu Burt (93) is a fungus "closely allied to *Corticium bombycinum* (Sommerf.) Karsten", fide Rogers and Jackson (567). Thus Burt's records are not included above. Rogers and Jackson (567) concluded that *Peniophora aspera* was the correct name for *Thelephora setigera* and their synonymy was followed by several North American mycologists. Hence, the reports of *P. aspera* in Conners (116), Harris (267), and Slysh (584) listing *T. setigera* as a synonym are compiled above.

Hyphoderma sibiricum (Parm.) Eriksson & Strid 1975
Cort. N. Europe 3:535.

Basionym *Radulomyces sibiricus* Parm., Consp. Syst. Cort., 223, 1968.

Distribution Canada: YT. United States: CO.

Hosts *Picea* sp., gymnosperm.

Ecology Log; associated with a white rot.

Culture Characters Not known.

References 232, 402.

Hyphoderma subtestaceum (Litsch.) Donk 1957
Fungus 27:15.

Basionym *Peniophora subtestacea* Litsch., Oesterr. Bot. Z. 77:132, 1928.

Distribution Canada: ON. United States: CT, IA, IL, MA, MN, NY, OH, PA, WI.

Hosts *Pinus banksiana, P. resinosa, Populus* sp., *P. tremuloides, Quercus alba, Ulmus* sp., angiosperm.

Ecology Not known.

Culture Characters
(1).2.3c.7.32.36.38.44-47.(50).54.55 (Ref. 474).

References 267, 406, 474, 584.

One entity of the *H. setigerum* complex that European mycologists (173, 326) have not distinguished is *H. subtestaceum*. In North America *H. subtestaceum* is separated from *H. setigerum* because it (*H. subtestaceum*) has a smooth to papillose hymenial surface which has pinkish or ochraceus tints.

Hyphoderma typhicola (Burt) Donk 1962
Persoonia 2:222.

Basionym *Peniophora typhicola* Burt, Ann. Missouri Bot. Gard. 12:319, 1926.

Distribution United States: NY.

Host *Typha latifolia*.

Ecology Not known.

Culture Characters Not known.

References 92, 584.

HYPHODERMELLA Eriksson & Ryv. 1976
Hyphodermataceae

Hyphodermella corrugata (Fr.) Eriksson & Ryv. 1976
Cort. N. Europe 4:579.
Basionym *Grandinia corrugata* Fr., Hym. Eur., 625, 1874.
Synonyms *Odontia corrugata* (Fr.) Bourd. & Galzin 1928, *Odontia coloradensis* Overh. 1930, fide Gilbertson and Larsen (210 as *Odontia pruni*), *Odontia livida* Bres. 1891, fide Ginns (243).
Distribution Canada: BC, MB, ON.
United States: AZ, CA, CO, IA, KS, NM, NY, WA.
Hosts *Acer* sp., *A. glabrum, Alnus* sp., *Artemisia vulgaris, Baccharis pilularis, B. sarothroides, Corylus* sp., *Cowania stansburiana, Fagus grandifolia, Fraxinus velutina, Hamamelis virginiana, Lycium* sp., *Malus* sp., *Olneya tesota, Physocarpus capitatus, Platanus wrightii, Populus* sp., *P. tremuloides, Prosopis juliflora, Prunus demissa, P. emarginata, Quercus* sp., *Q. arizonica, Ribes sanguineum, Rubus arizonensis, ?Tilia* sp., *Ulmus* sp., *Vitis arizonica*, angiosperm.
Ecology Small diameter (up to 3 cm), typically decorticated dead branches and stems on the ground; log; dead grape vines; associated with a white rot.
Culture Characters
(2).5.24.25.32.36.38.41-42.(54).(55) (Ref. 588).
References 116, 133, 183, 190, 207, 208, 209, 210, 216, 242, 243, 402, 406, 463, 515.

In five references (207, 208, 209, 210, 216) this fungus was mistakenly labelled *Odontia pruni*; see the discussion in Ginns (243).

HYPHODONTIA Eriksson 1958 *(nom. conserv.)*
Chaetoporellaceae

Synonyms are *Grandinia* Fr. 1838, fide Jülich (322), *Kneiffiella* Karsten 1889, fide Jülich (322), *Chaetoporellus* Bondartsev & Singer ex Singer 1944, fide Jülich (322), *Fibrodontia* Parm. 1968, fide Hjortstam (289).

Hyphodontia abieticola (Bourd. & Galzin) Eriksson 1958
Symb. Bot. Upsal. 16 (1):104.
Basionym *Odontia barba-jovis* subsp. *abieticola* Bourd. & Galzin, Hym. France, 426, 1928.
Synonym *Grandinia abieticola* (Bourd. & Galzin) Jülich 1982.
Distribution Canada: BC, NF. United States: AK, AZ, CO, ID, MT, NC-TN, NM, NY, WA.
Hosts *Abies* sp., *Picea* sp., *P. mariana, P. rubens, Pinus* sp., *P. ponderosa, Populus* sp., *Pseudotsuga menziesii*, gymnosperm.
Ecology Butt; associated with a white rot.
Culture Characters
(1).2.3c.(31ab).32.36.38.47.50.55.60 (Ref. 474).
References 174, 197, 204, 210, 216, 242, 327, 447, 474.

Hyphodontia alienata (Lundell) Eriksson 1958
Symb. Bot. Upsal. 16 (1):104.
Basionym *Peniophora alienata* Lundell in Lundell and Nannfeldt, Fungi Exs. Suec. No. 1043, Uppsala, 1941.
Distribution Canada: ON, PQ. United States: NC.
Hosts *Abies balsamea, Betula* sp., *Picea* sp., angiosperm, gymnosperm.
Ecology Not known.
Culture Characters
(1).2.3c.31b.32.36.38.45-47.(50).54.(55) (Ref. 474).
References 242, 391, 447, 474, 584.

Hyphodontia alutacea (Fr.) Eriksson 1958
Symb. Bot. Upsal. 16 (1):104.
Basionym *Hydnum alutaceum* Fr., Syst. Mycol. 1:417, 1821.
Synonyms *Odontia alutacea* (Fr.) Bres. 1897.
Distribution Canada: AB, BC, MB, NS, ON.
United States: AR, AZ, CO, IA, ID, MA, ME, MN, MT, NM, NY, WA, WI.
Hosts *Abies* sp., *A. balsamea, A. concolor, Fagus grandifolia, Picea* sp., *P. engelmannii, P. sitchensis, Pinus* sp., *P. contorta, P. monticola, P. ponderosa, P. resinosa, P. strobus, Populus* sp., *Quercus* sp., *Thuja occidentalis, T. plicata, Tsuga*? sp., angiosperm, gymnosperm.
Ecology Decayed wood; associated with a white rot.
Culture Characters
(1).2.3c.(31a).32.36.38.46-47.54.55.60 (Ref. 474).
References 116, 174, 197, 204, 208, 214, 216, 249, 356, 394, 402, 406, 447, 463, 474.

Hyphodontia alutaria (Burt) Eriksson 1958
Symb. Bot. Upsal. 16 (1):104.
Basionym *Peniophora alutaria* Burt, Ann. Missouri Bot. Gard. 12:332, 1925.
Synonym *Grandinia alutaria* (Burt) Jülich 1982
Distribution Canada: MB. United States: AZ, CO, MI, MN, MT, NC, NM, TN, VA, VT.
Hosts *Abies fraseri, Betula alleghaniensis, Larix* sp., *Picea* sp., *P. engelmannii, P. glauca, P. rubens, Pinus monticola, P. ponderosa, Populus* sp., *Thuja occidentalis*, angiosperm, gymnosperm.
Ecology Bark; branch; live tree trunk; trunk; stump; log; trail stairway; associated with a white rot.
Culture Characters
(1).2.3c.7.32.36.38.47.54.55.60 (Ref. 474);
2a.3c.7.32.36.38.44.54.60 (Ref. 255:354).
References 92, 197, 204, 208, 214, 216, 245, 279, 327, 402, 406, 447, 474.

Hyphodontia arguta (Fr.) Eriksson 1958
Symb. Bot. Upsal. 16 (1):104.
Basionym *Hydnum argutum* Fr., Syst. Mycol. 1:424, 1821.
Synonyms *Odontia arguta* (Fr.) Quél. 1888, *Grandinia arguta* (Fr.) Jülich 1982.

Distribution Canada: AB-NT, BC, MB, NB, NS, ON, PE. United States: AZ, CO, GA, IA, ID, IN, MD, MI, MN, MS, NC, NM, NY, PA, WI.
Hosts *Abies* sp., *A. balsamea, A. concolor, A. lasiocarpa* var. *arizonica, Acer negundo, A. saccharum, Alnus oblongifolia, Arbutus arizonica, Betula alleghaniensis, Juglans major, Juniperus deppeana, J. pachyphloea, Picea engelmannii, Pinus* sp., *P. contorta* var. *contorta, P. ponderosa, P. sylvestris, Pinus taeda, Populus* sp., *P. grandidentata, P. tremuloides, P. trichocarpa, Quercus* sp., *Q. gambelii, Robinia pseudoacacia, Salix* spp., *Thuja occidentalis, Tsuga heterophylla*, angiosperm, gymnosperm.
Ecology Butt; cut wood; dead wood; lowerside of logs; associated with a white rot.
Culture Characters
(1).2.3c.7.(13).32.36.38.47.(50).(53).54.55.60 (Ref. 474);
2.3c.7.32.36.38.47.54.58 (Ref. 255:356).
References 56, 96, 116, 119, 174, 197, 204, 208, 211, 214, 216, 242, 249, 255, 327, 383, 399, 402, 406, 463, 474.

Hyphodontia barba-jovis (Bull.) Eriksson 1958
Symb. Bot. Upsal. 16 (1):104.
Basionym *Hydnum barba-jobi* Bull., Histoire Champignons France, 303, 1791.
Synonyms *Odontia barba-jovis* (Bull.) Fr. 1838, *Hydnum nyssae* Berk. & Curtis 1873, fide Gilbertson (192) and L.W. Miller and Boyle (463).
Distribution Canada: BC, NS, ON, PQ. United States: IA, LA, MD, MI, NY, SC.
Hosts *Abies balsamea, Nyssa sylvatica, Picea* sp., *Pinus banksiana, P. strobus, Pseudotsuga menziesii, Thuja occidentalis*, angiosperm, gymnosperm.
Ecology Decorticated rotten log; decaying wood.
Culture Characters
2.3c.31a.(31b).32.36.38.46-47.(50).(53).(54).55.60 (Ref. 474).
References 116, 192, 242, 249, 255, 334, 447, 463, 474.

Hyphodontia breviseta (Karsten) Eriksson 1958
Symb. Bot. Upsal. 16 (1):104.
Basionym *Kneiffia breviseta* Karsten, Hedwigia 25:232, 1886.
Synonyms *Grandinia breviseta* (Karsten) Jülich 1982, *Odontia crustula* L.W. Miller 1934, fide Ginns (Isotype studied: Miller 82, BPI 265536), *Odontia lactea* Karsten sensu Aucts., fide Eriksson (165).
Distribution Canada: AB, BC, NB, NF, NS. United States: AR, AZ, CO, IA, MN, NC, NM, TN, VA.
Hosts *Abies* sp., *A. balsamea, A. amabilis, A. fraseri, A. lasiocarpa, Arbutus arizonica, Betula* sp., *B. alleghaniensis, Fagus grandifolia, Mortonia scabrella, Picea* sp., *P. glauca, P. mariana, P. rubens, Pinus contorta, P. ponderosa, P. strobus, Populus* sp., *P. tremuloides, Pseudotsuga menziesii, Rhododendron* sp., *R. catawbiense, Thuja plicata, Tilia* sp., *Tsuga heterophylla*, angiosperm, gymnosperm.
Ecology Decorticated wood and bark; moss-covered wood and bark; twig; branch; tree base; stump; exposed root; uprooted bottom; log; associated with a white rot.
Culture Characters Not known.
References 116, 191, 197, 204, 208, 214, 216, 242, 249, 255, 327, 383, 402, 406, 447, 463.

Hallenberg (255) reported several specimens as belonging to the *aspera-breviseta* group, from British Columbia, Ontario, and Quebec on *Abies balsamea, A. lasiocarpa, Pinus strobus*.

Hyphodontia burtii (Peck) Gilbn. 1971
P. 300. In *Evolution in the higher Basidiomycetes*. Ed., R.H. Petersen. Univ. Tenn. Press, Knoxville.
Basionym *Grandinia burtii* Peck, Annual Rep. New York State Mus. 53:847, 1900.
Synonym *Odontia burtii* (Peck) Gilbn. 1962.
Distribution United States: MI, NY.
Hosts *Fagus grandifolia*, gymnosperm.
Ecology Decorticated log.
Culture Characters Not known.
Reference 189.

Hyphodontia comptopsis Burdsall & Nakasone 1981
Mycologia 73:460 1981.
Distribution United States: FL.
Hosts Angiosperm.
Ecology Well-decayed wood.
Culture Characters Not known.

Known only from the type specimen.

Hyphodontia crustosa (Pers.:Fr.) Eriksson 1958
Symb. Bot. Upsal. 16 (1):104.
Basionym *Odontia crustosa* Pers., Obs. Mycol. 2:16, 1799.
Synonyms *Grandinia crustosa* (Pers.) Fr. 1838, *Odontia crustosa* (Pers.) Quél. 1888, now a superfluous combination.
Distribution Canada: AB, AB-NT, BC, MB, NB, NS, ON. United States: AK, AR, AZ, CA, CO, FL, GA, IA, ID, MD, MI, MN, MS, MT, NY, WI.
Hosts *Abies balsamea, Acer* sp., *A. rubrum, A. saccharum, Alnus* sp., *A. oblongifolia, A. rubra, A. rugosa, A. sinuata, Arbutus menziesii, Arctostaphylos patula, Betula* sp., *Cornus sericea, C. stolonifera, Corylus* sp., *Cupressus* sp., *Fagus grandifolia, Juniperus virginiana, Picea* sp., *P. engelmannii, P. mariana, P. pungens, Pinus* sp., *P. banksiana, P. ponderosa, Populus* sp., *P. tremuloides, Quercus* sp., *Q. gambelii, Q. garryana, Q. hypoleucoides, Salix* sp., *Sambucus pubens, Taxodium* sp., angiosperm, gymnosperm.
Ecology Bark or wood of dead branches; associated with a white rot.
Culture Characters
2.3c.(13).26.32.36.38.47.54.55.60 (Ref. 474).
References 116, 119, 120, 128, 197, 202, 204, 214, 216, 249, 383, 398, 402, 406, 447, 463, 474.

Hyphodontia fimbriaeformis (Berk. & Curtis) Ginns & Lefebvre, *comb. nov.*

Basionym *Irpex fimbriaeformis* Berk. & Curtis, Grevillea 1:145, 1873.

Synonyms *Hydnum pallidum* Cooke & Ellis 1881, fide Gilbertson (191), *Hydnum rimulosum* Peck 1896, fide Gilbertson (189), *Hydnum combinans* Peck 1900, fide Gilbertson (189), *Odontia stipata* (Fr.) Quél. 1888 sensu N. Amer. Aucts., *Hyphodontia stipata* (Fr.) Gilbn. 1971 sensu Gilbertson.

Distribution Canada: NS, ON. United States: AZ, IA, ID, MN, NJ, NY, OR, PA, WA.

Hosts *Abies balsamea, Acer* sp., *Alnus* sp., *A. tenuifolia, Fraxinus velutina, Pinus attenuata, Populus tremuloides, P. trichocarpa, Quercus* sp., *Ulmus* sp., angiosperm.

Ecology Decorticated log, old stump, associated with a white rot.

Culture Characters Not known.

References 183, 192, 214, 216, 242, 249, 463.

In these references this fungus is recorded as *Odontia stipata* (Fr.) Quél. or *Hyphodontia stipata* (Fr.) Gilbn. but Eriksson et al. (170) discarded the basionym, *Hydnum stipata* Fr., because Fries had applied the name to, at least, two fungi. They (170) adopted the name *Fibrodontia gossypina* Parm. for the European fungus previously labelled *Odontia (Hydnum) stipata*. *Fibrodontia gossypina* in Eriksson et al. (170) is dimitic, whereas Gilbertson's descriptions (189, 191, 192) are of a monomitic fungus. The oldest name for the North American fungus previously labelled "*stipata*" is *H. fimbriaeformis*. Thus we transfer the epithet to *Hyphodontia*.

Hyphodontia floccosa (Bourd. & Galzin) Eriksson 1958 Symb. Bot. Upsal. 16 (1):104.

Basionym *Odontia alutacea* subsp. *floccosa* Bourd. & Galzin, Hym. France, 423, 1928.

Distribution Canada: AB, BC, PQ. United States: AZ, CO, ID, MN, MT, NC, NM, NY.

Hosts *Cowania stansburiana, Larix occidentalis, Picea* sp., *Pinus* sp., *P. banksiana, P. ponderosa, P. resinosa, Quercus gambelii, Thuja plicata.*

Ecology Associated with a white rot.

Culture Characters 2.3c.31ab.32.36.38.46-47.53.(54).55.59 (Ref. 474).

References 116, 174, 197, 204, 210, 216, 391, 402, 447, 474.

Hyphodontia gossypina (Parm.) Hjort. 1990 Mycotaxon 39:416.

Basionym *Fibrodontia gossypina* Parm., Consp. Syst. Cort., 207, 1968.

Distribution United States: MD, MS, NC-TN.

Hosts *Fagus grandifolia, Ulmus* sp., angiosperm.

Ecology Branch.

Culture Characters (1).2.3c.(8).13.(23).32.36.38.46-47.54.59 (Ref. 474).

References 327, 474.

This species has a dimitic hyphal system, an unusual feature in the genus *Hyphodontia*. Jung (327) clearly illustrates and describes the skeletal hyphae, and Nakasone (474) observed skeletal (fiber) hyphae in cultures.

Hyphodontia granulosa (Pers.:Fr.) Ginns & Lefebvre, *comb. nov.*

Basionym *Thelephora granulosa* Pers., Synop. Meth. Fung., 576, 1801.

Synonyms *Grandinia aspera* Fr. 1874, fide Jülich (325), *Odontia aspera* (Fr.) Jϕst. 1937, *Hyphodontia aspera* (Fr.) Eriksson 1958.

Distribution Canada: BC, ON, PQ. United States: AK, AZ, CO, MN, NC-TN.

Hosts *Abies fraseri, Chamaecyparis nootkatensis, Picea* sp., *P. engelmannii, P. glauca, P. sitchensis, Pinus ponderosa, P. resinosa, Populus* sp., *P. tremuloides, Quercus gambelii, Robinia neomexicana, Thuja plicata*, gymnosperm.

Ecology Associated with a white rot.

Culture Characters Not known.

References 116, 204, 214, 216, 242, 273, 327, 402, 406, 447.

This fungus is best known as *Hyphodontia aspera*. The synonymy of *H. aspera* under *H. granulosa* was only proposed in 1984. Because we recognize the genus *Hyphodontia* rather than *Grandinia* it is necessary to transfer the epithet *granulosa* to *Hyphodontia*. Unfortunately, the names *Grandinia granulosa* and *Vararia granulosa* have been misapplied to a fungus with skeletal hyphae, see Pouzar (537). The name for the fungus with skeletal hyphae is *Dichostereum boreale*. Thus three reports (116, 332, 463) under the name *G. granulosa* from North America are included under *D. borealis*.

Hyphodontia hastata (Litsch.) Eriksson 1958 Symb. Bot. Upsal. 16 (1):104.

Basionym *Peniophora hastata* Litsch., Oesterr. Bot. Z. 77:130, 1928.

Distribution Canada: AB-NT, BC, ON. United States: AZ, CO, ID, NM, NY, WA, WI.

Hosts *Pinus* sp., *P. contorta, P. ponderosa, P. radiata, Quercus gambelii*, gymnosperm, rarely angiosperm.

Ecology Saprophytic on rotten wood; associated with a white rot.

Culture Characters (1).2.3c.19.(21).32.36.38.47.(50).55 (Ref. 474).

References 116, 119, 197, 204, 208, 216, 242, 255, 402, 474, 584.

Hyphodontia juniperi (Bourd. & Galzin) Eriksson & Hjort. 1976 Cort. N. Europe 4:666.

Basionym *Corticium serum* var. *juniperi* Bourd. & Galzin, Bull. Soc. Mycol. France 27:246, 1911.

Distribution United States: TX.

Host *Juniperus ashei.*
Ecology Dead branches; associated with a white rot.
Culture Characters Not known.
Reference 201.

Hyphodontia lanata Burdsall & Nakasone 1981 Mycologia 73:461.
Synonym *Grandinia lanata* (Burdsall & Nakasone) Nakasone 1990
Distribution United States: FL, MS, TN.
Hosts *Betula* sp., *Ligustrum* sp., *Liriodendron tulipifera, Sabal palmetto*, angiosperm.
Ecology Not known.
Culture Characters
2.3c.13.(31a).32.36.38.47.(50).54 (Ref. 474); 2.3.13.32.36.38.47.54 (Ref. 74).
References 74, 183, 474.

Hyphodontia latitans (Bourd. & Galzin) Ginns & Lefebvre, *comb. nov.*
Basionym *Poria latitans* Bourd. & Galzin, Bull. Soc. Mycol. France 41:226, 1925.
Synonym *Chaetoporellus latitans* (Bourd. & Galzin) Bondartsev & Singer 1944.
Distribution Canada: ON.
United States: AR, AZ, FL, GA, IA, IL, IN, MD, MI, MS, NC, NM, SC, TN, VA.
Hosts *Acer negundo, Pinus taeda*, angiosperm.
Ecology Associated with a white rot.
Culture Characters
(1).2.3c.31abc.32.36.38.45-46.50.54.55.60 (Ref. 474), 2.3.7.15.26.32.36.38.43.(48).55.60.61 (Ref. 270).
References 174, 217, 474.

The generic name *Chaetoporellus* was placed in synonymy under *Hyphodontia* by Jülich (322) and we agree. Thus the epithet *latitans* is transferred to *Hyphodontia*.

Hyphodontia macrescens (Banker) Ginns & Lefebvre, *comb. nov.*
Basionym *Hydnum macrescens* Banker in Peck, New York State Mus. Bull. 75:15, 1904.
Distribution United States: NY.
Host *Xylobolus frustulatus.*
Ecology Not known.
Culture Characters Not known.

Known only from the type specimen, which was redescribed by Gilbertson (191). He concluded it was very similar to *H. burtii* but differed in spore shape. Because the circumscription of *Hydnum* no longer encompasses this fungus and because it is similar to some species of *Hyphodontia*, we transfer the epithet *macrescens* to *Hyphodontia*.

Hyphodontia microspora Eriksson & Hjort. 1976 Cort. N. Europe 4:651.
Distribution United States: MS, TX, WI.
Hosts *Pinus* sp., *P. elliottii.*
Ecology Dead wood; plywood roof sheathing; associated with a uniform to pitted white rot.
Culture Characters
(1).2.3c.31b.32.36.38.44.53.55.59 (Ref. 474).
References 203, 474.

Hyphodontia nespori (Bres.) Eriksson & Hjort. 1976 Cort. N. Europe 4:655.
Basionym *Odontia nespori* Bres., Ann. Mycol. 18:43, 1920.
Distribution United States: FL.
Host *Juniperus virginiana.*
Ecology Decorticated dead branch.
Culture Characters Not known.
Reference 202.

This may be the fungus recorded as *Hyphodontia papillosa* in North America; see comment under *H. papillosa*.

Hyphodontia orasinusensis Gilbn. & Blackwell 1988 Mycotaxon 33:382.
Distribution United States: LA.
Host *Ailanthus altissima, Celtis laevigata, Pinus taeda, Quercus nigra.*
Ecology Dead wood; associated with a white pitted rot.
Culture Characters Not known.

Known only from the type specimens. Basidiomes similar to *H. microspora* but distinguished by broader, more ellipsoidal spores and simple septate basal hyphae.

Hyphodontia pallidula (Bres.) Eriksson 1958 Symb. Bot. Upsal. 16 (1):104.
Basionym *Gonatobotrys pallidula* Bres., Ann. Mycol. 1:127, 1903.
Synonyms *Peniophora pallidula* (Bres.) Bres. 1913, *Grandinia pallidula* (Bres.) Jülich 1982, *Peniophora laminata* Burt 1926, fide Rogers and Jackson (567).
Distribution Canada: BC, MB, NB, NS, ON, PQ.
United States: AZ, CA, CO, IL, MI, MN, NC, NM, NY, OR, PA, TN, VT, WI.
Hosts *Abies* sp., *A. balsamea, A. concolor, A. fraseri, A. magnifica, Alnus oblongifolia, Betula* sp., *Larix laricina, Picea* sp., *P. engelmannii, P. glauca, P. rubens, Pinus leiophylla* var. *chihuahuana, P. ponderosa, P. resinosa, P. strobiformis, P. strobus, P. taeda, Populus* sp., *P. tremuloides, Pseudotsuga menziesii, Salix* sp., *Thuja plicata, Tsuga canadensis*, angiosperm, gymnosperm.
Ecology Bark; bark and wood of fallen decaying trunk; rotten wood; decayed stem; log; exposed root; associated with a white rot.
Culture Characters
(1).2.3c.7.32.36.38.47.(50).54.55.58 (Ref. 474).
References 92, 116, 120, 128, 130, 197, 204, 208, 214, 216, 242, 249, 267, 315, 327, 388, 390, 391, 402, 406, 447, 474, 584.

Hyphodontia papillosa (Fr.) Eriksson sensu N. Amer. Aucts.
Basionym *Thelephora papillosa* Fr., Elenchus 1:212, 1828.
Distribution Canada: NS.
United States: AZ, MN, NM.
Hosts *Acer grandidentatum, Fagus grandifolia, Pinus engelmannii, P. ponderosa, P. strobus.*
Ecology Not known.
Culture Characters Not known.
References 197, 214, 216, 242.

The name is probably misapplied because Fries' concept is uncertain (174:682). Specimens from North America should be renamed. Although Eriksson (167) used the epithet in 1958, he later (174) realized the species concept was confused and preferred the name *H. verruculosa* Eriksson & Hjort. (now a synonym of *H. rimosissima*). Gilbertson's concept of *papillosa* seems to be distinct from *H. rimosissima.*

Hyphodontia pruni (Lasch) Eriksson & Hjort. 1976
Cort. N. Europe 4:663.
Basionym *Odontia pruni* Lasch in Rabenh., Klotzschii herb. viv. Mycol. no. 1514, 1851 and Bot. Zeit. (Berlin) 9 (36):644, 1851.
Synonyms *Hyphodontia bugellensis* (Ces.) Eriksson 1958, fide Eriksson and Ryvarden (174).
Distribution United States: AZ.
Hosts *Chilopsis linearis, Platanus wrightii, Populus* sp.
Ecology Not known.
Culture Characters
2.3.(24).(25).26.32.36.38.47.54 (Ref. 588).
References 216, 406.

These references refer to a fungus with clamp connections and relatively short spores, which agrees with the European concept. Several other records (see 243) of *Odontia pruni* were based on *Hyphodermella corrugata.*

Hyphodontia quercina (Pers.:Fr.) Eriksson 1958
Symb. Bot. Upsal. 16 (1):105.
Basionym *Odontia quercinum* Pers., Obs. Myc. 2:17, 1799.
Synonyms *Radulum quercinum* (Pers.) Fr. 1874.
Distribution Canada: AB-NT, BC, NS, YT.
United States: AK, AZ, IA.
Hosts *Acer* sp., *Alnus* sp., *A. rubra, Malus pumila, Prosopis juliflora, Salix* sp., angiosperm.
Ecology Not known.
Culture Characters Not known.
References 97, 116, 174, 207, 216, 242, 249, 255, 463.

Hyphodontia rimosissima (Peck) Gilbn. 1971
P. 300. In *Evolution in the higher Basidiomycetes.* Ed., R.H. Petersen. Univ. Tenn. Press, Knoxville.
Basionym *Odontia rimosissima* Peck, Annual Rep. New York State Mus. 50:114, 1897.
Synonyms *Grandinia rimosissima* (Peck) Jung 1987, *Hyphodontia verruculosa* Eriksson & Hjort. 1976, fide Hjortstam (278).
Distribution Canada: PQ.
United States: AZ, NC, NC-TN, NY, VA.
Hosts *Acer* sp., *Alnus* sp., *A. incana, A. oblongifolia, Betula alleghaniensis, Fagus grandifolia,* angiosperm.
Ecology Bark; moss-covered bark; broken wood; twig; branch; tree base; associated with a white rot.
Culture Characters Not known.
References 189, 216, 255, 327.

Listed as a possible synonym of *H. spathulata* by Jülich and Stalpers (326 as *Kneiffiella*).

Hyphodontia species A
Distribution Canada: BC, ON.
Hosts Not known.
Ecology Not known.
Culture Characters Not known.
Reference 174:597.

Similar to *H. alienata* but has cystidia which are more tapered (174).

Hyphodontia spathulata (Schrader:Fr.) Parm. 1968
Consp. Syst. Cort. 123.
Basionym *Hydnum spathulata* Schrader, Spic. Fl. German., 178, 1794.
Synonyms *Radulum spathulatum* (Schrader) Bres. 1903, *Odontia spathulata* (Schrader) Litsch. 1939, *Grandinia spathulata* (Schrader) Jülich 1982, *Hydnum pithyophilum* Berk. & Curtis 1849, *Hydnum caryophylleum* Berk. & Curtis 1873, *Hydnum velatum* Berk. & Curtis 1873, *Hydnum xanthum* Berk. & Curtis 1873, *Irpex ambiguus* Peck 1887. All fide Gilbertson (189, 192).
Distribution Canada: BC, MB, NB, NS, ON, PE.
United States: AZ, CO, FL, IA, IN, MD, MI, MN, NC, NM, NY, SC, WI.
Hosts *Abies* sp., *A. balsamea, A. fraseri, Acer negundo, A. rubrum, A. saccharum, Alnus oblongifolia, A. tenuifolia, Betula papyrifera, Fagus grandifolia, Fraxinus velutina, Juglans major, Juniperus deppeana, J. pachyphloea, J. virginiana, ?Liriodendron tulipifera, Malus pumila, Picea engelmannii, Pinus* sp., *P. palustris, P. ponderosa, P. resinosa, P. strobus, Platanus wrightii, Populus* sp., *P. tremuloides, Quercus* sp., *Thuja* sp. *T. occidentalis, Tsuga canadensis,* angiosperm, gymnosperm.
Ecology Bark; dead decorticated branches; branch; wood piece; associated with a white rot and a soft spongy white rot.
Culture Characters
2.3c.13.31b.32.36.38.45.47.55.60 (Ref. 474).
References 116, 174, 189, 192, 202, 204, 208, 211, 214, 216, 242, 249, 327, 402, 406, 463, 474.

Hyphodontia subalutacea (Karsten) Eriksson 1958
Symb. Bot. Upsal. 16(1):104.
Basionym *Corticium subalutacea* Karsten, Meddeland. Soc. Fauna Fl. Fenn. 9:65, 1882.

Synonyms *Peniophora subalutacea* (Karsten) Höhnel & Litsch. 1906, *Grandinia subalutacea* (Karsten) Jülich 1982.
Distribution Canada: AB-NT, BC, NS, ON, PQ. United States: AL, AZ, CO, GA, ID, IL, LA, MD, MI, MN, NC, NM, NJ, NY, UT, WA, WI.
Hosts *Abies balsamea, A. fraseri, A. lasiocarpa* var. *lasiocarpa, Acer saccharum, A. spicatum, Picea* sp., *P. engelmannii, P. pungens, Pinus* sp., *P. aristata* var. *longaeva, P. contorta, P. echinata, P. ponderosa, P. resinosa, Populus tremuloides, Pseudotsuga menziesii, Quercus* sp., *Q. gambelii*, angiosperm, gymnosperm.
Ecology Twig; decaying wood; log; associated with a white rot.
Culture Characters 2.3c.31a.32.36.38.47.(50).54.55.60 (Ref. 474); 2.3c.(7).26.32.36.38.47.54-55.60 (Ref. 255:366).
References 92, 96, 116, 133, 174, 197, 204, 214, 216, 242, 249, 255, 267, 327, 388, 391, 398, 402, 407, 474, 584.

HYPOCHNELLA Schröter 1889 Botryobasidiaceae

Hypochnella violacea Schröter 1888
As Auersw. ex Schröter in F. Cohn, Kryptogamen-Flora Schlesien 3 (1):420.
Distribution United States: IA.
Hosts Not known.
Ecology Not known.
Culture Characters Not known.
References 174, 435.

HYPOCHNICIELLUM Hjort. & Ryv. 1980 Coniophoraceae

Hypochniciellum molle (Fr.) Hjort. 1981
Mycotaxon 13:125.
Basionym *Thelephora mollis* Fr., Syst. Mycol. 1:443, 1821.
Synonym *Peniophora mollis* (Fr.) Bourd. & Galzin 1912.
Distribution United States: ID, OR, WA.
Host *Pinus* sp.
Ecology Rotten wood; underside pine guard rails aside highway.
Culture Characters Not known.
References 118, 119

In both references the name appears as *Peniophora mollis* (Fr. sensu Bres.) Bourd. & Galzin.

Hypochniciellum ovoideus (Jülich) Hjort. & Ryv. 1980
Mycotaxon 12:177.
Basionym *Leptosporomyces ovoideus* Jülich, Beih. Willdenowia 7:203-204, 1972.
Distribution Canada: ON. United States: CO, WV.
Hosts *Acer* sp., *A. saccharinum, Picea engelmannii.*
Ecology Mine timber.
Culture Characters 1.3c.(16).26.32.36.38.44-45.(53).54.55 (Ref. 474).
References 183, 312, 474.

Hypochniciellum subillaqueatum (Litsch.) Hjort. 1981
Mycotaxon 13:126.
Basionym *Corticium subillaqueatum* Litsch., Ann. Mycol. 39:128-129, 1941.
Synonym *Trechispora subillaqueata* (Litsch.) Gilbn. & Budington 1970.
Distribution United States: AZ, NM.
Hosts *Pinus ponderosa, Quercus hypoleucoides.*
Ecology Not known.
Culture Characters Not known.
References 57, 197, 204, 216.

HYPOCHNICIUM Eriksson 1958 Hyphodermataceae
Synonym is *Lagarobasidium* Jülich 1974, fide Eriksson and Ryvarden (174).

Hypochnicium analogum (Bourd. & Galzin) Eriksson 1958
Symb. Bot. Upsal. 16 (1):101.
Basionym *Gloeocystidium analogum* Bourd. & Galzin, Bull. Soc. Mycol. France 28:366, 1913.
Synonym *Corticium analogum* (Bourd. & Galzin) Burt 1926.
Distribution Canada: BC, ON. United States: AZ, ID, MD, ME, NC, WV.
Hosts *Acer macrophyllum, A. saccharum, Carya* sp., *Fraxinus velutina, Populus* sp., *P. trichocarpa, Quercus* sp., *Q. emoryi, Salix gooddingii, S. nigra*, angiosperm.
Ecology Mine timber; associated with a white rot.
Culture Characters (1).2.3c.15.(16).34.36.38.(39).44-47.48.(50).(51).54 (Ref. 474).
References 93, 116, 174, 204, 216, 242, 406, 474.

Type species for *Gloeohypochnicium* (Parm.) Hjort. (287).

Hypochnicium bombycinum (Sommerf.:Fr.) Eriksson 1958
Symb. Bot. Upsal. 16 (1):101.
Basionym *Thelephora bombycina* Sommerf., Fl. Lapp. Suppl., 284, 1826.
Synonym *Corticium bombycinum* (Sommerf.) Karsten 1893.
Distribution Canada: NS, NTM, ON, PQ. United States: AK, AZ, CO, FL, MA, MI, MN, NM, NY, TN, TX, VT, WA.
Hosts *Acer* sp., *A. saccharum, A. spicatum, Alnus* sp., *A. oblongifolia, Betula* sp., *Fagus* sp., *Juniperus virginiana, Larix* sp., *Picea engelmannii, Pinus engelmannii, P. ponderosa, Populus* sp., *P. grandidentata, P. tremuloides, Prosopis juliflora, Prunus serotina* var. *virens, Quercus* sp., *Q. arizonica, Q. emoryi, Q. gambelii, Q. reticulata, Salix* sp., *Sambucus* sp., angiosperm, gymnosperm.

Ecology Bark of live and dead species; dead decorticated branches; associated with a white rot.
Culture Characters
(1).2.3c.27.32.36.(38).(40).43-45.54.55.60 (Ref. 474); 2.3c.7.32.36.45.58.61 (Ref. 254).
References 93, 116, 133, 183, 197, 202, 207, 208, 216, 242, 249, 255, 402, 406, 464, 474.

Hypochnicium cymosum (D.P. Rogers & Jackson) Larsson & Hjort. 1977
Mycotaxon 5:477.
Basionym *Peniophora cymosa* D.P. Rogers & Jackson, Canad. J. Res., C, 26:133, 1945.
Distribution Canada: ON. United States: NC.
Hosts Gymnosperm.
Ecology Not known.
Culture Characters Not known.
References 291, 302, 584.

Hypochnicium detriticum (Bourd. & Galzin) Eriksson & Ryv. 1976
Cort. N. Europe 4:701.
Basionym *Peniophora detritica* Bourd. & Galzin, Rev. Sci. Bourbonnais Centr. France 23:15, 1910.
Synonym *Hyphodontia magnacystidiata* Lindsey & Gilbn. 1977, fide Eriksson, in litt.
Distribution United States: NY.
Host *Populus tremuloides*.
Ecology Dead, fallen trunks; associated with a white rot.
Culture Characters Not known.
References 405, 406.

Known in North America from only one collection.

Hypochnicium eichleri (Bres.) Eriksson & Ryv. 1976
Cort. N. Europe 4:707.
Basionym *Peniophora eichleri* Bres. in Saccardo and Sydow, Sylloge Fung. 16:194, 1902.
Distribution United States: AZ, LA, MI, NY.
Hosts *Acer rubrum, Juniperus virginiana, Pinus banksiana, P. ponderosa, P. resinosa*.
Ecology Dead branches; associated with a white rot.
Culture Characters
(1).2.3c.(7).(26).(31d).32.36.38.43-44.(50).(53).54.55.(57) (Ref. 474);
2.3c.(7).(26).32.36.38.47.54.55.(57) (Ref. 254).
References 174, 201, 474.

Hypochnicium geogenium (Bres.) Eriksson 1958
Symb. Bot. Upsal. 16 (1):101.
Basionym *Corticium geogenium* Bres., Ann. Mycol. 1:98, 1903.
Synonyms *Peniophora albostraminea* Bres. 1925, fide Eriksson and Ryvarden (174), *Corticium albostramineum* (Bres.) Overh. 1938.
Distribution Canada: AB-NT, MB, NS, NTM, PQ. United States: AZ, CA, ID, NC, NM.
Hosts *Abies balsamea, Alnus incana, A. tenuifolia, Juglans major, Picea* sp., *P. glauca, P. rubens, Pinus* sp., *P. ponderosa, P. strobus, Pseudotsuga menziesii, Quercus californica, Thuja plicata*.
Ecology Wood and bark; moss-covered wood.
Culture Characters Not known.
References 53, 92, 116, 174, 183, 197, 208, 216, 242, 249, 255, 327, 518.

Hypochnicium karstenii (Bres.) Hallenb. 1983
Mycotaxon 16:566.
Basionym *Corticium karstenii* Bres., Ann. Mycol. 9:427, 1911.
Distribution Canada: ON, PQ.
Hosts Angiosperm.
Ecology Not known.
Culture Characters
1.3c.7.32.36.38.44.55.60.61 (Ref. 254).
Reference 255.

Hypochnicium lundellii (Bourd.) Eriksson 1958
Symb. Bot. Upsal. 16 (1):101.
Basionym *Corticium lundellii* Bourd. in Eriksson, Svensk Bot. Tidskr. 43:56, 1949.
Distribution United States: AZ, LA, NM, NY, WI.
Hosts *Fraxinus velutina, Juniperus virginiana, Pinus ponderosa, P. resinosa, Populus* sp., *P. tremuloides*.
Ecology Dead branches; associated with a white rot.
Culture Characters
2.3c.(7).27.32.36.40.43-44.54.55.60 (Ref. 474).
References 183, 201, 204, 216, 406, 474.

Hypochnicium polonense (Bres.) Strid 1975
Wahlenbergia 1:68.
Basionym *Kneiffia polonensis* Bres., Ann. Mycol. 1:102, 1903.
Synonyms *Peniophora polonensis* (Bres.) Höhnel & Litsch. 1906, *Hyphoderma polonense* (Bres.) Donk 1957, *Peniophora canadensis* Burt 1926, fide Rogers and Jackson (567).
Distribution Canada: ON, PQ.
United States: CT, IL, MI, MN, NC, NY, PA.
Hosts *Abies balsamea, Fraxinus* sp., *F. nigra, Picea* sp., angiosperm, gymnosperm.
Ecology Bark and wood; log.
Culture Characters
1.3c.26.32.36.38.47.54.58.61 (Ref. 257).
References 92, 174, 214, 267, 447, 584.

Hypochnicium prosopidis Burdsall 1976
Mycotaxon 3:512.
Distribution United States: AZ.
Host *Prosopis juliflora*.
Ecology Not known.
Culture Characters Not known.

Known only from the type specimen.

Hypochnicium punctulatum (Cooke) Eriksson 1958
Symb. Bot. Upsal. 16 (1):101.
Basionym *Corticium punctulatum* Cooke, Grevillea 6:132, 1878.

Synonym *Peniophora peckii* Burt 1926, fide Rogers and Jackson (567).
Distribution Canada: ?PQ. United States: AZ, FL, ID, MA, MI, NJ, NM, NY, SC, WA, WI.
Hosts *Acer macrophyllum, Alnus* sp., *A. oblongifolia, Betula* sp., *Ceanothus* sp., *C. integerrimus, Juniperus virginiana, Pinus* sp., *P. ponderosa, Populus* sp., *P. fremontii, Pseudotsuga menziesii, Quercus* sp., *Sageretia wrightii, Simmondsia chinensis, Thuja plicata*, angiosperm, rarely gymnosperm.
Ecology Bark and wood; rotten logs; bare ground in woods; associated with a white rot.
Culture Characters
2.3c.7.32.36.(38).40.44-45.54.55.60 (Ref. 474); 2a.3c.26.32.36.38.45.54.55.60.61 (Ref. 254).
References 92, 93, 133, 174, 183, 197, 202, 204, 208, 216, 474, 567.

Hypochnicium sphaerosporum (Höhnel & Litsch.) Eriksson 1958
Symb. Bot. Upsal. 16 (1):101.
Basionym *Peniophora sphaerospora* Höhnel & Litsch., Sitzungsber. Kaiserl. Akad. Wiss., Math.-Naturwiss. Kl., Abt. 1, 115:1600, 1906.
Distribution United States: AZ, MN, NM, OR, TN.
Hosts *Abies* sp., *Acer negundo, Alnus oblongifolia, Pinus ponderosa, Tsuga* sp., *Ulmus* sp., angiosperm, gymnosperm.
Ecology Associated with a white rot.
Culture Characters
2.3c.7.(26).32.36.38.44-46.(50).54.55.60 (Ref. 474).
References 197, 204, 214, 216, 474 .

Hypochnicium stratosum Burdsall & Nakasone 1983
Mycotaxon 17:261.
Distribution United States: FL.
Host *Fagus grandifolia*.
Ecology Not known.
Culture Characters Not known.

Known only from the type specimen. The acyanophilous spores, bladder-like cystidia and basidial shape suggest this is a *Cystostereum* (284).

Hypochnicium vellereum (Ellis & Cragin) Parm. 1968
Consp. Syst. Cort., 116.
Basionym *Corticium vellereum* Ellis & Cragin, J. Mycol. 1:58, 1885 and Bull. Washburn College Lab. Nat. Hist. 1:66, 1885.
Distribution Canada: AB, BC, MB, NB, NS, ON, PQ. United States: AL, CA, GA, ID, IL, KS, MA, MI, MO, MN, MT, NH, NY, OH, PA, SD, TX, VT, WA, WI.
Hosts *Acer* sp., *A. rubrum, A. saccharum, Betula alleghaniensis, Celtis occidentalis, Fagus* sp., *F. grandifolia, Malus* sp., *Populus* sp., *P. balsamifera, P. tremuloides, P. trichocarpa, Thuja plicata, Ulmus* sp., *U. pumila*, angiosperm.
Ecology The most common fungus isolated from trunk rot in *Acer saccharum*; basidiomes produced on bark and wood decaying on the ground; associated with a white rot.
Culture Characters
1.3c.7.34.36.38.43-44.(48).(50).(53).54.60 (Ref. 474); 1.3.7.34.36.38.43.44.48.54.60 (Ref. 487).
References 93, 116, 183, 242, 249, 373, 406, 474, 490, 603.

Hypochnicium versatum (Burt) Ginns 1992
Mycotaxon 44:214.
Basionym *Peniophora versata* Burt, Ann. Missouri Bot. Gard. 12:305, 1926.
Synonym *Corticium versatum* (Burt) D.P. Rogers & Jackson 1943
Distribution United States: WA.
Hosts *Abies magnifica*, gymnosperm.
Ecology Slightly decayed planks and cross-ties; type of rot uncertain.
Culture Characters Not known.
References 92, 245.

The report (116) from British Columbia was based upon a collection of *Gloeocystidiellum ochraceum*, fide Ginns (245).

HYPOCHNOPSIS Karsten 1889 — Atheliaceae

Hypochnopsis mustialaensis (Karsten) Karsten 1889
Bidrag Kännedom Finlands Natur Folk 48:442.
Basionym *Hypochnus mustialaensis* Karsten, Not. Sällsk. Fauna Fl. Fenn. Förh. 8:222, 1870.
Synonyms *Coniophora mustialaensis* (Karsten) Massee 1889, *Coniophora cyanospora* D.P. Rogers 1935, fide Rogers and Jackson (567), and Ginns (243).
Distribution Canada: BC, MB, ON, PQ.
United States: AZ, IA, ID, MS, NJ, NM, OH.
Hosts *Abies balsamea, A. concolor, Acer saccharum, Alnus tenuifolia, Betula* sp., *Picea glauca, P. mariana, Pinus ponderosa, P. strobus, Quercus* sp., *?Thuja* sp., *Tsuga heterophylla*, angiosperm.
Ecology Well-rotted wood and bark of logs and fallen limbs; decayed leaves; apparently associated with a brown rot.
Culture Characters Not known.
References 57, 197, 208, 216, 242, 243, 563.

We suspect that this fungus is mycorrhizal.

INTEXTOMYCES Eriksson & Ryv. 1976 — Hyphodermataceae

Intextomyces contiguus (Karsten) Eriksson & Ryv. 1976
Cort. N. Europe 4:737.
Basionym *Corticium contiguum* Karsten, Meddeland. Soc. Fauna Fl. Fenn. 2(1):39, 1881.
Synonyms *Corticium crustaceum* (Karsten) Karsten 1903, fide Eriksson and Ryvarden (174), *Corticium crustaceum* (Karsten) Höhnel & Litsch. 1906 superfluous combination, *Corticium subcinereum*

Burt 1926, fide Rogers and Jackson (567).
Distribution Canada: AB-NT, BC, MB, NTM, ON, PQ.
United States: FL, KS, MA, MN, VT, WA, WV.
Hosts *Abies* sp., *Acer* sp., *Betula* sp., *B. papyrifera, Cornus* sp., *Crataegus* sp., *Fraxinus nigra, F. oregona, F. pennsylvanica, Populus* sp., *P. trichocarpa, Prunus* spp., *Quercus macrocarpa, Salix* sp., *Syringa* sp., *Ulmus* sp., *Viburnum* sp.
Ecology Dead bark.
Culture Characters
(1).2.3c.7.32.36.38.47.54 (Ref. 474).
References 93, 116, 133, 174, 242, 354, 474.

IRPEX Fr. 1825 Steccherinaceae

Irpex lacteus (Fr.:Fr.) Fr. 1825
Elenchus 1:145.
Basionym *Sistotrema lacteum* Fr., Obs. Mycol. 2:266, 1818.
Synonym *Irpiciporus lacteus* (Fr.) Murrill 1907, *Irpex rimosus* Peck 1890, fide Gilbertson (189), *Irpex tulipiferae* (Schw.) Schw. 1832, fide Maas Geesteranus (428), *Polyporus tulipiferae* (Schw.) Overh. 1915.
Distribution Canada: AB, BC, MB, NB, NS, NTM, ON, PE, PQ. United States: AL, AR, AZ, CA, CO, CT, DC, FL, GA, IA, ID, IN, KS, KY, LA, MA, MD, ME, MI, MN, MO, MT, NC, ND, NE, NH, NJ, NM, NY, OH, OK, OR, PA, SC, SD, TN, VA, VT, WA, WI, WV, WY.
Hosts *Abies grandis, Acer* sp., *A. negundo, A. nigrum, A. rubrum, A. saccharinum, A. saccharum, A. spicatum, Albizia julibrissin, Alnus* sp., *A. incana, A. rubra, A. rugosa* var. *americana, Amelanchier* sp., *Aralia spinosa, Baccharis halimifolia, Betula* sp., *B. alleghaniensis, B. papyrifera, B. populifolia, Buddleia* sp., *Calycanthus floridus, Caragana* sp., *Carpinus* sp., *C. caroliniana, Carya* sp., *Castanea crenata, C. mollissima, C. sativa, Catalpa* sp., *Celtis* sp., *Citrus* sp., *Cornus* sp., *Corylus* sp., *Fagus* sp., *F. grandifolia, F. sylvatica, Forsythia* sp., *Fraxinus* sp., *F. nigra, F. pennsylvanica, Ginkgo biloba, Gleditsia triacanthos, Gymnocladus* sp., *Juglans* sp., *J. cinerea, Liquidambar* sp., *Liriodendron tulipifera, Magnolia* sp., *Maclura* sp., *Malus* sp., *M. domestica, M. pumila, M. sylvestris, Morus* sp., *Nemopanthus mucronatus, Picea* sp., *Pinus* sp., *P. contorta, P. lambertiana, P. ponderosa, Platanus* sp., *Populus* sp., *P. balsamifera, P. deltoides, P. grandidentata, P. tremuloides, P. trichocarpa, Prunus* sp., *P. americana, P. armeniaca, P. avium, P. cerasus, P. emarginata, P. pensylvanica, P. persica, P. serotina, P. virginiana, Pseudotsuga menziesii, Pyrus* sp., *Quercus* sp., *Q. alba, Q. arizonica, Q. gambelii, Q. rubra* var. *borealis, Rhus* sp., *R. copallina, R. glabra, R. typhina, R. vernix, Ribes* sp., *Robinia pseudoacacia, Rosa* sp., *Rubus* sp., *Sabal* sp., *Salix* sp., *Sassafras* sp., *Sorbus* sp., *S. aria, Symphoricarpos occidentalis, Tilia americana, Tsuga canadensis, T. heterophylla, Ulmus americana, Vitis labrusca, V. vulpina*, angiosperm, gymnosperm.
Ecology Bark; decaying wood; fallen branches; slash; often on dead branches or trunks of standing dead trees; log; test stakes; wooden porch chairs; associated with a white rot.
Culture Characters
(1).2.6.8.(16).32.36.(37).40.42.(50).(53).54.(55).(57) (Ref. 474);
1.(2).6.7.(8).14.32.36.40.42-43.(48).53.54 (Ref. 487);
(1).2.6.8.12.32.36.38.41-42.54 (Ref. 150).
References 103, 127, 177, 179, 183, 189, 217, 242, 249, 327, 332, 333, 334, 402, 410, 464, 474, 484, 534, 609, 611.

JAAPIA Bres. 1911 Epitheleaceae

Jaapia argillacea Bres. 1911
Ann. Mycol. 9:428.
Distribution Canada: BC. United States: ID.
Host *Populus tremuloides*.
Ecology Wet wood, typically partially submerged, along shores of lakes and streams.
Culture Characters
1.3.7.32.(36).(39).40.43-45.(54).(55) (Ref. 588).
References 174, 563.

Jaapia ochroleuca (Bres.) Nannf. & Eriksson 1953
Svensk Bot. Tidskr. 47:184.
Basionym *Coniophora ochroleuca* Bres. in Brinkmann, Jahresber. Westfal. Prov.-Ver., Wiss. Kunst. 26: 130, 1898.
Synonym *Pellicularia ochroleuca* (Bres.) D.P. Rogers 1943, *Peniophora ochroleuca* (Bres.) Höhnel & Litsch. 1908.
Distribution United States: IL, MA, NC, OR.
Hosts *Abies fraseri, Acer* sp., *Pinus* sp., *Pseudotsuga mucronata*, angiosperm, gymnosperm.
Ecology Log.
Culture Characters Not known.
References 267, 327, 565.

KAVINIA Pilát 1938 Beenakiaceae

Kavinia alboviridis (Morgan) Gilbn. & Budington 1970
J. Arizona Acad. Sci. 6:95.
Basionym *Hydnum alboviride* Morgan, J. Cincinnati Soc. Nat. Hist. 10:12, 1887.
Synonym *Mycoacia alboviridis* (Morgan) L.W. Miller & Boyle 1943, *Oxydontia alboviride* (Morgan) L.W. Miller 1933.
Distribution Canada: BC, MB, NS, ON, PQ.
United States: AZ, CA, CO, CT, IA, ID, MA, MI, NM, NY, OH, TN.
Hosts *Abies balsamea, A. concolor, A. grandis, A. magnifica, Fagus* sp., *Picea* sp., *P. rubens, Pinus contorta, P. ponderosa, Populus* sp., *P. tremuloides, Pseudotsuga menziesii, Quercus gambelii, Thuja*

occidentalis, angiosperm, gymnosperm.
Ecology Rotted log; associated with a white rot.
Culture Characters
2.3c.26.(31d).32.36.39.47.54.55 (Ref. 474).
References 116, 120, 183, 197, 204, 208, 216, 242, 249, 402, 406, 463, 474, 528.

Kavinia himantia (Schw.) Eriksson 1958
Symb. Bot. Upsal. 16 (1):160.
Basionym *Hydnum himantia* Schw., Schriften Naturf. Ges. Leipzig 1:104, 1822.
Synonym *Mycoacia himantia* (Schw.) L.W. Miller & Boyle 1943, *Hydnum subfuscum* Peck 1887, fide Gilbertson (189), ?*Hydnum murraii* Berk. & Curtis 1873, fide Gilbertson (192).
Distribution Canada: BC, NS, ON. United States: AZ, CO, IA, MA, NC, NM, NY, OR, PA, TN, WA, WI.
Hosts *Abies concolor, Acer* sp., *Arbutus arizonica, A. menziesii, Ostrya virginiana, Pinus ponderosa, Quercus* sp., *Q. arizonica, Q. gambelii, Q. garryana, Tilia americana*, angiosperm, gymnosperm.
Ecology Leaves; old bark and wood; associated with a white rot.
Culture Characters
2.3c.26.31d.32.36.38.47.54.55 (Ref. 474).
References 116, 133, 183, 189, 192, 197, 204, 208, 216, 242, 249, 400, 402, 463, 474, 528.

LACHNELLA Fr. 1835 — Lachnellaceae

Lachnella alboviolascens (Alb. & Schw.:Fr.) Fr. 1849
Summa Veg. Scand. 2:365.
Basionym *Peziza alboviolascens* Alb. & Schw., Conspectus Fungorum 322, 1805.
Synonym *Cyphella alboviolascens* (Alb. & Schw.) Karsten 1869, *Cyphella pezizoides* Zopf in A.P. Morgan 1888, fide W.B. Cooke (125) and Donk (155).
Distribution Canada: BC, ON.
United States: AZ, CA, DE, ID, MA, MD, ME, MO, OH, PA, SC, VA, WA, WI.
Hosts *Arctostaphylos pungens, Ceanothus integerrimus, Cytisus scoparius, Equisetum arvense, Fouquieria splendens, Garrya wrightii, Helianthus* sp., *Juniperus monosperma, Lupinus arboreus, Lycium* sp., *Platanus wrightii, Salix* sp., *Sambucus* sp., *Syringa vulgaris*, angiosperm.
Ecology Bark and branches of trees and shrubs; woody herbaceous stems; litter and duff; associated with a white rot.
Culture Characters
1.3.24.26.34.36.40.47.53.54 (Ref. 477).
References 80, 93, 119, 125, 131, 211, 216, 242, 477, 545.

Lachnella manitobensis W.B. Cooke 1962
Beih. Sydowia 4:74.
Distribution Canada: MB.
Hosts Unidentified wood and fern.
Ecology Dead sticks and fern fronds.
Culture Characters Not known.

Known only from the type specimen, which could not be found (5). Cooke's description suggested, to Agerer (5), that *Henningsomyces* was a more appropriate genus for this fungus.

Lachnella tiliae (Peck) Donk 1951
Lilloa 22:345.
Basionym *Peziza tiliae* Peck, Annual Rep. New York State Mus. 24:96, 1872.
Synonym *Cyphella tiliae* (Peck) Cooke 1891.
Distribution Canada: MB, ON, PQ.
United States: IA, ME, MI, MN, MO, ND, NJ, NY, OH, OR, UT, VT, WI.
Hosts *Tilia* sp., *T. americana, Ulmus* sp.
Ecology Bark.
Culture Characters Not known.
References 80, 116, 125, 131, 242, 378.

Lachnella villosa (Pers.:Fr.) Gill. 1880
Champ. France Discomycetes, 80.
Basionym *Peziza villosa* Pers., Synop. Meth. Fung., 655, 1801.
Synonym *Cyphella villosa* (Pers.) Crouan 1867.
Distribution Canada: BC, ON. United States: CA, CO, DE, GA, OH, LA, MO, OR, SC, VA.
Hosts *Artemisia* sp., *Cirsium arvense, Encelia californica, Helianthus* sp., *Lupinus arboreus, Salix* sp., *Solidago* sp., *Urtica* sp.
Ecology Twigs of woody plants; herbaceous litter.
Culture Characters Not known.
References 80, 125, 131, 242, 402, 465, 545.

LAETISARIA Burdsall 1979 — Vuilleminiaceae

Laetisaria agavei Burdsall & Gilbn. 1982
Mycotaxon 15:336.
Distribution United States: AZ.
Hosts *Agave* sp., *A. chrysantha, A. parryi.*
Ecology Dead basal leaves.
Culture Characters
(1).2.6.16.32.36.(38).40.43.54 (Ref. 474).
References 67, 474.

Laetisaria arvalis Burdsall 1980
Mycologia 72:729.
Distribution United States: NE, OH.
Hosts *Moniliopsis (Rhizoctonia) solani, Pythium ultimum* Trow.
Ecology Apparently mycoparasitic on *M. solani* and possibly a biocontrol agent of *P. ultimum*; soil inhabiting, soil debris.
Culture Characters
2.5.16.23.36.38.41.(48).50.56 (Ref. 68).

Known only from the type specimens.

Laetisaria fuciformis (D. McAlpine) Burdsall 1979
Trans. Brit. Mycol. Soc. 72:422.
Basionym *Hypochnus fuciformis* D. McAlpine, Ann. Mycol. 4:549, 1906.
Synonym *Corticium fuciforme* ("Berk.") Wakef. 1916.
Anamorph *Isaria fuciformis* Berk. 1872.
Distribution United States: MD, RI.
Hosts Grasses.
Ecology Causes red thread disease of sports turf and agricultural grasses; attacks most of the common cool-season turf and agricultural grasses.
Culture Characters
(1).(2).6.16.35.36.(38).(40).42.43.56 (Ref. 592).
References 94, 183, 329, 592.

Because the name *Corticium fuciforme* has been applied to two fungi (*L. fuciformis* and *Limonomyces roseipellis*), we cite only those records which state that the fungus lacked clamp connections. Farr et al. (183), Smith et al. (585), and Stalpers and Loerakker (592) erred in citing Sprague's (586) records as *Laetisaria fuciformis*, because Sprague stated that the fungus had clamp connections.

LAURILIA Pouzar 1959 — Echinodontiaceae

Laurilia sulcata (Burt) Pouzar 1959
Ceská Mykol. 13:14.
Basionym *Stereum sulcatum* Burt in Peck, Annual Rep. New York State Mus. 54:154, 1901.
Synonym *Lloydella sulcata* (Burt) Lloyd 1916, *Echinodontium sulcatum* (Burt) H. Gross 1964.
Distribution Canada: AB, AB-NT, BC, ON, PQ.
United States: AK, CO, ID, LA, MI, MT, NC, NH, NY, OR, PA, TN, TX, UT, VT, WA, WI, WV.
Hosts *Abies* sp., *A. amabilis, A. grandis, A. lasiocarpa, Larix* sp., *L. lyallii, L. occidentalis, Picea* sp., *P. engelmannii, P. glauca, P. pungens, P. rubens, P. sitchensis, Pinus* sp., *P. contorta, P. monticola, P. ponderosa, P. strobus, Pseudotsuga* sp., *P. menziesii, Taxodium* sp., *Thuja plicata, Tsuga* sp., *T. canadensis, T. heterophylla, T. mertensiana*, gymnosperm.
Ecology Causes a minor trunk rot in *Picea engelmannii* and some root rot in *P. glauca*; basidiomes produced on bark; broken or cut wood; logs and stumps; butt; associated with a white pocket rot of dead wood.
Culture Characters
2.3c.8.(13).15.26.(27).33.(34).37.38.(39).46-47.(50).55.59 (Ref. 474);
2.3.8.(13).33.36.38.39.47.55.59 (Ref. 487);
2.3c.8.(13).33.36.(37).38.39.46-47.(50).55.59 (Ref. 104).
References 89, 97, 104, 116, 133, 145, 183, 242, 251, 327, 333, 379, 402, 448, 450, 462, 474, 511, 519.

Laurilia taxodii (Lentz & McKay) Parm. 1968
Consp. Syst. Cort., 180.
Basionym *Stereum taxodii* Lentz & McKay, Mycologia 52:262, 1960.
Synonym *Echinodontium taxodii* (Lentz & McKay) H. Gross 1964.
Distribution United States: FL, GA, LA, MI, MS, NC, TX.
Host *Taxodium distichum.*
Ecology Causes pocket rot, known as pecky cypress, in live trees, associated with a white rot.
Culture Characters
(1).2.3c.8.13.34.36.38.(39).47.(50).55.58 (Ref. 474).
References 145, 251, 474.

LAXITEXTUM Lentz 1955 — Gloeocystidiellaceae

Laxitextum bicolor (Pers.:Fr.) Lentz 1955
Agric. Monogr. U.S.D.A. 24:19.
Basionym *Thelephora bicolor* Pers., Synop. Meth. Fung., 568, 1801.
Synonym *Stereum bicolor* (Pers.) Fr. 1838, *Stereum coffeatum* Berk. & Curtis 1873, fide Lentz (379), *Stereum fuscum* (Schrader) Quél. (sic) 1888, fide Lentz (379).
Distribution Canada: BC, MB, NS, ON, PQ.
United States: AL, AR, CT, DC, DE, FL, GA, IA, ID, IN, KY, LA, MD, ME, MN, MO, MS, NC, NH, NJ, NY, OH, OR, PA, SC, TN, TX, VA, VT, WA, WI, WV.
Hosts *Acer* sp., *A. macrophyllum, A. negundo, A. nigrum, A. rubrum* var. *rubrum, A. saccharinum, A. saccharum, Aesculus hippocastanum, Alnus* sp., *A. oblongifolia, Betula alba, B. alleghaniensis, B. lenta, B. papyrifera, Castanea* sp., *Celtis* sp., *C. laevigata, Fagus* sp., *F. grandifolia, Fraxinus* sp., *F. americana, Liquidambar styraciflua, Liriodendron tulipifera, Magnolia virginiana, Malus* sp., *Nyssa* sp., *N. aquatica, Pinus* sp., *P. taeda, Platanus occidentalis, Populus* sp., *P. tremuloides, P. trichocarpa, Prunus pensylvanica, P. persica, Quercus* sp., *Q. chrysolepis, Q. nigra, Q. phellos, Q. rubra, Q. wislizenii, Salix* sp., *Tsuga heterophylla, Ulmus* sp., *U. americana*, usually angiosperm, rarely gymnosperm.
Ecology Rotting limbs; dead trees; log; associated with a white rot.
Culture Characters
2.3c.(15).(21).34.36.(37).(38).(39).(40).42-43.(50).(51).(53).54.(55).(57).(59).(60) (Ref. 474);
2.3c.15.(34).36.37.39.44-46.(48).54.(55).60.61 (Ref. 104).
References 89, 96, 104, 108, 116, 183, 242, 249, 333, 379, 474, 519, 611.

Laxitextum incrustatum Hjort. & Ryv. 1981
Mycotaxon 13:35.
Synonym *Gloeocystidiellum sinuosum* Freeman 1981 (see below).
Distribution United States: AL, AR, FL, LA, MS.
Hosts *Pinus* sp., *Quercus* sp., *Q. stellata, Salix* sp.

Ecology Branches on ground.
Culture Characters
2.3.15.34.36.43.54.55 (Ref. 76).
Reference 76.

Known only from the type specimens. We have seen the holotypes and all paratypes for *L. incrustatum* and *G. sinuosum*. They are the same fungus and *G. sinuosum* becomes a synonym because the epithet *incrustatum* was published in May 1981, whereas the epithet *sinuosum* was not published until December 1981.

LAZULINOSPORA Burdsall & Larsen 1974
Thelephoraceae

Lazulinospora cinnamomea Burdsall & Nakasone 1981
Mycologia 73:464.
Distribution United States: FL.
Hosts *Magnolia* sp., *Serenoa repens*.
Ecology Not known.
Culture Characters Not known.

Known only from the holotype and one paratype specimen.

Lazulinospora wakefieldii Burdsall & Larsen 1974
Mycologia 66:98.
Distribution United States: FL.
Host *Quercus nigra*.
Ecology Bark.
Culture Characters Not known.

Known only from the type specimen.

LEPTOSPOROMYCES Jülich 1972 Atheliaceae

Leptosporomyces fuscostratus (Burt) Hjort. 1987
Windahlia 17:58.
Basionym *Corticium fuscostratum* Burt, Ann. Missouri Bot. Gard. 13:299.
Synonyms *Athelia fuscostratum* (Burt) Donk 1957, *Confertobasidium olivaceo-album* (Bourd. & Galzin) Jülich sensu Jülich, and N. Amer. Aucts., fide Hjortstam (286).
Distribution Canada: AB-NT, BC, NS, ON, PQ. United States: AZ, CO, CT, FL, ID, IL, MA, MD, MI, MN, MO, NH, NM, NY, OR, PA, WI.
Hosts *Abies balsamea, A. concolor, A. lasiocarpa* var. *arizonica, Acer glabrum, A. saccharum, Picea* sp., *P. engelmannii, Pinus* sp., *P. banksiana, P. contorta, P. ponderosa, P. resinosa, P. strobus, P. taeda, Populus* sp., *P. tremuloides, Pseudotsuga menziesii, Thuja plicata*, angiosperm, gymnosperm.
Ecology A minor decay fungus in *Pinus strobus*; perhaps psychrophilic; basidiomes produced on woody debris; bark.
Culture Characters
1.(2).3c.21.35.36.39.47.(53).55.58 (Ref. 474); (1).(2a).3c.7.33.37.38.47.55.58.61 (Ref. 257).
References 30, 93, 116, 176, 183, 196, 197, 204, 208, 214, 216, 242, 249, 274, 312, 393, 406, 447, 474.

Leptosporomyces galzinii (Bourd.) Jülich 1972
Beih. Willdenowia 7:192.
Basionym *Corticium galzinii* Bourd., Rev. Sci. Bourbonnais Centr. France 23:11, 1910.
Synonym *Athelia galzinii* (Bourd.) Donk 1957, *Corticium canum* Burt 1926, fide Rogers and Jackson (567), *Athelia grisea* M. Christiansen 1960, fide Jülich (312).
Distribution Canada: AB, BC, MB, NTM, ON, PQ. United States: AZ, CA, CO, CT, ID, LA, MD, ME, MT, NH, NJ, NM, NY, WA.
Hosts *Abies balsamea, A. concolor, A. magnifica, Alnus oblongifolia, Betula* sp. *Heterobasidion (Fomes) annosum, Picea* sp., *P. engelmannii, P. glauca, Pinus ponderosa, Pseudotsuga menziesii, Quercus* sp., *Q. hypoleucoides, Tsuga mentensiana*, angiosperm, gymnosperm.
Ecology Perhaps psychrophilic; basidiomes produced primarily on decaying wood and bark; associated with a white rot.
Culture Characters
2.3.7.32.36.38.45-47.(54).55 (Ref. 588).
References 93, 128, 130, 183, 196, 197, 204, 208, 216, 242, 312, 402.

Leptosporomyces montanum (Jülich) Ginns & Lefebvre, *comb. nov.*
Basionym *Confertobasidium olivaceo-album* var. *montanum* Jülich, Beih. Willdenowia 7:174, 1972.
Synonym *Confertobasidium montanum* (Jülich) Jülich & Stalpers 1980.
Distribution United States: OR.
Hosts *Abies lasiocarpa, Larix occidentalis*.
Ecology Not known.
Culture Characters Not known.

Known in North America only from the holotype and a paratype. Concluding that the generic distinctions between *Confertobasidium* Jülich and *Leptosporomyces* were "vague", Hjortstam (286) tentatively incorporated the *Confertobasidum* epithets under *Leptosporomyces*. We propose this combination to conform with Hjortstam's arrangement of the species.

Leptosporomyces raunkiaerii (M. Christiansen) Jülich 1972
Beih. Willdenowia 7:206.
Basionym *Athelia raunkiaerii* M. Christiansen, Dansk Bot. Ark. 19:153, 1960.
Distribution Canada: ON. United States: CA, NJ, IA.
Hosts *Abies magnifica, Populus* sp., gymnosperm.
Ecology Rotting wood.
Culture Characters Not known.
References 128, 130, 312, 406.

LEUCOGYROPHANA Pouzar 1958 Coniophoraceae

Leucogyrophana arizonica Ginns 1978
Canad. J. Bot. 56:1955.
Distribution United States: AZ, MD, MT, NM, NY, TN.
Hosts *Abies concolor, Pinus* spp., *P. ponderosa, P. strobiformis, P. ?strobus, P. virginiana, Pseudotsuga menziesii,* gymnosperm.
Ecology Well decayed wood; rather solid dead wood; associated with a brown rot.
Culture Characters
1.(2).3c.7.(16).35.36.(37).38.(39).47.(53).55 (Ref. 474);
1.3.16.22.(32).35.36.38.47.55 (Ref. 228).
References 199, 228, 474.

Leucogyrophana mollusca (Fr.) Pouzar 1958
Ceská Mykol. 12:33.
Basionym *Merulius molluscus* Fr., Syst. Mycol. 1:329, 1821.
Synonym *Merulius subaurantiacus* Peck 1885, fide Ginns (228), *Leucogyrophana pseudomollusca* (Parm.) Parm. 1967, fide Ginns (228).
Distribution Canada: AB, BC, NS, NTM, ON, PQ. United States: AK, AL, AZ, CA, CO, ID, IN, MA, ME, MI, MT, NH, NM, NY, OR, PA, TN, VT, WA.
Hosts *Abies* sp., *A. balsamea, A. concolor, A. magnifica, Betula* sp., *Castanea dentata, Fagus* sp., *F. grandifolia, Larix* sp., *L. occidentalis, L. sibiricae, Picea* sp., *P. abies, P. engelmannii, P. glauca, P. sitchensis, Pinus* sp., *P. ayacahuite, P. contorta, P. densiflora, P. engelmannii, P. monticola, P. ponderosa, P. strobiformis, P. sylvestris, P. strobus, Populus* sp., *P. tremuloides, P. trichocarpa, Pseudotsuga* sp., *P. menziesii, Rhododendron* sp., *Thuja plicata, T. occidentalis, Tsuga* sp., *T. canadensis, T. heterophylla, T. mertensiana, Ulmus* sp., *U. americana*, angiosperm, gymnosperm.
Ecology Wood and bark, leaves, needles, sawdust, soil, discarded timbers, in buildings, greenhouses, and cellars, associated with a brown rot.
Culture Characters
1.(2).3c.(9).16.21.23.26.32.36.(38).(39).46-47.51.54.55 (Ref. 474);
1.3.16.22.(23).32.36.(37).38-39.(46).47.(53).(54).55 (Refs. 228 & 248).
References 57, 183, 197, 199, 204, 208, 216, 228, 242, 248, 249, 333, 402, 474.

Leucogyrophana montana (Burt) Domanski 1975
Mala fl. grzybow 1(2):57. Polska Akad. Nauk, Krakow.
Basionym *Merulius montanus* Burt, Ann. Missouri Bot. Gard. 4:354, 1917.
Distribution United States: ID, NY.
Hosts *Pinus contorta, P. monticola, P. strobus,* gymnosperm.
Ecology Well decayed wood, associated with a brown rot.
Culture Characters Not known.
References 199, 228.

Leucogyrophana olivascens (Berk. & Curtis) Ginns & Weresub 1976
Mem. New York Bot. Gard. 28:96.
Basionym *Corticium olivascens* Berk. & Curtis, Grevillea 1:179, 1873.
Synonym *Coniophora olivascens* (Berk. & Curtis) Massee 1889, *Corticium chlorinum* Berk. & Curtis 1873, *Corticium prasinum* Berk. & Curtis 1873, *Coniophora subochracea* Peck 1897. All fide Ginns (228).
Distribution Canada: NS, ON.
United States: AL, AZ, CO, CT, DE, FL, GA, ID, LA, MA, MD, ME, MI, NC, NH, NJ, NM, NY, NC, OH, PA, TX, VA, VT, WI.
Hosts *Picea pungens, Pinus* sp., *P. ponderosa, P. strobus, Platanus occidentalis, P. wrightii, Populus* sp., *P. deltoides, Quercus* sp., *Taxodium distichum, Tsuga canadensis, Ulmus* sp., angiosperm, gymnosperm.
Ecology Bark; slash; well rotted wood and debris; boards; associated with a brown rot.
Culture Characters
1.3c.(9).16.21.23.26.32.36.38.(39).45-46.54.55 (Ref. 474);
1.3.16.22.23.(26).32.36.(38).39.46.(47).(48).54-55 (Refs. 248 & 228).
References 183, 197, 199, 204, 208, 216, 228, 242, 248, 249, 382, 402, 474.

Leucogyrophana pinastri (Fr.:Fr.) Ginns & Weresub 1976
Mem. New York Bot. Gard. 28:96.
Basionym *Hydnum pinastri* Fr., Nov. Flora Suec. 2:38, 1814 (original not seen, the citation is from Fries, Obs. Myc. 1:149, 1815).
Synonym *Odontia pinastri* (Fr.) Quél. 1882, *Merulius pinastri* (Fr.) Burt 1917, *Serpula pinastri* (Fr.) W.B. Cooke 1957, *Merulius irpicinus* Peck 1894, *Merulius atrovirens* Burt 1917, *Serpula atrovirens* (Burt) W.B. Cooke 1957, *Merulius byssoideus* Burt 1917, *Serpula byssoidea* (Burt) W.B. Cooke 1957. All fide Ginns (228).
Distribution Canada: BC, ON, PQ. United States: AK, AR, AZ, CA, CO, DC, FL, IA, ID, IL, IN, KY, LA, MA, MD, ME, MI, MO, MT, NC, NE, NH, NM, NY, OH, OR, PA, SC, TN, VA, VT, WA, WI, WV.
Hosts *Abies* sp., *A. amabilis, A. concolor, A. pectinata, Alnus* sp., *Corylus californica, Larix americana, L. occidentalis, Liriodendron tulipifera, Picea* sp., *P. abies, P. engelmannii, P. excelsa, P. sitchensis, Pinus* sp., *P. edulis, P. ponderosa, P. silvestris, P. strobus, P. sylvestris, P. virginiana, Populus* sp., *P. tremuloides, P. trichocarpa, Pseudotsuga* sp., *P. menziesii, Quercus* sp., *Salix* sp. *Thuja plicata, Umbellularia californica,* rarely angiosperm, gymnosperm.
Ecology Wood and bark; underside of half-decayed log; pulpwood; discarded boards; compost; litter; soil; in

greenhouses; cellars; crawl spaces; mushroom bed; associated with a brown rot.

Culture Characters
1.(2).3c.(9).16.23.32.36.37.(38).(40).44-47.54.55 (Ref. 474);
1.3.16.22.23.32.36-37.(38).39.46-47.(54).55.56 (Refs. 228 & 248).

References 123, 133, 183, 197, 199, 208, 216, 228, 248, 242, 406, 442, 474, 582.

Leucogyrophana pulverulenta (Sowerby:Fr.) Ginns 1978
Canad. J. Bot. 56:1966.

Basionym *Auricularia pulverulenta* Sowerby, Brit. Fun., pl. 214, 1799.
Synonym *Merulius pulverulentus* (Sowerby) Fr. 1828.
Distribution Canada: BC, NS.
United States: CO, IN, ME, PA, TN.
Hosts *Picea* sp., *P. sitchensis, Pinus* sp., *P. sylvestris, P. ?strobus, Quercus* sp., *Thuja plicata*, angiosperm, gymnosperm.
Ecology Logs; stumps; live trees; boards and timbers in buildings; cellars; mines; associated with a brown rot.
Culture Characters
1.3.7.(26).32.36.38.47.(54).55 (Ref. 228).
References 199, 228, 242, 249.

Leucogyrophana romellii Ginns 1978
Canad. J. Bot. 56:1968.

Distribution Canada: BC, NS, ON, PE, PQ.
United States: AZ, ID, ME, MI, MT, NC, NH, NM, NY, SD, VT, WI.
Hosts *Abies balsamea, Betula* sp., *Picea* sp., *P. abies, P. engelmanni, P. jezoensis, Pinus* sp., *P. albicaulis, P. monticola, P. ponderosa, P. sibirica, P. sylvestris, Pseudotsuga menziesii, Tsuga canadensis*, gymnosperm.
Ecology Associated with a brown rot; rarely on live mosses.
Culture Characters
1.3c.7.(26).32.36.38.(39).47.53.(54).55 (Ref. 474);
1.3.7.32.(36).37.(38).39.47.55 (Ref. 228).
References 199, 228, 242, 249, 474.

Leucogyrophana sororia (Burt) Ginns 1976
Canad. J. Bot. 56:1970.

Basionym *Merulius sororius* Burt, Ann. Missouri Bot. Gard. 4:329, 1917.
Distribution United States: MD, WA.
Hosts *Pinus* sp., gymnosperm.
Ecology Well decayed wood; associated with a brown rot.
Culture Characters Not known.
References 199, 228.

LICROSTROMA Lemke 1964 Licrostromataceae

Licrostroma subgiganteum (Berk.) Lemke 1964
Canad. J. Bot. 42:763.

Basionym *Corticium subgiganteum* Berk., Grevillea 2:3, 1873.
Anamorph *Michenera artocreas* Berk. & Curtis 1868, fide Lemke (376).
Distribution Canada: ON, PQ. United States: AL, CT, ME, MN, NC, NH, NJ, NY, PA, SC, VA, VT, WI.
Hosts *Acer* sp., *A. rubrum, Fagus* sp., *Magnolia* sp., *Liriodendron* sp., angiosperm.
Ecology Principally on bark; associated with a white rot.
Culture Characters
2.6.(8).15.32.36.38.45-46.(50).54 (Ref. 474).
References 93, 183, 242, 376, 474, 511.

LIMONOMYCES Stalpers & Loerakker 1982 Vuilleminiaceae

Limonomyces culmigenus (J. Webster & D. Reid) Stalpers & Loerakker 1982
Canad. J. Bot. 60:536.

Basionym *Exobasidiellum culmigenum* J. Webster & D. Reid, Trans. Brit. Mycol. Soc. 52:20, 1969.
Synonym *Galzinia culmigena* (J. Webster & D. Reid) Johri & Bandoni 1975.
Distribution Canada: BC.
Hosts *Carex* sp., *Dactylus glomerata.*
Ecology Parasitic; basidiomes produced on overwintered culms and senescent leaves.
Culture Characters
2.3.(7).(16).26.32.36.38.47.48.56.(57).(58) (Ref. 592).
References 22, 592.

Limonomyces roseipellis Stalpers & Loerakker 1982
Canad. J. Bot. 60:534.

Synonym *Corticium fuciforme* ("Berk.") Wakef. sensu Wakef. 1917.
Distribution United States: MD, MS, OR, RI, WA.
Hosts *Agrostis alba, A. canina, A. palustris, A. tenuis, Briza media, Cynodon dactylon, Festuca myuros, F. ovina* var. *duriuscula, F. rubra, F. rubra* var. *commutata, Poa annua, P. pratensis.*
Ecology Causes pink patch disease of turf grasses.
Culture Characters
2.3.(20).(22).(25).(26).(35).36.38.41.42.56 (Ref. 592).
References 94, 184, 329, 586.

LINDTNERIA Pilát 1938 Stephanosporaceae

Synonym is *Cyanobasidium* Jülich 1979, fide Hjortstam (287).

Lindtneria baboquivariensis (Gilbn.) Gilbn. & Ryv. 1986
N. Amer. Polypores 1:429.

Basionym *Poria baboquivariensis* Gilbn., Mycotaxon 3:358, 1976.
Distribution United States: AZ.
Host *Prosopis juliflora.*
Ecology Dead wood; associated with a white rot.
Culture Characters Not known.

Known only from the type specimen.

Lindtneria chordulata (D.P. Rogers) Hjort. 1987
Mycotaxon 28:33.
Basionym *Pellicularia chordulata* D.P. Rogers, Farlowia 1: 98, 1943.
Synonym *Botryohypochnus chordulatus* (D.P. Rogers) Burdsall & Nakasone 1981.
Distribution Canada: AB-NT, NTM, ON. United States: FL, IA, OH.
Hosts *Picea glauca, Populus tremuloides, Quercus virginiana, Salix* sp., *Tilia americana.*
Ecology Dead bark and wood.
Culture Characters Not known.
References 74, 242, 565.

Type species of *Cyanobasidium* Jülich.

Lindtneria flava Parm. 1968
Izv. Akad. Nauk Estonsk. SSR, Ser. Biol. 17:408.
Distribution United States: NY.
Host *Fagus* sp.
Ecology Dead wood; associated with a white rot.
Culture Characters Not known.
Reference 217.

Lindtneria leucobryophila (Henn.) Jülich 1977
Persoonia 9:418.
Basionym *Thelephora leucobryophila* Henn., Verh. Bot. Vereins Prov. Brandenburg 39:96, 1898.
Synonym *Trechispora leucobryophila* (Henn.) Liberta 1966, *Trechispora variecolor* (Bourd. & Galzin) Liberta 1966, fide Liberta (395).
Distribution Canada: BC, ON. United States: IA, MN, OR.
Hosts *Alnus* sp., *Pinus resinosa.*
Ecology Not known.
Culture Characters Not known.
References 214, 242, 395.

Lindtneria rugospora (W.B. Cooke) Larsen 1986
Nova Hedwigia 43:258.
Basionym *Serpula rugospora* W.B. Cooke, Mycologia 49:214, 1957.
Distribution United States: KY.
Hosts Not known.
Ecology Rotten wood.
Culture Characters Not known.

Known only from the type specimen, which Larsen (366) redescribed. Our notes on the microscopic features of the type specimen (Lloyd 5804 at BPI) are: hyphae 2.5-5.5(-7) μm diam. with both single clamp connections and simple septa, hyaline, thin-walled; immature basidia clavate, almost pedicillate, 23-30 x 8-10 μm; basidiospores broadly ellipsoid, 6-7.5(-10) x 5.5-7 μm, the wall hyaline or very pale yellow, aculeate, the ornamentation of short ridges but some (?immature) spores were smooth.

Lindtneria thujatsugina Larsen 1986
Nova Hedwigia 43:258.
Distribution United States: ID.
Hosts Not known.
Ecology Growing in and on duff, around a brown-rotted gymnosperm stump under live standing *Thuja plicata* and *Tsuga heterophylla.*
Culture Characters Not known.

Known only from the type specimen.

Lindtneria trachyspora (Bourd. & Galzin) Pilát 1938
Stud. Bot. Cech. 1:72.
Basionym *Poria trachyspora* Bourd. & Galzin, Hym. France, 659, 1928.
Distribution Canada: ON. United States: FL, IA, IL, IN, MI, NC, NY, PA, TN.
Hosts Angiosperm, occasionally gymnosperm.
Ecology Very rotted wood.
Culture Characters Not known.
References 217, 416.

LITSCHAUERELLA Oberw. 1965 Tubulicrinaceae

Litschauerella clematidis (Bourd. & Galzin) Eriksson & Ryv. 1976
Cort. N. Europe 4:839.
Basionym *Peniophora clematitis* (sic) Bourd. & Galzin, Bull. Soc. Mycol. France 28:383, 1912.
Synonym *Peniophora abietis* (Bourd. & Galzin) Bourd. & Galzin 1928, fide Weresub (635).
Distribution Canada: ON. United States: FL, NY.
Hosts *Juniperus virginiana*, angiosperm.
Ecology Associated with a white rot.
Culture Characters Not known.
References 174, 202, 584.

Slysh (584) stated that L.K. Weresub had identified his New York collection of *P. abietis*. We follow Weresub's concept of *P. abietis* and list the name in synonymy.

LOPHARIA Kalchbr. & MacOwan 1881 Stereaceae

Lopharia cinerascens (Schw.) G.H. Cunn. 1956
Trans. Roy. Soc. New Zealand, Bot. 83:622.
Basionym *Thelephora cinerascens* Schw., Trans. Amer. Philos. Soc., N.S. 4:167, 1832.
Synonym *Stereum cinerascens* (Schw.) Massee 1890, *Lloydella cinerascens* (Schw.) Bres. 1901, *Stereum dissitum* Berk. 1873, fide Lentz (379), *Corticium ephebium* Berk. & Curtis 1873, fide Lentz (379) and Hjortstam (288), *Stereum neglectum* Peck 1880, fide Lentz (379), *Peniophora occidentalis* Ellis & Ev. 1897, fide Lentz (379), *Stereum purpurascens* Lloyd 1914, fide Burt (89) and Lentz (379), *Stereum moricola* Berk. 1873, fide Lentz (379).
Distribution Canada: MB, NB, ON, PQ. United States: AL, AR, CA, CO, CT, DC, FL, GA, IA, IL, IN, KS, KY, LA, MA, MD, MI, MN, MO,

MS, MT, ND, NE, NY, OH, OR, PA, RI, SC, SD, TN, TX, VA, VT.
Hosts *Acer* sp., *A. negundo, A. saccharum, Carya* sp., *C. illinoensis, Elaeagnus umbellata, Euonymus atropurpureus, E. europaeus, Magnolia* sp., *Morus* sp., *M. alba, Platanus* sp., *Populus* sp., *Prunus serotina, P. virginiana, Quercus* sp., *Q. agrifolia, Rhamnus frangula, Robinia* sp., *R. pseudoacacia, Salix* sp., *S. amygdaloides, Sambucus canadensis, Tilia* sp., *T. americana, Ulmus* sp., *U. americana, U. davidiana, Umbellularia californica, Vitis* sp., angiosperm.
Ecology Decaying wood; fallen limbs; logs.
Culture Characters
2.3c.8.(13).(26).32.(36).(37).(38).(40).44-45.(48).(53). 54.60 (Ref. 474);
2.3c.8.(13).(26).32.36.38.39.47.54.60.61 (Ref. 104).
References 89, 96, 104, 116, 183, 214, 242, 333, 379, 474, 519, 593, 620.

Lopharia papyrina (Mont.) Boidin 1959
Bull. Mens. Soc. Linn. Lyon 28:210.
Basionym *Stereum papyrinum* Mont. in de la Sagra, Hist. Cuba Pl. Cell., 374, 1845.
Synonym *Lloydella papyrina* (Mont.) Bres. 1901.
Distribution United States: FL.
Hosts Angiosperm.
Ecology Underside of fallen limbs.
Culture Characters Not known.
References 89, 593, 620.

Lopharia spadicea (Pers.:Fr.) Boidin 1959
Bull. Mens. Soc. Linn. Lyon 28:211.
Basionym *Thelephora spadicea* Pers., Synop. Meth. Fung. 568, 1801.
Synonym *Lloydella spadicea* (Pers.) Bres. 1901.
Distribution Specific localities not given.
Hosts Not known.
Ecology Not known.
Culture Characters Only a partial species code is known: 59 (Ref. 36).
References 174, 311.

Reported from "America" (174) and "Nordamerika" (311), and Stevenson and Cash (593) reported specimens in the Lloyd Herbarium from "United States, Canada" and other countries.

LUELLIA Larsson & Hjort. 1974 — Atheliaceae

Luellia recondita (Jackson) Larsson & Hjort. 1974
Svensk Bot. Tidskr. 68:60.
Basionym *Corticium reconditum* Jackson, Canad. J. Res., C, 26:154, 1948.
Distribution Canada: ON.
Host *Pinus strobus*.
Ecology Not known.
Culture Characters Not known.
References 116, 174, 303.

MAIREINA (Pilát) W.B. Cooke 1961 — 'Cyphellaceae'
Donk (156) treated *Maireina* as a section of *Cyphellopsis* but the following names await transfer to a suitable genus. Donk felt that *Maireina* was a "receptacle of several what I consider unrelated species,...."

Maireina cinerea (Burt) W.B. Cooke 1961
Beih. Sydowia 4:85.
Basionym *Solenia cinerea* Burt in Millspaugh and Nuttall, Flora Santa Catalina Island, 315, 1923.
Distribution United States: CA.
Host *Quercus* sp.
Ecology Bark.
Culture Characters Not known.

Known only from the type specimen, which has been redescribed by Burt (91) and Cooke (125).

Maireina fulvescens (Bourd. & Galzin) W.B. Cooke 1961
Beih. Sydowia 4:86.
Basionym *Cyphella fulvescens* Bourd. & Galzin, Hym. France, 162, 1928.
Distribution United States: NM.
Host *Populus tremuloides*.
Ecology Not known.
Culture Characters Not known.
Reference 208.

Maireina ilicis W.B. Cooke 1961
Beih. Sydowia 4:87.
Distribution United States: TN.
Host *Ilex* sp.
Ecology Not known.
Culture Characters Not known.

Known only from the type specimen.

Maireina jacksonii W.B. Cooke 1961
Beih. Sydowia 4:88.
Distribution Canada: ON.
Hosts *Osmunda* sp., *Pteridium* sp.
Ecology Fallen stems.
Culture Characters Not known.

Known only from the type specimen.

Maireina marginata (MacAlpine) W.B. Cooke 1961
Beih. Sydowia 4:89.
Basionym *Cyphella marginata* MacAlpine, Stone Fruit Dis. Australia 120, 1927.
Distribution United States: OR.
Hosts *Cytisus scoparius, Malus* sp., *Prunus* sp., *P. amygdalus, P. persica, Pyrus malus*.
Ecology Dead twigs which are typically on live plants.
Culture Characters Not known.
References 93, 125.

Maireina spiralis (Coker) W.B. Cooke 1961
Beih. Sydowia 4:94.
Basionym *Cyphella spiralis* Coker, J. Elisha Mitchell Sci. Soc. 43:137, 1927.
Distribution United States: NC.
Host *Rosa* sp.
Ecology Fallen canes.
Culture Characters Not known.

Known only from the type specimen.

Maireina subspiralis W.B. Cooke 1961
Beih. Sydowia 4:95.
Distribution United States: MA.
Hosts Not known.
Ecology Rotten wood.
Culture Characters Not known.

Known only from the type specimen.

Maireina texensis (Cooke) W.B. Cooke 1961
Beih. Sydowia 4:95.
Basionym *Cyphella texensis* Berk. & Curtis ex Cooke, Grevillea 20:9, 1893.
Distribution United States: TX.
Host *Quercus* sp.
Ecology Not known.
Culture Characters Not known.

Known only from the type specimen, which was redescribed by Burt (80) and Cooke (125).

Maireina thujae W.B. Cooke 1961
Beih. Sydowia 4:95.
Distribution United States: ID.
Host *Thuja plicata*.
Ecology Wood and bark of branch of windfall.
Culture Characters Not known.

Known only from the type specimen.

MELZERICIUM Hauerslev 1975 Hyphodermataceae

Melzericium udicolum (Bourd.) Hauerslev 1975
Friesia 10:316.
Basionym *Corticium udicolum* Bourd., Rev. Sci. Bourbonnais Centr. France 23:8, 1910.
Synonym *Corticium areolatum* Bres. 1925, fide L.K. Weresub (in litt., after study of type specimen BPI 280273 = Weir 23387).
Distribution Canada: BC, ON. United States: ID, WA.
Hosts *Alnus tenuifolia*, *Populus* sp., *Salix* sp., *Thuja occidentalis*.
Ecology Rotten wood.
Culture Characters Not known.
References 53, 116, 119, 174, 567.

Not *Corticium apiculatum* Bres. (= *Ceraceomyces tessulatus*) as listed by Burt but apparently a good species, fide Rogers and Jackson (567). The Ontario collections on *Thuja occidentalis* were referred, tentatively, to *C. areolatum* by Rogers and Jackson (567).

MERISMODES Earle 1909 Cyphellopsidacea
Synonym is *Phaeocyphellopsis* W.B. Cooke 1961, fide Reid (554).

Merismodes fasciculatus var. ***fasciculatus*** (Schw.) Earle 190
Bull. New York Bot. Gard. 5:406.
Basionym *Cantharellus fasciculatus* Schw., Trans. Amer. Philos. Soc., N.S. 4:153, 1832.
Synonym *Cyphella fasciculata* (Schw.) Berk. & Curtis 1856.
Distribution Canada: BC, MB, NF, NS, NT, ON, PQ, YT. United States: AL, CO, DE, GA, IN, LA, MA, MD, ME, MI, MO, NC, NH, NJ, NY, OR, PA, RI SC, VA, VT, WI, WY.
Hosts *Alnus* sp., *A. crispa* var. *mollis*, *A. oregana*, *A. rubra*, *A. rugosa*, *A. serrulata*, *Corylus* sp., *Populus* sp., *Prunus* sp., *P. virginiana*, *Pyrus malus*, *Salix* sp.
Ecology Bark of branches; dead twigs and branches.
Culture Characters Not known.
References 80, 93, 108, 116, 125, 183, 242, 249, 402, 511.

Merismodes fasciculatus var. ***caroliniensis*** W.B. Cooke 1961
Beih. Sydowia 4:103.
Distribution United States: SC.
Hosts Not known.
Ecology Not known.
Culture Characters Not known.

Known only from the type specimen.

Merismodes fasciculatus var. ***occidentalis*** W.B. Cooke 1961
Beih. Sydowia 4:102.
Distribution United States: CA, CO, ID, OR, WA.
Hosts *Alnus oregona*, *Salix* sp., *Umbellularia californica*.
Ecology Not known.
Culture Characters Not known.

Known only from the original description.

Merismodes fasciculatus var. ***oregonus*** W.B. Cooke 1961
Beih. Sydowia 4:102.
Distribution United States: OR, WA.
Host *Alnus oregona*.
Ecology Not known.
Culture Characters Not known.

Known only from the original description.

Merismodes ***ochraceus*** (Hoffm.:Fr.) D. Reid 1964
Persoonia 3:116.
Basionym *Solenia ochracea* Hoffm., Deutschl. Fl. (Cryptogamic) 2:tab. 8, 1795.

Synonym *Phaeocyphellopsis ochracea* (Hoffm.) W.B. Cooke 1962, *Cyphella mellea* Burt 1914, fide W.B. Cooke (125, 131).
Distribution Canada: BC, MB, NS, ON.
United States: AL, CA, CO, DE, FL, IA, IL, IN, KS, KY, LA, MA, MD, MI, MO, NE, NC, NJ, NY, OH, PA, TN, WA, WV.
Hosts *Acer macrophyllum, Fagus grandifolia, Populus* sp., *P. trichocarpa, Salix* sp., *S. nigra, S. scouleriana*, angiosperm.
Ecology Rotten wood; decayed stump; rotting litter.
Culture Characters Not known.
References 80, 120, 125, 128 as *Phaeosolenia* (sic), 116, 119, 131, 249, 378.

A number of genera of host plants were listed in Cooke (125) but it is not clear which records were from Canada or the United States.

MERULIOPSIS Bondartsev 1959 Phanerochaetaceae
Synonyms are *Byssomerulius* Parm. 1967, fide Ginns (226), *Ceraceomerulius* (Parm.) Eriksson & Ryv. 1973, fide Jülich and Stalpers (326).

Meruliopsis albostramineus (Torrend)Jülich & Stalpers 1980 Verh. Kon. Ned. Akad. Wetensch., Afd. Natuurk., Tweede Sect. 74:154.
Basionym *Merulius albostramineus* Torrend, Brotéria, Sér. Bot. 11:70, 1913.
Synonym *Ceraceomerulius albostramineus* (Torrend) Ginns 1978, *Merulius armeniacus* Bres. 1925, *Byssomerulius armeniacus* (Bres.) Gilbn. 1974, a later homonym of *Byssomerulius armeniacus* Parm. 1967, *Merulius atropurpureus* W.B. Cooke 1943, fide Ginns (226), *Merulius bellus* Berk. & Curtis sensu W.B. Cooke (120, 128, 133). All fide Ginns (226).
Distribution Canada: BC, PQ.
United States: AZ, CA, CO, ID, MA, MT, NC, NJ, NM, NV, NY, OR, PA, WA, WV, WY.
Hosts *Abies* sp., *A. balsamea, A. concolor, A. grandis, A. magnifica, Acer* sp., *Alnus* sp., *Cercocarpus ledifolius, Eucalyptus* sp., *Picea* sp., *Pinus* sp., *P. attenuata, P. controta* var. *latifolia, P.monticola, P. ponderosa, P. rigida, Populus* sp., *P. tremuloides, Pseudotsuga menziesii, Thuja* sp., *T. plicata, Tsuga mertensiana*, angiosperm, gymnosperm.
Ecology Rotting wood; logs; associated with a brown rot (226) or a white rot (402).
Culture Characters
1.3r.21.(22).26.32.36.(38).39.44-46.50.54.55.(57) (Ref. 474);
1.6.7.32.36.38 or 40.45-47.50.54.55.(57) (Ref. 226).
References 53, 120, 128, 130, 133, 197, 204, 208, 216, 223, 226, 242, 402, 406, 474.

Meruliopsis ambiguus (Berk.) Ginns 1976 Canad. J. Bot. 54:117.
Basionym *Merulius ambiguus* Berk., Grevillea 1:69, 1872.
Synonym *Byssomerulius ambiguus* (Berk.) Gilbn. & Budington 1970, *Merulius succineus* Lloyd 1924, fide Ginns (226).
Distribution Canada: AB, AB-NT, BC, MB, ON, PQ, SK, YT. United States: AK, AL, AR, AZ, CA, CO, CT, FL, GA, ID, MD, MN, MT, NC, NE, NJ, NM, NY, SC, TX, WI.
Hosts *Alnus rubra, Picea* sp., *P. glauca, Pinus* sp., *P. banksiana, P. contorta, P. echinata, P. engelmannii, P. ponderosa, P. taeda, Salix* sp., gymnosperm, primarily on *Pinus* spp., rarely angiosperm.
Ecology Saprophytic on bark and wood; associated with a white rot.
Culture Characters
1.(2).5.21.32.(36).37.38.(40).42-43.50.(54).55.(57) (Ref. 474);
1.(2).5.7.(22).(25).32.36.38.40.42.(48).50.(54).55.(57) (Ref. 226).
References 96, 116, 204, 216, 226, 242, 402, 474.

Meruliopsis bellus (Berk. & Curtis) Ginns 1976 Canad. J. Bot. 54:122.
Basionym *Merulius bellus* Berk. & Curtis, Grevillea 1:69, 1872.
Distribution United States: AL.
Hosts Angiosperm.
Ecology Apparently associated with a white rot.
Culture Characters Not known.

The type, from Alabama, is the only confirmed specimen known. Ginns (222, 226), after studying the type specimen, compared its features with the similar *Meruliopsis hirtellus*. Cooke's reports (120, 128, 133) are referred to *Meruliopsis albostramineus* because we have redetermined some of his collections labelled *M. bellus* and presume they are all the same.

Meruliopsis corium (Pers.:Fr.) Ginns 1976 Canad. J. Bot. 54:126.
Basionym *Thelephora corium* Pers., Synop. Meth. Fung., 574, 1801.
Synonym *Merulius corium* (Pers.) Fr. 1828, *Byssomerulius corium* (Pers.) Parm. 1967, *Merulius confluens* Schw. 1822, *Byssomerulius confluens* (Schw.) Lindsey & Gilbn. 1978, *Merulius pallens* Schw. 1832, *Merulius haedinus* Berk. & Curtis 1872, *Merulius ulmi* Peck 1906. All fide Ginns (226).
Distribution Canada: BC, MB, NS, ON, SK.
United States: AK, AL, AZ, CA, DC, FL, GA, LA, MD, MN, MT, NM, NY, OH, OR, PA, SC, TN, VA, WA, WI.
Hosts *Abies concolor, Acer* sp., *A. circinatum, A. macrophyllum, A. rubrum* var. *rubrum, A. pseudoplatanus, Alnus* sp., *A. oblongifolia, A. rubra, A. rugosa, A. sinuata, Amelanchier* sp., *A. alnifolia, A alnifolia* var. *semiintegrifolia, Arbutus* sp., *Baccharis halimifolia, B. sarothroides, Betula resinifera, Carya ovata, Celtis reticulata, Citrus* sp., *Cornus nuttallii, Corylus cornuta* var. *californica, Cytisus scoparius,*

Encelia californica, Ilex arenicola, Larix occidentalis, Ligustrum amurense, L. vulgare, Liquidambar sp., *Nerium oleander, Nyssa* sp., *N. sylvatica, Ostrya* sp., *Parthenocissus* sp., *Picea mariana, Pinus banksiana, P. contorta, P. muricata, P. ponderosa, Populus* sp., *P. tremuloides, P. trichocarpa, Prosopis juliflora, Prunus persica, P. virginiana, P. virginiana* var. *demissa, Quercus* sp., *Q. arizonica, Q. gambelii, Q. garryana, Q. hypoleucoides, Q. reticulata, Q. toumeyi, Q. virginiana, Rhamnus purshiana, Salix* sp., *S. alaxensis, S. scouleriana, S. reticulata, Simmondsia chinensis, Tilia* sp., *T. americana, Ulmus americana, Viburnum obovatum*, angiosperm, rarely on gymnosperm.

Ecology Saprophytic on bark and wood; dead twigs; dead branches on live trees; associated with a white rot.

Culture Characters
1.5.(7).(22).(25).32.38.43-44.(48).50.54.(55).(57) (Ref. 226).

References 96, 97, 116, 127, 133, 177, 183, 197, 204, 207, 208, 209, 214, 216, 226, 242, 249, 333, 406, 423, 465.

Meruliopsis hirtellus (Burt) Ginns 1976
Canad. J. Bot. 54:132.

Basionym *Merulius hirtellus* Burt, Ann. Missouri Bot. Gard. 4:335, 1917.

Synonym *Byssomerulius hirtellus* (Burt) Parm. 1967.

Distribution Canada: ON, PQ. United States: AZ, CO, MA, MI, NM, NY, OH, PA, UT, WI.

Hosts *Abies concolor, A. lasiocarpa* var. *arizonica, Acer* sp., *Betula* sp., *Buxus* sp., *Fagus* sp., *F. grandifolia, Ostrya* sp., *Picea* sp., *P. engelmannii, P. rubens, Pinus* sp., *P. ponderosa, Populus* sp., *P. tremuloides, Pseudotsuga menziesii, Quercus* sp., *Sambucus* sp., *Vitis* sp., angiosperm, gymnosperm.

Ecology Perhaps psychrophilic; basidiomes produced on twigs and branches up to 3 cm diam; associated with a white rot.

Culture Characters
1.5.7.(26).32.36.38.42.54.(57) (Ref. 226);
1.5.13.32.36.(39).40.41-42.53.54.55.(57) (Ref. 474).

References 183, 196, 197, 204, 216, 226, 242, 402, 406, 474.

Meruliopsis sulphureus (Burt) Ginns & Lefebvre, *comb. nov.*

Basionym *Merulius sulphureus* Burt, Ann. Missouri Bot. Gard. 4:333, 1917.

Synonym *Byssomerulius sulphureus* (Burt) Lindsey 1974, *Macrohyporia sulphurea* (Burt) Ginns & J. Lowe 1983, *Wolfiporia sulphureus* (Burt) Ginns 1984.

Distribution United States: AZ, FL.

Hosts *Olneya tesota, Opuntia fulgida, Prosopis juliflora, P. velutina, ?Quercus* sp.

Ecology Fallen trees; associated with a white rot.

Culture Characters
2.6.21.26.32.36.38.43-44.54 (Ref. 474).

References 207, 216, 218, 474.

Because the generic name *Byssomerulius* is, herein, accepted as a synonym of *Meruliopsis*, we propose the transfer of *sulphureus* to *Meruliopsis*.

Meruliopsis taxicola (Pers.:Fr.) Bondartsev 1959
In Parmasto, Izv. Akad. Nauk Estonsk SSR, Ser. Biol. 8:274.

Basionym *Xylomyzon taxicola* Pers., Mycol. Eur. 2:32, 1825.

Synonym *Merulius taxicola* (Pers.) Duby 1830, *Poria taxicola* (Pers.) Bres. 1897, *Gloeoporus taxicola* (Pers.) Gilbn. & Ryv. 1986, *Merulius ravenelii* Berk. 1872, fide Ginns (226), *Poria rufa* (Schrader:Fr.) Cooke 1886, fide Lowe (416).

Distribution Canada: AB, AB-NT, BC, MB, ON, PQ, SK. United States: AZ. CA, CO, CT, GA, ID, ME, MI, MS, MT, NH, NJ, NM, NY, OR, UT, WA, WY.

Hosts *Abies balsamea, A. concolor, A. lasiocarpa, A. magnifica, Picea* sp., *P. engelmannii, P. glauca, P. mariana, P. rubens, Pinus* sp., *P. banksiana, P. contorta, P. ponderosa, P. taeda*, rare on *Populus* sp., *Pseudotsuga menziesii, Tsuga heterophylla*, gymnosperm.

Ecology Saprophytic; associated with a white rot.

Culture Characters
1.5.7.32.36.38.42.55 (Ref. 487).

References 96, 120, 128, 217, 226, 242, 402, 484.

METULODONTIA Parm. 1968 — Hyphodermataceae

Metulodontia nivea (Karsten) Parm. 1968
Consp. Syst. Cort., 118.

Basionym *Kneiffia nivea* Karsten, Hedwigia 35:173, 1896.

Synonym *Peniophora nivea* (Karsten) Bourd. & Galzin 1912, *Hyphoderma karstenii* Jülich 1974, proposed as a *nom. nov.* for *K. nivea* because there already was a *Hyphoderma nivea* Fuckel 1869.

Distribution Canada: BC, NT, ON, PQ.
United States: MD, ME, MN, MT, NY, TN.

Hosts *Abies balsamea, A. lasiocarpa, Ostrya* sp., *Picea glauca, Pinus* sp., *Populus* sp., *P. grandidentata, P. tremuloides, Pseudotsuga menziesii, Quercus* sp., angiosperm, gymnosperm.

Ecology Associated with a white rot.

Culture Characters
2.3c.15p.32.36.38.45-46.54.55.60 (Ref. 474).

References 116, 174, 214, 242, 255, 406, 474, 584.

MUCRONELLA Fr. 1874 — Hericiaceae

Synonym is *Myxomycidium* Massee 1901, fide Petersen (530).

Mucronella aggregata Fr. 1863
Monogr. Hymen. Suec. 2:280.
Synonyms *Hydnum nudum* Berk. & Curtis 1873, fide Gilbertson (192), *Mucronella minutissima* Peck 1891, fide Corner (134), *Mucronella minutissima* var. *conferta* Peck 1902, fide Gilbertson (189), probably *Mucronella ramosa* Lloyd 1922, fide Corner (134).
Distribution Canada: BC, NS, ON.
United States: AL, AZ, CO, IA, ME, MN, NM, NY, OH, OR, SC, WA.
Hosts *Abies balsamea, A. concolor, Betula* sp., *B. alleghaniensis, Picea engelmannii, Pinus* sp., *P. ponderosa, Populus balsamifera, Pseudotsuga menziesii, Quercus* sp., angiosperm, gymnosperm.
Ecology Decaying wood; decorticated wood; logs; associated with a white rot.
Culture Characters Not known.
References 116, 119, 137, 183, 189, 190, 192, 197, 204, 208, 214, 216, 249, 402, 448, 463.

Probably the same as *M. calva*, see Corner (137), but in the absence of a critical comparison we recognize two species.

Mucronella bresadolae (Quél.) Corner 1970
Beih. Nova Hedwigia 33:172.
Basionym *Clavaria bresadolae* Quél., F. Mycol., 458, 1888.
Synonym *Myxomycidium pendulum* Massee 1901 (see below), *Mucronella pendula* (Massee) Petersen 1980, *Mucronella alba* Lloyd, 1919, fide Corner (137).
Distribution Canada: BC. United States: CA, WA.
Hosts Generally gymnosperm.
Ecology Associated with a brown rot.
Culture Characters Not known.
References 137, 530.

To our knowledge no one has compared the type specimens of *M. bresadolae* and *Myxomycidium pendulum*. However, Corner (137) and Petersen (530) having studied the type of *Mucronella alba* reduced the name to synonymy under *M. bresadolae* and *M. pendula*, respectively. It appears that *M. bresadolae* is the oldest, hence correct name for this fungus. Petersen (530) did not mention Corner's synonymy of *M. alba* under *M. bresadolae*.

Mucronella calva (Alb. & Schw.:Fr.)Fr. 1874
Hym. Eur., 629.
Basionym *Hydnum calvum* Alb. & Schw., Consp. Fung., 271, 1805.
Distribution Canada: AB-NT, BC, NS.
United States: LA.
Hosts *Abies balsamea, Picea engelmannii, Pinus contorta*.
Ecology Not known.
Culture Characters Not known.
References 242, 616.

Mucronella flava Corner 1953
Ann. Bot. (London), N.S. 17:356.
Distribution United States: AL, IA.
Hosts *Acer* sp., *Quercus* sp.
Ecology Large hewn oak beam; well decayed log deposited on river bank.
Culture Characters Not known.
References 137, 445.

MYCOACIA Donk 1931 — Meruliaceae

Mycoacia aurea (Fr.) Eriksson & Ryv. 1976
Cort. N. Europe 4:877.
Basionym *Hydnum aureum* Fr., Elenchus 1:137, 1828.
Synonym *Hydnum fasciculare* Berk. & Curtis 1873, fide Gilbertson (192), a later homonym of *Hydnum fasciculare* Alb. & Schw. 1805, *Odontia stenodon* (Pers.) Bres. 1897, fide Eriksson and Ryvarden (174), *Mycoacia stenodon* (Pers.) Donk 1931.
Distribution Canada: BC, ON.
United States: IA, MI, MN, NY, SC.
Hosts *Alnus rubra, ?Nyssa sylvatica, Populus* sp., *P. tremuloides, Ulmus* sp., angiosperm.
Ecology Rotted wood; dead wood.
Culture Characters
2.4.(26).32.36.37.(39).(40).43-44.48.54 (Ref. 588).
References 56, 116, 192, 214, 242, 406, 463.

Mycoacia austro-occidentale Canfield 1976
Mycotaxon 3:513.
Distribution United States: AZ.
Hosts *Fouquieria splendens, Prosopis juliflora*.
Ecology Associated with a white rot.
Culture Characters
2.3.7.13.21.32.36.38.47.50.54 (Ref. 477).
References 207, 477.

Mycoacia fuscoatra (Fr.:Fr.) Donk 1931
Medd. Nedl. Myc. Ver. 18-20:152.
Basionym *Hydnum fuscoatra* Fr., Nov. Flora Suec. 2:39, 1814 (original not seen).
Synonym *Odontia fuscoatra* (Fr.) Bres. 1897, *Steccherinum fuscoatrum* (Fr.) Gilbn. 1971, *Hydnum carbonarium* Peck 1887, fide Gilbertson (189).
Distribution Canada: MB, NS, ON. United States: AZ, CO, GA, IA, ID, MI, MN, NC, NY, OH, WA, WI.
Hosts *Acer* sp., *A. rubrum, Betula alleghaniensis, Castanea* sp., *Fagus grandifolia, Populus* sp., *P. tremuloides, Quercus arizonica, Q. emoryi*, angiosperm.
Ecology Slash; burnt log; logs; associated with a white rot.
Culture Characters
2.4.7.(27).34.36.(39).40.42-43.(50).54.58 (Ref. 474).
References 56, 116, 183, 189, 214, 216, 242, 249, 402, 406, 463, 474.

Mycoacia uda (Fr.) Donk 1931
Medd. Nedl. Myc. Ver. 18-20:151.
Basionym *Hydnum udum* Fr., Syst. Mycol. 1:422, 1821.
Synonym *Odontia uda* (Fr.) Bres. 1897.
Distribution Canada: BC, MB, ON. United States: IA, IN, LA, MI, MT, NY, OR, TN, WA, WI.
Hosts *Acer* sp., *A. macrophyllum*, *A. saccharinum*, *Alnus* sp., *A. rubra*, *Arbutus menziesii*, *Cytisus scoparius*, *Fagus grandifolia*, *Fraxinus nigra*, *Populus* sp., *Populus tremuloides*, *Pseudotsuga menziesii*, *Quercus* sp., *Salix* sp., angiosperm, gymnosperm.
Ecology Branch; associated with a white rot.
Culture Characters
2.4.14.(34).36.40.43-44.48.50.54.(55).59 (Ref. 474).
References 56, 116, 133, 183, 242, 406, 463, 474, 489, 616.

MYCOBONIA Pat. 1894 Mycoboniaceae
Synonym is *Grandinioides* Banker 1906, fide Donk (154).

Mycobonia flava (Swartz) Pat. 1894
Bull. Soc. Mycol. France 10:77.
Basionym *Peziza flava* Swartz, Nov. Gen. et Sp. Pl., 150, 1788.
Synonym *Grandinioides flavum* (Swartz) Banker 1906.
Distribution United States: FL, LA.
Hosts Not known.
Ecology Fallen branches; old logs.
Culture Characters Not known.
References 88, 319.

MYCOGLOEA Olive 1950 Cystobasidiaceae

Mycogloea carnosa Olive 1950
Mycologia 42:385.
Distribution United States: LA, SC.
Hosts *Acer rubrum*, angiosperm.
Ecology Dead corticated limb.
Culture Characters Not known.
References 183, 419, 500, 502.

MYCORRHAPHIUM Maas G. 1962 Climacodontaceae

Mycorrhaphium adustulum (Banker) Ryv. 1989
Mem. New York Bot. Gard. 49: 346.
Basionym *Steccherinum adustulum* Banker, Mem. Torrey Bot. Club 12:133, 1906.
Synonym *Steccherinum pusillum* (Fr.) Banker 1912 sensu N. Amer. Aucts., *Mycoleptodonoides pusillum* (Fr.) Harrison 1973, *Mycorrhaphium pusillum* (Fr.) Maas G. 1962 sensu N. Amer. Aucts.
Distribution Canada: NS.
United States: ?IA, MN, NJ, NY.
Hosts Angiosperm.
Ecology Rotten sticks on ground; associated with a white rot.
Culture Characters
2.3c.8.32.36.38.47.48.54 (Ref. 474).
References 25, 249, 428, 463, 474.

Maas Geesteranus (426) and Ryvarden (loc. cit.) recognized *M. adustulum* as distinct from the European *M. pusillum*. The generative hyphae of *M. adustulum* have clamp connections, whereas *M. pusillum* lacks them (428, 426); and they are geographically isolated. Reports (463, 474) of *pusillum* from North America seem to be based upon Banker's (26) placing *S. adustulum* in synonymy with the older *S. pusillum*.

Mycorrhaphium adustum (Schw.) Maas G. 1962
Persoonia 2:394.
Basionym *Hydnum adustum* Schw., Schriften Naturf. Ges. Leipzig 1:103, 1822.
Synonym *Steccherinum adustum* (Schw.) Banker 1906.
Distribution United States: AL, CT, IA, KY, MN, MO, NC, NJ, NY, OH, PA, TN.
Hosts *Quercus* sp., angiosperm.
Ecology Dead wood; logs; dead decaying, half buried branches.
Culture Characters
2.3c.7.32.36.38.46-47.(48).(51).54 (Ref. 474).
References 25, 333, 426, 463, 474, 610.

One report (214) from Minnesota was referred (474:200) to *M. adustulum*.

MYLITTOPSIS Pat. 1895 Auriculariaceae

Mylittopsis marmorata (Berk. & Curtis) D.P. Rogers & G.W. Martin 1955
Mycologia 47:891-894.
Basionym *Tremella marmorata* Berk. & Curtis, Grevillea 2:19, 1873.
Synonym *Mylittopsis langloisii* Pat. 1895, fide Rogers and Martin (loc. cit.).
Distribution United States: FL, LA, SC.
Host *Quercus* sp.
Ecology Logs in wet woods.
Culture Characters Not known.
References 419, 568.

MYXARIUM Wallr. 1833 Hyaloriaceae

Myxarium atratum (Peck) Ginns & Lefebvre, *comb. nov.*
Basionym *Naematelia atrata* Peck, Annual Rep. New York State Mus. 24:83, 1872.
Synonym *Tremella nucleata* Schw.:Fr., 1822, fide Bandoni (16), *Naematelia nucleata* (Schw.) Fr. 1822, *Exidia nucleata* (Schw.) Burt 1921.
Distribution Canada: BC, MB, NS, ON, PQ.
United States: AL, AZ, CA, GA, IA, LA, OH, ME, MI, MN, NC, NJ, NY, VT, WI, WV.
Hosts *Acer* sp., *A. macrophyllum*, *A. saccharum*, *Ampelopsis tricuspidata*, *Aralia spinosa*, *Juglans regia*,

Magnolia fraseri, Populus sp., *P. tremuloides, Quercus* sp., *Q. hypoleucoides, Q. tomentella, Salix* sp., *S. glauca, S. nigra, Sambucus* sp., *S. caerulea, Tilia americana, Vitis aestivalis*, scuppernong grape, almost all broad-leaved woody plants, angiosperm.
Ecology Bark; decaying wood; corticated wood; limbs; corticated limb; dead branches; decorticated branch on ground; branches and stems; dead grape vine.
Culture Characters Only a partial species code is known, 60 (Ref. 301).
References 90, 103, 107, 116, 183, 216, 242, 249, 301, 443, 465, 498, 499, 532, 626, 642.

Because there is a *Myxarium nucleatum* Wallr. 1833, Schweinitz's epithet *nucleatum* cannot be transferred to *Myxarium*. Peck's *atrata* is the next available epithet for this fungus and we propose the new combination. Although Hung and Wells (301), apparently following Neuhoff, list *Tremella nucleata* Schw. (≡ *Exidia nucleata* (Schw.:Fr.) Burt) as a synonym of *Myxarium nucleatum* Wallr., Donk (157:233-235) rejected the synonymy and recognized two species.

Myxarium grilletii (Boudier) D. Reid 1973
Persoonia 7:297.
Basionym *Tremella grilletii* Boudier, Bull. Soc. Bot. France 32:284, 1885.
Synonym *Stypella minor* Möller 1895 sensu G.W. Martin, fide Reid (557), *Sebacina sphaerospora* Bourd. & Galzin 1924, fide Donk (157) and Wells (625), *Tremella gangliformis* Linder 1933, fide Martin (433).
Distribution Canada: ON. United States: IA, LA, MA, MO, MN, MO, NC, OR, WI.
Hosts *Alnus* sp., *Pinus* sp., *Pseudotsuga* sp., *Quercus* sp. *Populus tremuloides, Ulmus* sp., angiosperm.
Ecology Dead wood; twigs and dead branches.
Culture Characters Not known.
References 214, 419, 433, 443, 451, 496, 502, 625, 641.

Donk (157) and Wells (625) concluded that *Stypella minor* sensu G.W. Martin was the same as *Sebacina sphaerospora*. McGuire's (451) report of *Sebacina sphaerospora* is *Stypella minor* sensu G.W. Martin, fide Wells (625).

Myxarium subhyalinum (A. Pearson) D. Reid 1970
Trans. Brit. Mycol. Soc. 55:426.
Basionym *Sebacina subhyalina* A. Pearson, Trans. Brit. Mycol. Soc. 13:70, 1928.
Synonym *Sebacina sublilacina* G.W. Martin 1934, fide Reid (556).
Distribution Canada: AB, ON, PQ. United States: CA, GA, IA, MA, MO, NC, NY, OH, OR.
Hosts *Liriodendron tulipifera, Populus* sp., *Prunus* sp., *Quercus* sp., *Salix* sp., *Tilia* sp., *Ulmus* sp., angiosperm, thelephoraceous fungus.
Ecology Rotten, decorticated wood; limb.
Culture Characters Not known.
References 434, 443, 451, 496, 498, 625.

Related to *Sebacina podlachica, Exidia nucleata, Stypella minor*, etc. (625). Wells (625) concluded that *Sebacina subhyalina* was a synonym of *S. podlachica* Bres. but Reid (556) pointed out that *S. subhyalina* has cystidia, and *S. podlachica* does not.

NOCHASCYPHA Agerer 1983 Lachnellaceae

Nochascypha filicina (Karsten) Agerer 1983
Mitt. Bot. Staatssamml. München 19:268.
Basionym *Cyphella filicina* Karsten, Not. Sällsk. Fauna Fl. Fenn. Förh. 11:220, 1870.
Synonym *Lachnella filicina* (Karsten) W.B. Cooke 1962.
Distribution Canada: ON. United States: NY.
Host Fern.
Ecology Fronds.
Culture Characters Not known.
Reference 125.

ODONTICIUM Parm. 1968 Chaetoporellaceae

Odonticium laxa (L.W. Miller) Ryv. 1978
Norw. J. Bot. 25:296.
Basionym *Odontia laxa* L.W. Miller, Mycologia 26:19, 1934.
Distribution Canada: NB.
United States: CO, IA, NC, OH.
Hosts *Abies balsamea, Acer* sp., *Picea engelmannii, Quercus* sp., *Ulmus* sp., angiosperm, gymnosperm.
Ecology Bark; associated with a white rot.
Culture Characters
2.6.7.(22).(26).32.36.38.44-45.(50).(53).54.55 (Ref. 474).
References 191, 242, 402, 463, 474, 571.

Odonticium romellii (Lundell) Parm. 1968
Consp. Syst. Cort., 127.
Basionym *Odontia romellii* Lundell in Eriksson, Symb. Bot. Upsal. 16 (1):124, 1958.
Distribution Canada: AB, AB-NT, BC.
United States: ID, MT, NC, WA.
Hosts *Abies fraseri, A. grandis, Acer circinatum, Picea engelmannii, Pinus banksiana, P. contorta, P. monticola, P. ponderosa, Pseudotsuga menziesii, Thuja plicata*, gymnosperm.
Ecology Log.
Culture Characters Not known.
References 210, 242.

OLIVEONIA Donk 1958 Ceratobasidiaceae
Synonym is *Heteromyces* Olive 1957, a later homonym of *Heteromyces* Müll. 1889, fide Donk (157).

Oliveonia fibrillosa (Burt) Donk 1958
Fungus 28:20.

Basionym *Sebacina fibrillosa* Burt, Ann. Missouri Bot. Gard. 13:335, 1926.
Synonyms *Heteromyces fibrillosum* (Burt) Olive 1957, *Peniophora heterobasidioides* D.P. Rogers 1935, fide Olive (507).
Distribution United States: IA.
Host *Populus* sp.
Ecology Sodden log.
Culture Characters Not known.
References 507, 563.

Known in the United States from one collection, Rogers 329, the type of *Peniophora heterobasidioides*.

Oliveonia pauxilla (Jackson) Donk 1958
Fungus 28:20.
Basionym *Corticium pauxillum* Jackson, Canad. J. Res., C, 28:724, 1950.
Synonym *Heteromyces pauxillum* (Jackson) Olive 1957.
Distribution Canada: ON.
Host *Pteretis pennsylvanica*.
Ecology Rachis.
Culture Characters Not known.

Known only from the type specimen, which was redescribed by Olive (507).

Oliveonia subviolacea (Peck) Larsen 1974
Mycologia Memoir 4:128.
Basionym *Hypochnus subviolaceus* Peck, New York State Bot. Rep. 1893:25, 1894.
Synonyms *Hydrabasidium subviolaceum* (Peck) Eriksson & Ryv. 1978, *Ceratobasidium atratum* (Bres.) D.P. Rogers, fide Larsen (362).
Distribution Canada: MB, ON, PQ.
United States: NC, OR, TN.
Hosts Gymnosperm.
Ecology Badly decayed wood; dead wood; exposed root of live tree; root bark.
Culture Characters Not known.
References 93, 169, 327, 443.

Eriksson et al. (169) felt that this fungus differed significantly from *O. fibrillosa*, the type species of *Oliveonia*, and placed it in *Hydrabasidium* Parker-Rhodes ex Eriksson & Ryv.

PAULLICORTICIUM Eriksson 1958 — Sistotremaceae

Paullicorticium ansatum Liberta 1962
Brittonia 14:220.
Distribution Canada: ON. United States: ME, OR.
Hosts *?Abies* sp., *Pinus rigida, Pseudotsuga menziesii, ?Thuja plicata*.
Ecology Much decayed wood.
Culture Characters Not known.

Known only from the type specimens.

Paullicorticium delicatissimum (Jackson) Liberta 1962
Brittonia 14:222.
Basionym *Corticium delicatissimum* Jackson, Canad. J. Res., C, 28:722, 1950.
Distribution Canada: ON, PQ.
Hosts *Picea* sp., *Tsuga canadensis*.
Ecology Much decayed, brown rotted wood.
Culture Characters Not known.
References 116, 242, 307, 386.

Paullicorticium jacksonii Liberta 1962
Brittonia 14:223.
Distribution Canada: ON. United States: IA.
Hosts *Quercus macrocarpa, Tsuga* sp.
Ecology Decayed wood.
Culture Characters Not known.

Known only from the type specimens. Possibly a synonym of *Sistotremastrum niveocremeum*, fide Jülich and Stalpers (326).

Paullicorticium pearsonii (Bourd.) Eriksson 1958
Symb. Bot. Upsal. 16 (1):67.
Basionym *Corticium pearsonii* Bourd., Trans. Brit. Mycol. Soc. 7:51, 1921.
Synonym *Corticium subinvisible* D.P. Rogers 1935, fide Liberta (386).
Distribution Canada: PQ.
United States: AZ, CO, IA, MA, NM, WI.
Hosts *Fraxinus velutina, Picea* sp., *Pinus* sp., *Pinus ponderosa, Tsuga canadensis*, angiosperm, gymnosperm.
Ecology Decayed wood.
Culture Characters Not known.
References 197, 204, 216, 242, 386, 388, 390, 563.

PELLIDISCUS Donk 1959 — Crepidotaceae

Synonyms are *Cyphellathelia* Jülich 1972, fide Hjortstam (283), *Phaeoglabrotricha* W.B. Cooke 1961, fide Reid (554).

Pellidiscus pallidus (Berk. & Br.) Donk 1959
Persoonia 1:90.
Basionym *Cyphella pallida* Berk. & Br., Ann. Mag. Nat. Hist., Ser. IV, 11:343, 1873.
Synonyms *Cyphella sessilis* Burt 1926, *Pellidiscus subiculosus* W.B. Cooke 1961. Both fide Reid (554).
Distribution Canada: BC, ON. United States: MI, NJ.
Hosts *Alnus* sp., *Cytisus scoparius, Populus* sp., *Pteridium aquilinum, ?Sambucus* sp., *Vicia gigantea*.
Ecology Decomposed litter.
Culture Characters Not known.
References 125, 242, 545.

Pellidiscus pezizoidea (Ellis & Ev.) Ginns & Lefebvre, *comb. nov.*
Basionym *Corticium pezizoideum* Ellis & Ev., J. Mycol. 4:74, 1888.

Synonym *Cyphellathelia pezizoidea* (Ellis & Ev.) Jülich 1972.
Distribution United States: LA, NJ.
Hosts *Arundinaria* sp., *Phaeolus* sp.
Ecology Bean vines.
Culture Characters Not known.
Reference 312.

Congeneric with *Pellidiscus pallidus* but the type specimen of *Corticium pezizoideum* seems distinct, fide Hjortstam (283). Thus the new combination in *Pellidiscus* is proposed to align *pezizoideum* with allied species. Neither *Corticium*, which is now restricted to quite a different group of fungi, nor *Cyphellathelia*, which is now accepted as a synonym of *Pellidiscus*, is available for the epithet *pezizoideum*.

PENIOPHORA Cooke 1879 Peniophoraceae

Peniophora aurantiaca (Bres.) Höhnel & Litsch. 1906 Sitzungsber. Kaiserl. Akad. Wiss., Math.-Naturwiss. Kl., Abt. I, 115:1583.
Basionym *Corticium aurantiaca* Bres., Fungi Tridentini 2:37, 1892.
Synonyms *Aleurodiscus zelleri* Burt 1926, fide Rogers and Jackson (567), *Peniophora lepida* Bres. 1925, fide Eriksson et al. (169), *Peniophora shearii* Burt 1926, fide Rogers and Jackson (567).
Distribution Canada: AB, BC, MB, NB, NF, NFL, NS, NTM, ON, PE, YT. United States: AK, ID, MA, ME, MI, MS, MT, NC, NH, NY, OR, PA, WA, WI.
Hosts *Abies grandis, Alnus* sp., *A. crispa, A. rubra, A. rugosa, A. rugosa* var. *americana, A. sinuata, Arbutus menziesii, Betula* sp., *Menziesia ferruginea, Populus trichocarpa, Salix* sp., *Ulmus* sp.
Ecology Dead wood; small dead twigs; dead stems and branches; associated with a white rot.
Culture Characters
2.3c.26.27.32.36.38.(39).43-44.(50).(53).54.58 (Ref. 474).
References 53, 92, 93, 97, 116, 169, 183, 242, 249, 340, 344, 474, 584.

Predominantly on species of *Alnus* and *Salix*. The report (183) on *Abies grandis* needs confirmation.

Peniophora borealis (Peck) Burt 1926 Ann. Missouri Bot. Gard. 12:295.
Basionym *Peniophora disciformis* (DC.) Cooke var. *borealis* Peck, Harriman Alaska Exped. 5. The Fungi of Alaska, 43, 1904.
Distribution United States: AK.
Hosts Angiosperm, perhaps *Alnus* sp.
Ecology Bark of small decaying twigs.
Culture Characters Not known.
References 92, 97.

Peniophora carnea (Berk. & Cooke) Cooke 1879 Grevillea 8:21.
Basionym *Corticium carneum* Berk. & Cooke, Grevillea 7:1, 1878.
Distribution United States: CA, TX, WA.
Hosts *Pinus contorta*, angiosperm.
Ecology Fallen, decaying limbs; logs.
Culture Characters Not known.
References 92, 133, 542.

Hjortstam (288) recently studied the two California collections cited in the prologue (1016 and 1025) and concluded "The first is indeterminable and the other represents a species of *Peniophora* Cooke s. str., although not *P. cinerea* (Pers.:Fr.) Cooke ...." Burt (92) cited Ravenel 78 (at K) as type, presumably because it was labelled *C. carneum* and because 1025 was, in Burt's opinion, *P. cinerea*. If 1025 can be used to characterize the species, then Burt's designation of 78 as type becomes irrelevant. The most detailed description associated with the name *P. carnea* is that of Burt, who based his concept on 78 and a collection from Cuba (Murrill & Earle 333), but it is uncertain whether they are conspecific with 1025.

Peniophora cinerea (Pers.:Fr.) Cooke 1879 Grevillea 8:20.
Basionym *Corticium cinereum* Pers., Neues Mag. Bot. 1:111, 1794.
Synonyms *Corticium fumigatum* Thümen 1876, fide Burt (92), *Thelephora lilacina* Schw. 1832, fide Burt (92), *Peniophora lilacina* (Schw.) Massee 1889, a later homonym of *Peniophora lilacina* (Berk. & Br.) Cooke 1879.
Distribution Canada: AB, BC, MB, NB, NF, NS, ON, PE, PQ. United States: AL, CA, CO, CT, FL, GA, ID, IL, IN, KY, LA, MA, MD, ME, MI, MN, MO, MS, NC, ND, NE, NH, NJ, NM, NY, OH, OR, PA, TN, VA, VT, WA, WI.
Hosts *Abies balsamea, A. concolor, A. lasiocarpa* var. *lasiocarpa, Acer* sp., *Acer macrophyllum, A. rubrum, A. saccharum, A. spicatum, Alnus* sp., *A. rubra, Amelanchier* sp., *A. alnifolia, Arbutus menziesii, Arctium minus, Betula* spp., *B. alleghaniensis, B. papyrifera, Carya* sp., *Cercocarpus ledifolius, Cornus* sp., *C. nuttallii, C. rugosa, C. stolonifera, Corylus cornuta* var. *californica, Cotoneaster* sp., *Eucalytpus* sp., *Fagus grandifolia, Fraxinus oregona, F. pennsylvanica, Holodiscus discolor, Juglans nigra, Lagerstroemia* sp., *Liquidambar styraciflua, Malus pumila, Picea* sp., *P. engelmannii, P. mariana, Pinus aristata, P. banksiana, P. ponderosa, P. strobus, Populus* sp., *P. grandidentata, P. tremuloides, Prunus* sp., *P. pensylvanica, P. virginiana, Pseudotsuga menziesii, Purshia tridentata, Quercus* sp., *Q. agrifolia, Q. alba, Q. coccinea, Q. garryana, Q. kelloggii, Q. macrocarpa, Q. rubra, Q. rubra* var. *borealis, Rhus diversiloba, Rubus* sp., *Salix* spp., *S. alba, S. nigra, Symphoricarpos occidentalis, Syringa vulgaris, Tilia americana, Tsuga heterophylla, Ulmus*

sp., *U. americana, Vitis* sp., *Weigelia florida*, southern yellow pine, angiosperm, gymnosperm.

Ecology Causes a minor trunk rot in yellow birch; basidiomes produced on bark of dead branch; bark of fallen limb; pulpwood; test crosses; associated with a white rot.

Culture Characters
2.3c.15.21.(27).31g.32.36.(37).(39).40.42-43.(50).54.55.60 (Ref. 474).

References 92, 96, 108, 116, 120, 127, 128, 133, 177, 178, 179, 183, 197, 208, 214, 242, 267, 274, 332, 333, 402, 406, 474, 486, 509, 511.

Several specimens reported (119, 249, 402) as belonging to the *Peniophora cinerea* complex have been found in Colorado, Idaho, Nova Scotia and Washington.

Peniophora colorea Burt 1926
Ann. Missouri Bot. Gard. 12:343.
Distribution United States: LA.
Hosts Angiosperm.
Ecology Bark of dead branches.
Culture Characters Not known.

Known only from the type specimen, which was redescribed by Liberta (392).

Peniophora decorticans Burt 1926
Ann. Missouri Bot. Gard. 12:344.
Distribution Canada: BC. United States: OR, WA.
Hosts *Acer macrophyllum, Quercus garryana, Rhus diversiloba.*
Ecology Not known.
Culture Characters Not known.
References 92, 133, 164, 169, 242, 579.

Peniophora erikssonii Boidin 1957
Bull. Soc. Hist. Nat. Toulouse 92:286.
Distribution Canada: NT, PQ, YT.
United States: AL, IN, NY, OR, PA, WA.
Hosts *Alnus* sp., *A. tenuifolia.*
Ecology Associated with a white rot.
Culture Characters
2.6.7.26.32.36.38.46-47.(50).(53).54.(57) (Ref. 474).
References 116, 169, 183, 474, 584.

Peniophora exima Jackson & Dearden 1951
Mycologia 43:60.
Distribution Canada: Specific localities not given.
United States: CA.
Host *Abies magnifica.*
Ecology Hand hewn boards; old bench log.
Culture Characters Not known.
References 120, 128, 183, 309.

Peniophora farlowii Burt 1926
Ann. Missouri Bot. Gard. 12:343.
Distribution Canada: ON. United States: NH.
Hosts Angiosperm.
Ecology Very rotten wood.
Culture Characters Not known.
References 92, 584.

Peniophora fuscomarginata Burt 1926
Ann. Missouri Bot. Gard. 12:335.
Synonym *Phanerochaete fuscomarginata* (Burt) Gilbn. in Gilbertson, Canfield and Cummins 1972.
Distribution United States: AZ, FL, LA, NM.
Hosts *Fraxinus velutina, Juglans major, Platanus wrightii, Quercus* sp., *Q. arizonica, Q. emoryi, Q. hypoleucoides, Q. rugosa, Q. toumeyi, Rubus arizonensis*, angiosperm.
Ecology Bark of fallen, decaying branches; associated with a white rot.
Culture Characters
2.3r.8.34.(36).37.(39).40.42.54 (Ref. 474).
References 92, 216, 392, 474.

Peniophora incarnata (Pers.:Fr.) Karsten 1889
Hedwigia 28:27.
Basionym *Thelephora incarnata* Pers., Synop. Meth. Fung., 573, 1801.
Distribution Canada: AB, AB-NT, BC, MB, NF, NS, ON. United States: AK, AL, CA, CO, DC, FL, GA, IA, IL, KS, KY, MA, MD, ME, MI, MO, ND, NH, NM, NY, OH, OR, PA, SC, VT, WA, WI, WV.
Hosts *Abies grandis, Acer* sp., *A. circinatum, A. macrophyllum, A. platanoides, Alnus* sp., *A. rubra, A. rugosa, Arbutus menziesii, Arctostaphylos patula, Baccharis* sp., *Bambusa* sp., *Cercocarpus ledifolius, Chrysanthemum* sp., *Corylus cornuta, Crataegus douglasii, Cytisus scoparius, Fagus* sp., *Fraxinus pennsylvanica, Holodiscus discolor, Juglans nigra, Magnolia x suolangiana, Pinus contorta, Platanus occidentalis, Populus* sp., *P. tremuloides, Prunus emarginata, Quercus* sp., *Q. garryana, Q. kellogii, Q. prinus, Rhus diversiloba, Rosa* sp., *Rubus spectabilis, Salix* sp., *Sambucus* sp., *S. caerulea, S. glaucus, Symphoricarpos occidentalis, Tilia americana, Vaccinium* sp., angiosperm.
Ecology Dead branches on live trees; wood and bark of fallen limbs; garden stake; associated with a white rot.
Culture Characters
2.3c.15.32.36.(37).(38).(39).(40).43.(50).54.(55).60 (Ref. 474).
References 92, 96, 97, 103, 116, 120, 127, 128, 130, 133, 183, 242, 249, 255, 267, 332, 402, 474, 584.

Peniophora junipericola Eriksson 1950
Symb. Bot. Upsal. 10 (5):52.
Distribution United States: FL, LA, MS.
Host *Juniperus virginiana.*
Ecology Dead branches; associated with a white rot.
Culture Characters
2.3.7.32.39.43.55.60 (Ref. 588) & 60 (Ref. 36).
References 201, 202.

Peniophora laeta (Fr.) Donk 1957
Fungus 27:17.
Basionym *Radulum laetum* Fr., Elenchus 1:152, 1828.
Synonym *Peniophora hydnoidea* (Pers.:Fr.) Donk 1933, a later homonym of *Peniophora hydnoides* Cooke & Massee 1889, fide Donk (152).
Distribution United States: CA.
Hosts *Amelanchier pallida, A. utahensis.*
Ecology Not known.
Culture Characters
2.3c.8.(26).32.36.(37).39.43-45.(50).54.60 (Ref. 474).
References 130, 474.

Peniophora laurentii Lundell 1946
In Lundell and Nannfeldt, Fungi Exs. Suec. No. 1342, Uppsala.
Distribution Canada: PQ.
Host *Populus tremuloides.*
Ecology Bark of log.
Culture Characters Not known.
Reference 243.

Peniophora lilacina (Berk. & Br.) Cooke 1879
Grevillea 8:20.
Basionym *Corticium lilacina* Berk. & Br., J. Linn. Soc., Bot. 14:70, 1875.
Distribution United States: AZ.
Host *Quercus gambelii.*
Ecology Dead branches; associated with a white rot.
Culture Characters Not known.
Reference 216.

Not to be confused with *Peniophora lilacina* (Schw.) Massee 1889, a later homonym and a synonym of *P. cinerea*, fide Burt (92).

Peniophora lycii (Pers.) Höhnel & Litsch. 1907
Sitzungsber. Kaiserl. Akad. Wiss., Math.-Naturwiss. Kl., Abt. I, 116:747.
Basionym *Thelephora lycii* Pers., Mycol. Eur. 1:148, 1822.
Synonym *Peniophora caesia* (Bres.) Bourd. & Galzin 1913, fide Eriksson et al. (169).
Distribution United States: DC, MO, VT.
Hosts *Betula* sp., *Quercus* sp., *Syringa* sp., angiosperm.
Ecology Fallen limbs.
Culture Characters
2.3.21.24.25.(26).32.36.(39).42-44.54.60 (Ref. 588).
Reference 92.

Peniophora malençonii Boidin & Lanquetin subsp. ***americana*** Chamuris 1987
Mycotaxon 28:547.
Distribution United States: AZ, CA, UT.
Hosts *Acer glabrum, Acacia* sp., *Aesculus californica, Adenostoma fasiculatum, Baccharis pilularis, Citrus limon, C. grandis, C. sinensis, Crossosoma* sp., *Encelia californica, Eucalyptus* sp., *E. globulus, E. robusta, Laurocerasus lyoni, Platanus* sp., *Quercus agrifolia, Rosa* sp., *Umbellularia californica.*
Ecology Dead, usually corticated, limbs and trunks of woody dicots; boards; vascular cylinder of cactus; associated with a white rot.
Culture Characters Not known.
Reference 102.

Peniophora nuda (Fr.) Bres. 1897
Atti Imp. Regia Accad. Rovereto, Ser. III, 3:114.
Basionym *Thelephora nuda* Fr., Syst. Mycol. 1:447, 1821.
Distribution Canada: MB, NB, NS, ON, PE. United States: AL, AZ, CO, FL, GA, IL, LA, MD, MI, MN, NJ, NY, OH, SC, TX, VT, VA, WA, WI.
Hosts *Acer* sp., *A. platanoides, A. saccharum, Betula* sp., *Celtis reticulata, Cercocarpus betuloides, Corylus californica, C. cornuta, Fagus grandifolia, Fraxinus velutina, Juglans major, Juniperus communis, Ostrya virginiana, Platanus wrightii, Populus* sp., *P. tremuloides, Prosopis* sp., *P. juliflora, Prunus serotina, Quercus* sp., *Q. arizonica, Q. gambelii, Q. hypoleucoides, Q. reticulata, Ribes divaricatum, Syringa vulgaris, Tilia americana, Weigelia florida,* angiosperm.
Ecology Branches; dead still attached branches; fallen limbs; slash; associated with a white rot.
Culture Characters
2.3c.15.21.(27).31g.32.36.(37).(39).40.42-43.(50). 54.60 (Ref. 474).
References 92, 96, 116, 127, 133, 183, 204, 207, 214, 216, 242, 249, 267, 398, 402, 474, 584.

Peniophora perexigua Jackson 1948
Canad. J. Res., C, 26:132.
Distribution Canada: ON. United States: AZ.
Hosts *Abies balsamea, Pseudotsuga menziesii, Tsuga canadensis,* gymnosperms.
Ecology Not known.
Culture Characters Not known.
References 116, 204, 216, 242, 302, 584.

Peniophora piceae (Pers.) Eriksson 1950
Symb. Bot. Upsal. 10 (5):49.
Basionym *Thelephora piceae* Pers., Mycol. Eur. 1:123, 1822.
Synonym *Peniophora separans* Burt 1926, fide Boidin and des Pomeys (47).
Distribution Canada: AB, AB-NT, BC, NB, NF. United States: MN.
Hosts *Abies balsamea, A. lasiocarpa, Picea* sp., *P. engelmannii, P. glauca, Pinus contorta, Pseudotsuga menziesii, Thuja plicata, Tsuga heterophylla,* gymnosperm.
Ecology Bark; log.
Culture Characters
2.3.(5).7.32.36.39.40.43.(48).55.60 (Ref. 487).
References 92, 116, 169, 214, 242, 486.

Peniophora pini (Schleich.:Fr.) Boidin
subsp. ***duplex*** (Burt) Weresub & S. Gibson 1960
Canad. J. Bot. 38:861.
Basionym *Peniophora duplex* Burt, Ann. Missouri Bot. Gard. 12:298, 1926.
Synonym *Corticium overholtsii* Burt 1926, fide Rogers and Jackson (567), and Weresub and Gibson (639).
Distribution Canada: ON. United States: CA, CT, FL, GA, MA, MD, NJ, NY, PA.
Hosts *Abies magnifica, Pinus* sp., *P. austriaca (cult.), P. banksiana, P. ponderosa, P. resinosa, P. rigida, P. strobus, P. sylvestris, P. taeda, P. virginiana*, gymnosperm.
Ecology Bark; dead branches.
Culture Characters
2.3.14.15.16.32.36.37.38.39.42.43.(48).55.60 (Ref. 487).
References 92, 93, 96, 116, 128, 242, 514, 584, 639.

The report (128) from California on *Abies magnifica* requires confirmation because it is beyond both geographic and host ranges of the fungus.

Peniophora pithya (Pers.) Eriksson 1950
Symb. Bot. Upsal. 10 (5):45.
Basionym *Thelephora pithya* Pers., Mycol. Eur. 1:146, 1822.
Distribution Canada: AB, AB-NT, BC, NB, NS, ON, PQ. United States: AZ, UT.
Hosts *Abies concolor, A. lasiocarpa, Acer macrophyllum, Alnus* sp., *Picea* sp., *P. engelmannii, P. glauca, P. mariana*, rarely on angiosperm, gymnosperm.
Ecology Bark and wood of logs; fallen branches; lumber.
Culture Characters
2.3.14.15.22.32.36.37.38.43.(48).52.55.60 (Ref. 487).
References 116, 183, 204, 216, 242, 249, 486, 584.

Peniophora polygonia (Pers.:Fr.) Bourd. & Galzin 1928
Hym. France, 320.
Basionym *Corticium polygonium* Pers., Tentamen dispositionis meth. fung., 30, 1797.
Synonym *Cryptochaete polygonia* (Pers.) Karsten 1889, *Gloeocystidium polygonium* (Pers.) Höhnel & Litsch. var. *fulvescens* Bres. 1925, fide Burt (93), and McKay and Lentz (452).
Distribution Canada: AB, BC, MB, NF, NFL, NS, NTM, ON, PQ, YT.
United States: AK, AZ, CO, ID, MN, NM, WA.
Hosts *Acer glabrum, Alnus* sp., *Populus* sp., *P. grandidentata, P. tremuloides, P. trichocarpa*.
Ecology Causes heart rot, top rot and incipient staining in aspen; saprophytic on rotten wood; basidiomes produced on dead bark; dead branches; especially fallen branches; slash; associated with a white rot.
Culture Characters
2.3.15.24.25.32.36.39.42-44.(48).54.60 (Ref. 588);
2.3.(15).26.32.36.(37).39.42.(48).50.54 (compiled from data in 452) & 60 (Ref. 36).
References 53, 93, 116, 119, 144, 180, 183, 204, 208, 214, 216, 242, 249, 255, 340, 349, 402, 406, 452, 603.

Peniophora pruinata (Berk. & Curtis) Burt 1926
Ann. Missouri Bot. Gard. 12:340.
Basionym *Stereum pruinatum* Berk. & Curtis, J. Linn. Soc., Bot. 10:332, 1868.
Distribution United States: AL, FL.
Hosts Angiosperm.
Ecology Rotten logs.
Culture Characters Not known.
Reference 92.

Peniophora pseudopini Weresub & S. Gibson 1960
Canad. J. Bot. 38:863.
Synonyms *Phlebia cervina* Overh. 1930, not validly published, fide Weresub and Gibson (639), *Stereum pini* N. Amer. Aucts., fide Weresub and Gibson (639).
Distribution Canada: AB, BC, MB, NTM, ON, PQ.
United States: MA, ME, NH, NY, OR, PA, WI.
Hosts *Abies balsamea, A. lasiocarpa, Picea glauca, P. mariana, Pinus* spp., *P. banksiana, P. contorta, P. ponderosa, P. resinosa, P. strobus, P. sylvestris, Pseudotsuga menziesii*.
Ecology Commonly isolated from mature lodgepole pine on which it was always associated with red heartwood stain; in one study caused 18% of the trunk rot in jack pine in Ontario; produces basidiomes on bark of dead branches; fallen limbs; bark of dead saplings; associated with a white rot.
Culture Characters
2.3c.15.26.31g.32.36.(38).(39).40.43.50.(53).55.60 (Ref. 474);
2.3.14.15.32.36.37.38.39.42.43.48.53.55.60 (Ref. 487).
References 89, 116, 242, 274, 414, 474, 486, 515, 516, 519, 639.

Peniophora pseudoversicolor Boidin 1965
Bull. Mens. Soc. Linn. Lyon 34:162.
Distribution United States: CO.
Host *Quercus gambelii*.
Ecology Slash; associated with a white rot.
Culture Characters
2.3.7.32.36.40.42-44.54 (Ref. 588).
References 398, 402.

Peniophora pusilla Jackson 1948
Canad. J. Res., C, 26:137.
Distribution Canada: ON, PQ. United States: NH.
Host *Tsuga canadensis*.
Ecology Bark.
Culture Characters Not known.

Known only from the holotype and two paratype specimens. The description of *P. pusilla* in Slysh (584) presumably was taken from Jackson's description.

Peniophora quercina (Pers.:Fr.) Cooke 1879
Grevillea 8:20.
Basionym *Thelephora quercina* Pers., Synop. Meth. Fung., 573, 1801.
Distribution United States: AZ, CA, NY.
Hosts *Mortonia scabrella, Quercus chrysolepis*, angiosperm.
Ecology Dead branches.
Culture Characters
2.3.(8).(21).(24).(25).32.36.(39).(40).42-44.54.60 (Ref. 588).
References 128, 216, 584.

Peniophora rufa (Fr.) Boidin 1959
Bull. Soc. Mycol. France 74:443.
Basionym *Thelephora rufa* Fr., Elenchus 1:187, 1828.
Synonyms *Stereum rufum* (Fr.) Fr. 1838, *Cryptochaete rufa* (Fr.) Karsten 1889, *Tubercularia pezizoidea* Schw. 1832, *Hypocrea richardsoni* Berk. & Mont. 1875. All fide Burt (89) and Lentz (379).
Distribution Canada: AB, BC, MB, NB, NF, NS, NTM, ON, PE, PQ, SK, YT. United States: AK, AZ, CO, IA, ID, MA, ME, MI, MN, MT, ND, NE, NH, NM, NY, OR, PA, VT, WI, WY.
Hosts *Populus* sp., *P. balsamifera, P. grandidentata, P. tremuloides, P. trichocarpa, Salix* sp.
Ecology Dead branches; slash; associated with a white rot.
Culture Characters
2.3.(13).15.24.25.32.36.(39).42-44.(48)54.60 (Ref. 588);
2.3.14.15.32.36.39.42.(48).(50).54 (compiled from data in 452) & 60 (Ref. 36).
References 89, 103, 116, 183, 204, 214, 216, 242, 334, 340, 379, 402, 406, 452, 509, 511 , 519.

Peniophora septentrionalis Laurila 1939
Ann. Bot. Soc. Zool.-Bot. Fenn. "Vanamo" 10(4):10.
Distribution Canada: AB, BC, NF, NS, NTM, ON, PQ, SK. United States: CO.
Hosts *Abies balsamea, A. lasiocarpa, Picea* sp., *P. engelmannii, P. glauca, P. glauca* var. *albertiana, P. mariana, P. rubens, Pinus banksiana, P. contorta, P. strobus, Pseudotsuga menziesii*, gymnosperm.
Ecology Causes a red heart rot in pines; commonly isolated from hail wounds; commonly enters trunk wounds; causing decay in *Picea glauca*; causes a minor trunk rot in *Abies lasiocarpa, P. engelmannii* and *P. mariana*; basidiomes typically produced on bark of fallen branches; logs; associated with a white rot.
Culture Characters
2.3.14.15.32.36.37.38.40.43.(48).55.60 (Ref. 487).
References 116, 169, 242, 249, 274, 402, 486, 539.

Peniophora seymouriana Burt 1926
Ann. Missouri Bot. Gard. 12:337.
Distribution United States: GA.
Hosts Angiosperm.
Ecology Decaying branches.
Culture Characters Not known.

Known only from the type specimen, which was redescribed by Liberta (392). A paratype from Cuba is not the same species (392).

Peniophora similis (Berk. & Curtis) Massee 1889
J. Linn. Soc., Bot. 25:147.
Basionym *Corticium simile* Berk. & Curtis, J. Linn. Soc., Bot. 10:337, 1868.
Distribution United States: FL.
Hosts Angiosperm.
Ecology Fallen limbs; underside of logs.
Culture Characters Not known.
Reference 92.

Peniophora sphaerocystidiata Burdsall & Nakasone 1983
Mycotaxon 17:261.
Distribution United States: MS.
Host *Vaccinium* sp.
Ecology Not known.
Culture Characters
2.3.15.32.36.37.(38).39.43.50.54 (Ref. 75);
2.3c.15.26.32.36.(37).(38).39.43.50.54 (Ref. 474).

Known only from the holotype specimen.

Peniophora tamaricicola Boidin & Malençon 1961
Rev. Mycol. (Paris) 26:153.
Distribution United States: AZ.
Hosts *Baccharis sarothroides, Carnegiea gigantea, Celtis reticulata, Cercidium microphyllum, Condalia mexicana, C. spathulata, Fouquieria splendens, Fraxinus velutina, Juglans major, Parkinsonia microphylla, Prosopis juliflora, P. velutina.*
Ecology Associated with a white rot.
Culture Characters
2.3c.8.15.32.36.(37).(38).39.44.50.54.60 (Ref. 474);
2.3.7.8.32.36.39.43.44.50.54 (Ref. 206);
2.3.9.15.32.36.39.42-43.50.54.60 (Ref. 477).
References 183, 206, 207, 216, 403, 474.

Gilbertson and Burdsall (206) inadvertently omitted number 32 from the species code. We have inserted it above.

Peniophora unica Jackson & Dearden 1949
Canad. J. Res., C, 27:148.
Distribution Canada: BC.
Host *Abies lasiocarpa.*
Ecology Not known.
Culture Characters Not known.

Known only from the holotype specimen.

Peniophora violaceolivida (Sommerf.) Massee 1890
J. Linn. Soc., Bot. 25:152.
Basionym *Thelephora violaceolivida* Sommerf., Suppl. Florae Lapponicae, 283, 1826.

Distribution Canada: AB, AB-NT, NS, ON, PQ. United States: AK, AZ, CA, CO, DC, GA, ID, IL, LA, MA, MD, MI, MN, NC, NJ, NY, VT, WI.
Hosts *Abies magnifica, Acer* sp., *A. saccharum, Alnus* sp., *A. rugosa, Baccharis* sp., *Betula alleghaniensis, Castanea* sp., *Fagus grandifolia, Fraxinus* sp., *Ligustrum* sp., *L. sinense, Liquidambar styraciflua, Ostrya* sp., *Picea engelmannii, Pinus* sp., *P. banksiana, P. contorta, Populus* sp., *P. tremuloides, Prunus* sp., *Quercus* sp., *Q. gambelii, Q. reticulata, Salix* sp., *S. alaxensis, Thuja plicata*, angiosperm, gymnosperm.
Ecology Rotting wood; decorticated branches; fallen limbs; slash; dead stem; associated with a white rot.
Culture Characters 2.3c.15.21.(27).31g.32.36.(37).(39).40.42-43.(50). 54.(55).60 (Ref. 474).
References 92, 96, 103, 108, 128, 130, 177, 183, 204, 214, 216, 242, 249, 255, 267, 344, 398, 402, 406, 474, 584.

PHAEOGLABROTRICHA W.B. Cooke 1961 Cyphellaceae
The type species of this genus is *Cyphella sessilis* Burt which Reid (554) placed in synonymy with *Pellidiscus pallidus*. Thus the following epithet is in limbo and awaiting transfer to an appropriate genus.

Phaeoglabrotricha globosa W.B. Cooke 1961 Beih. Sydowia 4:115.
Distribution United States: DE.
Host *Fraxinus americana*.
Ecology Not known.
Culture Characters Not known.

Known only from the type specimen.

PHAEOSOLENIA Speg. 1902 Crepidotaceae

Phaeosolenia betulae W.B. Cooke 1961 Beih. Sydowia 4:122.
Distribution Canada: ON.
Host *Betula alleghaniensis*.
Ecology Bark.
Culture Characters Not known.

Known only from the type specimen.

Phaeosolenia brenckleana (Sacc.) W.B. Cooke 1961 Beih. Sydowia 4:122.
Basionym *Solenia brenckleana* Sacc., Atti Reale Accad. Sci. Lett. Arti Padova, N.S. 33:163, 1917.
Distribution United States: KS, ND.
Host *Ulmus* sp.
Ecology Not known.
Culture Characters Not known.
Reference 125.

Phaeosolenia densa (Berk.) W.B. Cooke 1961 Beih. Sydowia 4:123.
Basionym *Cyphella densa* Berk., Flora N. Zealand 184, 1855.
Synonym *Cyphella endophila* Ces. in Rabenhorst 1872, fide Cooke (125), *Solenia endophila* (Ces.) Fr. 1874.
Distribution Canada: ON. United States: CO, DE, FL, ME, MA, PA, SC, VT.
Hosts *Populus* sp., *P. angustifoloia*.
Ecology Bark; decorticated wood.
Culture Characters Not known.
References 91, 125, 402.

A number of host plants were listed by Cooke (125) but it is not clear which records were from Canada and the United States.

Phaeosolenia ravenelii (Berk. & Curtis) W.B. Cooke 1961 Beih. Sydowia 4:127.
Basionym *Cyphella ravenelii* Berk. & Curtis, Grevillea 2:5, 1873.
Distribution United States: OH, SC.
Hosts *Carya* sp., *Juglans* sp.
Ecology Bark; decorticated wood.
Culture Characters Not known.
References 80, 125, 131.

PHANEROCHAETE Karsten 1889 Phanerochaetaceae

Phanerochaete affinis (Burt) Parm. 1968 Consp. Syst. Cort. 84.
Basionym *Peniophora affinis* Burt, Ann. Missouri Bot. Gard. 12:266, 1926.
Synonyms *Peniophora laevis* (Fr.) Burt in Peck sensu Burt 1902, *Phanerochaete laevis* (Fr.) Eriksson & Ryv. sensu Eriksson and Ryvarden 1978, fide Burdsall (64), Eriksson et al. (169), Jülich (325), and Jülich and Stalpers (326), *Peniophora gilvidula* Bres. 1925, fide Burdsall (64).
Distribution Canada: AB-NT, BC, MB, NB, NS, ON, PE, PQ. United States: AK, AL, AZ, CA, CT, ID, IL, KY, MA, MD, ME, MI, MN, MO, MT, NC, NE, NH, NJ, NY, OH?, OR, SC, TN, TX, VT, WA, WI.
Hosts *Abies magnifica, Acer* sp., *A. rubrum, A. saccharum, A. spicatum, Alnus* sp., *A. incana, A. sinuata, Betula* sp., *B. lenta, B. populifolia, Carpinus* sp., *Castanea* sp., *C. dentata, Cornus florida, Corylus* sp. *Crataegus* sp., *Fagus* sp., *F. grandifolia, Pinus* sp., *P. contorta* var. *contorta* and var. *latifolia, P. ponderosa, Populus* sp., *P. grandidentata, P. tremuloides, P. trichocarpa, Prunus pensylvanica, P. persica, P. virginiana, Quercus* sp., *Q. arizonica, Q. garryana, Q. rubra, Q. rubra* var. *borealis, Salix* sp., *Thuja plicata, Ulmus* sp., *U. americana*, angiosperm, gymnosperm.
Ecology Bark; slash and woody debris; decorticated logs and limbs; causing a white rot.

Culture Characters
2.5.31g.32.36.38.(40).42-43.(50).54.55.(57).(58).(59) (Ref. 474) & 59 (Ref. 36).
References 53, 64, 92, 97, 116, 119, 128, 133, 169, 183, 197, 214, 242, 249, 267, 308, 406, 474, 511, 542, 584.

Burt (93:280), and earlier authors, recognized that the name *Corticium laeve* had been used for two different fungi. He treated them as *Corticium laeve* Pers. and *Peniophora laevis* (Fr.) Burt (92:257). Both have the same basionym. Burt erred in attributing the combination in *Peniophora* to "Burt in R. Fries ... 1900." We did not find the name *Peniophora* nor any combination with *laevis* in R. Fries' article, Göteb. K. Vetensk. Vitterhets-samh. Hand. Sjatte Foljden, Ser. B, 4 (3):1-38, 1901. As Jülich (325), and Jülich and Stalpers (326) indicate, the earliest name for this fungus is *Peniophora affinis*.

Phanerochaete allantospora Burdsall & Gilbn. 1974
Mycologia 66:780.
Distribution United States: AZ.
Hosts *Fouquieria splendens, Fraxinus velutina, Juglans major, Platanus wrightii, Prosopis juliflora, P. velutina*, other desert trees and shrubs.
Ecology Dead branches; associated with a white rot.
Culture Characters
1.6.14.22.34.36.38.43-45.(48).53.54 (Ref. 477);
1.6.7.(14).32.36.(37).38.(40).43-45.(48).(53).54 (Ref. 474);
1.6.11.22.32.36.38.44.45.53.54 (Ref. 66).
References 64, 66, 183, 207, 216, 474, 477.

The listing (183) of *Pinus ponderosa* as a substrate was an error in compilation.

Phanerochaete arizonica Burdsall & Gilbn. 1974
Mycologia 66:785.
Distribution United States: AZ, CA, CO, MN, NM.
Hosts *Cercidium microphyllum, Cercocarpus breviflorus, Fouquieria splendens, Lemaireocereus thurberi, Mimosa biuncifera, Mortonia scabrella, Opuntia versicolor, Parkinsonia microphylla, Platanus wrightii, Populus fremontii, P. tremuloides, Prosopis juliflora, P. velutina, Prunus persica, Quercus gambelii*, angiosperm.
Ecology Twigs; slash; associated with a white rot.
Culture Characters
2.5.(7).21.26.32.36.38.43-44.(53).54 (Ref. 474);
2.5.11.32.36.40.43.44.53.54 (Ref. 66).
References 64, 66, 183, 207, 208, 216, 399, 402, 474, 477.

The cultural characters in Nakasone and Gilbertson (477) were based on cultures of *Phanerochaete sordida*, fide Nakasone (474), and they appear under that species.

Phanerochaete avellanea (Bres.) Eriksson & Hjort. 1981
Cort. N. Europe 6:1072.
Basionym *Corticium avellaneum* Bres. in Bourdot and Galzin, Bull. Soc. Mycol. France 27:236, 1911.
Distribution United States: AZ, CO, NC, OH, WI.
Hosts *Alnus* sp., *Platanus wrightii, Populus angustifolia, Quercus* sp., *Q. gambelii.*
Ecology Twigs; slash; log; associated with a white rot.
Culture Characters Not known.
References 64, 216, 327, 402.

Phanerochaete burtii (Romell) Parm. 1967
Izv. Akad. Nauk Estonsk. SSR, Ser. Biol. 16:388.
Basionym *Peniophora burtii* Romell in Burt, Ann. Missouri Bot. Gard. 12:279, 1926.
Distribution Canada: NB, ON. United States: AL, AZ, CA, DC, FL, IL, LA, MA, MD, MI, MS, MT, NJ, NM, NY, OH, OK, TN, VA, VT, WI, WV.
Hosts *Alnus oblongifolia, Fagus* sp., *Juglans* sp., *Liquidambar styraciflua, Nyssa sylvatica, Pinus palustris, P. ponderosa, P. strobus, P. taeda, Platanus wrightii, Populus fremontii, Prunus serotina, Pseudotsuga menziesii, Quercus* sp., *Q. arizonica, Q. emoryi, Q. gambelii, Q. nigra, Q. stellata, Q. virginiana, Sabal palmetto, Sequoia sempervirens, Ulmus alata*, angiosperm.
Ecology Fallen limbs; associated with a white rot.
Culture Characters
(1).2.5.(7).21.32.36.(39).(40).42-43.(53).54.(55) (Ref. 474).
References 57, 61, 64, 92, 183, 204, 216, 242, 267, 474, 584.

Eriksson et al. (169) after careful study placed *Peniophora burtii* and *P. limonea* in synonymy under *Corticium calotrichum*, whereas Burdsall (64) also having studied type specimens recognized *P. burtii* as distinct from *P. sanguinea*, and *C. calotrichum* and *P. limonea* were listed as synonyms of *P. sanguinea*.

Phanerochaete cacaina (Bourd. & Galzin) Burdsall & Gilbn. 1974
Mycologia 66: 781.
Basionym *Peniophora cacaina* Bourd. & Galzin, Bull. Soc. Mycol. France 28:397, 1913.
Distribution United States: AZ, NM.
Host *Pinus ponderosa.*
Ecology Not known.
Culture Characters Not known.
References 64, 197, 216.

Phanerochaete cana (Burt) Burdsall 1985
Mycologia Memoir 10:50.
Basionym *Peniophora cana* Burt, Ann. Missouri Bot. Gard. 12:227, 1926.
Distribution United States: FL.
Host *Liquidambar styraciflua.*
Ecology Brittle wood humus; woody debris.
Culture Characters Not known.
References 64, 92, 392.

Known from only two specimens.

Phanerochaete carnosa (Burt) Parm. 1967
Izv. Akad. Nauk Estonsk. SSR, Ser. Biol. 16:388.
Basionym *Peniophora carnosa* Burt, Ann. Missouri Bot. Gard. 12:325, 1926.
Synonym *Peniophora firma* Burt 1926, fide Burdsall (64).
Distribution Canada: BC, MB, NF, NS, ON, PQ. United States: AK, AZ, CA, CO, ID, MA, ME, MI, MN, MT, NH, NM, NY, OR, SD, VT, WA.
Hosts *Abies* sp., *A. amabilis, A. balsamea, A. concolor, A. grandis, A. lasiocarpa, A. magnifica, Acer* sp., *A. spicatum, Alnus* sp., *A. oblongifolia, Arctostaphylos patula, Betula* sp., *B. pumila, Fagus grandifolia, Juniperus communis, J. nana, J. virginiana, Larix occidentalis, ?Picea* sp., *P. engelmannii, P. mariana, P. rubens, Pinus monticola, P. ponderosa, P. ponderosa* var. *arizonica, P. rigida, P. strobus, Platanus wrightii, Populus* sp., *P. grandidentata, P. tremuloides, P. trichocarpa, Pseudotsuga menziesii, Quercus gambelii, Robinia neo-mexicana, Thuja occidentalis, T. plicata, Umbellularia californica*, rarely angiosperm, gymnosperm.
Ecology Bark and wood; rotten wood; underside of prostrate rotten limb; logs; associated with a white rot.
Culture Characters
(1).2.5.(7).21.32.36.(39).40.43.(53).(54).55 (Ref. 474).
References 64, 92, 116, 119, 120, 128, 183, 197, 204, 208, 214, 216, 242, 249, 392, 399, 402, 406, 474, 511, 584.

Phanerochaete chrysorhiza (Torrey) Budington & Gilbn. 1973
Southw. Naturalist 17:417.
Basionym *Hydnum chrysorhizon* Torrey in Eaton, Manual Bot. 309, 1822.
Synonyms *Hydnum chrysocomum* Underw. 1897, fide Burdsall (64), *Hydnum fragilissimum* Berk. & Curtis 1873, fide Burdsall (64), *Mycoacia fragilissima* (Berk. & Curtis) L.W. Miller & Boyle 1943, *Oxydontia fragilissima* (Berk. & Curtis) L.W. Miller 1933, *Odontia fragilissima* (Berk. & Curtis) C.A. Brown 1935.
Distribution United States: AL, AZ, FL, GA, IA, IL, LA, MD, MI, MS, NC, NY, OH, SC, TN, WI.
Hosts *Acer* sp., *Carnegiea gigantea, Carpinus caroliniana, Citrus* sp., *Cornus florida, Fouquieria splendens, Fraxinus* sp., *F. velutina, Gossypium hirsutum, Haplopappus laricifolius, Juglans* sp., *J. major, Liquidambar sytraciflua, Lycium* sp., *Nectandra coriacea, Nyssa sylvatica, Olneya tesota, Ostrya virginiana, Pinus* sp., *P. taeda, Platanus wrightii, Populus* sp., *P. grandidentata, P. tremuloides, Prosopis juliflora, Quercus* sp., *Q. arizonica, Q. emoryi*, angiosperm.
Ecology Dead branches; rotten logs; roots; associated with a white rot.
Culture Characters
2.5.(16).21.32.36.40.42-43.54.55.57 (Ref. 474).
References 56, 61, 64, 73, 191, 192, 207, 216, 403, 406, 463, 474.

Phanerochaete chrysosporium Burdsall 1974
Mycotaxon 1:124.
Anamorph *Sporotrichum pulverulentum* Novobr. 1972, fide Burdsall (62a).
Distribution Canada: BC, NB. United States: AZ, CA, GA, IL, MD, ME, MS, NY, WI.
Hosts *Abies balsamea, Fagus grandifolia, Heteromeles arbutifolia, Liriodendron tulipifera, Picea* sp., *Pinus* sp., *P. contorta, P. resinosa, Platanus wrightii, Populus fremontii, Sequoia sempervirens, Ulmus americana*, angiosperm.
Ecology Fallen branches; piles of wood chips; causing a white rot.
Culture Characters
1.6.(7).14.33.34.35.36.40.41.(48).(51).54.55.(57).(59) (Ref. 474);
1.6.14.24.32.33.34.35.36.40.41.(48).54.55 (Ref. 65);
1.6.(7).14.(25).26.33.34.35.36.40.41-42.(48).54.55 (Ref. 589).
References 64, 177, 183, 242, 474.

Phanerochaete corymbata (G.H. Cunn.) Burdsall 1985
Mycologia Memoir 10:65.
Basionym *Corticium corymbatum* G.H. Cunn., Trans. Roy. Soc. New Zealand 82:324, 1954.
Distribution United States: LA.
Host *Ulmus rubra*.
Ecology Branches; associated with a white rot.
Culture Characters Not known.
Reference 64.

Phanerochaete crassa (Lév.) Burdsall 1985
Mycologia Memoir 10:67.
Basionym *Thelephora crassa* Lév., Ann. Sci. Nat. Bot., Sér. III, 2:209, 1844.
Synonyms *Laxitextum crassum* (Lév.) Lentz 1955, *Lopharia crassa* (Lév.) Boidin 1959, *Stereum umbrinum* Berk. & Curtis 1873, a later homonym of *Stereum umbrinum* Fr. 1847, *Hymenochaete multispinulosa* Peck 1882, *Hymenochaete scabriseta* Cooke 1883, *Hymenochaete purpurea* Cooke & Morgan 1883, *Peniophora intermedia* Massee 1889. All facultative synonyms fide Burt (89) and Lentz (379).
Distribution United States: AL, AR, AZ, CT, FL, GA, IA, ID, IL, IN, KS, KY, LA, MD, MO, MS, NC, NH, OH, PA, SC, TN, TX, VA.
Hosts *Acacia* sp., *Agave palmeri, Alnus rubra, Canotia holocantha, Carpinus caroliniana, Carya* sp., *Castanea* sp., *Dasylirion wheeleri, Fagus grandifolia, Fouquieria splendens, Ilex opaca, Juglans major, Liquidambar styraciflua, Liriodendron tulipifera, Platanus wrightii, Prosopis juliflora, P. velutina, Prunus persica, Quercus* sp., *Q. alba, Q. arizonica, Q. emoryi, Vitis labrusca*, angiosperm.

Ecology Dead trunks and branches; dead fallen branches; test crosses; associated with a white rot.
Culture Characters
(1).2.5.7.32.36.(37).38.(40).43.(53).54.(57) (Ref. 474);
2.5.7.34.36.40.41-42.54 (Ref. 477).
References 64, 89, 96, 177, 179, 183, 204, 207, 216, 333, 379, 474, 477, 519, 620.

Phanerochaete ericina (Bourd.) Eriksson & Ryv. 1978 Cort. N. Europe 5:1011.
Basionym *Peniophora ericina* Bourd., Rev. Sci. Bourbonnais Centr. France 23:14, 1910.
Distribution United States: GA, IL, MD, NC, TN.
Hosts *Acer* sp., *Carya* sp., *Fagus grandifolia, Liriodendron tulipiferae, Quercus* sp., *Robinia pseudoacacia*, angiosperm.
Ecology Twigs; fallen branches; associated with a white rot.
Culture Characters
2.5.21.31g.32.36.38.42.54.(57) (Ref. 474).
References 64, 474.

Phanerochaete exilis (Burt) Burdsall 1985 Mycologia Memoir 10:74.
Basionym *Peniophora exilis* Burt, Ann. Missouri Bot. Gard. 12:239, 1926.
Distribution United States: FL.
Hosts Angiosperm.
Ecology Fallen woody debris; log; associated with a white rot.
Culture Characters
1.(2).5.7.(22).32.36.40.42-43.(50).(51).54 (Ref. 474).
References 64, 474.

Phanerochaete filamentosa (Berk. & Curtis) Burdsall 1975 Dist. Hist. Biota S. Appalachians Part IV: 278.
Basionym *Corticium filamentosum* Berk. & Curtis, Grevillea 1:178, 1873.
Synonyms *Peniophora filamentosa* (Berk. & Curtis) Burt in Coker 1921, fide Eriksson et al. (169), *Peniophora unicolor* Peck 1890, fide Burdsall (64).
Distribution Canada: PQ.
United States: AL, AR, AZ, CA, FL, IL, KY, MA, MD, MI, MO, NC, NH, NJ, NY, OH, TN, VA, VT.
Hosts *Abies concolor, Acer* sp., *A. saccharum, Betula alleghaniensis, Castanea dentata, Cornus* sp., *Fagus* sp., *F. grandifolia, Liriodendron tulipifera, Platanus wrightii, Populus* sp., *P. tremuloides, Prunus serotina, Quercus* sp., *Q. arizonica, Q. gambelii, Q. nigra, Robinia pseudoacacia, Tilia* sp., *T. americana, Ulmus americana*, angiosperm, gymnosperm.
Ecology Bark; decorticated and rotten wood; decaying wood; fallen limbs; live and dead trees; logs; associated with a white rot.
Culture Characters
2.3i.13.21.32.(36).(37).38.(40).(44).45-47.54.55 (Ref. 474).
References 61, 64, 92, 108, 128, 130, 183, 216, 267, 327, 406, 474, 584.

The unusual records on gymnosperms need confirmation.

Phanerochaete galactites (Bourd. & Galzin) Eriksson & Ryv. 1978
Cort. N. Europe 5:1008.
Basionym *Corticium rhodoleucum* Bourd. subsp. *galactites* Bourd. & Galzin, Hym. France, 189, 1928.
Distribution United States: NC.
Host *Betula alleghaniensis.*
Ecology Bark; twig.
Culture Characters Not known.
Reference 327.

Phanerochaete hiulca (Burt) Welden 1980 Mycotaxon 10:441.
Basionym *Peniophora hiulca* Burt, Ann. Missouri Bot. Gard. 12:272, 1926.
Distribution United States: FL, GA, LA, MS.
Hosts *Quercus* sp., *Q. laurifolia, Q. virginiana*, angiosperm.
Ecology Usually on fallen small branches and twigs, occasionally on larger branches and logs; associated with a white rot.
Culture Characters
2.3r.7.(22).32.36.(37).40.42-43.54 (Ref. 474).
References 64, 474.

Phanerochaete jose-ferreirae (D. Reid) D. Reid 1975 Acta Bot. Croat. 34:135.
Basionym *Corticium jose-ferreirae* D. Reid, Rev. Biol. 5: 140, 1965.
Distribution United States: WI.
Hosts *Quercus* sp., *Q. alba.*
Ecology Associated with a white rot.
Culture Characters
1.3r.7.(22).32.36.38.(40).43.53.55 (Ref. 474).
Reference 474.

Phanerochaete magnoliae (Berk. & Curtis) Burdsall 1985 Mycologia Memoir 10:95.
Basionym *Radulum magnoliae* Berk. & Curtis, Hooker's J. Bot. Kew Gard. Misc. 1:236, 1849.
Synonym *Phanerochaete cumulodentata* (Nikolaeva) Parm. 1968 nom. nudum, fide Burdsall (64).
Distribution Canada: ON.
United States: FL, MN, NY, SC.
Hosts *Betula alleghaniensis, B. pumila, Magnolia virginiana, Populus* sp., *Quercus* sp., *Salix nigra.*
Ecology Slash and woody debris; associated with a white rot.
Culture Characters
(1).2.6.7.34.36.(38).40.41.54 (Ref. 474).
References 64, 183, 192, 214, 474.

Phanerochaete omnivora (Shear) Burdsall & Nakasone 1978 Mycotaxon 7:17.

Basionym *Hydnum omnivorum* Shear, J. Agric. Res. 30:476, 1925.

Distribution United States: AZ, FL, TX.

Hosts *Acacia* sp., *Baccharis* sp., *Carnegiea gigantea*, *Carya* sp., *Chilopsis linearis*, *Fouquieria splendens*, *Maclura pomifera*, *Platanus* sp., *P. wrightii*, *Prosopis juliflora*, *P. velutina*, native angiosperms.

Ecology Twigs and small branches; railroad ties; associated with a white rot.

Culture Characters
2.5.16.20.24.32.37.40.42-43.54 (Ref. 73);
2.5.7.(16).32.36.(37).40.43-44.54 (Ref. 474).

References 64, 73, 191, 474, 477.

Phanerochaete salmoneolutea Burdsall & Gilbn. 1974 Mycologia 66:787.

Distribution United States: AZ, FL, LA.

Hosts *Carya* sp., *Quercus arizonica*, angiosperm.

Ecology Fallen branches; associated with a white rot.

Culture Characters Not known.

References 64, 66.

Phanerochaete sanguinea (Fr.) Pouzar 1973 Ceská Mykol. 27:26.

Basionym *Thelephora sanguinea* Fr., Elenchus 1:203, 1828.

Synonyms *Peniophora sanguinea* (Fr.) Höhnel & Litsch. 1906, *Peniophora limonea* Burt 1926, fide Burdsall (64), and Rogers and Jackson (567), *Peniophora miniata* sensu Burt 1926, fide Rogers and Jackson (567).

Distribution Canada: AB-NT, BC, NS, ON, PQ. United States: AL, AZ, CA, CO, FL, GA, ID, LA, MA, MD, ME, MI, MN, MS, MT, NC, NH, NJ, NM, NY, OR, PA, SC, TN, VA, WI, WA.

Hosts *Abies grandis*, *Acer glabrum*, *A. saccharum*, *Betula alleghaniensis*, *B. papyrifera*, *Castanea dentata*, *Fagus grandifolia*, *Picea* sp., *P. engelmannii*, *P. glauca*, *P. rubens*, *Pinus* sp., *P. banksiana*, *P. contorta*, *P. ponderosa*, *P. resinosa*, *P. strobus*, *Populus* sp., *P. tremuloides*, *Pseudotsuga menziesii*, *Quercus* sp., *Q. arizonica*, *Q. emoryi*, *Q. gambelii*, *Q. garryana*, *Q. hypoleucoides*, *Q. nigra*, *Q. rubra*, *Robinia neomexicana*, *Sequoia sempervirens*, *Thuja occidentalis*, *T. plicata*, *Tsuga canadensis*, *T. heterophylla*, angiosperm, gymnosperm.

Ecology Bark; decayed wood; twig; branch; bark on root; slash; fallen limbs; log; associated with a white rot.

Culture Characters
1.2.5.(16).21.(26).32.36.(37).38.42-44.(45).(50).(53).54.55 (Ref. 474).

References 64, 92, 116, 119, 120, 128, 183, 197, 204, 208, 214, 216, 242, 249, 327, 335, 400, 402, 406, 474, 584, 615.

See comment under *P. burtii*.

Phanerochaete sordida (Karsten) Eriksson & Ryv. 1978 Cort. N. Europe 5:1023.

Basionym *Corticium sordida* Karsten, Meddeland. Soc. Fauna Fl. Fenn. 9:65, 1882.

Synonyms *Peniophora sordida* (Karsten) Burt 1926, *Corticium glabrum* Berk. & Curtis 1873, fide Hjortstam (288), *Peniophora cremea* (Bres.) Sacc. & Syd. 1902, fide Burdsall (64), and Rogers and Jackson (567), *Phanerochaete cremea* (Bres.) Parm. 1968, *Peniophora arachnoidea* Burt 1926, fide Burdsall (64), and Rogers and Jackson (567).

Distribution Canada: AB, BC, MB, NB, NS, NTM, ON, PE, PQ. United States: AK, AL, AZ, CA, CO, DC, FL, GA, ID, IL, LA, MA, MD, MI, MS, MT, NC, NH, NJ, NM, NY, OR, TN, VT, WA, WI.

Hosts *Abies balsamea*, *A. fraseri*, *A. lasiocarpa*, *Acer* sp., *A. macrophyllum*, *A. rubrum*, *A. saccharum*, *A. spicatum*, *Alnus* sp., *A. oblongifolia*, *A. rugosa*, *A. rubra*, *A. sinuata*, *A. tenuifolia*, *Betula* sp., *B. alleghaniensis*, *B. occidentalis*, *B. papyrifera*, *Carya illinoensis*, *Corylus* sp., *Crataegus* sp., *Delonix regia*, *Elaeagnus commutata*, *Fagus grandifolia*, *Juglans cinerea*, *Heteromeles arbutifolia*, *Larix occidentalis*, *Liquidambar styraciflua*, *Liriodendron tulipifera*, *Nyssa* sp., *Picea* sp., *P. glauca*, *Pinus* sp., *P. banksiana*, *P. contorta*, *P. ponderosa*, *P. resinosa*, *P. taeda*, *Piptoporus betulinus*, *Platanus wrightii*, *Populus* sp., *P. tremuloides*, *P. trichocarpa*, *Quercus* sp., *Q. gambelii*, *Q. kelloggii*, *Q. nigra*, *Q. rubra*, *Rhododendron* sp., *Thuja occidentalis*, *Tsuga heterophylla*, *Viburnum obovatum*, angiosperm, gymnosperm, herb.

Ecology Corticated and decorticated fallen branches; slash; decorticated wood on the ground; creosote or penta-treated poles; associated with a white rot.

Culture Characters
(1).2.(5).6.7.(14).32.(34).36.38.(40).42.(48).(50).(53).54.(55).(57) (Ref. 474);
2.6.14.34.36.40.41.48.54 (Ref. 477);
2.(5).7.(13).36.38.41-42.(50).54.(55).57 (Ref. 609).

References 64, 92, 116, 119, 128, 133, 177, 183, 197, 204, 208, 216, 242, 249, 267, 327, 400, 402, 406, 474, 584, 609, 654.

Nakasone (474) pointed out that the cultural characters under *Phanerochaete arizonica* in Nakasone and Gilbertson (477) were based on cultures of *P. sordida*. Thus they are listed here.

Phanerochaete subceracea (Burt) Burdsall 1985 Mycologia Memoir 10:128.

Basionym *Corticium subceracea* Burt, Ann. Missouri Bot. Gard. 13:239, 1926.

Distribution United States: AL, DC, IL, KY, MD, NC, OH, PA.

Hosts *Quercus* sp., *Rhus* sp.?, rarely on *Pinus* sp., angiosperm.

Ecology Bark and wood; associated with a white rot.

Culture Characters Not known.

References 64, 93, 393.

Phanerochaete tuberculata (Karsten) Parm. 1968
Consp. Syst. Cort., 83.
Basionym *Corticium tuberculatum* Karsten, Hedwigia 35:45, 1896.
Distribution Canada: BC, ON.
United States: AZ, CO, MI, MN, PA, WI.
Hosts *Acer saccharum, Agave palmeri, A. parryi, Betula papyrifera, Carnegiea gigantea, Celtis reticulata, Chilopsis linearis, Condalia mexicana, Fouquieria splendens, Fraxinus* sp., *Juglans major, Lycium* sp., *Platanus wrightii, Populus* sp., *P. fremontii, P. trichocarpa, Prosopis juliflora, Quercus emoryi, Q. gambelii, Q. oblongifolia, Vauquelinia californica, Yucca brevifolia*, angiosperm.
Ecology Decayed wood; slash; associated with a white rot.
Culture Characters
1.5.(6).7.21.32.36.40.41-42.(43).48.53.54.57 (Ref. 477).
References 64, 93, 116, 207, 214, 216, 242, 398, 402, 403, 477.

Phanerochaete velutina (DC.:Fr.) Karsten 1898
Kritisk öfversigt Finlands Basidsvampar Tillägg III, 33.
Basionym *Thelephora velutina* DC., Flora France 6:33, 1815.
Synonyms *Peniophora velutina* (DC.) Cooke 1879, *Peniophora rhodochroa* Bres. 1925, *Corticium rhodellum* Peck 1889. All fide Burdsall (64).
Distribution Canada: AB, BC, MB, NS, NTM, ON, PQ. United States: AL, AZ, CA, CO, CT, ID, IL, LA, MA, MD, ME, MI, MN, MT, NC, NH, NJ, NM, NY, OR, PA, TN, VT, WA, WI.
Hosts *Abies concolor, A. lasiocarpa, Acer* sp., *A. saccharum, Alnus incana, A. oblongifolia, A. tenuifolia, Amelanchier* sp., *Arbutus menziesii, Arctostaphylos pungens, Betula papyrifera, Carya* sp., *Castanea* sp., *Corylus cornuta, C. cornuta* var. *californica, Crossosoma* sp., *C. californicum, Fagus* sp., *F. grandifolia, Juniperus deppeana* var. *pachyphloea, Liquidambar styraciflua, Liriodendron tulipifera, Magnolia* sp., *Picea englemannii, Pinus* sp., *P. cembroides, P. engelmannii, P. ponderosa, P. resinosa, P. strobus, Platanus wrightii, Populus* sp., *P. tremuloides, Pseudotsuga menziesii, Quercus* sp., *Q. arizonica, Q. emoryi, Q. gambelii, Q. hypoleucoides, Q. reticulata, Q. robur, Q. rubra, Thuja plicata*, angiosperm, gymnosperm.
Ecology Decaying limbs; slash; decaying logs; associated with a white rot.
Culture Characters
2.5.(16).(21).(26).32.36.38.40.43.(53).54.55.(57) (Ref. 474).
References 53, 64, 92, 116, 133, 183, 197, 204, 208, 214, 216, 242, 245, 249, 267, 333, 334, 398, 402, 406, 465, 474, 584.

Phanerochaete viticola (Schw.) Parm. 1967
Izv. Akad. Nauk Estonsk. SSR, Ser. Biol. 16:389.
Basionym *Thelephora viticola* Schw., Schriften Naturf. Ges. Leipzig 1:107, 1822.
Synonyms *Corticium viticola* (Schw.) Fr. 1838, *Peniophora viticola* (Schw.) Höhnel & Litsch. 1907, *Corticium crocicreas* Berk. & Curtis 1873, fide Burt (93) and Hjortstam (288).
Distribution United States: AL, AR, KY, MD, MS, NC, NH, NY, TN, VT.
Hosts *Abies* sp., *A. balsamea, A. fraseri, Acer* sp., *A. saccharum, A. spicatum, Betula* sp., *B. alleghaniensis, Fagus* sp., *F. grandifolia, Picea* sp., *Thuja occidentalis, Viburnum alnifolium, Vitis* sp., angiosperm, gymnosperm.
Ecology Bark and wood; bark covered with moss and lichens; dead bark of live vine; twig; log; slash and debris.
Culture Characters Not known.
References 61, 64, 92, 108, 183, 327, 584.

Phanerochaete xerophila Burdsall 1985
Mycologia Memoir 10:141.
Distribution United States: AZ.
Hosts *Baccharis* sp., *Carnegiea gigantea, Opuntia versicolor, Platanus wrightii, Prosopis velutina, Quercus arizonica.*
Ecology Dead, fallen branches.
Culture Characters Not known.

Known only from the type specimen.

PHLEBIA Fr. 1821 Meruliaceae
Synonym is *Merulius* Boehmer:Fr. 1760, fide Nakasone and Burdsall (474a).

Phlebia albida Post 1863
In Fries, Monogr. Hym. Suec. II:280.
Synonym *Phlebia canadensis* W.B. Cooke 1956, fide Eriksson et al. (170).
Distribution Canada: BC, NS, ON. United States: AZ, AK, CA, CO, ID, IL, KS, KY, ME, MI, MT, NH, NM, NY, OR, PA, SC, TN, VT, WA, WV, WI.
Hosts *Abies alba, A. amabilis, A. arizonica, A. balsamea, A. concolor, A. grandis, A. lasiocarpa, A. lasiocarpa* var. *arizonica, A. lasiocarpa* var. *lasiocarpa, A. magnifica, Acer* sp., *A. saccharum, A. spicatum, Alnus* sp., *A. incana, A. tenuifolia, Betula* sp., *B. alba* var. *papyrifera, B. alleghaniensis, Carya* sp., *Corylus avellana, Fagus* sp., *Fagus grandifolia, F. silvatica, Larix occidentalis, Picea* sp., *P. canadensis, P. engelmannii, P. glauca, P. parryana, P. pugens, Pinus* sp., *P. albicaulis, P. contorta* var. *latifolia, P. engelmannii, P. leiophylla* var. *chihuahuana, P. monticola, P. ponderosa, Populus* sp., *P. tremuloides, Prunus virginiana* var. *melanocarpa, Pseudotsuga menziesii, Quercus* sp., *Q. gambelii, Q. hypoleucoides, Q. kelloggii, Salix scouleriana, Sequoia gigantea, Thuja plicata, Tsuga heterophylla, T. mertensiana,*

angiosperm, usually gymnosperm.
Ecology Rotting wood; slash; underside of logs; associated with a white rot.
Culture Characters
2.4.14.16.26.27.32.(36).(37).(39).40.42-43.48.55.59 (Ref. 474);
2a.4.7.32.36.38.42.54.(55).58 (Ref. 253).
References 116, 119, 120, 122, 128, 130, 133, 170, 183, 192, 197, 204, 208, 216, 249, 333, 398, 402, 406, 474.

Many of the herbarium specimens labelled *P. albida* that we have seen are, in fact, specimens of *P. centrifuga*. Thus, many of the above reports of *P. albida* may have been based upon misidentified specimens of *P. centrifuga*. Eriksson et al. (170) mentioned the confusing labelling of *P. centrifuga* specimens as *P. albida*. Hjortstam (pers. comm. 1993) is of the opinion that the name *P. albida* may be a synonym of *Grandinia tuberculata* Berk. & Curtis 1849. See also *P. pallida* (below).

Phlebia brevispora Nakasone 1981
Mycologia 73:805.
Distribution United States: FL, IL, LA, MD, MS, OK, SC, VA.
Hosts *Betula alleghaniensis, Pinus* sp., *P. elliottii, P. elliottii* var. *densa, P. resinosa, Pseudotsuga menziesii, Quercus* sp., *Q. lyrata, Q. nuttallii*, yellow pine.
Ecology Causing internal decay in creosote or penta-treated southern pine poles; slash; associated with a white rot.
Culture Characters
2.4.(13).21.33.34.36.40.41.48.54.55.59 (Ref. 476);
2.4.13.14.21.(27).33.34.36.40.41.48.54.55.59 (Ref. 474);
2.4.(13).21.34.36.40.41.(48).54.55.59 (Ref. 609).
References 177, 474, 476, 609, 654.

Phlebia brunnescens Burdsall & Nakasone, *ined.*
Synonym *Odontia brunnescens* Overh., Bull. Torrey Bot. Club 65:171, 1938, not validly published because a Latin description was lacking.
Distribution United States: LA.
Host *Liquidambar* sp.
Ecology Dead branch; associated with a white rot.
Culture Characters Not known.

The type specimen, designated by Overholts, was redescribed by Gilbertson (190).

Phlebia celtidis W.B. Cooke 1956
Mycologia 48:394.
Distribution United States: ID, WA.
Hosts *Celtis* sp., *C. reticulata*.
Ecology Dead wood.
Culture Characters Not known.
References 122, 183.

Phlebia centrifuga Karsten 1881
Meddeland. Soc. Fauna Fl. Fenn. 6:10, 1881.
Synonyms *Phlebia mellea* Overh. 1930, *Phlebia subalbida* W.B. Cooke 1956. Both fide Eriksson et al. (170).
Distribution Canada: BC, NB, ON.
United States: AZ, CA, CO, MI, MT, NY.
Hosts *Abies* sp., *A. amabilis, A. balsamea, A. grandis, A. lasiocarpa, A. magnifica, Acer saccharum, Arbutus menziesii, Betula alleghaniensis, ?Picea* sp., *P. engelmannii, P. glauca, P. mariana, Pinus banksiana, P. ponderosa, Pseudotsuga menziesii, Thuja plicata, Tsuga heterophylla*, gymnosperm.
Ecology Decaying sticks; rotting log; associated with a white rot.
Culture Characters
2.4.7.(16).34.36.(37).(39).40.41-42.(54).55.59 (Ref. 474);
2.4.7.32.36.38.42.55.58 (Ref. 253).
References 122, 170, 242, 255, 332, 402, 474, 515.

It is quite likely that many of the records of *Phlebia albida* (above) were misidentified specimens of *P. centrifuga*. See the discussion under *P. albida*.

Phlebia chiricahuaensis (Gilbn. & Budington) Wu 1990
Acta Bot. Fennica 142:34.
Basionym *Resinicium chiricahuaensis* Gilbn. & Budington, Mycologia 62:675, 1970.
Distribution Canada: AB, NTM.
United States: AZ, CO, MT, NM.
Hosts *Abies concolor, Picea* sp., *P. engelmannii, Pinus ponderosa, Pseudotsuga menziesii*, gymnosperm.
Ecology Associated with a white rot.
Culture Characters
2.3c.7.(26).32.36.38.47.55 (Ref. 474).
References 170, 197, 205, 208, 216, 242, 402, 474.

Phlebia coccineofulva Schw. 1832
Trans. Amer. Philos. Soc., N.S., 4:165.
Synonyms *Peniophora coccineofulva* (Schw.) Burt 1926, *Phlebia atkinsoniana* W.B. Cooke 1956, fide Eriksson et al. (170), *Corticium martianum* Berk. & Curtis 1873, fide Burdsall (64) and Hjortstam (288), *Peniophora martiana* (Berk. & Curtis) Burt 1926, *Phlebia martiana* (Berk. & Curtis) Parm. 1967.
Distribution Canada: BC, NB, ON, PQ. United States: AL, AZ, CA, CO, FL, ID, MA, ME, MD, MI, NH, NJ, NT, NY, OH, PA, WI, WV.
Hosts *Betula* sp., *Carya* sp., *Quercus rubra*, angiosperm.
Ecology Rotting wood and bark; very rotten wood.
Culture Characters Not known.
References 64, 92, 98, 122, 170, 220.

Harris's (267) and Slysh's (584) descriptions of *Peniophora coccineofulva* as a species with simple septate hyphae are in error and their records are not cited. However, the specimen Slysh illustrated is *P.*

coccineofulva, fide Burdsall (64). Slysh's concept of *Peniophora martiana* as a simple septate species is also in error, fide Burdsall (64).

Phlebia cretacea (Bourd. & Galzin) Eriksson & Hjort. 1981 Cort. N. Europe 6:1105.
Basionym *Peniophora cretacea* Bourd. & Galzin, Hym. France, 288, 1928.
Synonyms *Peniophora romellii* Litsch. in Bourdot 1932, fide Eriksson et al. (170), *Phlebia romellii* (Litsch.) Parm. 1967.
Distribution Canada: BC, ON.
United States: MN, NY.
Hosts *Thuja occidentalis*, gymnosperm.
Ecology Not known.
Culture Characters
1.(3c).(7).(26).(34).36.38.47.51.55 (Ref. 260).
References 170, 214, 584.

Phlebia deflectens (Karsten) Ryv. 1971 Rep. Kevo Subarc. Res. Station 8:150.
Basionym *Grandinia deflectens* Karsten, Bidrag Kännedom Finlands Natur Folk 37:239, 1882.
Synonyms *Corticium deflectens* (Karsten) Karsten 1889, *Phlebia lilacea* M. Christiansen 1960, fide Eriksson et al. (170).
Distribution Canada: BC, NS. United States: AZ.
Hosts *Acer saccharum, Arbutus menziesii, Quercus* sp., *Q. hypoleucoides, Salix* sp.
Ecology Not known.
Culture Characters Not known.
References 57, 116, 204, 216, 242, 249.

Phlebia georgica Parm. 1967 Izv. Akad. Nauk Estonsk. SSR, Ser. Biol. 16:390.
Synonym *Rogersella eburnea* Hjort. & Høgholen 1980, fide Eriksson et al. (170).
Distribution Canada: MB, NS, ON. United States: TN.
Hosts *Juniperus* sp., *Tsuga* sp., gymnosperm.
Ecology Not known.
Culture Characters Not known.
References 170, 242, 290.

Phlebia hydnoidea Schw. 1832 Trans. Amer. Philos. Soc., N.S., 4:165.
Synonyms *Odontia hydnoidea* (Schw.) Peck 1903, *Gloeoradulum hydnoideum* (Schw.) Lloyd 1917, *Radulum hydnoideum* (Schw.) Lloyd 1917, *Odontia lateritia* Berk. & Curtis 1873, fide Burdsall (61).
Distribution United States: AL, DC, FL, GA, NC, MA, NC, NY, OH, PA, VA, WV.
Hosts *Castanea dentata, Quercus* sp., angiosperm.
Ecology Bark; associated with a white rot.
Culture Characters
2.3c.21.31e.35.36.(37).38.47.53.54.(55) (Ref. 474);
2.4.21.31e.35.36.(37).38.47.53.54 (Ref. 474).
References 61, 192, 220, 474, 593.

Perhaps *Steccherinum*, fide Gilbertson (195) or *Veluticeps*, fide Ginns (220). Berkeley and Curtis, in the original description of *O. lateritia*, state this seems to be the same as *Phlebia hydnoidea* Schw. Subsequently several authors, e.g., Peck (526), Cooke (122), Gilbertson (195), Ginns (220), either listed or suggested *O. lateritia* as a synonym of *P. hydnoidea*, but only Burdsall (61) has compared type specimens.

Phlebia incarnata (Schw.) Nakasone & Burdsall 1984 Mycotaxon 21:245.
Basionym *Merulius incarnatus* Schw., Schriften Naturf. Ges. Leipzig 1:92, 1822.
Synonym *Byssomerulius incarnatus* (Schw.) Gilbn. 1974, *Merulius rubellus* Peck 1882, fide Ginns (226).
Distribution United States: AL, AR, DC, DE, FL, GA, IL, IN, KY, LA, MA, MD, MO, MS, NC, OH, OK, PA, SC, TN, TX, VA, WV.
Hosts *Acer* sp., *Quercus* spp., *Q. alba, Q. falcata, Q. prinus, Q. rubra* var. *borealis*, angiosperm.
Ecology Saprophytic, associated with a white rot.
Culture Characters
1.3.7.(26).32.36.38.40.47.54.(55) (Ref. 226).
References 96, 183, 226.

The specimen reported (197) from Arizona was subsequently determined (204) to be *Meruliopsis ambiguus*. Presumably the report in Gilbertson et al. (216) was based on the same collection. The reports from New Mexico (208) and Alberta (242) probably were specimens of *M. ambiguus*. Some of these misdeterminations were compiled by Farr et al. (183).

Phlebia lilascens (Bourd.) Eriksson & Hjort. 1981 Cort. N. Europe 6:1123.
Basionym *Corticium lilascens* Bourd., Rev. Sci. Bourbonnais Centr. France 23:13, 1910.
Distribution United States: CO.
Host *Quercus gambelii.*
Ecology Not known.
Culture Characters Not known.
References 399, 402.

Known in North America from one collection.

Phlebia livida (Pers.:Fr.) Bres. 1897 Atti Imp. Regia Accad. Rovereto, Ser. III, 3:105.
Basionym *Corticium lividum* Pers., Obs. Mycol. 1:38, 1796.
Synonyms ?*Corticium hepaticum* Berk. & Curtis 1873, fide Burt (93) and Hjortstam (288), ?*Corticium siparium* Berk. & Curtis 1873, fide Hjortstam (288).
Distribution Canada: AB, BC, NS, ON, PQ.
United States: AL, AZ, CA, CO, ID, LA, MD, MT, NC, NH, NM, NY, OR, PA, TX, VT, WA, WI.
Hosts *Abies* sp., *A. balsamea, Betula* sp., *Picea* sp., *P. engelmannii, Pinus* sp., *P. contorta, P. ponderosa, P. strobus, Pseudotsuga menziesii, Thuja plicata, Tsuga* sp., *T. heterophylla*, gymnosperm.
Ecology Rotting logs; associated with a white rot.
Culture Characters
2.3(sic).12?.26.32.36.(37).38.42.(50).55 (Ref. 415).

References 93, 116, 170, 183, 197, 204, 208, 216, 242, 249, 402, 415.

Corticium siparium has been listed as a synonym of *Chondrostereum purpureum*, and *Corticium hepaticum* as a synonym of *Punctularia strigoso-zonata*, see Cooke (122), Hjortstam (288) and Lentz (379).

Phlebia ludoviciana (Burt) Nakasone & Burdsall 1982 Mycotaxon 14:3.

Basionym *Peniophora ludoviciana* Burt, Ann. Missouri Bot. Gard. 12:244, 1926.

Synonym *Peniophora flammea* Burt 1925, fide Rogers and Jackson (567).

Distribution Canada: ON. United States: AL, AZ, CO, FL, IL, LA, MI, MT, NM, NY, TX, WI.

Hosts *Acer rubrum, A. saccharinum, A. saccharum, Alnus* sp., *Baccharis sarothroides, Betula* sp., *B. nigra, Celtis* sp., *Fraxinus* sp., *Liquidambar styraciflua, Lycium* sp., *Picea engelmannii, Platanus wrightii, Populus tremuloides, Quercus* sp., *Q. hypoleucoides, Sabal palmetto*, angiosperm, gymnosperm.

Ecology Rotting decorticated wood; associated with a white rot.

Culture Characters 2.4.(7).14.(34).36.40.41.42.48.54.(55).59 (Ref. 475).

References 92, 116, 183, 204, 208, 216, 267, 474, 475, 542, 584.

Listed as a synonym of *Phlebia subochracea* in Jülich and Stalpers (326), and Eriksson et al. (170) but Nakasone et al. (475) itemized the differences between the two and reported incompatibility in matings. Some of the above reports may have been based upon specimens of *P. subochracea*.

Phlebia murrillii W.B. Cooke 1956 Mycologia 48:398.

Distribution United States: FL.

Host *Pinus palustris*.

Ecology Fallen branch.

Culture Characters Not known.

Known only from the type specimen.

Phlebia nitidula (Karsten) Ryv. 1971 Rep. Kevo. Subarc. Res. Sta. 8:151.

Basionym *Corticium nitidulum* Karsten, Meddeland. Soc. Fauna Fl. Fenn. 6:11, 1881.

Distribution NORTH AMERICA, specific localities not given.

Hosts Not known.

Ecology Not known.

Culture Characters Not known.

Reference 170.

Phlebia pallida (Berk. & Curtis) Ginns 1993 Mycotaxon 46:322.

Basionym *Radulum pallidum* Berk. & Curtis, Grevillea 1:145, 1873.

Distribution Canada: PEI. United States: AL, CT, IA, IL, KY, MD, MI, MS, NJ, NY, OH, SC.

Hosts *Abies balsames, Carya* sp., *Prunus pensylvanica, Quercus* sp., angiosperm.

Ecology Decaying wood and bark; burnt logs; charred brush.

Culture Characters Not known.

References 192, 242, 410, 463.

Most of the distribution records are from Ginns (loc. cit.). The report (412) from Maine was based upon a specimen of *Meruliopsis corium* (Ginns, loc. cit.). The reports (242) from Prince Edward Island, Canada, on *Abies balsamea* and *Prunus pensylvanica*, were based upon field observations (loc. cit.) and cannot be confirmed. Ginns concluded they were probably misidentifications. Hjortstam (pers. comm. 1993) is of the opinion that *Grandinia tuberculata* Berk. & Curtis 1849 is an earlier name for this fungus.

Phlebia radiata Fr Syst. Mycol. 1:427.

Synonyms *Phlebia merismoides* (Fr.):Fr. 1821, fide Eriksson et al. (170), *Phlebia cinnabarina* Schw. 1832, fide Ginns (220), *Phlebia cystidiata* Jackson ex W.B. Cooke 1956, fide Ginns (226).

Distribution Canada: AB, BC, MB, NB, NF, NS, NTM, ON, PQ. United States: AL, AK, AR, AZ, CA, CO, CT, DC, DE, FL, IA, ID, IL, IN, KY, MA, ME, MD, MI, MN, MO, MT, NC, NH, NJ, NY, OH, OR, PA, RI, TN, TX, VA, VT, WA, WI, WV.

Hosts *Abies* sp., *A. balsamea, A. grandis, A. lasiocarpa, A. magnifica, Acer* sp., *A. macrophyllum, A. platanoides, A. rubrum, A. saccharinum, A. saccharum, A. spicatum, Alnus* sp., *A. glutinosa, A. incana, A. oblongifolia, A. rubra, A. rugosa, A. rugosa* var. *americana, A. sinuata, A. tenuifolia, Betula* sp., *B. alba, B. alleghaniensis, B. nigra, B. occidentalis, B. papyrifera, B. populifolia, B. pumila, B. verrucosa, Carpinus caroliniana, Carya* sp., *Castanea dentata, Corylus cornuta* var. *californica, Fagus* sp., *F. grandifolia, Fraxinus* sp., *Ilex opaca, Juglans cinerea, J. nigra, J. regia, Liriodendron tulipifera, Magnolia* sp., *M. fraseri, M. glauca, Malus domestica, Nyssa* sp., *Picea glauca, Pinus* sp., *P. banksiana, P. contorta, P. monticola, P. resinosa, P. taeda, P. virginiana, Platanus wrightii, Populus* sp., *P. acuminata, P. balsamifera, P. grandidentata, P. tremuloides, P. trichocarpa, Prunus* sp. (wild & cult.), *P. avium, P. emarginata, P. pensylvanica, P. serotinum, Pseudotsuga menziesii, Ostrya virginica, Quercus* sp., *Q. agrifolia, Q. borealis, Q. chrysolepis, Q. coccinea,*

Q. falcata, Q. gambelii, Q. garryana, Q. hypoleucoides, Q. pedunculata, Q. robur, Q. rubra, Q. velutina, Rhus typhina, Salix sp., *Sorbus aucuparia, Tilia* sp., *T. americana, Tsuga heterophylla, Ulmus* sp., *U. americana, Viburnum* sp., angiosperm, gymnosperm.

Ecology Saprophytic; bark; rotting wood; log; creosote or penta-treated southern pine poles; associated with a white rot.

Culture Characters
2.4.13.(14).21.(27).32.(35).36.40.42.(48).54.55.59 (Ref. 474);
2.4.15.(35).36.38.41-42.(48).54.(55).59 (Ref. 226);
2.4.(13).15b.(35).36.40.41-42.(48).(50).54.(55).59.65 (Ref. 609).

References 116, 122, 133, 170, 183, 204, 216, 226, 242, 249, 255, 332, 333, 335, 402, 406, 474, 489, 609, 654.

Phlebia rufa (Pers.:Fr.) M. Christiansen 1960
Dansk Bot. Ark. 19 (2):164.

Basionym *Merulius rufus* Pers., Synop. Meth. Fung. 498, 1801.

Synonyms *Phlebia acerina* Peck 1889, *Merulius pruni* Peck 1906, *Phlebia merulioidea* Lloyd 1915, *Merulius pilosus* Burt 1922, *Merulius interruptus* Bres. 1925. All fide Ginns (226).

Distribution Canada: AB, BC, MB, ON, PQ, SK.
United States: AZ, CA, DC, IL, KY, MA, MD, MI, MN, NC, NH, NJ, NY, OR, PA, VA, VT, WA, WI, WV.

Hosts *Abies balsamea, Acer* sp., *A. macrophyllum, A. saccharum, A. spicatum, Alnus* sp., *Betula papyrifera, Castanea dentata, Fagus* sp., *Fraxinus nigra, Juglans cinerea, Liriodendron* sp., *L. tulipifera, Malus domestica, Nyssa sylvatica, Pinus* sp., *Populus* sp., *P. balsamifera, Prunus* sp., *P. emarginata, P. pensylvanica, P. serotina, Quercus* sp., *Q. agrifolia, Q. arizonica, Q. garryana, Q. hypoleucoides, Q. tomentella, Salix* sp., *Tilia americana, Tsuga* sp., *Ulmus* sp., angiosperm, rarely gymnosperm.

Ecology Creosote or penta-treated southern pine poles; associated with a white rot.

Culture Characters
2.4.13.(14).21.32.36.(37).(38).(40).42.(48).(53).54.(55).59 (Ref. 474);
2.4.14.15.(35).36.38.41-42.(48).54.59 (Ref. 226);
2.4.13.15b.36.40.41-42.(48).(50).54.(55).59.65 (Ref. 609).

References 170, 183, 214, 216, 226, 204, 242, 255, 406, 465, 474, 654.

Phlebia segregata (Bourd. & Galzin) Parm. 1967
Izv. Akad. Nauk Estonsk. SSR, Ser. Biol. 16:393.

Basionym *Peniophora segregata* Bourd. & Galzin, Hym. France, 284, 1928.

Synonym *Peniophora livida* Burt 1926, fide Eriksson and Ryvarden (170) and Weresub (635).

Distribution Canada: BC, MB, ON, PQ.
United States: LA.

Hosts *Picea* sp., *Pseudotsuga menziesii, Tsuga heterophylla*, rarely angiosperm, gymnosperm.

Ecology Not known.

Culture Characters
2a.(3c).7.32.36.38.47.55.60.61 (Ref. 260).

References 92, 116, 170, 242, 584.

The earliest name for this fungus is *Peniophora livida*, but the epithet cannot be transferred to *Phlebia* because there already exists a *P. livida* (Pers.:Fr.) Bres., see above. The epithet *segregata* is the next available epithet for this fungus. Circumboreal, according to Eriksson et al. (170).

Phlebia separata (Jackson & Dearden) Parm. 1967
Izv. Akad. Nauk Estonsk. SSR, Ser. Biol. 16:393.

Basionym *Corticium separatum* Jackson & Dearden, Canad. J. Res., C, 27:154, 1949.

Distribution Canada: BC.

Host *Abies grandis*.

Ecology Not known.

Culture Characters Not known.

References 116, 308.

Phlebia serialis (Fr.) Donk 1957
Fungus 27:12.

Basionym *Thelephora serialis* Fr., Syst. Mycol. 1:445, 1821.

Synonyms *Peniophora serialis* (Fr.) Höhnel & Litsch. 1907, *Phlebia flavoferruginea* (Karsten) Parm. 1967, fide Eriksson et al. (170), *Peniophora flavoferruginea* (Karsten) Litsch. 1938.

Distribution Canada: BC, ON.
United States: NY, WA.

Hosts *Pinus strobus, Thuja plicata*, gymnosperm.

Ecology Decaying logs.

Culture Characters Not known.

References 92, 170, 242, 584.

Phlebia setulosa (Berk. & Curtis) Nakasone 1990
Mycologia Memoir 15:262.

Basionym *Hydnum setulosum* Berk. & Curtis, Grevillea 1:100, 1873.

Synonyms *Hyphodontia setulosa* (Berk. & Curtis) Maas G. 1974, *Steccherinum setulosum* (Berk. & Curtis) L.W. Miller 1935.

Distribution Canada: MB, NS, NTM.
United States: AL, FL, IA, IL, MI, MS, NY.

Hosts *Acer* sp., *A. saccharinum, Fagus grandifolia, Fomes* sp., *Liquidambar styraciflua, Nyssa sylvatica, Platanus* sp., *Populus tremuloides, Quercus* sp., *Sabal palmetto, Ulmus* sp., angiosperm.

Ecology Dead wood; old polypore; creosote or penta-treated southern pine poles; associated with a white rot.

Culture Characters
2.4.13.14.21.34.36.40.42.48.(50).54.59 (Ref. 474).

References 192, 242, 428, 463, 464, 474, 609, 654.

Phlebia subochracea (Bres.) Eriksson & Ryv. 1976 Cort. N. Europe 4:873.

Basionym *Grandinia subochracea* Bres., Hedwigia 33:206, 1894.

Synonyms *Corticium granulatum* Burt 1926, fide Nakasone et al. (475), a later homonym of *Corticium granulatum* (Bonorden) Karsten 1882, *Phlebia ochraceofulva* (Bourd. & Galzin) Donk 1957, fide Nakasone et al. (475).

Distribution Canada: ON. United States: AR, AZ, CO, ID, MS, MT, NY.

Hosts *Betula papyrifera, Picea engelmannii, Platanus* sp., *P. wrightii, Populus* sp., *P. tremuloides, P. trichocarpa, Prosopis juliflora, Quercus* sp., *Q. arizonica, Q. hypoleucoides, Salix* sp., *Ulmus* sp., angiosperm.

Ecology Rotting wood and bark; log; associated with a white rot.

Culture Characters
2.4.7.(14).34.35.36.38.40.41.42.(48).54.59 (Ref. 475)
2.4.7.(14).34.36.(38).(40).41.(42).(48).54.59 (Ref. 474).

References 93, 183, 207, 255, 393, 402, 474, 475.

Basidiomes are similar to those of *Phlebia ludoviciana*. Because the name *P. subochracea* has only been applied to North American specimens since 1982 (475), it is possible that some early records of *P. ludoviciana* are a mixture.

Phlebia subserialis (Bourd. & Galzin) Donk 1957 Fungus 27:12.

Basionym *Corticium subseriale* Bourd. & Galzin, Hym. France, 219, 1928.

Synonym *Peniophora subserialis* (Bourd. & Galzin) Slysh 1960, *Peniophora phlebioides* Jackson & Dearden 1949, fide Nakasone et al. (475), *Phlebia phlebioides* (Jackson & Dearden) Donk 1957

Distribution Canada: AB, BC, NB, NTM, ON, PQ. United States: AZ, CO, FL, ID, MS, MT, NM, NY, WA, WI.

Hosts *Abies balsamea, A. concolor, Acer glabrum, Juniperus deppeana, Liquidambar styraciflua, Picea* sp., *P. engelmannii, P. pungens, Pinus* sp., *P. banksiana, P. contorta, P. echinata, P. engelmannii, P. leiophylla* var. *chihuahuana, P. ponderosa, P. strobus, Populus* sp., *Pseudotsuga menziesii, Quercus* sp., *Q. nigra, Tsuga heterophylla*, gymnosperm.

Ecology Caused 18% of the decay in lodgepole pine logging slash in Alberta; isolated from creosote or penta-treated southern pine poles; basidiomes produced on logs; associated with a white rot.

Culture Characters
2.4.(7).14.34.36.38.41.48.(50).(54).55.59 (Ref. 475);
2.4.(7).14.34.36.38.41.(48).54.55.59 (Ref. 609).

References 116, 119, 170, 177, 183, 197, 204, 208, 216, 242, 308, 402, 406, 413, 474, 475, 489, 584, 609, 654.

Slysh's (584) description resembles the description of *Phlebia serialis* in Eriksson et al. (170).

Phlebia tremellosus (Schrader:Fr.) Nakasone & Burdsall 1984 Mycotaxon 21:245.

Basionym *Merulius tremellosus* Schrader, Spicilegium Fl. German., 139, 1794.

Distribution Canada: AB, AB-NT, BC, MB, NB, NS, ON, PQ, SK. United States: AR, CA, CT, FL, GA, IA, ID, IN, LA, MA, MD, MI, MN, MO, MT, NC, NY, OH, OR, TN, VA, WA, WI, WV.

Hosts *Abies amabilis, A. concolor, Acer* sp., *A. macrophyllum, A. rubrum, A. saccharinum, A. saccharum, Alnus* sp., *A. serrulata, Betula* sp., *B. alleghaniensis, B. occidentalis, B. papyrifera, Carya glabra, Castanea dentata, Fagus* sp., *F. grandifolia, Fraxinus americana, F. nigra, Malus* sp., *M. domestica, M. pumila, Pinus* sp., *P. monticola, P. ponderosa, Populus* sp., *P. tremuloides, P. trichocarpa, Pseudotsuga* sp., *P. menziesii, Quercus* sp., *Q. alba, Q. falcata, Q. garryana, Q. nigra, Q. nuttallii, Q. prinus, Q. velutina, Salix* sp., *Sorbus* sp., *Thuja* sp., *Tsuga* sp., *T. heterophylla, Ulmus* sp., angiosperm, gymnosperm.

Ecology Saprophytic; rotting wood; fallen branches; logs; associated with a white rot.

Culture Characters
2.4.21.26.34.36.38.41.54.55.59 (Ref. 226);
2.4.7.(26).(27).31e.34.36.(37).40.42.(48).54.55.59 (Ref. 474).

References 96, 116, 119, 127, 130, 183, 226, 242, 249, 255, 332, 333, 334, 406, 423, 474, 611.

This fungus produces the antibiotic Merulidial (543).

Phlebia vinosa Burdsall, *ined.* In Nakasone, Mycologia Memoir 15: 267, 1990.

Synonym *Radulum vinosum* Overh., Bull. Torrey Bot. Club 65:172, 1938, not validly published, lacking a Latin description.

Distribution United States: FL, GA, LA, MS, NC, SC.

Hosts *Carya* sp., *Liriodendron tulipifera, Magnolia* sp., *Pinus* sp., *Quercus* sp., angiosperm.

Ecology Dead wood; associated with a white rot.

Culture Characters
(1).2.4.(26).27.(34).36.(37).(38).(39).(40).42-43.(48).(50).54.(55).59 (Ref. 474).

References 190, 474.

PHLEBIELLA Karsten 1890 — Sistotremaceae

Synonym is *Xenasmatella* Oberw. 1965, fide Hjortstam and Larsson (293).

Phlebiella californica (Liberta) Larsson & Hjort. 1987 Mycotaxon 29:316.

Basionym *Xenasma californica* Liberta, Mycologia 57:967, 1965.

Synonym *Xenasmatella californica* (Liberta) Hjort. 1983.

Distribution United States: CA.

Hosts Angiosperm.

Ecology Not known.
Culture Characters Not known.

Known only from the holotype and two paratype specimens, see Liberta (389) and Hjortstam (281).

Phlebiella christiansenii (Parm.) Larsson & Hjort. 1987 Mycotaxon 29:316.
Basionym *Cristella christiansenii* Parm., Izv. Akad. Nauk Estonsk. SSR, Ser. Biol. 14 (2):222, 1965.
Synonym *Trechispora christiansenii* (Parm.) Liberta 1966.
Distribution Canada: PQ. United States: VT.
Hosts Not known.
Ecology Not known.
Culture Characters Not known.
Reference 395.

Phlebiella filicina (Bourd.) Larsson & Hjort. 1987 Mycotaxon 29:317.
Basionym *Corticium filicinum* Bourd., Rev. Sci. Bourbonnais Centr. France 23:12, 1910.
Synonyms *Xenasma filicinum* (Bourd.) M. Christiansen 1960, *Xenasmatella filicina* (Bourd.) Oberw. 1965.
Distribution Canada: BC, MB, PQ. United States: CA, CO, ID, MA, NY, OR, PA.
Hosts *Abies balsamea, A. magnifica, Picea* sp., *Pinus* sp., *P. strobus, Polystichum munitum, Vitis ?riparia*, gymnosperm.
Ecology Fern stems; rotting wood; log; associated with a white rot.
Culture Characters Not known.
References 116, 128, 130, 232, 255, 384, 402.

The reports (116, 128, 130, 255, 384, 402) of this species on wood may have been based upon misidentified specimens of *P. pseudotsugae*, see the discussion under that epithet.

Phlebiella gaspesica (Liberta) Larsson & Hjort. 1987 Mycotaxon 29:317.
Basionym *Xenasma gaspesica* Liberta, Mycologia 58:932, 1966.
Synonym *Xenasmatella gaspesica* (Liberta) Hjort. 1983.
Distribution Canada: PQ.
Host *Picea* sp.
Ecology Not known.
Culture Characters Not known.
Reference 281.

Phlebiella grisella (Bourd.) Larsson & Hjort. 1987 Mycotaxon 29:318.
Basionym *Corticium grisellum* Bourd., Rev. Sci. Bourbonnais Centr. France 35:17, 1922.
Synonym *Xenasma grisellum* (Bourd.) Liberta 1960.
Distribution Canada: PQ. United States: IL.
Hosts Angiosperm.
Ecology Not known.
Culture Characters Not known.
Reference 394.

Phlebiella inopinata (Jackson) Larsson & Hjort. 1987 Mycotaxon 29:316.
Basionym *Corticium inopinatum* Jackson, Canad. J. Res., C, 28:718, 1950.
Synonym *Xenasmatella inopinata* (Jackson) Hjort. & Ryv. 1979.
Distribution Canada: BC, ON.
Hosts *Pseudotsuga menziesii, Tsuga canadensis.*
Ecology Not known.
Culture Characters Not known.
References 116, 295, 307.

Phlebiella insperata (Jackson) Ginns & Lefebvre, *comb. nov.*
Basionym *Corticium insperatum* Jackson, Canad. J. Res., C, 28:718, 1950.
Synonym *Xenasma insperatum* (Jackson) Donk 1957.
Distribution Canada: ON.
Hosts Angiosperm.
Ecology Bark of branches.
Culture Characters Not known.

Known only from the type specimen, which was redescribed by Liberta (384). The combination in *Phlebiella* was proposed by Oberwinkler (493) in 1977 but not validly published. We agree with Oberwinkler and transfer the epithet to *Phlebiella*.

Phlebiella lloydii (Liberta) Larsson & Hjort. 1987 Mycotaxon 29:318.
Basionym *Xenasma lloydii* Liberta, Mycologia 52:906, 1960.
Distribution Canada: ON.
Host *Pinus banksiana.*
Ecology Not known.
Culture Characters Not known.

Known only from the type specimen.

Phlebiella pseudotsugae (Burt) Larsson & Hjort. 1987 Mycotaxon 29:317.
Basionym *Corticium pseudotsugae* Burt, Ann. Missouri Bot. Gard. 13: 246, 1926.
Distribution Canada: BC. United States: ID, NY.
Hosts *Pseudotsuga menziesii, Tsuga canadensis.*
Ecology Decorticated, decaying wood.
Culture Characters Not known.
References 93, 116.

Hjortstam and Larsson (293) do not accept *C. pseudotsugae* as a synonym of *P. filicina*, as did both Liberta (384) and Oberwinkler (492), because the former has clamp connections and occurs primarily on wood, whereas *P. filicina* lacks clamp connections and "seems to be obligate on ferns."

Phlebiella ralla (Jackson) Larsson & Hjort. 1987 Mycotaxon 29:318.
Basionym *Corticium rallum* Jackson, Canad. J. Res., C, 28:723, 1950.

Synonym *Xenasma rallum* (Jackson) Liberta 1960.
Distribution Canada: BC, ON.
United States: IA, MA, NY, OR, RI.
Hosts *Acer* sp., *Castanea grandidentata, Fraxinus* sp., *Picea sitchensis, Pseudotsuga menziesii, Quercus macrocarpa, Tsuga canadensis, Vitis vulpinus.*
Ecology Decayed wood and bark.
Culture Characters Not known.
References 307, 384.

Phlebiella subnitens (Bourd. & Galzin) Larsson & Hjort. 1987
Mycotaxon 29:317.
Basionym *Corticium subnitens* Bourd. & Galzin, Hym. France, 224, 1928.
Synonym *Xenasma subnitens* (Bourd. & Galzin) Liberta 1960.
Distribution United States: IA, OR.
Hosts *Pseudotsuga menziesii, Quercus* sp.
Ecology Decayed wood.
Culture Characters Not known.
Reference 384.

Phlebiella sulphurea (Pers.:Fr.) Ginns & Lefebvre, *comb. nov.*
Basionym *Corticium sulphureum* Pers., Obs. Mycol. 1:38, 1796.
Synonyms *Phlebia vaga* Fr. 1821, fide Donk (152), *Hypochnus vaga* (Fr.) Kauffm. 1915, *Phlebiella vaga* (Fr.) D.P. Rogers 1944, *Trechispora vaga* (Fr.) Liberta 1966, *Hypochnus filamentosus* Burt 1926 (a later homonym of *Hypochnus filamentosus* Pat. 1891), fide Liberta (395), *Odontia tenuis* Peck 1891, fide Gilbertson (189), *Odontia fusca* Cooke & Ellis 1881, fide Rogers and Jackson (567), *Grandinia tabacina* Cooke & Ellis 1881, fide Gilbertson (191), ?*Hypochnus fumosus* Fr. 1818, fide Donk (152), *Tomentella fumosa* (Fr.) Pilát 1936.
Distribution Canada: AB, BC, MB, NB, NS, NT, ON, PQ. United States: AK, AL, AZ, CA, CO, ID, IL, IN, MA, MD, MI, MN, MO, MT, NC, NH, NJ, NM, NY, OH, PA, TN, UT, VT, WA, WI.
Hosts *Abies amabilis, A. balsamea, A. concolor, A. fraseri, A. grandis, Acer* sp., *A. macrophyllum, Alnus oblongifolia, Arbutus menziesii, Betula* sp., *B. alleghaniensis, B. papyrifera, B. pumila, Fagus,* sp., *Juglans major, Juniperus virginiana, Liquidambar styraciflua, Picea engelmannii, P. glauca, P. pungens, P. rubens, P. sitchensis, Pinus aristata* var. *longaeva, P. banksiana, P. engelmannii, P. leiophylla* var. *chihuahuana, P. monticola, P. ponderosa, P. strobus, P. taeda, Populus* spp., *P. tremuloides,* ?*Pseudotsuga* sp., *P. menziesii, Quercus* sp., *Q. gambelii, Thuja occidentalis, T. plicata, Tsuga canadensis,* angiosperm, gymnosperm.
Ecology Bark; moss-covered wood; rotten wood; branch; slash; logs; associated with a white rot.
Culture Characters
2.3c.(21).31d.32.36.(37).(38).39.46-47.54.55.60 (Ref. 474).
References 93, 116, 128, 130, 133, 183, 191, 197, 204, 208, 214, 216, 249, 327, 332, 334, 395, 399, 402, 406, 407, 474, 516.

The oldest epithet for this fungus is *sulphurea*, see Donk (152). It dates from 1796, whereas the commonly seen epithet *vaga* was proposed in 1821. Thus the epithet *sulphurea* is transferred to *Phlebiella*.

Phlebiella tulasnelloideum (Höhnel & Litsch.) Ginns & Lefebvre, *comb. nov.*
Basionym *Corticium tulasnelloideum* Höhnel & Litsch., Sitzungsber. Kaiserl. Akad. Wiss., Math.-Naturwiss. Kl., Abt. I, 117:1118, 1908.
Synonyms *Xenasma tulasnelloideum* (Höhnel & Litsch.) Donk 1957, *Xenasmatella tulasnelloidea* (Höhnel & Litsch.) Oberw. 1965, *Corticium incanum* Burt 1926, fide Rogers and Jackson (567), and Liberta (384).
Distribution Canada: BC, NS, ON.
United States: FL, GA, IA, IL, IN, MA, MO, NC, NH, NJ, NY, OH, OR, RI, VT.
Hosts *Acer* sp., *Betula alleghaniensis, Liquidambar styraciflua, Pinus strobus, Pseudotsuga menziesii, Quercus* sp., *Q. alba, Q. nigra*, angiosperm, gymnosperm.
Ecology Bark and wood of dead limbs; decorticated wood; decayed wood and bark; vegetable debris.
Culture Characters
2.3c.7.35.36.38.46-47.(50).54.55.58 (Ref. 474);
2a.3c.11.34.36.38.47.48.54.58.61 (Ref. 259).
References 93, 96, 116, 249, 384, 387, 474, 562.

The combination in *Phlebiella* was proposed by Oberwinkler (493) in 1977 but not validly published. We agree with Oberwinkler and transfer the epithet to *Phlebiella*.

PHLEBIOPIS Jülich 1978 Phanerochaetaceae

Phlebiopsis flavido-alba (Cooke) Hjort. 1987
Windahlia 17:58.
Basionym *Peniophora flavido-alba* Cooke, Grevillea 8:21, 1879.
Synonyms *Phanerochaete flavido-alba* (Cooke) S.S. Rattan 1977, *Peniophora texana* Burt 1926, *Peniophora vernicosa* Ellis & Ev. ex Burt 1926. Both fide Burdsall (64).
Distribution Canada: NS.
United States: AL, AR, AZ, FL, GA, IN, KY, LA, MD, MS, NC, NJ, NY, OH, OR, SC, TX, WV.
Hosts *Acer* sp., *A. rubrum, A. saccharum, Carpinus caroliniana, Celtis* sp., *Citrus sinensis, Cornus* sp., *Fagus* sp., *F. grandifolia, Fraxinus* sp., *Gleditsia triacanthos, Juniperus mexicana, J. sabinoides, J. virginiana, Libocedrus* sp., *Liquidambar* sp., *L. styraciflua, Liriodendron* sp., *L. tulipifera, Magnolia* sp., *Malus domestica, Myrica cerifera, Nyssa sylvatica, Pinus* sp. (southern pine), *Pinus ponderosa, P. taeda,*

Populus sp., *P. deltoides, Quercus* sp., *Q. alba, Q. laurifolia, Salix* sp., *S. nigra, Vitis* sp., angiosperm.

Ecology Bark; occasionally on dead branches on live trees; limb; slash; log; dead pieces of wood; pulpwood; wood chips; test crosses; test post rail units; creosote or penta-treated poles; associated with a white rot.

Culture Characters
2.5.7.32.36.40.41-42.54.55 (Ref. 179);
2.5.7.32.36.40.41-42.54.(55) (Ref. 474);
2.5.7.32.36.40.42.54.55 (Ref. 609).

References 64, 92, 177, 178, 179, 183, 201, 242, 249, 335, 392, 474, 584, 609, 654.

Phlebiopsis gigantea (Fr.:Fr.) Jülich 1978
Persoonia 10:137.

Basionym *Thelephora gigantea* Fr., Obs. Mycol. 1:152, 1818.

Synonyms *Peniophora gigantea* (Fr.) Massee 1889, *Phlebia gigantea* (Fr.) Donk 1957, *Phanerochaete gigantea* (Fr.) S.S. Rattan, Abdullah & Ismail 1977, *Peniophora globifera* Ellis & Ev. 1897, fide Burdsall (64).

Distribution Canada: BC, NB, NS, ON, PQ.
United States: AK, AR, AZ, CA, CO, FL, GA, ID, LA, MA, MD, MI, MN, MS, MT, NC, NH, NJ, NM, NY, OR, PA, SC, SD, TX, VA, WI.

Hosts *Abies* sp., *A. balsamea, A. concolor, A. lasiocarpa, A. lasiocarpa* var. *arizonica, Larix occidentalis, Picea* sp., *P. engelmannii, P. glauca, P. mariana, P. rubens, Pinus* sp., *P. banksiana, P. contorta, P. echinata, P. elliottii, P. leiophylla, P. leiophylla* var. *chihuahuana, P. monticola, P. ponderosa, P. radiata, P. resinosa, P. strobus, Populus* sp., *Pseudotsuga menziesii, Thuja plicata, Tsuga* sp., *T. canadensis, T. heterophylla*, angiosperm, gymnosperm.

Ecology Causes a root rot in *Pinus contorta*; basidiomes produced on bark; slash; dead trees; logs; ground side of new sills; leaves which had been on ground four months; associated with a white rot. Used in biological control of Annosus root rot (p. 6).

Culture Characters
2.5.7.35.36.38.42-43.(48).(50).(53).(54).55.(57) (Ref. 474);
1.(2).5.(6).7.35.36.38.42.43.53.54 (Ref. 487).

References 64, 92, 96, 97, 108, 116, 183, 197, 204, 208, 216, 249, 274, 335, 402, 406, 474, 584.

Phlebiopsis ravenelii (Cooke) Hjort. 1987
Windahlia 17:58.

Basionym *Peniophora ravenelii* Cooke, Grevillea 8:21, 1879.

Synonyms *Phanerochaete ravenelii* (Cooke) Burdsall 1985, *Peniophora roumeguerii* (Bres.) Höhnel & Litsch. 1906, fide Burdsall (64), *Metulodontia roumeguerii* (Bres.) Parm. 1968, *Phlebiopsis roumeguerii* (Bres.) Jülich & Stalpers 1980, *Peniophora stratosa* Burt 1926, fide Burdsall (64).

Distribution Canada: BC.
United States: AL, AR, CA, CO, DC, FL, GA, ID, LA, MN, MO, MT, OR, PA, SC.

Hosts *Acer rubrum, Alnus* sp., *A. rugosa, Carya* sp., *Crataegus* sp., *Diospyros* sp., *Eucalyptus* sp., *Ligustrum amurense, Lithocarpus densiflora, Picea engelmannii, Quercus* sp., *Q. agrifolia, Q. chrysolepis, Q. virginiana*, angiosperm.

Ecology Dead wood; bark; fallen branches; decaying logs; associated with a white rot.

Culture Characters
2.5.(16).24.25.(26).32.36.38.44-46.54 (Ref. 588).

References 64, 92, 96, 183, 214, 402, 584, 615.

PHLEOGENA Link 1833 Ecchynaceae

Synonym is *Martindalia* Sacc. & Ellis 1885, fide Barr and Bigelow (27).

Phleogena faginea (Fr.) Link 1833
Handbuch Erkennung ... Gewächse 3:396.

Basionym *Onygena faginea* Fr., Symb. Gast., 25, 1818.

Synonyms *Pilacre faginea* (Fr.) Berk. & Br. 1850, *Pilacre petersii* Berk. & Curtis in Berk. & Br. 1859, fide Donk (157), *Martindalia spironema* Sacc. & Ellis 1885, fide Barr and Bigelow (27).

Distribution Canada: NB, NS, ON, PQ.
United States: AL, LA, MA, NC, NJ, NY, PA.

Hosts *Acer rubrum, A. saccharum, Carpinus caroliniana, Fagus grandifolia, Pinus* sp., *Quercus* sp., angiosperm.

Ecology Decorticated log.

Culture Characters Not known.

References 27, 116, 157, 159, 183, 242, 249, 332, 419, 443, 512.

PHYSALACRIA Peck 1882 Physalacriaceae

Physalacria andina (Pat. & Lagerh.) Pat. 1900
Essai taxonomique, 50.

Basionym *Physalacria orinocensis* Pat. & Gaill. var. *adina* Pat. & Lagerh., Bull. Soc. Mycol. France 9:36, 1893.

Distribution United States: FL, TN.

Hosts Lauraceae, *Rhus* sp., *Tilia* sp.

Ecology Petioles; branches.

Culture Characters Not known.

References 32, 530 as *Physalacria concinna* Sydow, see Berthier (32).

Physalacria cryptomeriae Berthier & C.T. Rogerson 1981
Mycologia 63:643.

Distribution United States: NY.

Host *Cryptomeria japonica.*

Ecology Fallen dead needles.

Culture Characters Not known.

Known only from the holotype and a paratype, collected a year later at the same site. Berthier (32), in a monograph of the genus, compared this species with several similar species.

Physalacria inflata (Schw.:Fr.) Peck 1882
Bull. Torrey Bot. Club 9:2.
Basionym *Mitrula inflata* Schw. in Fries, Elenchus 1:234, 1828.
Distribution Canada: NS, ON, PQ. United States: GA, IA, MD, MI, MN, NE, NH, NY, PA, VT, WI.
Hosts *Malus* sp., *Salix* sp.
Ecology Dead wood; decorticated fallen branches; rotted log; stump; leaves.
Culture Characters Not known.
References 9, 32, 249, 331, 332, 532, 562, 610.

Physalacria langloisii Ellis & Ev. 1888
J. Mycol. 4:73.
Distribution United States: LA.
Hosts Not known.
Ecology Rotten wood.
Culture Characters Not known.

Known only from the specimen at FH, which was designated type by Baker (9). Baker also referred a Florida collection to this name, but Berthier (32) concluded it was not conspecific.

Physalacria luttrellii Baker 1946
Mycologia 38:636.
Distribution United States: GA.
Host *Lespedeza bicolor.*
Ecology Dead stems.
Culture Characters Not known.

Known only from the type specimen, which was redescribed by Berthier (32).

PILODERMA Jülich 1969 — Atheliaceae

Piloderma byssinum var. ***byssinum*** (Karsten) Jülich 1969
Ber. Deutsch. Bot. Ges. 81:418.
Basionym *Lyomyces byssinus* Karsten, Meddeland. Soc. Fauna Fl. Fenn. 11:137, 1884.
Synonyms *Athelia byssina* (Karsten) Parm. 1967, *Corticium bicolor* Peck 1873, *Piloderma sphaerosporum* Jülich 1972. Both fide Eriksson et al. (170).
Distribution Canada: AB, BC, NS, NT, ON. United States: AZ, CO, ID, IL, MA, MN, MT, ND, NM, NY, OR, UT, WA, WY.
Hosts *Abies balsamea, Picea* sp., *P. engelmannii, P. glauca, Pinus aristata* var. *longaeva, P. contorta, P. ponderosa, Populus* sp., *P. tremuloides, Symphoricarpos occidentalis*, angiosperm, gymnosperm.
Ecology Mycorrhizal, at least, with *Pseudotsuga menziesii* (187); reported to be associated with a white rot.
Culture Characters Not known.
References 170, 183, 197, 204, 214, 216, 242, 249, 312, 402, 406, 407.

Piloderma byssinum var. ***minutum*** Jülich 1972
Beih. Willdenowia 7:232.
Distribution Canada: ON. United States: NY.
Hosts Not known.
Ecology Not known.
Culture Characters Not known.

Known only from the type specimens.

Piloderma fallax (Libert) Stalpers 1984
Studies in Mycol. (Baarn) 24:53.
Basionym *Sporotrichum fallax* Libert, Plantae Cryptog. Ardem. No. 187, 1832.
Synonyms *Corticium bicolor* Peck 1873 sensu Aucts., *Athelia bicolor* (Peck) Parm. 1967 sensu Aucts., *Piloderma bicolor* (Peck) Jülich 1972 sensu Aucts., *Piloderma croceum* Eriksson & Hjort. 1981. All fide Stalpers (589).
Distribution Canada: AB, BC, MB, ON, PQ, SK. United States: AZ, CA, CO, ID, MA, ME, MI, MN, MT, NC, NH, NJ, NM, NY, OR, PA, WA, WI.
Hosts *Abies balsamea, A. concolor, A. lasiocarpa, A. magnifica, Picea* sp., *P. glauca, Pinus contorta, P. ponderosa, Populus* sp., *P. tremuloides, Pseudotsuga menziesii, Thuja plicata, Tsuga canadensis, T. heterophylla*, gymnosperm.
Ecology Mycorrhizal (188, 348, 656); basidiomes produced on needles; twigs; rotten wood, etc. on ground; associated with a white rot.
Culture Characters Not known.
References 93, 116, 119, 120, 128, 197, 204, 208, 214, 216, 312, 388, 390, 402, 406.

Piloderma lanatum (Jülich) Eriksson & Hjort. 1981
Cort. N. Europe 6:1207.
Basionym *Piloderma byssinum* var. *lanatum* Jülich, Beih. Willdenowia 7:230, 1972.
Distribution United States: IA, PA.
Hosts Not known.
Ecology Not known.
Culture Characters Not known.
Reference 312.

Piloderma olivaceum (Parm.) Hjort. 1984
Windahlia 14:25.
Basionym *Athelia bicolor* forma *olivacea* Parm., Izv. Akad. Nauk Estonsk. SSR, Ser. Biol. 16:380, 1967.
Distribution Canada: BC. United States: AZ.
Hosts Not known.
Ecology Not known.
Culture Characters Not known.
References 170, 282.

Piloderma reticulatum Jülich 1969
Ber. Deutsch. Bot. Ges. 81:417.
Basionym Proposed as a *nom. nov.* for *Corticium reticulatum* Litsch., Ann. Mycol. 39:124, 1941, which is a later homonym of *Corticium reticulatum* (Fr.) Fr. 1874.
Distribution United States: AZ.
Host *Picea engelmannii.*
Ecology Not known.
Culture Characters Not known.
Reference 447.

In the reference (447) the name appeared as *Athelia reticulata* (Litsch.) Parm.

PIREX Hjort. & Ryv. 1985 Cylindrobasidiaceae

Pirex concentricus (Cooke & Ellis) Hjort. & Ryv. 1985
Mycotaxon 24:289.
Basionym *Radulum concentricum* Cooke & Ellis, Grevillea 14:13, 1885.
Synonyms *Phlebia concentrica* (Cooke & Ellis) Kropp & Nakasone 1985, *Irpex owensii* Lloyd 1916, fide Gilbertson (191), *Radulum owensii* (Lloyd) Lloyd 1917.
Distribution Canada: BC.
United States: CA, MT, OR, WA.
Hosts *Acer* sp., *A. macrophyllum, Alnus* sp., *A. rubra, Betula papyrifera, Corylus* sp., *C. cornuta* var. *californica, Heteromeles arbutifolia, Prunus* sp., *Pseudotsuga menziesii, Quercus* sp., *Q. garryana, Salix* sp., angiosperm, myrtle.
Ecology Associated with a white rot.
Culture Characters
2.4.(7).21.34.36.(38).(39).40.41-42.(50).54.55.59 (Ref. 474);
1.3i.25.34.36.39.42.54.58.63 (Ref. 264).
References 116, 133, 190, 191, 242, 335, 474, 593.

PLATYGLOEA Schröter 1887 Cystobasidiaceae

Platygloea abdita Bandoni 1959
Mycologia 51:94.
Distribution United States: IA.
Host *Exidiopsis ?sublilacina* (G.W. Martin).
Ecology Mycoparasitic within basidiomes on wood of angiosperm trees.
Culture Characters Not known.

Known only from the type specimen.

Platygloea arrhytidiae Olive 1951
Bull. Torrey Bot. Club 78:103.
Distribution Canada: AB-NT, ON.
United States: IA, NC.
Hosts Dacrymycetaceae, *Arrhytidia enata, Pinus contorta* var. *contorta.*
Ecology Mycoparasitic within basidiomes.
Culture Characters Not known.
References 13, 242, 501.

Platygloea caroliniana Coker 1920
J. Elisha Mitchell Sci. Soc. 35:123.
Distribution United States: NC.
Host *Lagerstroemia* sp.
Ecology Corticated and decorticated dead branches.
Culture Characters Not known.

Known only from the type and one paratype specimen, see Bandoni (13).

Platygloea fimetaria (Schum.) Höhnel 1917
Ann. Mycol. 15:293.
Basionym *Tremella fimetaria* Schum., Enumerario Plant. Saellandiae 2:440, 1803.
Distribution Canada: BC, MB.
Host Associated with *Saccobolus violascens* Boudier on herbivore dung.
Ecology Presumably mycoparasitic.
Culture Characters Not known.
References 13, 19, 443.

Platygloea lagerstroemiae Coker 1920
J. Elisha Mitchell Sci. Soc. 35:124.
Distribution United States: NC.
Host *Lagerstroemia* sp.
Ecology Dead branches; bark of dead limb.
Culture Characters Not known.

Known only from the type specimen, see Bandoni (13).

Platygloea laplata Lindsey 1986
Mycotaxon 27:333.
Distribution United States: CO.
Host *Peniophora nuda.*
Ecology Presumably mycoparasitic on basidiomes on *Quercus gambelii.*
Culture Characters Not known.

Known only from the type specimen.

Platygloea longibasidia Lowy 1954
Mycologia 46:100.
Distribution United States: LA.
Hosts Angiosperm.
Ecology Dead, partly decorticated wood.
Culture Characters Not known.

Known only from the type specimen, see Bandoni (13) and Lowy (419).

Platygloea mycophila Burdsall & Gilbn. 1974
Mycologia 66:703.
Distribution United States: AZ.
Host *Peniophora tamaricicola.*
Ecology Mycoparasitic on the hymenial surface.
Culture Characters Not known.

Known only from the holotype and several paratype specimens. Also described in Gilbertson et al. (207).

Platygloea peniophorae* var. *peniophorae Bourd. & Galzin 1909
Bull. Soc. Mycol. France 25:17.
Distribution Canada: BC, ON, PQ. United States: AZ, GA, IA, IL, MN, NC, OR, TN, WA.
Hosts *Corticium* sp., *Dacrymyces* sp., *D. minor, Hyphoderma argillaceum, Peniophora* sp.
Ecology Mycoparasitic.
Culture Characters Not known.
References 13, 19, 207, 214, 216, 267, 437, 443, 496, 498, 504.

Platygloea peniophorae* var. *interna Olive 1954
Bull. Torrey Bot. Club 81:331.
Distribution United States: GA, NC.
Host *Dacrymyces* sp.
Ecology Mycoparasitic on basidiomes.
Culture Characters Not known.

Known only from the type specimen.

Platygloea pustulata G.W. Martin & Cain 1940
Mycologia 32:691.
Distribution Canada: BC, ON, PQ.
Hosts *Abies balsamea*, gymnosperm.
Ecology Dead wood and bark.
Culture Characters Not known.
References 13, 19, 116, 437, 443.

Platygloea sebacea (Berk. & Br.) McNabb 1965
Trans. Brit. Mycol. Soc. 48:188.
Basionym *Dacrymyces sebaceus* Berk. & Br., Ann. Mag. Nat. Hist., Ser. IV, 7:430, 1871.
Synonym *Platygloea miedzyrzecensis* Bres. 1903, fide McNabb (455).
Distribution Canada: BC. United States: CA, GA.
Hosts *Botryodiplodia* sp., *Botryosphaeria* sp., *B. quercuum* (Schw.) Sacc., *Phialophorophoma*-like fungus, sphaeriaceous fungus.
Ecology Mycoparasitic on sphaeriaceous fungi on dead wood, growing from the perithecial stromata; on or in fructifications of other fungi.
Culture Characters Not known.
References 13, 19, 455, 498.

Laeticorticium simplicibasidium Lindsey & Gilbn. 1977
Mycotaxon 5:317.
Distribution United States: AZ.
Hosts *Populus* sp., *P. tremuloides*.
Ecology Associated with a white rot.
Culture Characters
(1).2.3c.21.32.(36).37.38.47.(50).54 (Ref. 474).
References 405, 406, 474.

The unusual basidia of this species make it difficult to determine its generic affinities. However, it is misplaced in *Laeticorticium*. We include it with the species of *Platygloea* because it seems most closely related to them, and because we do not recognize the generic name *Laeticorticium*. Larsen (365) concluded that it probably belonged in the Auriculariaceae and should be compared with *Platygloea unispora* and *Itersonilia perplexans* Derx.

Platygloea sphaerospora G.W. Martin 1934
Mycologia 26:261.
Distribution United States: NJ.
Hosts *Quercus falcata, Q. rubra*.
Ecology Rotten wood.
Culture Characters Not known.

Known only from the type specimen, see Bandoni (13).

Platygloea subvestita Olive 1953
Bull. Torrey Bot. Club 80:33.
Distribution United States: NC.
Hosts Angiosperm.
Ecology Decorticated wood.
Culture Characters Not known.

Known only from the type specimen.

Platygloea unispora Olive 1958
J. Elisha Mitchell Sci. Soc. 74:41.
Distribution United States: GA, NC.
Hosts *Chamaecyparis thyoides, Juniperus* sp.
Ecology Dead branches on the ground.
Culture Characters Not known.
References 13, 498.

Detailed description and illustrations in Olive (495).

Platygloea vestita Bourd. & Galzin 1923
Bull. Soc. Mycol. France 39:261.
Distribution Canada: BC. United States: IA.
Hosts *Quercus* sp., angiosperm.
Ecology Dead branches; litter.
Culture Characters Not known.
References 13, 19, 436, 443.

PLICATURA Peck 1872 — Atheliaceae

Synonym is *Plicaturopsis* D. Reid 1964, fide Ginns (226).

Plicatura crispa (Pers.:Fr.) Rea 1922
British Basidiomycetes, 626.
Basionym *Cantharellus crispus* Pers., Neues Mag. Bot. 1:106, 1794.
Synonyms *Trogia crispa* (Pers.) Fr. 1863, *Plicaturopsis crispa* (Pers.) D. Reid 1964, *Plicatura faginea* (Fr.) Karsten, fide Reid (554).
Distribution Canada: AB, AB-NT, BC, NB, NF, NS, ON, PQ, YT. United States: AK, IA, MI, MN, MT, NC-TN, NY, WI.
Hosts *Acer* sp., *A. rubrum, A. saccharum, Alnus* sp., *A. rubra, A. rugosa, Betula* sp., *B. alleghaniensis, B.*

papyrifera, B. occidentalis, Fagus sp., *F. grandifolia, Picea glauca, Populus* sp., *Prunus* sp., *P. pensylvanica, Pseudotsuga menziesii, Quercus* sp., *Tilia americana.*

Ecology Bark and wood; limbs; branch; trunk; associated with a white rot.

Culture Characters
(1).2.3c.7.(16).32.36.(37).38.44-47.50.54.(55).60 (Ref. 474).

References 97, 103, 214, 242, 249, 327, 340, 442, 474.

Plicatura nivea (Fr.) Karsten 1889
Bidrag Kännedom Finlands Natur Folk 48:342.

Basionym *Merulius niveus* Fr., Elenchus 1:59, 1828.

Synonym *Plicatura alni* Peck 1872, *Merulius rimosus* Berk. 1891, *Cantharellus candidus* Peck 1898, fide Ginns (in litt., type studied, NYS), *Radulum cuneatum* Lloyd 1916. All fide Ginns (221), unless stated otherwise.

Distribution Canada: AB, BC, MB, NB, NF, NS, NT, NTM, ON, PQ, SK.
United States: AK, CA, CT, ID, MA, ME, MI, MN, MT, NH, NY, OR, PA, VT, WA.

Hosts *Abies grandis, Acer pensylvanicum, Alnus* sp., *A. incana, A, rubra, A. rugosa, A. rugosa* var. *americana, A. sinuata, A. tenuifolia, Betula* sp., *B. alleghaniensis, B. occidentalis, B. papyrifera, Populus* sp., *P. tremuloides, Salix* sp., *Tilia* sp., *T. americana.*

Ecology Principally on dead branches and stems of small diameter of *Alnus* species; associated with a white rot.

Culture Characters
(1).2.3c.7.34.36.(37).39.44-45.(50).(51).(53).54.60 (Ref. 474);
(1).2.3.26.32.34.36.(37).(38).39.(44).46.53.54.60 (Ref. 221).

References 97, 116, 130, 183, 221, 226, 242, 249, 334, 335, 474.

PODOSCYPHA Pat. 1900 — Podoscyphaceae

Podoscypha aculeata (Berk. & Curtis) Boidin 1959
Rev. Mycol. (Paris) 24:210.

Basionym *Thelephora aculeata* Berk. & Curtis, Grevillea 1:149, 1873.

Synonyms *Cotylidia aculeata* (Berk. & Curtis) Lentz 1955, *Stereum aculeatum* (Berk. & Curtis) Burt 1926, a later homonym of *Stereum aculeatum* Velen. 1922.

Distribution United States: MO, SC.

Hosts Not known.

Ecology On the ground, we suspect that the basidiomes are attached to buried wood.

Culture Characters Not known.

References 93, 379, 555.

Podoscypha cristata (Berk. & Curtis) D. Reid 1965
Beih. Nova Hedwigia 18:174.

Basionym *Stereum cristata* Berk. & Curtis, Grevillea 1:163, 1873.

Synonym *Cymatoderma cristatum* (Berk. & Curtis) Welden 1960.

Distribution United States: SC.

Host *Vitis* sp.

Ecology Dead plant; dead wood; vines.

Culture Characters Not known.

References 89, 555, 617.

In the United States known from only two collections from the Santee River area of South Carolina.

Podoscypha ravenelii (Berk. & Curtis) Pat. 1900
Essai taxonomique, 71.

Basionym *Stereum ravenelii* Berk. & Curtis, Grevillea 1:162, 1873.

Distribution United States: AL, LA, NC, OH, SC.

Hosts *Carpinus caroliniana, Pinus* sp.

Ecology Usually on the ground and then growing from buried wood; sometimes on rotting wood or roots of trees; rarely on wood humus.

Culture Characters
2.3.8.34.36.38.41.56.60 (Ref. 621).

References 89, 555, 612, 621.

Podoscypha thozetii (Berk.) Boidin 1959
Rev. Mycol. (Paris) 24:208.

Basionym *Stereum thozetii* Berk., J. Linn. Soc., Bot. 18:385, 1881.

Distribution United States: AL.

Host Perhaps grasses.

Ecology Ground amongst grass.

Culture Characters Not known.

Reference 555.

POROTHELEUM Fr. 1818 — Stromatoscyphaceae

Synonym is *Stromatoscypha* Donk 1951, see Cooke (131).

Porotheleum fimbriatum (Pers.) Fr.:Fr. 1818
Obs. Mycol. 1:272.

Basionym *Poria fimbriata* Pers., Neues Mag. Bot. 1:109, 1794.

Synonym *Stromatoscypha fimbriata* (Pers.) Donk 1951, *Porotheleum papillatum* Peck 1887, fide W.B. Cooke (124).

Distribution Canada: MB, NB, NS, ON, PE, PQ.
United States: AL, AZ, CA, CO, FL, IA, ID, IL, IN, KY, MA, MD, ME, MI, MN, MO, MT, NC, NH, NM, NY, OH, PA, SC, TN, VA, VT, WA, WI, WV.

Hosts *Abies concolor, ?Aesculus* sp., *Alnus incana, Betula* sp., *B. alba, B. alleghaniensis, Fagus* sp., *F. grandifolia, Juglans cinerea, Juniperus deppeana, Populus* sp., *P. tremuloides, Quercus alba, Q. hypoleucoides*, angiosperm, gymnosperm.

Ecology Bark; moss-covered bark; dead fallen tree; log; associated with a white rot.
Culture Characters
2.3.(24).(25).32.36.(39).46-47.(54).(55) (Ref. 588).
References 124, 183, 208, 212, 214, 216, 242, 249, 327, 333, 334, 378, 402, 406.

Cooke (124) listed numerous host plants but it is uncertain which were from Canada and the United States.

Porotheleum perenne W.B. Cooke 1957
Mycologia 49:687.
Distribution United States: NY.
Host *Acer saccharum.*
Ecology Not known.
Culture Characters Not known.

Known only from the type specimen.

Porotheleum revivescens Berk. & Curtis 1869
J. Linn. Soc., Bot. 10:324.
Distribution United States: LA.
Hosts Not Known.
Ecology Not known.
Culture Characters Not known.
Reference 124.

Porotheleum subiculosa W.B. Cooke 1957
As Overh. ex W.B. Cooke, Mycologia 49:685.
Distribution United States: WV.
Host *Liquidambar styraciflua.*
Ecology Not known.
Culture Characters Not known.

Known only from the type specimen.

PROTODONTIA Höhnel 1907 Hyaloriaceae

Protodontia oligacantha G.W. Martin 1953
J. Wash. Acad. Sci. 43:16.
Distribution Canada: BC.
Host *Populus trichocarpa.*
Ecology Slash.
Culture Characters Not known.

Known only from the holotype and three paratype specimens.

Protodontia piceicola (Kühner) G.W. Martin 1952
Stud. Nat. Hist. Iowa Univ. 19:63.
Basionym *Protohydnum lividum* var. *piceicolum* Kühner, Botaniste 17:30, 1926.
Distribution Canada: ON. United States: WI.
Hosts *Tsuga canadensis*, gymnosperm.
Ecology Decayed wood.
Culture Characters Not known.
References 388, 443.

Protodontia subgelatinosum (Karsten) Pilát 1952
Sborn. Nár. Mus. v Praze, Rada B, Prír. Vedy 13:200.
Basionym *Hydnum subgelatinosum* Karsten, Meddeland. Soc. Fauna Fl. Fenn. 9:50, 1882.
Synonym *Protodontia uda* Höhnel 1907, fide Strid (596).
Distribution Canada: ON. United States: IA, LA, MO, NC, NEW ENGLAND, OR.
Hosts *Pinus* sp., angiosperm, ?gymnosperm.
Ecology Old log; inside rotting stump.
Culture Characters Not known.
References 432, 443, 503.

PSEUDOHYDNUM Karsten 1868 Hyaloriaceae
Synonym is *Tremellodon* Fr. 1874, fide Donk (157).

Pseudohydnum gelatinosum (Scop.:Fr.) Karsten 1868
Not. Sällsk. Fauna Fl. Fenn. Förh. 9, N.S. 6:374.
Basionym *Hydnum gelatinosum* Scop., Flora Carniolica, Ed. 2, 2:472, 1772.
Synonym *Tremellodon gelatinosum* (Scop.) Fr. 1874.
Distribution Canada: BC, MB, NS, ON, PQ. United States: AK, AZ, CA, CO, ID, LA, MI, MN, NC, NH, NY, OH, OR, PA, TN, WA, WI.
Hosts *Abies concolor, A. lasiocarpa, A. magnifica, Chamaecyparis nootkatensis, Picea* sp., *P. engelmannii, Pinus* sp., *Pseudotsuga menziesii, Quercus* sp., *Q. stellata, Tsuga canadensis, T. heterophylla*, gymnosperm, in the tropics on angiosperm.
Ecology Wet wood; decaying stumps; logs; rotting log; litter.
Culture Characters Not known.
References 107, 116, 119, 120, 130, 133, 183, 216, 242, 249, 273, 332, 333, 335, 402, 419, 443, 448, 510, 511.

PSEUDOMERULIUS Jülich 1979 Coniophoraceae

Pseudomerulius aureus (Fr.) Jülich 1979
Persoonia 10:330.
Basionym *Merulius aureus* Fr., Elenchus 1:62, 1828.
Synonyms *Plicatura aurea* (Fr.) Parm. 1967, *Merulius vastator* Fr. sensu W.B. Cooke 1955, 1974, fide Ginns (in litt.).
Distribution Canada: AB, AB-NT, BC, MB, NB, NS, ON, PQ. United States: AL, AZ, CA, CT, GA, ID, IN, LA, MA, MD, ME, MI, MN, MT, NC, NH, NJ, NM, NY, PA, SC, VA, VT, WA.
Hosts *Picea* sp., *P. engelmannii, P. glauca, Pinus* sp., *P. banksiana, P. contorta* var. *contorta, P. engelmannii, P. ponderosa, P. strobus, Pseudotsuga menziesii, Tsuga heterophylla*, gymnosperm.
Ecology Associated with a brown rot.
Culture Characters
1.3c.(9).21.27.32.36.(37).39.47.51.55.58 (Ref. 474); 1.3.7.(26).32.36.37.39.47.55.58 (Ref. 226).
References 119 as *M. testator* (sic), 120, 128, 197, 199, 208, 216, 226, 242, 249, 474.

Pseudomerulius curtisii (Berk.) Redhead & Ginns 1985
Trans. Mycol. Soc. Japan 26:372.
Basionym *Paxillus curtisii* Berk., Ann. Mag. Nat. Hist., Ser. II, 12:423-424, 1853.
Synonyms *Merulius crassus* Lloyd 1925, fide Ginns (type studied, Lloyd 26987 at BPI), *Serpula crassa* (Lloyd) W.B. Cooke 1957, *Paxillus corrugatus* Atk., Studies Amer. Fungi, 170, 1900, fide Ginns (type studied, CUP 3332).
Distribution Canada: ON, PQ. United States: GA, KY, LA, MA, MN, NC, NH, NY, SC, TN, VA, VT.
Hosts *Fagus* sp., *Pinus* sp., *Quercus* sp., *Tsuga* sp.
Ecology Decorticated log; decaying logs; stump.
Culture Characters Not known.
References 33, 123, 219, 275, 611 and Ginns, in litt.

PSEUDOTOMENTELLA Svrcek 1958 Thelephoraceae

Pseudotomentella atrocyanea (Wakef.) Burdsall & Larsen 1974
Mycologia 66:97.
Basionym *Tomentella atrocyanea* Wakef., Trans. Brit. Mycol. Soc. 49:357, 1966.
Distribution United States: AZ.
Hosts Not known.
Ecology Not known.
Culture Characters Not known.
Reference 69.

Pseudotomentella atrofusca Larsen 1972
Bull. Torrey Bot. Club 98:39.
Distribution Canada: AB, BC. United States: AZ, ID.
Hosts *Pinus ponderosa, P. strobus.*
Ecology Not known.
Culture Characters Not known.
References 216, 242, 361.

Pseudotomentella flavovirens (Höhnel & Litsch.) Svrcek 1958
Ceská Mykol. 12:68.
Basionym *Tomentella flavovirens* Höhnel & Litsch., Sitzungsber. Kaiserl. Akad. Wiss., Math.-Naturwiss. Kl., Abt. 1, 116:831, 1907.
Distribution Canada: ON. United States: CO, ID, NY.
Hosts *Acer* sp., *Betula alleghaniensis, Picea mariana.*
Ecology Not known.
Culture Characters Not known.
References 242, 361, 358.

Pseudotomentella fumosa Larsen 1983
Mycologia 75:558.
Distribution United States: OR.
Host *Pseudotsuga menziesii.*
Ecology Not known.
Culture Characters Not known.

Known only from the type specimen.

Pseudotomentella griseopergamacea Larsen 1971
Bull. Torrey Bot. Club 98:38.
Distribution Canada: ON. United States: NY.
Hosts *Acer* sp., *Pinus* sp., *P. resinosa, P. strobus, Quercus* sp., *Tsuga* sp., *T. canadensis*, angiosperm.
Ecology Not known.
Culture Characters Not known.
References 183, 242, 361.

Pseudotomentella griseoveneta Larsen 1974
Mycologia 66:165.
Distribution United States: OR.
Hosts *Pseudotsuga menziesii, Tsuga* sp., gymnosperm.
Ecology Not known.
Culture Characters Not known.

Known only from the holotype and two paratype specimens.

Pseudotomentella humicola Larsen 1968
Mycologia 60:547.
Distribution Canada: BC, ON. United States: WA.
Hosts *Picea mariana, Thuja occidentalis.*
Ecology Not known.
Culture Characters Not known.
References 242, 361, 448.

Pseudotomentella kaniksuensis Larsen 1983
Mycologia 75:560.
Distribution United States: ID.
Host *Pseudotsuga menziesii.*
Ecology Not known.
Culture Characters Not known.

Known only from the type specimen.

Pseudotomentella longisterigmata Larsen 1967
Canad. J. Bot. 45:1298.
Distribution United States: CO, MN, NY, WA.
Hosts *Abies* sp. *A. balsamea, Tsuga* sp., gymnosperm.
Ecology Not known.
Culture Characters Not known.
References 214, 361.

Pseudotomentella molybdea Larsen 1983
Mycologia 75:556.
Distribution United States: MT.
Host *Pseudotsuga menziesii.*
Ecology Not known.
Culture Characters Not known.

Known only from the type specimen.

Pseudotomentella mucidula (Karsten) Svrcek 1958
Ceská Mykol. 12:68.
Basionym *Hypochnus mucidulus* Karsten, Bidrag Kännedom Finlands Natur Folk 37:163, 1882.

Distribution Canada: AB, BC, ON.
United States: AZ, ID, NM, NY, WA.
Hosts *Castanea dentata, Picea* sp., *Pinus contorta, P. ponderosa, Thuja occidentalis, T. plicata.*
Ecology Not known.
Culture Characters Not known.
References 197, 216, 242, 361, 359, 448.

Pseudotomentella nigra (Höhnel & Litsch.) Svrcek 1960
Sydowia 14:178.
Basionym *Tomentella nigra* Höhnel & Litsch., Wiesner Festschrift, Wien, 78, 1908.
Distribution Canada: AB, BC, ON.
United States: CO, ID, NY.
Hosts *Picea* sp., *Pinus strobus*, gymnosperm.
Ecology Charred wood; log.
Culture Characters Not known.
References 359, 361, 402, 448.

Pseudotomentella tristis (Karsten) Larsen 1972
Nova Hedwigia 22:613.
Basionym *Hypochnus subfuscus* subsp. *tristis* Karsten, Meddeland. Soc. Fauna Fl. Fenn. 9:71, 1882.
Synonyms *Tomentella tristis* (Karsten) Höhnel & Litsch. 1906, *Hypochnus rhacodium* Berk. & Curtis ex Burt 1926, fide Larsen (359), *Pseudotomentella umbrina* (Fr.) Larsen 1967 sensu Aucts., fide Larsen (361).
Distribution Canada: AB, BC, MB, NS, ON.
United States: AZ, CO, ID, MA, MI, MN, MT, NJ, NM, NY, OH, PA, SD, WA.
Hosts ?*Abies* sp., *A. balsamea, A. concolor, A. grandis, Acer* sp., *A. negundo, Amelanchier* sp., *Betula* sp., *B. alleghaniensis, Picea* sp., *P. rubens, Pinus* sp., *P. banksiana, P. contorta, P. ponderosa, P. strobus, P. sylvestris, Populus* spp., *P. tremuloides, Poria* sp., *Quercus* sp., *Q. gambelii, Thuja occidentalis, Tsuga* sp., *T. canadensis, Ulmus* sp., angiosperm, gymnosperm.
Ecology Decaying logs; stone.
Culture Characters Not known.
References 93, 116, 197, 208, 214, 216, 242, 249, 359, 361, 402, 448.

Pseudotomentella vepallidospora Larsen 1967
Canad. J. Bot. 45:1299.
Distribution Canada: BC. United States: OR, WA.
Hosts *Pinus* sp., *Thuja* sp., *T. plicata*, gymnosperm.
Ecology Rotten wood.
Culture Characters Not known.
References 242, 359, 361, 363.

PULCHERRICIUM Parm. 1968 Vuilleminiaceae

Pulcherricium caeruleum (Lamarck:Fr.) Parm. 1968
Consp. Syst. Cort. 133.
Basionym *Byssus caerulea* Lamarck, Flora France, Ed. 2, 1 (Meth. Anal., 103), 1795.
Synonyms *Corticium caeruleum* (Lamarck) Fr. 1838, *Thelephora indigo* Schw. 1822, fide Burt (93) and Fries, Epicrisis, 562, 1838.
Distribution Canada: ON. United States: AL, AR, FL, GA, IL, IN, NC, SC, TN.
Hosts *Acer* sp., *Castanea dentata, C. pumila, Cornus florida, Fagus grandifolia, Juglans sieboldiana, Lagerstroemia* sp., *Ligustrum* spp., *Liriodendron tulipifera, Malus sylvestris, Quercus* sp., *Q. alba, Q. nigra, Q. velutina, Vitis rotundifolia.*
Ecology Wood and bark; fallen limbs; underside decaying limbs; dead branches.
Culture Characters
1.3c.(7).(9).32.(36).(37).40.43.54.58.63 (Ref. 260).
References 93, 96, 108, 183, 315.

The Ontario report is based upon our identification of specimen DAOM 191490.

PUNCTULARIA Pat. 1895 Punctulariaceae
Synonym is *Phaeophlebia* W.B. Cooke 1956, fide Talbot (598).

Punctularia atropurpurascens (Berk. & Br.) Petch 1916
Ann. Roy. Bot. Gard. (Peradeniya) 6:160.
Basionym *Thelephora atropurpurascens* Berk. & Br., J. Linn. Soc., Bot. 14:64, 1875.
Synonyms *Punctularia tuberculosa* (Pat.) Pat. 1895, *Corticium conigenum* Shear & R.W. Davidson 1944. Both fide Talbot (598).
Anamorph *Ptychogaster rubescens* Boudier 1887, fide Nakasone (474).
Distribution United States: FL.
Host *Quercus* sp.
Ecology Bark of decaying stump; associated with a white rot.
Culture Characters
2.3c.27.33.37.(38).40.43.54.59 (Ref. 474 as *P. tuberculosa*).
References 474, 598.

Punctularia strigoso-zonata (Schw.) Talbot 1958
Bothalia 7:143.
Basionym *Merulius strigoso-zonata* Schw., Trans. Amer. Philos. Soc., N.S., 4:160, 1832.
Synonyms *Phlebia strigoso-zonata* (Schw.) Lloyd 1913, *Phaeophlebia strigoso-zonata* (Schw.) W.B. Cooke 1956, *Phlebia orbicularis* Berk. & Curtis 1849, fide Cooke (122), *Phlebia anomala* Berk. & Rav. 1872, fide Cooke (122), *Phlebia rubiginosa* Berk. & Rav. 1872, fide Burt (90) and Cooke (122), *Phlebia zonata* Berk. & Curtis 1872, fide Cooke (122), *Phlebia pileata* Peck 1878, fide Burt (90) and Cooke (122), *Phlebia spilomea* Berk. & Curtis ex Cooke 1891, fide Cooke (122).

Distribution Canada: AB, AB-NT, BC, MB, NB, NF, NS, ON, PE, PQ, SK. United States: AL, AR, CO, CT, DE, GA, IA, IL, IN, KS, KY, LA, MA, ME, MI, MN, MO, MT, ND, NH, NY, OH, PA, RI, SC, SD, TN, VT, WI.

Hosts *Acer* sp. *A. saccharum, Alnus* sp., *A. incana, A. rugosa, Betula papyrifera, B. populifolia, Carya* sp., *C. ?ovata, ?Castanea dentata, Corylus* sp., *Fagus* sp., *Fraxinus* sp., *Malus baccata, Picea* sp., *Populus* sp., *P. acuminata, P. acutissima, P. alba, P. deltoidea, P. grandidentata, P. tremula, P. tremuloides, P. trichocarpa, Prunus* sp., *P. pensylvanica, P. serotina, Quercus* sp., *Q. minor, Q. stellata, Q. tinctoria, Salix* sp., *?Tilia* sp., *Ulmus* sp., angiosperm.

Ecology Infrequently isolated from decay behind wounds in trembling aspen; decaying wood of trees; basidiomes produced on limbs; associated with a white rot.

Culture Characters 2.3c.21.27.35.36.(37).38.43.(50).54.(55).59 (Ref. 474).

References 90, 103, 116, 122, 214, 242, 249, 332, 334, 349, 406, 439, 448, 474, 511.

RADULODON **Ryv. 1972** Hyphodermataceae

Radulodon americanus Ryv. 1972
Canad. J. Bot. 50:2074.

Synonym *Radulum casearium* Aucts., fide Gilbertson (191) and Ryvarden (569).

Distribution Canada: AB, BC, MB, NB, NS, NTM, ON, PQ, SK. United States: AZ, CO, ID, MI, MN, MT, NH, NY, WI.

Hosts *Betula papyrifera, Populus* sp., *P. balsamifera, P. grandidentata, P. grandifolia, P. tremuloides, P. trichocarpa*, angiosperm.

Ecology Heart rot in live aspen; associated with a white stringy rot.

Culture Characters 2.4.7.34.36.40.41-42.(48).(53).54.59 (Ref. 474).

References 116, 183, 214, 242, 249, 402, 406, 474, 489, 511, 569, 603.

Radulodon casearium (Morgan) Ryv. 1972
Canad. J. Bot. 50:2075.

Basionym *Hydnum casearium* Morgan, J. Cincinnati Soc. Nat. Hist. 10:11, 1887.

Synonym *Radulum casearium* (Morgan) Lloyd 1917.

Distribution United States: OH.

Host *Carya* sp.

Ecology Not known.

Culture Characters Not known.

References 191, 569.

Known only from the type specimen. Reports of *Radulum casearium*, invariably, were based upon misidentified specimens of *Radulodon americanus*.

RADULOMYCES **M. Christiansen 1960** Hyphodermataceae

Radulomyces confluens (Fr.:Fr.) M. Christiansen 1960
Dansk Bot. Ark. 19:230.

Basionym *Thelephora confluens* Fr., Obs. Mycol. 1:152, 1815.

Synonyms *Corticium confluens* (Fr.) Fr. 1838, *Cerocorticium confluens* (Fr.) Jülich & Stalpers 1980, *Coniophora avellanea* Burt 1917, *Corticium rubellum* Burt 1926. All fide D.P. Rogers and Jackson (567).

Distribution Canada: MB, NF, NS, NTM, ON. United States: AL, AZ, CO, DC, FL, IA, IL, LA, MA, MD, NH, NJ, NM, NY, OH, PA, VT, WA, WI.

Hosts *Acer* sp., *A. saccharum, Alnus* sp., *A. oblongifolia, Betula* sp., *B. papyrifera, Castanea mollissima, Fouquieria splendens, Juglans major, Picea* sp., *Populus* sp., *P. tremuloides, Quercus* sp., *Q. gambelii, Q. macrocarpa, Q. nigra, Robinia neomexicana, Salix* sp., *Sambucus* sp., *Thuja occidentalis, Tilia* sp., *?T. americana, Vitis* sp., angiosperm.

Ecology Decorticated wood; limbs; dead branches; slash; associated with a white rot.

Culture Characters (1).2.3c.26.32.36.38.44-45.(51).(53).54.60 (Ref. 474).

References 93, 103, 116, 183, 204, 216, 242, 249, 255, 334, 393, 402, 474.

Radulomyces cremoricolor (Berk. & Curtis) Ginns & Lefebvre, *comb. nov.*

Basionym *Corticium cremoricolor* Berk. & Curtis, Grevillea 1:180, 1873.

Synonym *Cerocorticium cremoricolor* (Berk. & Curtis) Ginns 1984.

Distribution Canada: BC, NS. United States: AL, CA, DC, FL, IL, IN, MA, MD, MI, MN, MO, NJ, NM, NY, OH, PA, TX, WI.

Hosts *Abies magnifica, Ilex* sp., *Prunus* sp., *Quercus* sp., angiosperm.

Ecology Dead bark; rotten wood.

Culture Characters Not known.

References 93, 120, 242, 249.

The type (Alabama, on *Ilex*, Peters 5205 at K) "is obviously an extreme form of the familiar and variable *Radulomyces confluens*", fide Hjortstam (288). The new combination is proposed to keep *cremoricolor* in the same genus as *R. confluens*.

Radulomyces fuscus (Lloyd) Ginns 1976
Canad. J. Bot. 54:131.

Basionym *Merulius fuscus* Lloyd, Mycol. Writ. 7: 1348, 1925.

Distribution United States: PA.

Host *Quercus* sp.

Ecology Bark.

Culture Characters Not known.

Reference 226.

Radulomyces notabilis (Jackson) Parm. 1968
Consp. Syst. Cort., 110.
Basionym *Corticium notabile* Jackson, Canad. J. Res., C, 26:156, 1948.
Distribution Canada: AB, BC, MB, NF, NS, ON. United States: CA, CT, MI, MN.
Hosts *Castanea dentata, Picea* sp., *P. glauca, P. mariana, Pinus banksiana, P. resinosa, Thuja occidentalis, T. plicata*, angiosperm, gymnosperm.
Ecology Associated with a white rot.
Culture Characters
2.3c.8.(26).34.36.38.44.(51).(52).(54).55 (Ref. 474).
References 116, 242, 249, 303, 474.

RAMARICIUM Eriksson 1954 Beenakiaceae

Synonym is *Phlyctibasidium* Jülich 1974, fide Ginns (229).

Ramaricium alboflavescens (Ellis & Ev.) Ginns 1979
Bot. Not. 132:94.
Basionym *Corticium alboflavescens* Ellis & Ev., Proc. Acad. Nat. Sci. Philadelphia 1894, 324, 1894.
Synonyms *Coniophora corticola* Overh. 1938, *Serpula illudens* Overh. ex W.B. Cooke 1957, *Serpula imperfecta* Overh. ex W.B. Cooke 1957. All fide Ginns (229).
Distribution Canada: ON (the southern peninsula near Lakes Erie and Ontario). United States: NY and the Appalachian Mountains in PA, TN, VA, WV.
Hosts *Kalmia latifolia, Thuja* sp., *T. occidentalis, Tsuga canadensis*, gymnosperm (probably *Tsuga canadensis*).
Ecology Dead twigs and branches on ground; logs; dry, decaying wood often raised off the ground; associated with a white rot.
Culture Characters
2.3c.7.(31c).32.36.38.47.50.54.55 (Ref. 474);
(1).2.3.7.11.26.32.36.39a.47.(54).55 (Ref. 229).
References 123, 229, 242, 474, 518.

Ramaricium albo-ochraceum (Bres.) Jülich 1977
Persoonia 9:417.
Basionym *Corticium albo-ochraceum* Bres., Ann. Mycol. 1:96, 1903.
Synonym *Trechispora albo-ochracea* (Bres.) Liberta 1973.
Distribution Canada: BC, MB, NB, NS, ON. United States: MI, OH.
Hosts *Abies* sp., *A. balsamea, Betula* sp., *Dactylis glomerata, Picea* sp., *P. mariana, Pinus* sp., *P. resinosa, Populus* sp., *Tsuga* sp., *T. canadensis*.
Ecology Bark; associated with a white rot.
Culture Characters
1.3c.26.(31d).32.36.38.47.54.55 (Ref. 474).
References 116, 229, 242, 395, 474.

Ramaricium flavomarginatum (Burt) Ginns 1979
Bot. Not. 132:94.
Basionym *Coniophora flavomarginata* Burt, Ann. Missouri Bot. Gard. 13:311, 1926.
Distribution Canada: BC. United States: WA.
Host *Quercus garryana*.
Ecology Bark of large dead branches; associated with a white rot.
Culture Characters Not known.
References 93, 133, 229, 242.

Ramaricium polyporoideum (Berk. & Curtis) Ginns 1979
Bot. Not. 132:98.
Basionym *Corticium polyporoideum* Berk & Curtis, Grevillea 1:177, 1873.
Synonyms *Hypochnus polyporoides* (Berk. & Curtis) Overh. 1938, *Phlyctibasidium polyporoideum* (Berk. & Curtis) Jülich 1974, *Trechispora polyporoidea* (Berk. & Curtis) Liberta 1966.
Distribution Canada: PQ. United States: AL, AR, CT, FL, GA, KY, MI, NC, NY, OH, TN, TX, VA, VT.
Hosts *Acer* sp., *Fagus grandifolia, Pinus strobus, Quercus* sp., *Robinia pseudoacacia, Vitis* sp., angiosperm, rarely gymnosperm.
Ecology Needles; twigs; dead fallen limbs; woody debris and live plant parts; log; rock surface; associated with a white rot.
Culture Characters
2.3c.20.(26).27.32.36.38.47.54 (Ref. 474).
References 59, 229, 314, 327, 333, 474, 528.

RECTIPILUS Agerer 1973 Schizophyllaceae

Rectipilus confertus (Burt) Agerer 1973
Persoonia 7:417.
Basionym *Solenia conferta* Burt, Ann. Missouri Bot. Gard. 11:17, 1924.
Distribution United States: AL, GA, KY, MO, SC, VA.
Hosts Not known.
Ecology Rotten log.
Culture Characters Not known.
References 1, 91, 96, 125.

Cooke's (125) records may be of a different species; see Agerer (1).

Rectipilus davidii (D. Reid) Agerer 1973
Persoonia 7:417.
Basionym *Calathella davidii* D. Reid, Persoonia 3:127, 1962.
Distribution Canada: MB, NS, ON, PQ. United States: DE, IA, LA, MA, MD, ME, MI, NC, NH, NJ, NY, OH, PA, SC, TN, VA, VT, WV.
Hosts Angiosperm.
Ecology Rotting wood.
Culture Characters Not known.
References 1, 125, 131, 554.

Apparently this is the fungus Cooke (125, 131) mislabelled as *Solenia fasciculata*, fide Reid (554).

Rectipilus fasciculatus (Pers.:Fr.) Agerer 1973
Persoonia 7:419.
Basionym *Solenia fasciculata* Pers., Mycol. Eur. 1:335, 1822.
Distribution Canada: NS, ON. United States: FL, IA, LA, NJ, NY, PA, SC, VA, VT.
Hosts *Abies balsamea, Betula* sp., gymnosperm.
Ecology Dead wood.
Culture Characters Not known.
References 91, 116, 242, 378, 512.

Cooke's (125, 131) records of *fasciculata* are cited under *Calathella davidii* and see explanation there.

Rectipilus idahoensis (W.B. Cooke) Agerer 1973
Persoonia 7:424.
Basionym *Solenia idahoensis* W.B. Cooke, Beih. Sydowia 4:24, 1961.
Distribution United States: ID.
Host *Pinus monticola.*
Ecology Not known.
Culture Characters Not known.

Known only from the type specimen.

Rectipilus sulphureus (Sacc. & Ellis) W.B. Cooke 1989
Mem. New York Bot. Gard. 49:168.
Basionym *Solenia sulphurea* Sacc. & Ellis, Michelia 2:564, 1882.
Distribution United States: NJ, OH.
Host *Magnolia glauca.*
Ecology Fallen leaves; rotten wood; dead places in live trunk.
Culture Characters Not known.
References 91, 125, 131.

REPETOBASIDIUM Eriksson 1958 — Sistotremaceae

Repetobasidium americanum Eriksson & Hjort. 1981
Cort. N. Europe 6:1251.
Distribution United States: MN.
Hosts Gymnosperm.
Ecology Not known.
Culture Characters Not known.

Known only from the type specimen.

Repetobasidium canadense Eriksson & Hjort. 1981
Cort. N. Europe 6:1253.
Distribution Canada: BC.
Hosts Not known.
Ecology Not known.
Culture Characters Not known.

Known only from the type specimen.

Repetobasidium conicum (Oberw.) Eriksson & Hjort. 1981
Cort. N. Europe. 6:1255.
Basionym *Repetobasidium mirificum* var. *conicum* Oberw., Sydowia 19:61, 1965.
Distribution Canada: BC.
Host *Picea* sp.
Ecology Not known.
Culture Characters
2a.3c.7.32.38.47.55.58.61 (Ref. 261).
References 170, 261.

Repetobasidium macrosporum (Oberw.) Eriksson & Hjort. 1981
Cort. N. Europe 6:1257.
Basionym *Repetobasidium vile* var. *macrosporum* Oberw., Sydowia 19:60, 1965.
Distribution Canada: BC.
Hosts Not known.
Ecology Not known.
Culture Characters Not known.
Reference 170.

Repetobasidium mirificum Eriksson 1958
Symb. Bot. Upsal. 16 (1):70.
Distribution Canada: BC, PQ. United States: NM.
Hosts *Abies concolor, Picea* sp., *Pseudotsuga menziesii.*
Ecology Not known.
Culture Characters
2a.(2b).3c.11.34.36.38.47.55.59.61 (Ref. 261).
References 170, 208, 242, 261.

Repetobasidium vile (Bourd. & Galzin) Eriksson 1958
Symb. Bot. Upsal. 16 (1):67.
Basionym *Peniophora vilis* Bourd. & Galzin, Hym. France, 282, 1928.
Distribution Canada: ON, PQ.
Hosts *Picea* sp., angiosperm, gymnosperm.
Ecology Not known.
Culture Characters Not known.
References 170, 242, 584.

RESINICIUM Parm. 1968 — Meruliaceae

Resinicium bicolor (Alb. & Schw.:Fr.) Parm. 1968
Consp. Syst. Cort., 98.
Basionym *Hydnum bicolor* Alb. & Schw., Conspectus Fungorum, 270, 1801.
Synonyms *Odontia bicolor* (Alb. & Schw.) Quél. 1888, *Hydnum serratum* Peck 1897, fide Gilbertson (189), *Hydnum balsameum* Peck 1904, fide Brown (56) and Gilbertson (189).
Distribution Canada: AB, BC, MB, NB, NF, NS, ON, PQ, SK.
United States: AZ, CO, FL, GA, IA, ID, LA, MA, MI, MN, MS, MT, NC, NM, NY, OR, TN.
Hosts *Abies* sp., *A. amabilis, A. balsamea, A. fraseri, A. grandis, A. lasiocarpa, Acer* sp., *Betula* sp., *B. alleghaniensis, B. papyrifera, Juniperus* sp., *Larix laricina, L. gmelinii, L. occidentalis, Liquidambar styraciflua, Oxydendrum arboreum, Picea* sp., *P. engelmannii, P. glauca, P. mariana, P. rubens, Pinus* sp., *P. banksiana, P. contorta, P. monticola, P.*

ponderosa, P. resinosa, P. strobus, P. taeda, Populus sp., *P. tremuloides, Pseudotsuga menziesii, Rhododendron catawbiense, Thuja plicata, T. occidentalis, Tsuga* sp., *T. canadensis, T. heterophylla*, gymnosperm.

Ecology Isolated from root rot in black spruce saplings; causes a root and butt rot in *Abies amabilis* and *Thuja plicata*; root rot in *A. lasiocarpa*; trunk rot in *A. balsamea* and young *A. grandis*; one of the three principal white rot fungi in roots of *A. balsamea*; causes a white stringy butt rot in *Pinus strobus*; basidiomes produced on bark; branch; slash; log; associated with a white rot.

Culture Characters
2.3c.13.26.32.(35).36.(38).40.44-46.(50).(52).54.55.59 (Ref. 474);
2.3.13.32.36.38.40.44-47.54.55.58 (Ref. 149);
2.3.13.32.(35).36.40.43-46.54.55.59 (Refs. 487 & 582).

References 56, 96, 116, 119, 177, 183, 197, 204, 214, 216, 242, 249, 255, 274, 327, 402, 406, 463, 474, 485, 645.

Resinicium furfuraceum (Bres.) Parm. 1968
Consp. Syst. Cort., 98.

Basionym *Corticium furfuraceum* Bres., Mycologia 17:69, 1925.

Distribution Canada: BC, NS, ON, PQ.
United States: AZ, CO, FL, ID, MA, MI, MN, MT, NC, NH, NM, TN, VA, WA, WI.

Hosts *Abies* sp., *A. balsamea, A. fraseri, A. grandis, A. lasiocarpa, Betula alleghaniensis, Fagus grandifolia, Juniperus* sp., *Larix occidentalis, Picea* sp., *P. engelmannii, P. rubens, Pinus* sp., *P. contorta, P. monticola, P. ponderosa, P. resinosa, P. strobus, P. taeda, Pseudotsuga menziesii, Thuja plicata, T. occidentalis, Tsuga canadensis, T. heterophylla*, gymnosperm.

Ecology Moss-covered bark; charred wood; decaying wood of logs; log; associated with a white rot.

Culture Characters
2.3c.7.(13).(16).33.36.38.(39).47.(53).55.58 (Ref. 474);
2.3.7.33.36.39.47.52.55 (Ref. 487);
2.3.7.33.36.39.47.52.55 (Ref. 450).

References 53, 93, 116, 119, 183, 197, 204, 214, 216, 242, 249, 255, 327, 388, 402, 450, 474.

Resinicium meridionale (Burdsall & Nakasone) Nakasone 1990
Mycologia Memoir 15:285.

Basionym *Mycoacia meridionalis* Burdsall & Nakasone, Mycologia 73:465, 1981.

Distribution United States: FL, GA, MS, SC, TX.

Hosts *Carpinus* sp., *Carya* sp., *Persea borbonia, Pinus* sp., *P. palustris, P. taeda, Platanus occidentalis, Quercus* sp., angiosperm.

Ecology Associated with a white rot.

Culture Characters
2.3c.(16).(26).28.32.36.38.44-47.(48).(53).54.55.59 (Ref. 474);
2.3.(26).28.32.36.38.44-47.(48).54.55.59 (Ref. 74).

References 74, 474.

Resinicium praeteritum (Jackson & Dearden) Ginns & Lefebvre, *comb. nov.*

Basionym *Corticium praeteritum* Jackson & Dearden, Canad. J. Res., C, 27:152, 1949.

Distribution Canada: BC.

Host *Alnus rubra*.

Ecology Not known.

Culture Characters Not known.

Known only from the type specimen. An isotype (DAOM 31350) seems to be a young specimen because the hymenium is not a distinct palisade. However, in the hymenial area there are numerous broadly clavate cells (9 µm diam) with pale yellow, slightly refractive, sulfo-negative contents. The contents apparently recede from the apical half of the cell and the cells look like the halocystidia which are typical of this genus. The presence of this feature and because the other features are within the limits of the genus, we transfer the epithet to *Resinicium*.

SARCODONTIA S. Schulzer 1866 Hyphodermataceae

Sarcodontia crocea (Schw.:Fr.) Kotlaba 1953
Ceská Mykol. 7:117.

Basionym *Sistotrema croceum* Schw., Schriften Naturf. Ges. Leipzig 1:102, 1822.

Synonyms *Hericium croceum* (Schw.) Banker 1906, *Mycoacia setosa* (Pers.) Donk 1931, fide Kotlaba (346), *Hydnum amplissimum* Berk. & Curtis 1873, fide Gilbertson (192 as *Mycoacia setosa*), *Hydnum subvelutinum* Berk. & Curtis 1873, fide Gilbertson (192).

Distribution Canada: NS, ON.
United States: IA, MA, ME, NJ, NY, PA, WV.

Hosts *Crataegus* sp., *Fagus grandifolia, Malus* sp., *Prunus* sp.

Ecology Dead branches on live tree; live but diseased tree; dying tree.

Culture Characters
(2).4.(21).(24).(25).32.36.38.(40).44-47.48.50.54 (Ref. 588).

References 25, 170, 192, 249, 463.

SCHIZOPHYLLUM Fr.:Fr. 1815 Schizophyllaceae

Schizophyllum commune Fr.:Fr. 1815
Obs. Mycol. 1:103.

Synonym *Schizophyllum radiatum* (Swartz) Fr. 1855, fide W.B. Cooke (126).

Distribution Canada: AB, BC, MB, NB, NF, NS, NTM, ON, PE, PQ, SK.

United States: AR, AL, AZ, CA, CT, DC, FL, GA, IA, ID, IL, IN, KS, KY, LA, MA, MD, ME, MI, MN, MO, MS, NC, NH, NY, OH, OK, OR, PA, SC, SD, TN, TX, VA, WA, WV.
Hosts Farr et al. (183) list 60 genera of host plants. In addition, the references below report the fungus on *Abies*, *Acer*, *Aesculus*, *Lindera*, *Paulownia*, *Rhamnus* and *Sambucus*.
Ecology Twigs; dead branch; causes sapwood rot behind wounds in *Carya* sp.; associated with a white rot.
Culture Characters
(1).2.3c.8.20.34.36.38.(40).43.48.53.54.55.60 (Ref. 474);
1.2.3.7.20.32.34.36.38.43.48.54.55.60 (Ref. 487).
References 96, 116, 127, 128, 130, 131, 178, 183, 225, 242, 249, 274, 396, 468, 474, 484.

Schizophyllum fasciatum Pat. 1887
J. Bot. (Morot) 1:170.
Distribution United States: FL, TX.
Hosts Not known.
Ecology Not known.
Culture Characters Not known.
References 126, 396.

Schizophyllum umbrinum Berk. 1851
Hooker's J. Bot. Kew Gard. Misc. 3:15.
Distribution United States: FL.
Hosts Not known.
Ecology Not known.
Culture Characters Not known.
Reference 126.

SCOPULOIDES (Massee) Hjort. & Ryv. 1979 Phanerochaetaceae

Scopuloides rimosa (Cooke) Jülich 1982
Persoonia 11:422.
Basionym *Peniophora rimosa* Cooke, Grevillea 9:94, 1881.
Synonyms *Phanerochaete rimosa* (Cooke) Burdsall 1985, *Peniophora hydnoides* Cooke & Massee in Massee 1890, fide Burdsall (64), *Phlebia hydnoides* (Cooke & Massee) M. Christiansen 1960, a later homonym of *Phlebia hydnoidea* Schw. 1832, *Odontia hydnoides* (Cooke & Massee) Höhnel 1909, *Scopuloides hydnoides* (Cooke & Massee) Hjort. & Ryv. 1979.
Distribution Canada: BC, NS, ON.
United States: AL, AZ, DC, FL, GA, IA, IL, LA, MI, MN, NC, NY, PA, TN, VT, WI.
Hosts *Acer* sp., *A. saccharum*, *Castanea dentata*, *Fagus grandifolia*, *Fraxinus* sp., *F. nigra*, *Nyssa* sp., *Populus* sp., *P. tremuloides*, *Quercus* sp., *Q. garryana*, *Salix* sp., *Tilia americana*, *Ulmus americana*, angiosperm, gymnosperm.
Ecology Bark and wood; charred wood; dead branches; rotten trunks; associated with a white rot.
Culture Characters
(1).2.5.7.32.36.(38).(39).(40).42-43.(48).(50).54.55.(57) (Ref. 474).
References 56, 61, 64, 116, 183, 214, 216, 249, 327, 406, 463, 474.

Massee used the Greek suffix for the epithet and spelled it *hydnoides*, whereas Schwienitz used the Latin and spelled it *hydnoidea*.

SCYTINOSTROMA Donk 1956 Lachnocladiaceae

Scytinostroma arachnoideum (Peck) Gilbn. 1962
Mycologia 54:660.
Basionym *Hydnum arachnoideum* Peck, Annual Rep. New York State Mus. 44:133, 1891.
Synonym *Corticium quaesitum* Jackson & Dearden 1949, fide Gilbertson (189).
Distribution Canada: BC. United States: AK, AZ, FL, ID, NC, NH, NM, NY, OR, WA.
Hosts *Picea* sp., *P. engelmannii*, *P. rubens*, *P. sitchensis*, *Pinus* sp., *P. ponderosa*, *Pseudotsuga menziesii*, *Tsuga canadensis*, *T. heterophylla*, gymnosperm.
Ecology Rotted log; log; associated with a white rot.
Culture Characters Not known.
References 116, 183, 189, 197, 216, 242, 308, 448.

With ornamented spores, probably *Scytinostromella*, fide Hallenberg (258).

Scytinostroma duriusculum (Berk. & Br.) Donk 1956
Fungus 26:20.
Basionym *Stereum duriusculum* Berk. & Br., J. Linn. Soc., Bot. 14:66, 1873.
Distribution United States: TN, TX.
Hosts Angiosperm.
Ecology Dead wood.
Culture Characters Not known.
References 44, 354, 518 as *S. duriusculum* Bres. (sic).

Scytinostroma fulvum (Berk. & Curtis) Hjort. 1990
Mycotaxon 39:420.
Basionym *Kneiffia fulva* Berk. & Curtis, J. Linn. Soc., Bot. 10:327, 1868.
Synonym *Grandinia alutacea* Berk. & Rav., 1873, fide Hjortstam (289).
Distribution United States: GA, presumably SC.
Host *Platanus occidentalis*.
Ecology Bark.
Culture Characters Not known.
References 96, 289.

Scytinostroma galactinum (Fr.) Donk 1956
Fungus 26:20.
Basionym *Thelephora galactina* Fr., Nova Acta Regiae Soc. Sci. Upsal., Ser. III, 1:136, 1851.
Synonym *Corticium galactinum* (Fr.) Burt 1926.
Distribution Canada: AB, BC, MB, NB, NF, NS, ON, PE, PQ, SK.

United States: AK, AL, AR, AZ, CA, CO, CT, DE, FL, GA, ID, IL, IN, KY, LA, MA, MD, ME, MI, MN, MO, MS, MT, NC, NH, NM, NY, OK, OR, SC, SD, TN, TX, VA, VT, WA, WI, WV, WY.

Hosts *Abies balsamea, A. grandis, A. lasiocarpa, A. magnifica, Acer* sp., *A. saccharum, Alnus rubra, Baptisia australis, Betula* sp., *B. alleghaniensis, B. papyrifera, B. populifolia, Carpinus caroliniana, Castanea* sp., *Ceanothus thyrsiflorus, Clethra alnifolia, Cornus florida, Cytisus scoparius, Exochorda racemosa, Fagus grandifolia, Fomes* sp., *Hibiscus syriacus, Ilex* sp., *Ilex opaca, Iris* sp., *Jasminum* sp., *J. nudiflorum, Juniperus* sp., *Kalmia* sp., *K. latifolia, Larix* sp., *L. laricina, L. occidentalis, Liquidambar styraciflua, Lychnis alba, Magnolia virginiana, Malus* sp., *M. sieboldi, M. sylvestris, Nerium oleander, Paeonia officinalis, Picea* sp., *P. abies, P. canadensis, P. engelmannii, P. glauca, P. glauca* var. *albertiana, P. mariana, P. rubens, P. rubra, Pinus* sp., *P. banksiana, P. contorta, P. monticola, P. ponderosa, P. resinosa, P. strobus, P. taeda, Platanus occidentalis, Populus* sp., *P. balsamifera, P. tremuloides, P. trichocarpa, Prunus* sp., *P. glandulosa, P. persica, P. triloba, Pseudotsuga* sp., *P. menziesii, Quercus* sp., *Q. alba, Q. nuttallii, Q. palustris, Q. rubra* var. *borealis, Q. velutina, Rhus glabra, Rubus* sp., *Rubus allegheniensis, R. flagellaris, R. phoenicolasius, Silene alba, Spiraea thunbergii, Thuja occidentalis, T. plicata, Tilia americana, Tsuga* sp., *T. canadensis, T. heterophylla, Viburnum carlesi*, fruit trees, angiosperm, gymnosperm.

Ecology Causes a root and butt rot of woody plants; basidiomes produced on slash, rotten logs, stumps, wood and roots, old conk; associated with a white rot.

Culture Characters
2.3.8.32.36.38.42.(50).54.55.60 (Ref. 381);
2.3c.8(d).15p.21.32.36.38.(40).44-45.(54).55.60 (Ref. 474).

References 44, 93, 116, 120, 128, 183, 197, 204, 208, 214, 216, 242, 249, 381, 402, 406, 448, 474, 479, 509, 511, 644, 645.

Scytinostroma jacksonii Boidin 1981
Naturaliste Canad. 108:199.

Distribution Canada: BC. United States: MT.
Hosts *Abies lasiocarpa, Tsuga* sp., gymnosperm.
Ecology Associated with a white rot.
Culture Characters
2a.3c.8d.15a.15p.32.36.38.47.55.58.(61) (Ref. 44);
2.3c.(8).15.26.32.36.38.47.50.55.58 (Ref. 474).
References 39, 44, 474.

Scytinostroma ochroleucum (Bres. & Torrend) Donk 1956
Fungus 26:20.

Basionym *Gloeocystidium ochroleucum* Bres. & Torrend, Brotéria, Sér. Bot. 11:81, 1913.
Synonym *Corticium abeuns* Burt 1926, fide Jülich and Stalpers (326), and Boidin and Lanquetin (44).
Distribution Canada: BC.
United States: AL, AZ, CA, NH, NY.
Hosts *Pinus ponderosa, Pseudotsuga menziesii, Thuja plicata, Tsuga canadensis, T. heterophylla*, gymnosperm.
Ecology Bark; decaying wood.
Culture Characters
2.6.8d.15a.15p.32.36.39.47.54.55.58.62 (Ref. 44).
References 93, 183, 216, 242, 567.

Rogers and Jackson (567) hesitated in placing *C. abeuns* in synonymy because they had not seen the type of *S. ochroleucum*. They, having examined all the other collections cited by Burt, found only one New York and one British Columbia collection to "resemble the type". Jülich and Stalpers (326), and Boidin and Lanquetin (44) accept *C. abeuns* as a synonym but do not give supporting evidence.

Scytinostroma odoratum (Fr.:Fr.) Donk 1956
Fungus 26:20.

Basionym *Thelephora odorata* Fr., Obs. Mycol. 1:151, 1815.
Synonym *Corticium odoratum* (Fr.) Bourd. & Galzin 1928.
Distribution Canada: ON.
Hosts *Abies* sp., *Pinus* sp., angiosperm, gymnosperm.
Ecology Not known.
Culture Characters Not known.
References 116, 644.

Scytinostroma portentosum (Berk. & Curtis) Donk 1956
Fungus 26:20.

Basionym *Corticium portentosum* Berk. & Curtis, Grevillea 2:3, 1873.
Synonyms *Corticium portentosum* var. *crystallophorum* Ellis & Ev. 1897, fide Burt (93), *Corticium diminuens* Berk. & Curtis 1873, fide Burt (93) and Hjortstam (288), ?*Scytinostroma hemidichophyticum* Pouzar 1966, fide Hallenberg (258).
Distribution Canada: BC, ON.
United States: AK, AL, AZ, CA, FL, ID, IN, KY, LA, ME, MO, NY, OH, PA, TX, VA, VT.
Hosts *Acacia flexicaulis, Acer rubrum* var. *rubrum, Aleurites fordii, Alnus* sp., *Betula pumila, Bougainvillea spectabilis, Carya tomentosa, Crossosoma bigelovii, Gleditsia* sp., *Juniperus virginiana, Mortonia scabrella, Prosopis juliflora, Pseudotsuga menziesii, Quercus* sp., *Q. arizonica, Q. emoryi, Q. gambelii, Q. hypoleucoides, Sambucus* sp., *Sapindus saponaria* var. *drummondii*, angiosperm.
Ecology Logs; associated with a white rot.
Culture Characters
2a(b).6.8d.12.15.26.34.36.38.47.50.53.54.60.62 (Ref. 354);
2.6.8.15.26.32.36.38.46-47.(51).(53).54.60 (Ref. 474).
References 44, 93, 183, 201, 204, 216, 354, 474.

Because Burt's (93) records from Louisiana (Langlois 244 & 2098), Wisconsin (Harper 848), and New York

(Atkinson 3406) are not *S. portentosum*, fide Boidin and Lanquetin (44), we have not cited them above.

Scytinostroma praestans (Jackson) Donk 1956
Fungus 26:20.
Basionym *Corticium praestans* Jackson, Canad. J. Res., C, 26:148, 1948.
Distribution Canada: AB, ON.
United States: IA, MN, WI.
Hosts *Osmunda claytoniana, Picea* sp., *Pinus controta, Quercus alba, Rubus* spp., angiosperm.
Ecology Associated with a white rot; ground.
Culture Characters
2.3c.(8).15.26.32.36.38.45-46.53.54.(55) (Ref. 474).
References 44, 116, 214, 303, 474.

Scytinostroma protrusum subsp. ***protrusum*** (Burt) Nakasone 1988
Mycologia 80:555.
Basionym *Corticium protrusum* Burt, Ann. Missouri Bot. Gard. 13:260, 1926.
Distribution United States: GA, FL, IL, IN, KY, MD, MS, VA.
Hosts *Alnus* sp., *Carya* sp., *Liquidambar styraciflua, Malus* sp., *Populus tremuloides, Prunus* sp., *Quercus* sp., *Q. rubra*, angiosperm.
Ecology Often associated with a root rot.
Culture Characters
2.3c.8(d).15.21.32.36.38.(40).44-45.54.(55).60 (Ref. 479).
Reference 479.

Scytinostroma protrusum subsp. ***septentrionale*** Nakasone 1988
Mycologia 80:555.
Distribution United States: ID, MI, MN, NY, WI.
Hosts *Acer* sp., *Betula papyrifera, Fagus grandifolia, Pinus* sp., *Populus* sp., *P. tremuloides*, angiosperm.
Ecology Not known.
Culture Characters
2.3c.8(d).15.21.32.36.38.(40).44-45.54.(55).60 (Ref. 479).
Reference 479.

SCYTINOSTROMELLA Parm. 1968 Gloeocystidiellaceae
Synonym is *Confertobasidium* Jülich 1972, see *S. olivaceo-album* below.

Scytinostromella heterogenea (Bourd. & Galzin) Parm. 1968
Consp. Syst. Cort., 171.
Basionym *Peniophora heterogenea* Bourd. & Galzin, Bull. Soc. Mycol. France 28:293, 1913.
Synonym *Gloeocystidiellum heterogeneum* (Bourd. & Galzin) Donk 1956.
Distribution Canada: BC, ON.
United States: AZ, CA, MI, NM.
Hosts *Pinus ponderosa*, angiosperm, gymnosperm.
Ecology Not known.
Culture Characters Not known.
References 185, 197, 208, 584.

Scytinostromella humifaciens (Burt) Freeman & Petersen 1979
Mycologia 71:86.
Basionym *Peniophora humifaciens* Burt, Ann. Missouri. Bot. Gard. 12:225, 1926.
Distribution Canada: BC. United States: ID, MT, WA.
Hosts *Pseudotsuga menziesii, Thuja plicata*, gymnosperm.
Ecology Very rotten log.
Culture Characters Not known.
References 92, 116, 171, 185, 242, 392.

Scytinostromella nannfeldtii (Eriksson) Freeman & Petersen 1979
Mycologia 71:90.
Basionym *Gloeocystidiellum nannfeldtii* Eriksson, Svensk. Bot. Tidskr. 52:14, 1958.
Distribution United States: AK.
Host *Picea* sp.
Ecology Decayed wood, debris, leaves, etc.
Culture Characters Not known.
References 171, 185.

Scytinostromella olivaceo-album (Bourd. & Galzin) Ginns & Lefebvre, *comb. nov.*
Basionym *Corticium olivaceo-album* Bourd. & Galzin, Bull. Soc. Mycol. France 27:239, 1911.
Synonyms *Confertobasidium olivaceo-album* (Bourd. & Galzin) Jülich 1972, ?*Scytinostromella fallax* Burdsall & Nakasone 1981, fide Hjortstam (286:64).
Distribution United States: FL, GA, MS.
Hosts *Pinus* sp., *P. taeda*.
Ecology Associated with a white rot.
Culture Characters
2.3c.(8).13.15p.(16).31e.32.37.38.44-47.55 (Ref. 474);
2.3.(8).15.21.26.32.37.38.44.54.55 (Ref. 74).
References 74, 474.

The lectotype (Bourdot 7509 at UPS) has skeletal hyphae and amyloid spores (286). Because these are characteristics of the genus *Scytinostromella*, *olivaceo-album* is transferred to that genus. The transfer of the type species *olivaceo-album* means *Confertobasidium* becomes a synonym of *Scytinostromella*. The species concept has been confused. Hjortstam (286) concluded that the description of *C. olivaceo-album* in Jülich (312), and subsequently Jülich and Stalpers (326), was *Leptosporomyces fuscostratus*.

SEBACINA Tul. 1871 Sebacinaceae

Sebacina arctica Y. Kobayasi 1967
Ann. Rept. Inst. Fermentation, Osaka 3:62.
Distribution United States: AK.
Host *Salix alaxensis*.

Ecology Superficial on bark.
Culture Characters Not known.

Known only from the type specimen.

Sebacina cinnamomea Burt 1915
Ann. Missouri Bot. Gard. 2:763.
Distribution United States: MD.
Host *Magnolia glauca.*
Ecology Limbs on dead tree.
Culture Characters Not known.

Known only from the type specimen.

Sebacina epigaea var. ***epigaea*** (Berk. & Br.) Neuhoff 1931
Z. Pilzk. 10:71.
Basionym *Tremella epigaea* Berk. & Br., Ann. Mag. Nat. Hist., Ser. II, 2:266, 1848.
Synonyms *Sebacina atrata* Burt 1915, fide McGuire (451), *Sebacina cokeri* Burt 1926, fide McGuire (451), *Tremella gyroso-alba* Lloyd ex J.A. Stevenson & Cash 1936, fide Bandoni (15).
Distribution Canada: NS, ON. United States: CA, GA, IA, LA, MA, MO, NC, NH, NY, OR, WI.
Hosts *Acer* sp., *Betula* sp., *Elaeagnus* sp., *E. commutata, Fraxinus nigra, Juniperus virginiana, Mimosa* sp., *Populus* sp., *Quercus* sp., angiosperm, rarely gymnosperm.
Ecology Rotten wood; branches; old hard heart of branch; bark at the bases of live trees; lower side of logs; fallen leaves and debris; underside of some very old cow dung; soil.
Culture Characters Not known.
References 15, 81, 93, 116, 183, 201, 202, 242, 249, 406, 443, 451, 498, 499, 643 .

Sebacina epigaea var. ***bicolor*** Olive 1958
J. Elisha Mitchell Sci. Soc. 74:41.
Distribution United States: NC.
Host *Quercus* sp.
Ecology Very rotten log.
Culture Characters Not known.

Known only from the type specimen. The name was proposed (495) with a detailed description and illustrations in 1944 but not validly published with a Latin description until 1958.

Sebacina helvelloides (Schw.) Burt 1915
Ann. Missouri Bot. Gard. 2:756.
Basionym *Thelephora helvelloides* Schw., Schriften Naturf. Ges. Leipzig 1:108, 1822.
Synonym *Sebacina chlorascens* Burt 1915, fide McGuire (451).
Distribution Canada: ON. United States: CT, FL, IA, IN, MA, NC, NH, NY, PA.
Hosts *Fagus grandifolia, Fraxinus nigra, Kalmia* sp., *Ostrya virginiana, Picea* sp., *Prunus* sp., *Quercus* sp., *Q. alba*, angiosperm.
Ecology Incrusting soil and bark at base of live trees and shrubs, but does not affect the inner bark or cambium.
Culture Characters Not known.
References 81, 107, 183, 274, 443, 451.

Sebacina incrustans (Pers.:Fr.) Tul. 1871
J. Linn. Soc., Bot. 13:36.
Basionym *Corticium incrustans* Pers., Obs. Mycol. 1:39, 1796.
Synonyms *Merisma cristata* Pers. 1797, *Sebacina cristata* (Pers.) Lloyd 1925, *Sebacina amesii* Lloyd 1916, *Ptychogaster subiculoides* Lloyd 1922. All fide Martin (443).
Distribution Canada: AB-NT, NS, NTM, ON, PQ. United States: CT, DC, FL, IA, IL, LA, MA, MD, ME, MI, MN, MO, NC, NH, NY, OH, PA, SC, TN, VA, VT, WI, WV.
Hosts *Acer* sp., *Adiantum pedatum, Aster* sp., *Cornus* sp., *Corylus americana, Elephantopus carolinianus, Erigeron annuus, Glechoma hederacea, Hypericum prolificum, Juniperus virginiana, Nyssa* sp., *Onoclea sensibilis, Picea glauca, Poa pratensis, Polystichum acrostichoides, Potentilla* sp., *Quercus macrocarpa, Thelypteris noveboracensis, Tilia americana, Viola* sp., *Vitis* sp., ferns.
Ecology Debris or bases of live plants; resupinate on logs; at base of live tree; roots of trees and base of fronds; soil.
Culture Characters Not known.
References 81, 107, 183, 202, 242, 333, 443, 451, 593.

Sebacina podlachica Bres. 1903
Ann. Mycol. 1: 117.
Synonym *Exidiopsis podlachica* (Bres.) Ervin 1957.
Distribution Canada: ON.
United States: CA, GA, IA, LA, MA, NC, OR, TN.
Hosts *Amelanchier* sp., *Elaeagnus commutata, Lyonothamnus floribundus, Pinus strobus, Prunus* sp., *P. serotina, Quercus* sp., *Ulmus* sp., angiosperm, rarely gymnosperm.
Ecology Decorticated wood; decaying wood; fallen branches; dead branches.
Culture Characters Not known.
References 81, 183, 419, 443, 451, 465, 498, 499, 622, 625.

More closely related to *S. sublilacina, Exidia nucleata, Stypella minor*, etc., than to *Exidiopsis* species, fide Wells (625) but Lowy (422) placed the species in *Exidiopsis.*

Sebacina polyschista Burt 1926
As Berk. & Curtis ex Burt, Ann. Missouri Bot. Gard. 13:338.
Distribution United States: SC.
Host *Pyrus malus.*
Ecology Underside of limb on dead tree.
Culture Characters Not known.

Known only from the type specimen.

Sebacina sordida Olive 1958
J. Elisha Mitchell Sci. Soc. 74:41.
Distribution United States: NC.
Hosts *Pinus* sp., *Platanus* sp.
Ecology Decorticated limb on the ground.
Culture Characters Not known.

Known only from the type specimens. The name was proposed (495) in 1944 but lacked a Latin description. See *S. sordida* in Excluded Names (p. 190).

SERPULA S.F. Gray 1821 Coniophoraceae

Serpula himantioides (Fr.:Fr.) Karsten 1889
Bidrag Kännedom Finlands Natur Folk 48:344.
Basionym *Merulius himantioides* Fr., Obs. Mycol. 2:238, 1818.
Synonyms *Serpula lacrimans* var. *himantioides* (Fr.) W.B. Cooke 1957, *Merulius brassicaefolius* Schw. 1822, fide Ginns and Lefebvre (herein), *Merulius lacrimans* var. *tenuissimum* Berk. ex Cooke 1878, fide Ginns (226), *Merulius tenuis* Peck 1894, fide Ginns (220), *Merulius americanus* Burt 1917, fide Ginns (218), *Serpula americana* (Burt) W.B. Cooke 1957.
Distribution Canada: AB, BC, MB, NB, NF, NS, ON, PQ. United States: AL, AZ, CA, CO, DC, FL, GA, IA, ID, IL, KY, LA, MA, MD, ME, MI, MO, MT, NC, NH, NJ, NM, NY, NV, OH, OR, PA, SC, TN, VA, VT, WA, WI, WV, WY.
Hosts *Abies amabilis, A. balsamea, A. concolor, A. lasiocarpa, A. magnifica, Chamaecyparis nootkatensis, Larix laricina, Picea* sp., *P. engelmannii, P. glauca, P. mariana, Pinus* sp., *P. ayacahuite, P. banksiana, P. contorta, P. monticola, P. ponderosa, P. resinosa, P. strobiformis, P. strobus, Populus* sp., *P. tremuloides, Pseudotsuga menziesii, Quercus* sp., *Thuja plicata, Tsuga canadensis, T. heterophylla, T. mertensiana*, angiosperm, gymnosperm.
Ecology Causes significant butt rot in *Abies balsamea*; a trunk rot in *Chamaecyparis nootkatensis*; and heart rot in *Picea glauca* and *P. rubens*; basidiomes produced on litter and rotten wood; dead trees; associated with a brown rot.
Culture Characters
1.3.7.8.32.(35).36.38.43-47.53.54.55.60 (Ref. 146 and others).
References 116, 119, 120, 123, 128, 130, 146, 183, 197, 199, 204, 208, 216, 242, 249, 266, 402, 406, 448, 582.

Authentic specimens of *M. brasicaefolius* are not known in North America (220). We have seen a specimen in Fries' herbarium (UPS), labelled *Merul. brassicaefolius* Sz., and it is here designated neotype. Harmsen (266) suggested it was a specimen of *Serpula himantioides* and we agree. Cooke (123) included a worldwide list of woody substrates but records from Canada and the United States could not be segregated.

Serpula incrassata (Berk. & Curtis) Donk 1948
Bull. Jard. Bot. Buitenzorg, Sér. III, 17:474.
Basionym *Merulius incrassatus* Berk. & Curtis, Hooker's J. Bot. Kew Gard. Misc. 1:234, 1849.
Synonyms *Meruliporia incrassata* (Berk. & Curtis) Murrill 1942, *Poria incrassata* (Berk. & Curtis) Burt 1917, *Merulius spissus* Berk. 1872, fide Burt (83) and Ginns (222).
Distribution Canada: BC, NB, ON.
United States: AL, CA, CT, DC, NY, FL, GA, ID, IL, KY, LA, MI, MS, NC, NE, NY, OK, OR, PA, SC, TN, TX, VA, WA.
Hosts *Libocedrus decurrens, Magnolia* sp., *Picea* sp., *Pinus* spp., *P. palustris, P. ponderosa, Pseudotsuga menziesii, Quercus* sp., *Q. borealis* var. *maxima, Robinia pseudoacacia, Sequoia sempervirens, Sequoiadendron giganteum, Taxodium* sp., *T. distichum, Tsuga canadensis*, angiosperm, gymnosperm.
Ecology Structural timbers; flooring and other wood in buildings; railroad ties; lumber in lumber yards; "gum wood".
Culture Characters
1.3.(5).7.35.36.38.45-46.53.54.55 (Ref. 146).
References 123, 183, 222, 242, 416, 582.

Serpula lacrimans var. ***lacrimans*** (Jacq.:Fr.) Schröter 1888
In F. Cohn, Kryptogamen-Flora Schlesien 3 (1):466.
Basionym *Boletus lacrimans* Jacq., Miscellanea Austriaca 2:111, 1781.
Synonyms *Merulius lacrimans* (Jacq.) Schum. 1803, *Merulius lacrimans* var. *terrestris* Peck 1896, fide Ginns (226), *Merulius terrestris* (Peck) Burt 1917.
Distribution Canada: AB, BC, MB, NS, ON, PQ, SK. United States: AK, AL, AZ, CA, CO, CT, DC, FL, GA, IA, ID, IL, KS, KY, LA, MA, MD, ME, MI, MO, MT, NC, NE, NH, NJ, NM, NV, NY, OR, PA, SC, UT, VA, VT, WA, WV, WY.
Hosts *Abies lasiocarpa, A. magnifica, Picea* sp., *Pinus* sp., *P. contorta, P. ponderosa, Pseudotsuga menziesii, Thuja plicata, Tsuga mertensiana*, gymnosperm.
Ecology This fungus is one of the principal dry rot fungi. Rotten wood; cordwood; planks; beams in gymnosperm log shed; structural timbers; restricted to buildings; clay bank in cellar; associated with a brown rot.
Culture Characters
1.3.(7).8.32.35.36.37.38.45.46.53.(54).55.60 (Ref. 487).
References 96, 116, 123, 128, 130, 183, 199, 242, 249, 266, 484, 582.

Reported (242) on *Betula* sp. in the forest which strongly suggests that the fungus was *S. himantioides*. Cooke (123) included a worldwide list of substrates but records from Canada and the United States could not be segregated.

Serpula lacrimans var. ***shastensis*** (L. Harmsen) Ginns & Lefebvre, *comb. nov.*
Basionym *Merulius lacrimans* var. *shastensis* L. Harmsen, Friesia 6:273, 1960.
Distribution United States: CA.
Hosts *Abies magnifica, Tsuga martensiana.*
Ecology Logs.
Culture Characters Not known.
Reference 130.

The variety *shastensis* is transferred to *Serpula* to keep the variety in the same genus as the species. In addition the generic name *Merulius* is now accepted as a synonym of *Phlebia* and the variety *shastensis* is not a *Phlebia*.

SIROBASIDIUM Lagerh. & Pat. 1892 Tremellaceae

Sirobasidium brefeldianum Möller 1895
Protobasidiomyceten. In A.F.W. Schimper (Ed.), Botanischen Mittheilungen aus den Tropen 8:165. G. Fischer, Jena.
Distribution United States: GA, NC.
Hosts *Albizzia julibrissin, Mimosa* sp.
Ecology Corticated limbs; decorticated branch.
Culture Characters Not known.
References 107, 498.

Sirobasidium sanguineum Lagerh. & Pat. 1892
J. Bot. (Morot) 6:469.
Distribution United States: LA, MD, NC.
Hosts *Fraxinus nigra, Ligustrum* sp., angiosperm.
Ecology Emerging from bark; dead branch; decorticated limb on the ground; dead privet shoot.
Culture Characters Not known.
References 14, 111, 420, 497, 502.

SISTOTREMA Fr. 1821 Sistotremaceae

Sistotrema adnatum Hallenb. 1984
Mycotaxon 21:397.
Distribution Canada: BC.
Hosts Gymnosperm.
Ecology Log.
Culture Characters Not known.

Known only from the type specimen.

Sistotrema athelioides Hallenb. 1984
Mycotaxon 21:400.
Distribution Canada: BC.
Hosts Gymnosperm.
Ecology Log.
Culture Characters Not known.

Known only from the type specimen.

Sistotrema biggsiae Hallenb. 1984
Mycotaxon 21:401.
Distribution Canada: ON, PQ. United States: AZ, MD, NY, WI.
Hosts *Abies* sp., *Acer* sp., *A. saccharinum, Betula* sp., *Trametes (Coriolus) versicolor, Populus* sp., *P. grandidentata, Salix gooddingii*, angiosperm.
Ecology Not known.
Culture Characters 1.3c.23.31d.32.36.37.(38).39.43-44.(48).(51).54.55.60 (Ref. 474).
References 256, 474.

Sistotrema binucleosporum Hallenb. 1984
Mycotaxon 21:409.
Distribution Canada: BC.
Hosts Gymnosperm.
Ecology Log.
Culture Characters Not known.

Known only from the type specimen.

Sistotrema brinkmannii (Bres.) Eriksson 1948
Kungl. Fysiogr. Sällsk. Lund Förh. 18:17.
Basionym *Odontia brinkmannii* Bres., Ann. Mycol. 1:88, 1903.
Synonyms *Grandinia brinkmannii* (Bres.) Bourd. & Galzin 1914, *Trechispora brinkmannii* (Bres.) D.P. Rogers & Jackson 1943.
Distribution Canada: AB, BC, MB, NB, NF, NS, NTM, ON, PQ. United States: AZ, CA, CO, CT, GA, IA, ID, LA, MA, MI, MO, NC, NM, NY, OH, OR, RI, TN, VA.
Hosts *Abies amabilis, A. balsamea, A. concolor, A. grandis, A. lasiocarpa, A. magnifica, Acer* sp., *A. macrophyllum, A. rubrum, A. saccharum, A. spicatum, Alnus oregana, A. rubra, A. tenuifolia, Arctostaphylos columbiana, Armillaria mellea, Betula* sp., *B. alleghaniensis, B. nigra, B. papyrifera, Carya ovata, Fagus grandifolia, Ganoderma* sp., *Gloeotulasnella pinicola, Larix* sp., *Liquidambar styraciflua, Liriodendron tulipifera, Peniophora* sp., *Phellinus (Hapalopilus) gilvus, Picea* sp., *P. glauca, P. glauca* var. *albertiana, P. mariana, P. rubens, P. sitchensis, Pinus* sp., *P. banksiana, P. contorta, P. ponderosa, P. resinosa, P. strobus, Piptoporus betulinus, Populus* sp., *P. balsamifera, P. tremuloides, P. trichocarpa, Prunus* sp., *P. emarginata, Pseudotsuga menziesii, P. mucronata, Pyrus malus, Quercus* sp., *Q. gambelii, Q. garryana, Q. macrocarpa, Q. nigra, Rhus ?glabra, Salix* sp., *Sphagnum* sp., *Tilia americana, Thuja plicata, Tsuga canadensis, Tsuga heterophylla, Ulmus americana, Vitis vulpina, Zea mays*, angiosperm, gymnosperm.
Ecology Limbs; slash; rotten logs; under side of very rotten log; isolated from southern pine utility poles; burlap covering of bale of imported peat moss; associated with a brown rot.

Culture Characters 1.3.32.36.(37).38.42.43.(48).50.54.55 (Ref. 581); 1.3.(22).32.36.38.43.(48).50.54.55.59 (Ref. 609); and a partial code 57.59.60 (Ref. 36).
References 56, 96, 103, 116, 120, 127, 128, 183, 197, 204, 208, 216, 242, 249, 256, 390, 398, 402, 406, 463, 495, 566, 609, 615.

For decades *S. brinkmannii* was reported, e.g. Boidin (36), to exhibit three mating systems, bipolar (unifactorial), tetrapolar (bifactorial) and homothallic. In 1984, Hallenberg (256), using mating tests and morphological characters, segregated six species from the *S. brinkmannii* complex. Thus prior reports of *S. brinkmannii* were, no doubt, based upon mixed collections. The report (24) that *Phymatotrichum omnivorum* Duggar is an anamorph is an error.

Sistotrema confluens Pers.:Fr. 1801
Synop. Meth. Fung., 551.
Distribution Canada: AB, NS, PQ.
United States: AZ, MI, MT, NC, NH, NY, VT, WI.
Hosts *Pinus* sp., *P. contorta, Populus* sp., gymnosperm.
Ecology Needles and wood on ground.
Culture Characters Not known.
References 109, 114, 183, 204, 216, 249, 331, 406, 532.

Sistotrema coroniferum (Höhnel & Litsch.) Donk 1956
Fungus 26:4.
Basionym *Gloeocystidium coroniferum* Höhnel & Litsch., Sitzungsber. Kaiserl. Akad. Wiss., Math.-Naturwiss. Kl., Abt.1, 116:825, 1907.
Synonyms *Trechispora coronifera* (Höhnel & Litsch.) D.P. Rogers & Jackson 1943, *Corticium atkinsonii* Burt 1926, fide Rogers and Jackson (567).
Distribution Canada: ON. United States: AZ, CO, IA, LA, MA, NY, OH, OR, VT.
Hosts ?*Abies grandis, Acer* sp., *Picea engelmannii, Pinus strobus, P. sylvestris, Populus* sp., *P. grandidentata, P. tremuloides, Quercus* sp., *Q. emoryi, Ulmus* sp., angiosperm, gymnosperm.
Ecology Fallen cones and dead wood; decaying, charred wood; associated with a white rot.
Culture Characters Not known.
References 93, 116, 127, 216, 402, 406, 566.

Sistotrema coronilla (Höhnel & Litsch.) Donk 1935
Stud. Nat. Hist. Iowa Univ. 7:23.
Basionym *Corticium coronilla* Höhnel & Litsch., Ann. Mycol. 4:291, 1906.
Distribution Canada: ON, PQ.
United States: GA, IA, MI, MN, OH, NY.
Hosts *Acer* sp., *A. saccharum, Betula alleghaniensis, B. nigra, Fagus* sp., *Picea glauca, P. mariana, Pinus strobus, Populus grandidentata, Quercus nigra, Salix* sp., *Taxus* sp., angiosperm.
Ecology Discarded fertilizer bags; fiberboard; insulating board.
Culture Characters 1.3c.16.23.26.31d.32.36.37.38.44-45.(51).54.55.60 (Ref. 474).
References 96, 256, 474, 563.

Sistotrema diademiferum (Bourd. & Galzin) Donk 1956
Fungus 26:4.
Basionym *Corticium diademiferum* Bourd. & Galzin, Bull. Soc. Mycol. France 27:244, 1911.
Synonym *Trechispora diademifera* (Bourd. & Galzin) D.P. Rogers 1944.
Distribution Canada: PQ.
United States: CO, IA, MA, NY, OR.
Hosts *Gossypium barbadense, Picea* sp., *P. sitchensis, Pinus contorta, P. strobus, Populus* sp., *P. tremuloides, Pyrus malus, Quercus* sp.
Ecology Associated with a white rot.
Culture Characters Not known.
References 242, 402, 566.

Sistotrema exima (Jackson) Ryv. & Solheim 1977
Mycotaxon 6:380.
Basionym *Corticium eximum* Jackson, Canad. J. Res., C, 26:152, 1948.
Distribution Canada: ON.
Host *Pinus strobus*.
Ecology Not known.
Culture Characters Not known.

Known only from the type specimen.

Sistotrema farinaceum Hallenb. 1984
Mycotaxon 21:406.
Distribution Canada: BC.
Host *Populus* sp.
Ecology Not known.
Culture Characters Not known.

Known only from the type specimen.

Sistotrema hirschii (Donk) Donk 1956
Fungus 26:4.
Basionym *Corticium hirschii* Donk, Medd. Nedl. Myc. Ver. 18-20:139, 1931.
Synonym *Trechispora hirschii* (Donk) D.P. Rogers 1944.
Distribution Canada: AB-NT, PQ.
United States: AZ, IA, OR.
Hosts *Betula* sp., *Pinus contorta, Populus grandidentata, Quercus garryana, Salix gooddingii*.
Ecology Not known.
Culture Characters Not known.
References 216, 242, 566.

Sistotrema muscicola (Pers.) Lundell 1947
In Lundell and Nannfeldt, Fungi Exs. Suec. No. 145, Uppsala.
Basionym *Hydnum muscicola* Pers., Mycol. Eur. 2:181, 1825.
Distribution United States: AK, AZ, CO.

Hosts *Populus* sp., *P. tremuloides*, gymnosperm.
Ecology Not known.
Culture Characters Not known.
References 183, 216, 217, 402, 406.

Sistotrema oblongisporum M. Christiansen & Hauerslev 1960
Dansk Bot. Ark. 19:82.
Distribution Canada: BC.
Hosts *Picea* sp., *Pinus* sp., angiosperm, gymnosperm.
Ecology Fallen or dead branches.
Culture Characters Not known.
References 171, 256.

Sistotrema populicola Ginns 1988
Mycologia 80:67.
Distribution Canada: YT.
Host *Populus tremuloides*.
Ecology Rotted decorticated branch on ground.
Culture Characters Not known.

Known only from the type specimen.

Sistotrema porulosum Hallenb. 1984
Mycotaxon 21:407.
Distribution Canada: BC, ON, PQ.
Hosts *Acer* sp., *Populus* sp.
Ecology Fallen branches.
Culture Characters Not known.

Known only from the holotype and two paratype specimens.

Sistotrema raduloides (Karsten) Donk 1956
Fungus 26:4.
Basionym *Hydnum raduloides* Karsten, Meddeland. Soc. Fauna Fl. Fenn. 9:110, 1883.
Synonyms *Grandinia raduloides* (Karsten) Bourd. & Galzin 1928, *Trechispora raduloides* (Karsten) D.P. Rogers 1944, *Hydnum populinum* Peck 1900, fide Gilbertson (189).
Distribution Canada: AB, BC, NB, NS, NTM, ON, PQ. United States: AK, AZ, CO, IA, MI, MN, MT, NC, NY, OR, TN, WA, WI.
Hosts *Abies balsamea*, *A. lasiocarpa*, *Acer rubrum*, *A. saccharum*, *Betula alleghaniensis*, *Picea* sp., *P. engelmannii*, *P. glauca*, *P. mariana*, *P. rubens*, *Populus* sp., *P. balsamifera*, *P. grandidentata*, *P. tremuloides*, *P. trichocarpa*, *Tilia americana*, *Tsuga heterophylla*, angiosperm, gymnosperm.
Ecology Occasionally isolated from decay behind wounds on trembling aspen; basidiomes produced on bark; decayed, decorticated wood; logs; associated with a white rot.
Culture Characters
1.3c.23.31d.33.36.37.38.44-45.(50).(53).54.(55).59 (Ref. 474);
1.2.3.7.23.33.36.38.45.46.54.55 (Ref. 487).
References 116, 171, 183, 189, 214, 216, 242, 327, 349, 402, 406, 450, 463, 474, 566.

Sistotrema resinicystidium Hallenb. 1980
Mycotaxon 11:466.
Distribution Canada: BC.
Hosts Not known.
Ecology Decayed wood.
Culture Characters Not known.
Reference 171.

Sistotrema sernanderi (Litsch.) Donk 1956
Fungus 26:4.
Basionym *Gloeocystidium sernanderi* Litsch., Svensk Bot. Tidskr. 25:437, 1931.
Synonym *Trechispora sernanderi* (Litsch.) D.P. Rogers 1944.
Distribution Canada: ON.
Host *Fagus grandifolia*.
Ecology Not known.
Culture Characters Not known.
Reference 566.

Sistotrema species A
In Currah, R.S., E.A. Smreciu, and S. Hambleton, Canad. J. Bot. 68:1180, 1990.
Distribution Canada: AB.
Hosts *Piperia unalascensis, Platanthera obtusata*.
Ecology Mycorrhizal with orchids.
Cultural Characters: See Currah et al. (loc. cit.).

Sistotrema subtrigonospermum D.P. Rogers 1935
Stud. Nat. Hist. Iowa Univ. 17:22.
Synonym *Trechispora subtrigonosperma* (D.P. Rogers) D.P. Rogers & Jackson 1943.
Distribution Canada: ON. United States: AZ, IA, WI.
Hosts *Abies concolor, Acer* sp., *Pinus ponderosa, Populus* sp., *P. tremuloides, Quercus* sp., *Ulmus* sp., angiosperm.
Ecology Not known.
Culture Characters Not known.
References 204, 216, 406, 563, 566.

SISTOTREMASTRUM Eriksson 1958 Sistotremaceae

Sistotremastrum niveocremeum (Höhnel & Litsch.) Eriksson 1958
Symb. Bot. Upsal. 16 (1):62.
Basionym *Corticium niveocremeum* Höhnel & Litsch., Sitzungsber. Kaiserl. Akad. Wiss., Math.-Naturwiss. Kl., Abt. 1, 117:1117, 1908.
Synonym *Paullicorticium niveocremeum* (Höhnel & Litsch.) Oberw. 1962.
Distribution Canada: BC, ON, PQ. United States: AZ, FL, IA, MA, MD, MO, NY, PA, RI, VA, WI.
Hosts *Abies balsamea, Acer* sp., *Alnus rubra, Betula* sp., *B. alleghaniensis, B. papyrifera, ?Fagus grandifolia, Pinus* sp.?, *Populus grandidentata, Quercus* sp., *Q. hypoleucoides, Q. rubra, Q. virginiana, Sassafras* sp., angiosperm.
Ecology Not known.

Culture Characters
2.3c.7.35.36.38.(39).45-47.(53).54.(55).59 (Ref. 474).
References 216, 255, 474, 492, 566.

Sistotremastrum suecicum Eriksson 1958
As Litsch. ex Eriksson, Symb. Bot. Upsal. 16 (1):62.
Synonym *Corticium suecicum* Litsch. in Lundell and Nannfeldt 1937, *nomen nudum*, fide Eriksson (167).
Distribution Canada: NS, ON, PQ.
United States: MA, NH, NY, OH, OR, PA.
Hosts *Abies balsamea, Acer saccharum, Alnus oregana, Betula* sp., *Castanea dentata, Juniperus* sp., *Picea* sp., *Pinus* sp., *P. resinosa, P. strobus, Populus* sp., *Pseudotsuga mucronata, Thuja occidentalis*, angiosperm, gymnosperm.
Ecology Not known.
Culture Characters
(1).(2a).(7).(26).32.36.38.47.55.(59).61 (Ref. 257).
References 171, 242, 249, 406, 492, 566.

SKVORTZOVIA V. Bononi & Hjort. 1987 Meruliaceae

Skvortzovia furfurella (Bres.) V. Bononi & Hjort. 1987
Mycotaxon 28:12.
Basionym *Odontia furfurella* Bres., Mycologia 17:71, 1925.
Synonym *Resinicium furfurella* (Bres.) Nakasone 1990.
Distribution United States: GA, IN, KY, LA, MD, MS, NC, NY, RI, TN, TX, VA.
Hosts *Carya* sp., *Pinus* spp., *P. glabra, P. palustris, P. taeda, P. virginiana, Quercus* sp., *Q. virginiana*, angiosperm.
Ecology Associated with a pitted to uniform white rot.
Culture Characters
2.3c.(27).28.32.36.38.44-47.(53).54.55 (Ref. 474).
References 53, 190, 203, 474.

SOLENIA Pers.:Fr. 1794 Cyphellaceae
Solenia Pers. is a later homonym of *Solenia* Lour. 1790, thus cannot be used in the Fungi. Many of the species have been transferred to *Henningsomyces* Kuntze. These five species await transfer to a valid genus.

Solenia andropogonis W.B. Cooke 1961
Beih. Sydowia 4:17.
Distribution United States: LA.
Host *Andropogon virginicus*.
Ecology Base of dead culm.
Culture Characters Not known.

Known only from the type specimen.

Solenia canadensis W.B. Cooke 1961
Beih. Sydowia 4:19.
Distribution Canada: NB, ON.
Hosts *Abies balsamea, Fagus* sp., *F. grandifolia, Picea* sp.
Ecology Not known.
Culture Characters Not known.

Known only from the holotype and three paratype specimens.

Solenia filicina Peck 1876
Annual Rep. New York State Mus. 28:52.
Distribution Canada: MB. United States: NY.
Host *Pteretis pensylvanica.*
Ecology Base of live stem.
Culture Characters Not known.
References 91, 116.

Solenia sphaerospora W.B. Cooke 1961
As Ellis ex W.B. Cooke, Beih. Sydowia 4:27.
Distribution United States: LA.
Host *Maclura pomifera.*
Ecology Not known.
Culture Characters Not known.

Known only from the type specimen.

Solenia subvillosa W.B. Cooke 1961
Beih. Sydowia 4:28.
Distribution United States: IL.
Hosts Not known.
Ecology Rotten wood.
Culture Characters Not known.

Known only from the type specimen.

SPARASSIS Fr. 1821 Sparassidaceae

Sparassis crispa Wülfen:Fr. 1781
In Jacquin, Miscellanea Austriaca 2:100.
Synonym *Sparassis radicata* Weir 1917, fide Martin and Gilbertson (446).
Distribution Canada: BC. United States: AZ, CA, ID, MT, NC, NM, OR, WA.
Hosts *Abies concolor, Carex occidentalis, Larix occidentalis, Picea engelmannii, Pinus* sp., *P. ayacahuite, P. monticola, P. ponderosa, P. strobiformis, P. muricata, Pseudotsuga menziesii.*
Ecology Base of live trees; roots; stumps; logs; ground; causes a brown cubical root and butt rot.
Culture Characters
1.3.26.34.36.38.47.53.55.59 (Ref. 446).
References 71, 116, 183, 242, 446.

Sparassis spathulata (Schw.) Fr. 1828
Elenchus 1:227.
Basionym *Merisma spathulata* Schw., Schriften Naturf. Ges. Leipzig 1:110, 1822.
Synonyms *Stereum caroliniense* Cooke & Rav. 1885, *Sparassis herbstii* Peck 1895, *Sparassis crispa* sensu N. Amer. Aucts. All fide Burdsall and Miller (71, 72).
Distribution United States: AR, MD, NC, PA, SC.
Hosts *Pinus* sp., *P. virginiana, Quercus* sp., *Q. alba.*

Ecology Base of stump; stump; ground under oak and pine.
Culture Characters 1.4.26.32.36.38.43.51.54.55.58 (Ref. 446).
References 71, 446.

SPHAEROBASIDIUM Oberw. 1965 Sistotremaceae

Sphaerobasidium minutum (Eriksson) Jülich 1979
As Oberw. ex Jülich, Persoonia 10:335.
Basionym *Xenasma minutum* Eriksson, Symb. Bot. Upsal. 16 (1):65, 1958.
Synonym *Paullicorticium minutum* (Eriksson) Liberta 1965.
Distribution Canada: PQ. United States: AZ, CO, CT, IL, MA, MT, NM, OR, WI.
Hosts *Abies* sp., *A. balsamea, Picea* sp., *P. abies, P. engelmannii, Pinus contorta, P. ponderosa, P. rigida, P. strobus, Pseudotsuga menziesii, Pteridium aquilinum, Tsuga* sp., *T. canadensis*, angiosperm, gymnosperm.
Ecology Bark; decayed wood; stems.
Culture Characters Not known.
References 183, 197, 204, 216, 242, 387, 388, 390, 402.

Sphaerobasidium subinvisible Liberta 1966
Mycologia 58:931.
Distribution Canada: PQ.
Host *Picea* sp.
Ecology Not known.
Culture Characters Not known.

Known only from the holotype, see Hjortstam (281).

STECCHERICIUM D. Reid 1963 Hericiaceae

Steccherinum seriatum (Lloyd) Maas G. 1966
Verh. Kon. Ned. Akad. Wetensch., Afd. Natuurk., Tweede Sect. C, 69:325.
Basionym *Hydnum seriatum* Lloyd, Mycol. Writ. 7:1196, 1923.
Synonym *Steccherinum westii* Murrill 1940, fide Maas Geesteranus (428).
Distribution United States: FL.
Hosts *Quercus* sp., angiosperm.
Ecology Not known.
Culture Characters 2a.3c.15.34.36.38.47.48.53.58.60.61 (Ref. 429); 2.3c.15.34.36.38.47.48.(50).54.60 (Ref. 474).
References 75, 428, 429, 474.

STECCHERINUM S.F. Gray 1821 Steccherinaceae

Steccherinum alaskense Lindsey & Gilbn. 1979
Mycologia 71:1264.
Distribution United States: AK.
Host *Alnus incana*.
Ecology Dead fallen stem, associated with a white rot.
Culture Characters Not known.

Known only from the type specimen.

Steccherinum ciliolatum (Berk. & Curtis) Gilbn. & Budington 1970
J. Arizona Acad. Sci. 6:97.
Basionym *Hydnum ciliolatum* Berk. & Curtis, Hooker's J. Bot. Kew Gard. Misc. 1:235, 1849.
Synonym *Odontia ciliolata* (Berk. & Curtis) L.W. Miller 1934.
Distribution Canada: AB, BC, NB, ON, PE, PQ. United States: AZ, CO, IA, MA, MD, MI, MN, MT, NC, NM, NY, SC.
Hosts *Abies concolor, A. lasiocarpa* var. *arizonica, A. lasiocarpa* var. *lasiocarpa, Acer* sp., *A. saccharum, Betula* sp., *B. alleghaniensis, B. papyrifera, Fagus* sp., *F. grandifolia, Malus* sp., *Picea* sp., *P. engelmannii, Populus* sp., *P. balsamifera, P. grandidentata., P. tremuloides, P. trichocarpa, Pseudotsuga menziesii, Quercus* sp., angiosperm, gymnosperm.
Ecology Decayed wood; slash; logs; associated with a white rot.
Culture Characters 2.3c.8.12.32.36.(38).40.46-47.54.55 (Ref. 474).
References 116, 183, 192, 204, 208, 214, 216, 242, 402, 406, 428, 448, 463, 474.

Steccherinum fimbriatum (Pers.:Fr.) Eriksson 1958
Symb. Bot. Upsal. 16 (1):134.
Basionym *Odontia fimbriata* Pers., Obs. Mycol. 1:88, 1796.
Distribution Canada: BC, MB, NB, NS, NT, ON, PE. United States: AZ, CA, CO, GA, IA, MI, MN, MT, NC, NM, NY, OH, OR, PA, WA, WI.
Hosts *Abies concolor, Acer* sp., *A. macrophyllum, A. saccharum, Alnus* sp., *A. rubra, Amelanchier* sp., *A. alnifolia, Betula alleghaniensis, B. papyrifera, Castanea* sp., *Fagus grandifolia, Juniperus* sp., *Liriodendron tulipifera, Lithocarpus densiflora, Ostrya virginiana, Picea* sp., *Pinus ponderosa, P. resinosa, P. strobus, Populus* sp., *P. tremuloides, P. trichocarpa, Prunus virginiana, Quercus* sp., *Q. gambelii, Rhus* sp., *R. typhina, Salix* sp., *Umbellularia californica*, angiosperm, gymnosperm.
Ecology Causes a minor decay, mostly associated with wounds in *Umbellaria californica*; prefers oak; saprophytic on rotten wood; basidiomes produced on bark and decorticated wood; decaying wood; slash; logs; associated with a white rot.
Culture Characters 2.3c.8.(16).(31c).32.36.(37).38.(40).44-45.54.55.60 (Ref. 474).
References 56, 96, 116, 119, 133, 171, 183, 197, 204, 208, 214, 216, 242, 249, 274, 327, 335, 398, 402, 406, 428, 463, 474.

Steccherinum laeticolor (Berk. & Curtis) Banker 1912
Mycologia 4:316.
Basionym *Hydnum laeticolor* Berk. & Curtis, Grevillea 1:99, 1873.
Synonyms *Hydnum floridanum* Berk. & Cooke 1878, *Hydnum parasitans* Berk. & Curtis 1873. Both fide Maas Geesteranus (428).
Distribution United States: AL, AZ, FL, IA, ID, LA, MI, NY, TX.
Hosts *Acer* sp., *A. saccharinum, Celtis reticulata, Pinus* sp., *Platanus* sp., *Prunus serotina* subsp. *virens, Quercus* sp., *Q. alba, Q. arizonica, Q. emoryi, Q. lyrata, Q. reticulata, Q. toumeyi, Salix* sp., *Ulmus* sp., *U. americana*, angiosperm.
Ecology Dead wood; associated with a white rot.
Culture Characters
2.3.8.24.25.26.32.36.38.40.46-47.54 (Ref. 588).
References 183, 192, 204, 209, 216, 428, 463.

The type of *H. laeticolor* is similar to *Steccherinum ochraceum* but lacks hymenial detail (192).

Steccherinum ochraceum (Pers.:Fr.) S.F. Gray 1821
Natural Arrangement British Plants 1:651.
Basionym *Hydnum ochraceum* Pers. in J.F. Gmelin, Caroli à Linné, ... Systema Naturae 2:1440, 1792.
Synonyms *Hydnum rhois* Schw.:Fr. 1822, fide Maas Geesteranus (428), *Steccherinum rhois* (Schw.) Banker 1906.
Distribution Canada: BC, MB, NS, ON, PQ.
United States: AK, AL, AR, AZ, CA, FL, GA, IA, ID, IL, IN, KY, LA, MD, ME, MI, MN, MO, MS, NC, NH, NJ, NY, OH, OK, OR, PA, SC, TN, TX, VA, WA, WI, WV.
Hosts *Acer* sp., *A. circinatum, A. rubrum* var. *rubrum, A. macrophyllum, Alnus* sp., *A. incana, A. oblongifolia, A. rubra, A. rugosa, A. sinuata, A. tenuifolia, Betula* spp., *B. alleghaniensis, B. papyrifera, B. pendula, B. pumila, Carpinus caroliniana, Carya* sp., *Castanea dentata, Cupressus macrocarpa, Fagus* sp., *F. grandifolia, Juglans major, Juniperus virginiana, Liquidambar* sp., *L. styraciflua, Magnolia* sp., *Malus sylvestris, Nyssa* sp., *Ostrya* sp., *Picea pungens, Populus* sp., *P. balsamifera, P. deltoides, P. tremuloides, P. trichocarpa, Prunus pensylvanica, P. persica, Pseudotsuga menziesii, Quercus* sp., *Q. agrifolia, Q. alba, Q. coccinea, Q. falcata, Q. garryana, Q. muhlenbergii, Q. nigra, Q. prinus, Q. shumardii, Q. stellata, Q. tomentella, Rhus* sp., *R. copallina, R. glabra, R. typhina, R. vernix, Salix* sp., *Tilia americana, Ulmus* sp., angiosperm, gymnosperm.
Ecology Causes a trunk rot in *Betula* spp. and *Carpinus caroliniana*; basidiomes produced on bark; dead wood; decaying wood; log; associated with a white rot.
Culture Characters
2a.3c.8.9.32.36.38.43.53.58.61 (Ref. 429);
2.3c.8.26.27.32.36.(37).(39).(40).43-44.(48).(50).54.(55).60 (Ref. 474).
References 25, 96, 97, 116, 183, 202, 204, 214, 216, 242, 249, 274, 327, 333, 340, 406, 428, 463, 465, 474, 511, 532, 611.

Steccherinum oreophilum Lindsey & Gilbn. 1977
Mycologia 69:194.
Distribution Canada: AB, NT, NTM, ON, PQ.
United States: AZ, CO, UT.
Hosts *Populus* sp., *P. tremuloides, Prunus* sp., *Tilia americana.*
Ecology Logs and slash; associated with a white rot.
Culture Characters
2.3c.(8).(13).(26).32.36.(38).(40).44.54.60 (Ref. 474);
2.3.8.13.22.26.34.36.38.44.54.60 (Ref. 233).
References 233, 242, 402, 404, 406, 474.

Steccherinum peckii Banker 1912
Mycologia 4:314.
Distribution United States: NY.
Host *Acer* sp.
Ecology Not known.
Culture Characters Not known.

Known only from the type specimen, which was redescribed by Maas Geesteranus (428).

Steccherinum queletii (Bourd. & Galzin) Hallenb. & Hjort. 1988
Mycotaxon 31:443.
Basionym *Odontia queletii* Bourd. & Galzin, Bull. Soc. Mycol. France 30:270, 1914.
Synonym *Phlebia queletii* (Bourd. & Galzin) M. Christiansen 1960.
Distribution Canada: MB.
United States: CA, FL, IA, NC, NY, TN, WA.
Hosts *Acer rubrum, A. saccharinum*, angiosperm.
Ecology Associated with a white rot.
Culture Characters
2.4.13.14.(32).(34).(36).37.40.43-45.48.(52).53.54.59 (Ref. 474) and 60 (Ref. 43).
References 463, 474.

The mating system, having been reported as unifactorial (474) and bifactorial (43), needs to be confirmed.

Steccherinum reniforme (Berk. & Curtis) Banker 1906
Mem. Torrey Bot. Club 12:127.
Basionym *Hydnum reniforme* Berk. & Curtis, J. Linn. Soc., Bot. 10:325, 1868.
Synonyms *Hydnum glabrescens* Berk. & Rav. 1873, *Steccherinum morganii* Banker 1906. Both fide Maas Geesteranus (428).
Distribution United States: IA, IN, OH, SC.
Host *Carya* sp.
Ecology Rotting wood; dead forest trees.
Culture Characters Not known.
References 428, 463.

Errorneously listed (26, 463) as a synonym of *S. rawakense*. Maas Geesteranus (428) itemized the differences between *S. reniforme* and *S. rawakense*. And see the comments under *S. rawakense* in the section on Excluded Names.

Steccherinum subcrinale (Peck) Ryv. 1978
Norw. J. Bot. 25:294.
Basionym *Hydnum subcrinale* Peck, New York State Mus. Bull. 167:27, 1913.
Synonym *Odontia subcrinale* (Peck) Gilbn. 1962.
Distribution Canada: ON.
United States: ID, MN, MT, NC, NY, WA.
Hosts *Abies balsamea, Acer* sp., *A. glabrum* var. *douglasii, Betula alleghaniensis, Castanea dentata, Pinus* sp., *Tsuga canadensis*, angiosperm, gymnosperm.
Ecology Rotting logs; associated with a white rot.
Culture Characters
2.6.(16).26.32.36.(38).(40).45-46.(50).54.55 (Ref. 474).
References 189, 214, 474, 571.

Maas Geesteranus (428) excluded this fungus from *Steccherinum* but did not suggest another genus.

Steccherinum subrawakense Murrill 1940
Bull. Torrey Bot. Club 67:275.
Distribution United States: FL.
Hosts *Magnolia grandiflora, Nyssa sylvatica, Liquidambar styraciflua*, angiosperm.
Ecology Log; associated with a white rot.
Culture Characters
2.3c.8.26.27.32.36.38.43-44.48.50.54 (Ref. 474).
References 428, 474.

Steccherinum tenue Burdsall & Nakasone 1981
Mycologia 73:472.
Distribution United States: MD, NC, TN.
Hosts *Acer* sp., *Fagus grandifolia, Vitis* sp., angiosperm.
Ecology Associated with a white rot.
Culture Characters
2.3.7.32.36.38.42.53.54 (Ref. 74);
2.3c.(8).(16).(26).32.36.(38).(40).43-44.(53).54 (Ref. 474).
References 74, 474.

Steccherinum vagum Burdsall & Nakasone 1981
Mycologia 73:473.
Distribution United States: TN.
Hosts *Liriodendron tulipifera*, angiosperm.
Ecology Log; associated with a white rot.
Culture Characters
2.3i.8.13.16.32.36.(37).40.42.54 (Ref. 474).

Known only from the type specimen.

STEREOPSIS D. Reid 1965 — Podoscyphaceae

Stereopsis burtianum (Peck) D. Reid 1965
Beih. Nova Hedwigia 18:292.
Basionym *Stereum burtianum* Peck, New York State Mus. Bull. 75:21, 1904.
Distribution United States: MA, NC, NH, NY, PA, TN, VA, VT.
Hosts Not known.
Ecology The ground; sandy banks; bushy places; angiosperm woods.
Culture Characters Not known.
References 89, 519, 555.

Stereopsis hiscens (Berk. & Rav.) D. Reid 1965
Beih. Nova Hedwigia 18:298.
Basionym *Thelephora hiscens* Berk. & Rav., Grevillea 1:148, 1873.
Synonyms *Thelephora ravenelii* Berk. 1873, *Stereum insolitum* Lloyd 1917. Both fide Reid (555).
Distribution United States: KY, SC, TN, VA.
Hosts Not known.
Ecology Arising from buried wood or roots.
Culture Characters Not known.
Reference 555.

Stereopsis humphreyi (Burt) Redhead & D. Reid 1983
Canad. J. Bot. 61:3088.
Basionym *Craterellus hymphreyi* Burt, Ann. Missouri Bot. Gard. 1:344.
Distribution Canada: BC. United States: WA.
Hosts *Picea* sp., *P. sitchensis, Thuja plicata, Tsuga heterophylla*, ferns, mosses.
Ecology Fallen needles; cones and twigs; fern fronds.
Culture Characters Not known.
Reference 549.

STEREUM Pers. 1794 — Stereaceae
Synonym is *Haematostereum* Pouzar 1959, fide Eriksson et al. (171), and Jülich and Stalpers (326).

Stereum ardoisiacum Lloyd 1923
Mycol. Writ. 7:1197.
Distribution Canada: ?PQ.
Hosts Not known.
Ecology Not known.
Culture Characters Not known.

Known only from the type specimen.

Stereum atrorubrum Ellis & Ev. 1890
Proc. Acad. Nat. Sci. Philadelphia 1890:219.

Distribution Canada: BC.
Hosts Not known.
Ecology Old logs.
Culture Characters Not known.

Known only from the type specimen, which was redescribed by Burt (93).

Stereum australe Lloyd 1913
Mycol. Writ. 4 (Letter 48:10).
Distribution United States: FL, MS, NH, WA.
Hosts Angiosperm.
Ecology Logs.
Culture Characters Not known.
References 89, 593.

Stereum avellanaceum J.A. Stevenson & Cash 1936
As Lloyd ex J.A. Stevenson & Cash, Bull. Lloyd Libr. Bot. 35:51.
Distribution United States: MN.
Hosts Not known.
Ecology Not known.
Culture Characters Not known.

Known only from the type specimen.

Stereum complicatum (Fr.) Fr. 1838
Epicrisis, 548.
Basionym *Thelephora complicata* Fr., Elenchus 1:179, 1828.
Synonyms *Thelephora ramealis* Schw. 1822, fide Lentz (379), *Stereum rameale* (Schw.) Burt 1920, a later homonym of *Stereum rameale* (Berk.) Massee 1889.
Distribution Canada: AB-NT, BC, ON. United States: AL, CO, DC, FL, GA, IA, IN, KY, LA, MD, ME, MI, MN, MS, NC, NH, NY, OH, PA, TN, TX, VA.
Hosts *Acer* sp., *A. rubrum, A. saccharum, Alnus* sp., *A. rubra, Betula alleghaniensis, B. papyrifera, Fagus* sp., *F. grandifolia, Juniperus* sp., *Nyssa* sp., *N. sylvatica, Picea rubens, Populus deltoides, Prunus serotina, Pseudotsuga menziesii, Quercus* sp., *Q. coccinea, Q. rubra, Salix nigra, Thuja* sp., *Weigelia florida*, angiosperm.
Ecology Considered to be a wound parasite; isolated from Douglas-fir utility poles; basidiomes produced on lichen covered bark; dead twigs and stumps; relatively common in broken branch stubs of cherry and maple; dead branch; corticated branch; butt; cut wood; cedar poles with bark; logs and bark; associated with a white rot.
Culture Characters
2.5.(8).21.24.32.36.(37).(38).39.42.54.(57) (Ref. 474); 2.5.8.(21).32.37.(39).43.54.57.66 (Ref. 609).
References 89, 95, 96, 100, 108, 116, 127, 178, 242, 327, 333, 379, 474, 509, 511, 519, 609, 611.

Stereum contrastum Lloyd 1924
Mycol. Writ. 7:1261.
Distribution United States: FL.
Hosts Not known.
Ecology Not known.
Culture Characters Not known.

Known only from the type specimen.

Stereum craspedium (Fr.) Burt 1920
Ann. Missouri Bot. Gard. 7:113.
Basionym *Thelephora craspedia* Fr., Nova Acta Regiae Soc. Sci. Upsal., Ser. III, 1:108, 1851.
Distribution United States: NC.
Hosts Angiosperm.
Ecology Decayed log.
Culture Characters Not known.
Reference 112.

Known only from the type specimen. Burt (89) quoted the original description. Apparently he did not see the type.

Stereum gausapatum (Fr.) Fr. 1874
Hym. Eur., 638.
Basionym *Thelephora gausapata* Fr., Elenchus 1:171, 1828.
Synonyms *Haematostereum gausapatum* (Fr.) Pouzar 1959, *Stereum spadiceum* var. *plicatum* Peck 1898, fide Chamuris (104), *Stereum occidentale* Lloyd 1919, fide Burt (89) and Lentz (379).
Distribution Canada: BC, MB, ON, PE. United States: AK, AL, AR, AZ, CA, CT, DC, DE, FL, GA, IA, IL, KS, LA, MA, MD, MO, MS, NC, ND, NJ, NY, OH, OK, OR, PA, TN, TX, VA, WA, WV.
Hosts *Acer* sp. *A. saccharum, Alnus* sp., *A. sinuata, Baccharis pilularis, Betula* sp., *Carpinus caroliniana, Castanea dentata, C. mollissima, C. pumila, Fagus* sp., *Lithocarpus densiflora, Lyonothamnus floribundus, Malus sylvestris, Prunus pensylvanica, Pseudotsuga menziesii, Quercus* sp., *Q. agrifolia, Q. alba, Q. coccinea, Q. emoryi, Q. falcata* var. *pagodifolia, Q. garryana, Q. laurifolia, Q. macrocarpa, Q. marilandica, Q. montana, Q. nigra, Q. nuttallii, Q. palustris, Q. prinus, Q. rubra, Q. rubra* var. *borealis, Q. velutina*, angiosperm.
Ecology Causes trunk rot in *Castanea* and *Carpinus*; the principal fungus causing sprout rot in *Quercus*; responsible for 29% of the decay entering via dead branches in southern angiosperms; basidiomes produced on bark; limb; trunk; dead stems; base of rotten stump; log; stumps and wounds; associated with a white rot.
Culture Characters
2.5.8.32.36.(37).39.42.(50).(53).54.(57).(59) (Ref. 474);
2.3i.8.21.32.(36).37.39.42-44.54.57.66 (Ref. 104);
2.5.8.21.32.(36).37.39.43.54.(55).57.66 (Ref. 609).
References 89, 96, 97, 100, 104, 108, 116, 142, 183, 216, 242, 274, 327, 340, 379, 465, 474, 519, 609.

Stereum hirsutum (Willd.:Fr.) S.F. Gray 1821
Natural Arrangement British Plants 1:653.

Basionym *Thelephora hirsuta* Willd., Flora Berolinensis Prodromus, 397, 1787.

Synonyms *Thelephora ochracea* Schw. 1822, fide Lentz (379), *Stereum variicolor* Lloyd 1914, fide Burt (89) and Lentz (379).

Distribution Canada: AB, BC, NB, NF, NS, NTM, ON, PE, PQ, SK. United States: AK, AL, AR, AZ, CA, CO, CT, DC, FL, GA, IA, ID, IN, LA, MA, MD, ME, MI, MN, MO, MS, MT, NC, NH, NJ, NM, NY, OH, OK, OR, PA, RI, SC, SD, TN, TX, VA, VT, WA, WI, WV, WY.

Hosts *Abies balsamea, A. fraseri, Acacia* sp., *Acer* sp., *A. circinatum, A. grandidentatum, A. macrophyllum, A. pensylvanicum, A. rubrum, A. rubrum* var. *rubrum, A. saccharinum, A. saccharum, A. spicatum, Adenostoma fasiculatum, Aesculus californica, Alnus* sp., *A. incana, A. rubra, A. rugosa, A. rugosa* var. *americana, A. tenuifolia, Amelanchier alnifolia, Arbutus menziesii, Arctostaphylos patula, Betula* sp., *B. alleghaniensis, B. fontinalis, B. lenta, B. nigra, B. occidentalis, B. papyrifera, B. populifolia, Calocedrus decurrens, Carpinus* sp., *C. caroliniana, Carya* sp., *C. cordiformis, Castanea* sp., *C. crenata, C. dentata, C. mollissima, C. sativa, Castanopsis chrysophylla, Catalpa* sp., *Ceanothus velutinus, Cephalanthus* sp., *C. occidentalis, Cercocarpus* sp., *C. betuloides, Corylus avellana, C. cornuta, C. cornuta* var. *californica, Eucalyptus globulus, Fagus* sp., *F. grandifolia, Hamamelis virginiana, Heteromeles* sp., *Ilex opaca, Juniperus virginiana, Kalmia latifolia, Larix kaempferi, Libocedrus decurrens, Ligustrum* sp., *L. lucidum, L. vulgare, Liquidambar* sp., *L. styraciflua, Liriodendron tulipifera, Lithocarpus densiflora, Lyonothamnus floribundus, Magnolia* sp., *Malus pumila, M. sylvestris, Myrica cerifera, Nyssa* sp., *N. sylvatica, Ostrya virginiana, Physocarpus capitatus, Picea rubens, P. sitchensis, Pinus contorta, Platanus occidentalis, Populus* sp., *P. balsamifera, P. deltoides, P. grandidentata, P. tremuloides, P. trichocarpa, Prunus* sp., *P. americana, P. cerasus, P. domestica, P. pensylvanica, P. persica, P. serotina, Pseudotsuga menziesii, Pyrus communis, Quercus* sp., *Q. agrifolia, Q. alba, Q. arizonica, Q. chrysolepsis, Q. dumosa, Q. emoryi, Q. falcata, Q. gambelii, Q. garryana, Q. kelloggii, Q. laurifolia, Q. nigra, Q. nuttallii, Q. prinus, Q. rubra, Q. rubra* var. *borealis, Q. stellata, Rhus copallina, R. toxicodendron, Rosa nutkana, Salix* sp., *S. lasiandra, S. nigra, S. purpurea, S. scouleriana, Schinus molle, Sequoia sempervirens, Sequoiadendron giganteum, Tilia americana, Tsuga heterophylla, Ulmus* sp., *U. americana, Umbellularia californica, Vitis* sp., angiosperm, less commonly on gymnosperm.

Ecology Causes a trunk rot in *Nyssa aquatica*; basidiomes produced on bark; cutwood; broken wood; twig; branch; dead branches; sapwood of dead trees and slash; butt; trunk; logs and stumps; associated with a white rot.

Culture Characters
2.5.8.21.32.(36).37.(39).(40).42.54.55.(57).(59) (Ref. 474);
2.5.8.(21).32.36.(37).39.40.42.50.54.57 (Ref. 581);
(2).3i.8.21.32.(36).37.(38).(39a).42-44.54.57.66 (Ref. 104);
2.5.7.32.37.39.44.(50).53.54 (Ref. 651).

References 89, 97, 104, 116, 119, 120, 128, 130, 133, 183, 214, 216, 242, 249, 274, 327, 333, 334, 335, 340, 379, 399, 402, 448, 465, 474, 511, 519, 579, 593.

Wright and Deschamps (651) inadvertently omitted number 32 from the species code. Thus we have inserted it above.

Stereum ochraceoflavum (Schw.) Peck 1869
Annual Rep. New York State Mus. 22:86.

Basionym *Thelephora ochraceoflava* Schw., Trans. Amer. Philos. Soc., N.S., 4:167, 1832.

Synonyms *Stereum striatum* var. *ochraceoflavum* (Schw.) Welden 1971, *Stereum sulphuratum* Berk. & Rav. in Berkeley and Curtis 1868, fide Jülich and Stalpers (326) and tentatively accepted by Hjortstam (289).

Distribution Canada: BC, NF, NS, ON.
United States: AL, AR, AZ, CA, CO, CT, DE, FL, GA, IA, IN, KS, KY, LA, MA, MD, ME, MI, MN, MO, MS, NC, NH, NJ, NY, OH, PA, SC, TN, TX, VA, VT, WA, WI.

Hosts *Acer macrophyllum, Adenostoma fasciculatum, Ailanthus altissima, Alnus* sp., *A. rugosa, A. serrulata, Amelanchier* sp., *Andromeda* sp., *Aronia arbutifolia, Betula* sp., *B. lenta, B. papyrifera, B. populifolia, Castanea* sp., *C. dentata, Cercis canadensis, Corylus americana, Eucalyptus globulus, Fagus* sp., *F. grandifolia, Liquidambar styraciflua, Myrica* sp., *M. cerifera, M. pusilla, Nicotiana glauca, Picea engelmannii, Populus* sp., *Prunus* sp., *P. armeniaca, P. pensylvanica, P. persica, Quercus* sp., *Q. alba, Q. arizonica, Q. garryana, Q. hypoleucoides, Q. ilicifolia, Q. muhlenbergii, Q. palustris, Q. rubra, Q. rubra* var. *borealis, Rhus copallina, R. vernix, Salix* sp., *Symphoricarpos* sp., *Vaccinium* sp., *Vitis* sp., angiosperm.

Ecology Isolated from decay in Douglas-fir utility poles; basidiomes produced on bark; twigs; dead branches; exposed root; tree; dead vine; associated with a white rot.

Culture Characters
2.5.8.21.32.36.37.39.42-43.(53).54 (Ref. 474);
2.3i.8.(21).32.36.37.38.(41).42.54.57.66 (Ref. 104);
2.5.8.(21).32.37.(39).43.54.57.66 (Ref. 609).

References 89, 96, 104, 108, 116, 183, 216, 242, 249, 327, 333, 379, 402, 465, 474, 519, 609.

Stereum ostrea (Blume & Nees:Fr.) Fr. 1838
Epicrisis, 547.

Basionym *Thelephora ostrea* Blume & Nees, Nova Acta Acad. Caes. Leop.-Carol. 13:13, 1826.

Synonyms *Thelephora fasciata* Schw. 1822, *Stereum fasciatum* (Schw.) Fr. 1838, *Stereum lobatum* (Kuntze:Fr.) Fr. 1830, *Thelephora mollis* Lév. 1846, a later homonym of *Thelephora mollis* Fr. 1821. All fide Lentz (379).

Distribution Canada: BC, MB, NB, NF, NS, ON, PE, PQ. United States: AL, AZ, CA, CO, CT, DC, FL, GA, IA, ID, IL, IN, KS, KY, LA, MA, MD, ME, MI, MN, MS, MT, NC, ND, NM, NJ, NY, OH, OK, OR, PA, SC, SD, TN, TX, VA, VT, WA, WI.

Hosts *Abies balsamea, A. lasiocarpa, Acer* sp., *A. macrophyllum, A. pensylvanicum, A. rubrum, A. saccharinum, A. saccharum, A. spicatum, Alnus* sp., *A. incana, A. rubra, Arbutus arizonica, A. menziesii, Betula* sp., *B. alleghaniensis, B. occidentalis, B. papyrifera, Carpinus caroliniana, Carya* sp., *C. aquatica, C. cordiformis, C. glabra, C. ovata, Castanea* sp., *C. dentata, Catalpa* sp., *Celtis laevigata, Fagus* sp., *F. grandifolia, Fraxinus* sp., *Juglans nigra, Liquidambar styraciflua, Liriodendron tulipifera, Myrica cerifera, Nyssa* sp., *N. aquatica, N. sylvatica, Picea rubens* or *Abies fraseri, P. mariana, Platanus occidentalis, Populus* sp., *P. grandidentata, P. tremuloides, P. trichocarpa, Pseudotsuga menziesii, Quercus* sp., *Q. agrifolia, Q. alba, Q. garryana, Q. hypoleucoides, Q. nigra, Q. palustris, Q. prinus, Q. rubra, Q. rubra* var. *borealis, Q. stellata, Q. velutina, Q. virginiana, Sequoia sempervirens, Tilia* sp., *T. americana, Tsuga heterophylla, Ulmus americana*, angiosperm.

Ecology Dead twigs and bark; branches; trunk; logs and stumps; associated with a white rot.

Culture Characters
2.5.8.21.32.36.(37).(38).39.42-44.(53).54.(57) (Ref. 474);
2.3i.8.(13).21.32.36.(37).38.(39).42.44.54.57.66 (Ref. 104).

References 89, 96, 104, 108, 116, 119, 133, 147, 183, 214, 216, 242, 249, 327, 333, 334, 335, 379, 402, 406, 474, 509, 519, 611.

Stereum pergamenum Berk. & Curtis 1873
Grevillea 1:161.

Distribution United States: AL, LA, NC, OH, SC.

Hosts Not known.

Ecology Stumps or buried wood, rarely on the ground.

Culture Characters Not known.

References 89, 612.

Stereum pubescens Burt 1920
Ann. Missouri Bot. Gard. 7:178.

Distribution United States: MT.

Hosts Angiosperm.

Ecology Dead limbs.

Culture Characters Not known.

Known only from the type specimen.

Stereum rugosum Pers.:Fr. 1794
Neues Mag. Bot. 1:110.

Synonym *Haematostereum rugosum* (Pers.) Pouzar 1959.

Distribution Canada: AB-NT, BC, NB, NF, NS, ON, PE, PQ. United States: AK, ID, MI, MN, NC, NH, NY, PA, TN.

Hosts *Abies lasiocarpa, Acer* sp., *A. rubrum, A. saccharum, Aesculus hippocastanum, Alnus* sp., *A. crispa, A. tenuifolia, Betula* sp., *B. papyrifera, B. populifolia, Corylus* sp., *Fagus grandifolia, Malus* sp., *Picea glauca, Populus* sp., *P. tremuloides, Prunus pensylvanica, P. virginiana, Pseudotsuga menziesii, Rhododendron* sp., *R. catawbiense, R. maximum, Quercus* sp., *Salix* sp., angiosperm, gymnosperm.

Ecology Stumps.

Culture Characters
2.3i.8.32.36.38.42.43.(50).54.57.66 (Ref. 104);
2.5.8.32.36.38.40.42.43.50.54 (Ref. 581).

References 89, 97, 104, 116, 183, 242, 249, 334, 519.

Stereum sanguinolentum (Alb. & Schw.:Fr.) Fr. 1838
Epicrisis, 549.

Basionym *Thelephora sanguinolenta* Alb. & Schw., Conspectus Fungorum, 274-275, 1805.

Synonyms *Haematostereum sanguinolentum* (Alb. & Schw.) Pouzar 1959, *Stereum balsameum* Peck 1875, *Stereum balsameum* forma *reflexum* Peck 1894. Both fide Lentz (379).

Distribution Canada: AB, BC, MB, NB, NF, NFL, NS, NT, ON, PE, PQ, SK, YT. United States: AK, AZ, CA, CO, CT, ID, MA, ME, MI, MT, NC, NH, NM, NY, OR, PA, TN, UT, VA, VT, WA, WI, WY.

Hosts *Abies* sp., *A. amabilis, A. balsamea, A. concolor, A. fraseri, A. grandis, A. lasiocarpa* var. *arizonica, A. lasiocarpa* var. *lasiocarpa, Alnus* sp., *A. tenuifolia, Amelanchier alnifolia* var. *cusickii, Fagus grandifolia, Larix* sp., *L. decidua, L. kaempferi, L. laricina, L. occidentalis, Picea* sp., *P. abies, P. engelmannii, P. glauca, P. glauca* var. *albertiana, P. mariana, P. rubens, P. sitchensis, Pinus* sp., *P. aristata, P. aristata* var. *longaeva, P. banksiana, P. contorta, P. contorta* var. *latifolia, P. monticola, P. ponderosa, P. resinosa, P. rigida, P. strobus, Populus balsamifera, Pseudotsuga* sp., *P. menziesii, Thuja plicata, Tsuga* sp., *T. canadensis, T. heterophylla, T. mertensiana*, gymnosperm.

Ecology Common as a sapwood rot, heartrot and root rot in balsam fir where it enters only through injuries to live stems and branches; caused 14% of decay in lodgepole pine logging slash in one survey; common fungus causing decay behind logging scars in *Picea glauca* and *Abies lasiocarpa*; basidiomes produced on bark; moss-covered bark; dead branches; rotting branches; dead sapling; live tree trunk; stumps and logs; associated with a white rot.

Culture Characters
2.5.8.(13).32.36.37.(38).(39).43-46.(50).(54).55.(57) (Ref. 474);

2.5.(7).(13).(21).32.36.38.43-45.55.57 (Ref. 487); 2.5.13.21.32.36.39.43-45.53.55 (Ref. 659); 2.3i.(8).(13).21.32.(36).37.(38).39.43-46.55.57.66 (Ref. 104).
References 30, 31, 89, 97, 104, 108, 116, 119, 128, 130, 132, 133, 141, 183, 197, 208, 216, 242, 249, 274, 327, 333, 335, 379, 391, 402, 407, 413, 448, 466, 474, 484, 511, 519, 521, 532, 595, 645.

Stereum spumeum Burt 1920
Ann. Missouri Bot. Gard. 7:208.
Distribution United States: LA, NY, PA, SC.
Hosts *Fagus* sp., *Quercus* sp., angiosperm.
Ecology Dead bark and wood.
Culture Characters Not known.

Known only from the type specimens.

Stereum striatum (Fr.) Fr. 1838
Epicrisis, 548.
Basionym *Thelephora striata* Fr., Elenchus 1:179, 1828.
Synonyms *Thelephora sericea* Schw. 1822, fide Fries (loc. cit.) and Lentz (379), *Stereum sericeum* (Schw.) Sacc. 1888.
Distribution Canada: NF, ON, PQ.
United States: AL, AR, CT, DC, FL, GA, IA, IN, KY, LA, MA, MD, MI, MN, MO, MS, NC, NJ, NY, OH, PA, SC, TN, TX, VA, VT, WV.
Hosts *Alnus* sp., *Betula* sp., *B. lenta, B. nigra, B. papyrifera, Carpinus* sp, *C. caroliniana, Castanea* sp., *Fagus grandifolia, Ilex opaca, Liquidambar styraciflua, Nyssa* sp., *N. multiflora, N. sylvatica, Ostrya virginiana, Prunus* sp., *P. pensylvanica, P. serotina, Quercus* sp., *Q. alba, Q. laurifolia, Rhus copallina*, angiosperm.
Ecology Dead wood; underside of dead twigs and branches; associated with a white rot.
Culture Characters
2.5.6.(21).32.36.(37).(38).(39).(40).43-44.(45).(53).54.(57) (Ref. 474);
2.3i.8.(13).(21).32.36.(37).38.42-44.54.57.66 (Ref. 104).
References 89, 104, 108, 183, 333, 379, 474, 519.

Stereum styracifluum (Schw.:Fr.) Fr. 1838
Epicrisis, 549.
Basionym *Thelephora styracifluum* Schw., Schriften Naturf. Ges. Leipzig 1:105, 1822.
Distribution United States: AL, NC.
Hosts *Carpinus* sp., *Liquidambar* sp.
Ecology Underside of dead fallen limbs; mossy dead trunk.
Culture Characters Not known.
Reference 89.

Stereum subtomentosum Pouzar 1964
Ceská Mykol. 18:147-148.
Distribution Canada: NT, ON, PQ.
United States: NEW ENGLAND, MI, NY.
Hosts *Acer* sp., *A. saccharum, Betula* sp., *B. alleghaniensis, Fagus* sp., *Tilia* sp.
Ecology Log.
Culture Characters Not known.
References 171, 536.

We suspect that this is a common species and specimens have been misnamed *S. ostrea*.

Stereum tennebrosum Lloyd 1918
Mycol. Writ. 7 (Letter 67:16).
Distribution United States: FL.
Hosts Not known.
Ecology Not known.
Culture Characters Not known.
Reference 593.

Known in North America from only the one Florida collection, but Lloyd placed several collections from Africa and Ecuador under this name.

Stereum versicolor (Swartz:Fr.) Fr. 1838
Epicrisis, 547.
Basionym *Helvella versicolor* Swartz, Novae genera et species plantarum, seu prodromus, 149, 1788.
Distribution United States: FL.
Hosts Not known.
Ecology Dead wood.
Culture Characters Not known.
References 89, 100.

STIGMATOLEMMA Kalchbr. 1882 Tricholomataceae

Synonym is *Rhodocyphella* W.B. Cooke 1961, fide Donk (156).

Stigmatolemma poriaeforme (Pers.:Fr.) Singer 1962
Sydowia 15:52.
Basionym *Peziza anomala* var. *poriaeformis* Pers., Synop. Meth. Fung., 656, 1801.
Synonyms *Peziza poriaeformis* (Pers.) DC. 1815, *Porotheleum poriaeforme* (Pers.) W.B. Cooke 1957, *Stigmatolemma poriaeforme* (Pers.) W.B. Cooke 1961 (not validly published, lacking citation of the basionym), *Peziza pruinata* Schw. 1822, fide Cooke (131), *Solenia tephrosia* (Pers.) Lentz 1947, fide Lentz (378).
Distribution Canada: BC, ON, PQ. United States: AL, AZ, DE, GA, IA, LA, MA, ME, MD, MN, MO, NC, NH, NJ, NY, OH, PA, SC, VA, WI.
Hosts *Juglans major, Platanus occidentalis, P. wrightii, Populus* sp., *P. tremuloides, Quercus emoryi, Q. hypoleucoides, Q. reticulata, Vitis* sp., *V. labrusca, V. rotundifolia*, angiosperm.
Ecology Bark of live grape vine; decaying limbs and logs; log (charred); often associated with a soft brown rot.
Culture Characters Not known.
References 91, 96, 108, 124, 183, 204, 214, 216, 242, 378, 406, 545, 604.

Cooke (131) lists *S. poriaeformis* as a synonym of *S. poroides* (Alb. & Schw.:Fr.) W.B. Cooke but in the absence of evidence for this synonymy we prefer to use "*poriaeformis*". Cooke (124) lists numerous substrates but it is uncertain which were from Canada and the United States.

Stigmatolemma taxi (Lév.) Donk 1962
Persoonia 2:342.
Basionym *Cyphella taxi* Lév., Ann. Sci. Nat. Bot., Sér. II, 8:336, 1837.
Synonyms *Cyphella cupulaeformis* Berk. & Rav. 1873, fide Donk (156), *Rhodocyphella cupulaeformis* (Berk. & Rav.) W.B. Cooke 1961.
Distribution United States: FL, GA, LA, NC, OH, SC, TN.
Hosts *Juniperus virginiana, Robinia pseudoacacia, Thuja occidentalis.*
Ecology Bark of live trees; decaying cedar limb.
Culture Characters Not known.
References 80, 108, 114, 125, 131, 156, 202.

Cyphella cupulaeformis is not a synonym of *C. ampla* Lév., fide Donk (156:344).

STROMATOCYPHELLA W.B. Cooke 1961 Cyphellaceae

Stromatocyphella aceris W.B. Cooke 1961
Beih. Sydowia 4:104.
Distribution United States: VT.
Host ?*Acer* sp.
Ecology Branches.
Culture Characters Not known.

Known only from the type specimen.

Stromatocyphella conglobata (Burt) W.B. Cooke 1961
Beih. Sydowia 4:104.
Basionym *Cyphella conglobata* Burt, Ann. Missouri Bot. Gard. 1:375, 1914.
Distribution Canada: BC, NB, ON, PQ.
United States: MI, NH, NY, PA.
Hosts *Alnus* sp., *A. incana, Betula* sp., *Juglans cinerea.*
Ecology Dead branches.
Culture Characters Not known.
References 80, 116, 125, 242, 554, 604.

STYPELLA Möller 1895 Hyaloriaceae

Stypella vermiformis (Berk. & Br.) D. Reid 1974
Trans. Brit. Mycol. Soc. 62:473.
Basionym *Dacrymyces vermiformis* Berk. & Br., Ann. Mag. Nat. Hist., Ser. V, 1:25, 1879.
Synonyms *Stypella papillata* Möller 1895, fide Reid (558), *Protomerulius farlowii* Burt 1919, fide Martin (443).
Distribution Canada: ON.
United States: MA, NC, NH.
Hosts *Abies balsamea, Pinus* sp., *P. strobus, Thuja* sp., gymnosperm.
Ecology Very rotten, decorticated wood.
Culture Characters Not known.
References 87, 116, 424, 433, 443.

SUBULICIUM Hjort. & Ryv. 1979 Tubulicrinaceae

Subulicium lautum (Jackson) Hjort. & Ryv. 1979
Mycotaxon 9:513.
Basionym *Peniophora lauta* Jackson, Canad. J. Res., C, 26:129, 1948.
Distribution Canada: ON. United States: OR.
Hosts *Thuja occidentalis, Tsuga canadensis*, angiosperm, gymnosperm.
Ecology Not known.
Culture Characters Not known.
References 116, 242, 295, 302, 584.

Subulicium rallum (Jackson) Jülich & Stalpers 1980
Verh. Kon. Ned. Akad. Wetensch., Afd. Natuurk., Tweede Sect. 74:223.
Basionym *Peniophora ralla* Jackson, Canad. J. Res., C, 26:136, 1948.
Synonym *Subulicystidium rallum* (Jackson) Hjort. & Ryv. (295).
Distribution Canada: ON.
Host *Abies balsamea.*
Ecology Not known.
Culture Characters Not known.
References 116, 242, 295, 302, 584.

Known only from the holotype and three paratype specimens, all collected the same day at the same locality.

SUBULICYSTIDIUM Parm. 1968 Tubulicrinaceae

Subulicystidium longisporum (Pat.) Parm. 1968
Consp. Syst. Cort., 121.
Basionym *Hypochnus longisporus* Pat., J. Bot. (Morot) 8:221, 1894.
Synonyms *Peniophora longispora* (Pat.) Höhnel 1905, *Peniophora asperipilata* Burt 1926, fide Rogers and Jackson (567).
Anamorph *Aegerita tortuosa* Bourd. & Galzin 1928.
Distribution Canada: AB-NT, BC, MB, ON, PQ.
United States: AZ, CO, FL, GA, ID, IL, LA, ME, MI, MN, MT, NC, NY, TX, WA.
Hosts *Abies balsamea, Acer negundo, Alnus* sp., *A. oblongifolia, Betula* sp., *B. alleghaniensis, Fraxinus nigra, Liriodendron tulipifera, Picea mariana, Platanus* sp., *Populus* sp., *P. tremuloides, P. trichocarpa, Quercus* sp., *Q. gambelii, Q. garryana, Salix* sp., *Ulmus americana*, angiosperm, rarely gymnosperm.

Ecology Bark and decaying wood; decorticated wood; bark of rotten limb; slash; associated with a white rot.
Culture Characters Not known.
References 92, 96, 108, 116, 183, 204, 214, 216, 242, 267, 315, 400, 402, 406, 584.

SUILLOSPORIUM Pouzar 1958 Botryobasidiaceae

Suillosporium cystidiatum (D.P. Rogers) Pouzar 1958
Ceská Mykol. 12:31.
Basionym *Pellicularia cystidiata* D.P. Rogers, Farlowia 1:101, 1943.
Synonym *Botryobasidium cystidiatum* (D.P. Rogers) Eriksson 1948.
Distribution Canada: ON, PQ. United States: CT, IA.
Hosts *Picea* sp., gymnosperm.
Ecology Decayed wood; trees.
Culture Characters Not known.
References 166, 171, 242, 448, 535, 565.

THANATEPHORUS Donk 1956 Ceratobasidiaceae

Thanatephorus cucumeris (A.B. Frank) Donk 1956
Reinwardtia 3:376.
Basionym *Hypochnus cucumeris* A.B. Frank, Ber. Deutsch. Bot. Ges. 1:62, 1883.
Synonyms *Hypochnus filamentosus* Pat. 1891, fide Talbot (600), *Pellicularia filamentosa* (Pat.) D.P. Rogers 1943 sensu D.P. Rogers, fide Donk (157) and Talbot (599), *Corticium microsclerotium* Weber 1951, fide Talbot (600), *Corticium praticola* Kotila 1929, fide Talbot (600), *Pellicularia praticola* (Kotila) Fletje 1952, *Corticium vagum* Berk. & Curtis sensu Burt, fide Rogers (563) and Donk (157), *Corticium vagum* var. *solani* Burt in Rolfs 1903, fide Donk (157) and Talbot (600).
Anamorph *Moniliopsis solani* (Kühn) R.T. Moore 1987 (≡ *Rhizoctonia solani* Kühn 1858, = *Rhizoctonia microsclerotia* Matz 1917, fide Talbot (600), = *Rhizoctonia dichotoma* Saksena & Vaartaja 1960, fide Talbot (600), = *Rhizoctonia praticola* Saksena & Vaartaja 1961, not validly published, fide Donk (157)).
Distribution Canada: AB, BC, MB, NB, NS, NT, NTF, ON, PQ, SK. United States: AK, AL, AR, AZ, CA, CO, CT, DC, DE, FL, GA, IA, ID, IL, IN, KY, LA, MA, MD, ME, MI, MN, MO, MS, MT, NC, ND, NE, NJ, NY, OH, OK, OR, PA, RI, SC, SD, TX, UT, VA, WA, WI, WV, WY.
Hosts Cosmopolitan, Farr et al. (183) compiled over 400 genera of host plants.
Ecology Causes a variety of diseases, see Baker (11); including Rhizoctonia root rot; brown patch of turf; web blight of seedlings of *Aleurites fordii, Catalapa* spp., and *Ficus* sp.; basidiomes typically produced on live stems; wood and bark lying on the ground.
Culture Characters
(2).6.19.(23).24.25.26.32.36.(39).41-42.56.(57) (Ref. 588). See also Boidin (36) and Parmeter and Whitney (524).
References 84, 93, 116, 183, 242, 249, 274, 565, 585.

Thanatephorus pennatus Currah 1987
Canad. J. Bot. 65:1958.
Distribution Canada: AB.
Host *Calypso bulbosa*.
Ecology Endophyte, isolated from roots of live plant.
Culture Characters Not known.

Known only from the type specimen and one additional isolate.

Thanatephorus species A
Anamorph *Moniliopsis anomala* Currah 1990.
As Burgeff ex Currah, Canad. J. Bot. 68:1180.
Distribution Canada: AB.
Hosts *Coeloglossum viride, Platanthera hyperborea.*
Ecology Endophyte, isolated from roots of live plants.
Culture Characters See Currah (loc. cit.).

The vegetative hyphae are similar in many respects to that of *Thanatephorus cucumeris* and *T. pennatus*. Thus the teleomorph is most likely a species of *Thanatephorus*.

THELEPHORA Fr. 1821 Thelephoraceae

Thelephora terrestris is known, from laboratory tests, to be mycorrhizal, and the other species of *Thelephora* are probably mycorrhizal. However, the plants listed below as hosts refer to the occurrence of basidiomes and not, unless specified, to proven mycorrhizal associations.

Thelephora albidobrunnea Schw. 1832
Trans. Amer. Philos. Soc., N.S., 4:166.
Synonyms *Stereum micheneri* Berk. & Curtis 1873, *Stereum spongiosum* Massee 1889, *Thelephora odorifera* Peck 1891. All fide Burt (78).
Distribution Canada: ON, PQ. United States: IA, LA, MI, NC, NY, OH, PA, SC, VT, WI.
Hosts *Acer rubrum, Amelanchier* sp., *A. arborea, A. canadensis, A. laevis, Carpinus carolinana, Fraxinus americana, Juniperus virginiana, Quercus alba, Viburnum rufidulum.*
Ecology Running up and encircling twigs on the ground and against base of shrubs; soil and sticks, base of young live trees and shrubs, base of tree, around a rotten twig, on damp soil; rich soil under cedars; surrounds the stems and root collars of young trees of *Amelanchier*, but does little damage; ground encrusting twigs, stems, etc.; crowns and stem of young trees.
Culture characters: Not known.
References 78, 108, 136, 183, 274, 377a, 533.

Thelephora americana Lloyd 1915
Mycol. Writ. 4 (Letter 59:4).
Distribution Canada: BC, ON. United States: NC, NH, NJ, NY, OH, PA, TN, VA, WA, WI, WV.
Host Probably *Tsuga heterophylla.*
Ecology Possibly ectomycorrhizal with hemlock; basidiomes produced on ground and small twigs.
Culture characters: Not known.
References 136, 348.

Thelephora anthocephala var. ***anthocephala*** (Bull.):Fr. 1836
Epicrisis, 535.
Basionym *Clavaria anthocephala* Bull., Herb. France 2:197, 1789.
Distribution Canada: PQ. United States: AK, IA, KY, LA, MA, MO, NC, NY, OH, PA.
Hosts Not known.
Ecology Ground in woods, especially of *Fagus*, *Quercus*, also conifers; ground in deciduous stands; wet peat soil in tundra.
Culture characters: Not known.
References 78, 136, 344, 377a, 533.

Corner (136) suggested, without seeing specimens, that the Alaska report was based upon specimens of *T. caryophyllea*.

Thelephora anthocephala var. ***americana*** (Peck) Corner 1968
Beih. Nova Hedwigia 27:40.
Basionym *Thelephora palmata* var. *americana* Peck, Annual Rep. New York State Mus. 53:857, 1900.
Distribution Canada: Specific localities not given. United States: NY.
Hosts Not known.
Ecology Ground in woods (? not coniferous).
Culture characters: Not known.
References 136.

Corner (136) reported the distribution simply as "Canada, U.S.A., Mexico, Japan, China".

Thelephora anthocephala var. ***clavularis*** (Fr.) Quél. 1886
Enchiridion Fung., 203.
Basionym *Merisma clavulare* Fr., Obs. Mycol. 1:156, 1815.
Synonym *Thelephora palmata* var. *clavularis* (Fr.):Fr. 1821.
Distribution Canada and United States, but specific localities not given.
Hosts Not known.
Ecology Not known.
Culture characters: Not known.
References 136.

Corner (136) gave the distribution simply as "Europe, Canada, U.S.A., Japan."

Thelephora caespitulans Schw. 1832
Trans. Amer. Philos. Soc., N.S., 4:166.
Distribution Canada: Specific localities not given. United States: PA, SC, VT, WA.
Hosts Not known.
Ecology Ground in mixed and coniferous woods.
Culture characters: Not known.
References 78, 133, 136, 553.

Corner (136) gave the Canadian distribution simply as "Canada (common)." He suggested this might be a young state of *T. anthocephala*.

Thelephora caryophyllea Fr. 1821
Syst. Mycol. 1:565.
Distribution Canada: NB, ON, PQ. United States: AK, CA, CO, CT, DC, GA, IA, ID, KY, MA, ME, MI, MN, MT, NC, NH, NY, OH, PA, VT, WA.
Hosts *Betula nigra, Larix occidentalis, Pinus contorta, Salix sitchensis, Tsuga heterophylla.*
Ecology Sandy ground in coniferous woods; ground under pines; soil in pine and deciduous woods; at base of large white pine, growing on soil and exposed roots; sand under willow; soil along creek; seedling gridle and inhabiting wood or bark; ground in forests and on mountain slopes; growing around stems and foliage of coniferous seedlings, occasionally smothering some by its copious thallus development.
Culture characters: Not known.
References 78, 116, 119, 133, 136, 183, 214, 274, 444, 509, 533.

Thelephora cervicornis Corner 1968
Beih. Nova Hedwigia 27:49.
Distribution United States: AZ, MD.
Hosts Not known.
Ecology Not known.
Culture characters: Not known.

Known only from three collections; the type from the Bahamas and two paratypes from Arizona and Maryland.

Thelephora cuticularis Berk. 1847
London J. Bot. 6:324.
Distribution United States: DE, FL, IA, MO, NC, OH, PA, RI, TX, VT, WI.
Hosts *Juniperus virginiana, Quercus* sp.
Ecology Bark of oak tree; mossy bark at the base of trees and on fallen twigs.
Culture characters: Not known.
References 78, 108, 136, 183, 377a.

Thelephora griseozonata Cooke 1891
Grevillea 19:104.
Distribution United States: AL, IA, LA, MS, NC, NJ, SC.

Host *Pinus* sp.
Ecology Sandy ground in pine woods; hillside pasture.
Culture characters: Not known.
References 78, 108, 136, 377a.

Thelephora intybacea Pers.:Fr. 1801
Synop. Meth. Fung., 567.
Distribution Canada: ON. United States: CT, DC, IA, ID, MA, ME, MI, NC, NH, NY, OH, VT, WA.
Hosts Not known.
Ecology Ground in pine woods, growing up from the layer of fallen leaves; saprobic on rotten woody material of all kinds; in mixed woods.
Culture characters: Not known.
References 78, 119, 136, 377a.

Thelephora lutosa Schw. 1832
Trans. Amer. Philos. Soc., N.S., 4:166.
Distribution United States: NC, VA.
Hosts Not known.
Ecology Ground in roads and in woods; clay ground in mixed woods with fragment of wood attached to short radicating base.
Culture characters: Not known.
References 78, 93, 136, 553.

Thelephora magnispora Burt 1914
Ann. Missouri Bot. Gard. 1:211.
Distribution United States: FL.
Hosts Not known.
Ecology Mossy ground.
Culture characters: Not known.
References 136.

In the United States known from only the one collection.

Thelephora palmata (Scop.) Fr. 1821
Syst. Mycol. 1:432.
Basionym *Clavaria palmata* Scop., Fl. Carn. 2:483, 1760.
Distribution Canada: PE, PQ.
United States: CA, CO, CT, DC, DE, IA, IL, MO, NC, NH, NJ, NY, OH, PA, ?VT, WA.
Hosts Not known.
Ecology Moist ground in coniferous woods, also in grassy fields, on soil in conifer woods.
Culture characters: Not known.
References 78, 108, 128, 377a, 509, 511, 533.

Thelephora penicillata var. ***penicillata*** Fr. 1821
Syst. Mycol. 1:434.
Synonym *Thelephora spiculosa* Fr. sensu Burt 1915, fide Corner (136).
Distribution United States: IA, MI, NJ, OH, WI.
Hosts Not known.
Ecology Leaves, including partially decayed *Fagus* leaves, on ground in moist groves.
Culture characters: Not known.
References 78, 136, 377a.

Thelephora penicillata var. ***byssoideofimbriata*** (Bourd. & Galzin) Corner 1968
Beih. Nova Hedwigia 27:77.
Basionym *Phylacteria mollissima* forma *byssoideofimbriata* Bourd. & Galzin, Hym. France, 468, 1928.
Distribution United States: ID.
Host *Pinus contorta* var. *latifolia.*
Ecology Not known.
Culture characters: Not known.
Reference 136.

Known in North America from only one collection.

Thelephora penicillata forma ***subfimbriata*** (Bourd. & Galzin) Corner 1968
Beih. Nova Hedwigia 27:77.
Basionym *Phylacteria mollissima* forma *subfimbriata* Bourd. & Galzin, Hym. France, 468, 1928.
Synonym *Merisma fimbriata* Schw. 1822, fide Corner (136), *Thelephora fimbriata* (Schw.) Fr. 1825, *Thelephora scoparia* Peck 1889, fide Burt (78) and Corner (136).
Distribution Canada: PQ. United States: IA, IL, IN, MA, NC, NJ, NY, PA, SC.
Host *Pinus ponderosa*.
Ecology Moist places, encrusting small twigs and moss; encrusting herbaceous plants; encrusting stems.
Culture characters: Not known.
References 78, 274, 377a, 533.

Thelephora regularis var. ***regularis*** Schw. 1822
Schriften Naturf. Ges. Leipzig 1:105.
Distribution Canada: ON, PQ. United States: AL, DE, IA, IL, KS, MA, ME, MO, NC, NH, PA, SC, WI.
Hosts Not known.
Ecology In moss in wet places and in humus; damp, mossy earth, damp sandy soil; soil among mosses in moist places.
Culture characters: Not known.
References 78, 108, 136, 533, 553.

Thelephora regularis var. ***multipartita*** (Schw.) Corner 1968
Beih. Nova Hedwigia. 27:83.
Basionym *Thelephora multipartita* Schw. in E.M. Fries, Elenchus 1:166, 1828.
Distribution United States: DC, IA, IL, NC, NJ, NY, OH, PA.
Hosts Not known.
Ecology Soil under deciduous trees, especially oaks; on earth, upland rocky woods (mixed oak and pine), damp woods, in rich humus near base of oak.
Culture characters: Not known.
References 78, 108, 136, 377a, 533.

Thelephora scissilis Burt 1914
Ann. Missouri Bot. Gard. 1:204-205.
Distribution United States: MI, WA.
Hosts Not known.

Ecology Ground, in woods.
Culture characters: Not known.
References 78, 133, 136.

Thelephora terrestris Ehrenb.:Fr. 1785
Crypt. Exsicc. No. 178.
Distribution Canada: AB, BC, MB, NF, ON, PQ. United States: AK, AL, AZ, CA, CT, IA, ID, IN, MA, ME, MI, MN, MT, NC, ND, NH, NJ, NY, OH, OR, PA, SC, SD, WA, WY.
Hosts *Abies grandis, A. lasiocarpa, Dicentra spectabilis, Larix occidentalis, Picea* sp., *P. engelmannii, P. mariana, Pinus* sp., *P. banksiana, P. contorta, P. contorta* var. *latifolia, P. echinata, P. elliottii, P. monticola, P. nigra, P. ponderosa, P. resinosa, P. strobus, P. sylvestris, P. taeda, P. virginiana, Pseudotsuga menziesii, Quercus* sp., *Q. alba, Ribes* sp., *Spiraea* sp., *Thuja plicata, Tsuga heterophylla, Vaccinium* sp., *V. corymbosum*.
Ecology Ectomycorrhizal with, at least, hemlock, slash pine and Virginia pine; causes smothering disease, by its copious thallus development, it harms only slow growing seedlings; basidiomes sometimes encircling canes and stems of seedlings; basidiomes produced on sandy ground in bare fields and at base of trunks and from fallen twigs and leaves in pine woods; growing around stems and foliage of coniferous seedlings.
Culture characters: Not known.
References 78, 97, 108, 116, 119, 128, 136, 183, 197, 216, 274, 348, 377a, 448, 511, 533.

Causes smothering disease of coniferous seedlings in nature and in nurserys, but it does little harm. It is of benefit to seedlings in its role as a mycorrhiza-former.

Thelephora vialis Schw. 1832
Trans. Amer. Philos. Soc., N.S., 4:165.
Synonym *Thelephora tephroleuca* Berk. & Curtis 1873, fide Burt (78).
Distribution United States: DC, IL, NC, NJ, OH, PA, SC, VT.
Hosts Not known.
Ecology Ground in angiosperm woods.
Culture characters: Not known.
References 78, 108, 136.

THUJACORTICIUM Ginns 1988 — Hyphodermataceae

Thujacorticium mirabile Ginns 1988
Mycologia 80:69.
Distribution Canada: BC.
Host *Thuja plicata*.
Ecology Old, decorticated, decayed logs.
Culture Characters Not known.

Known only from the type specimen.

TOMENTELLA Pat. 1887 — Thelephoraceae

Tomentella angulospora Larsen 1975
Beih. Nova Hedwigia 51:172.
Distribution United States: NY.
Host *Acer* sp.
Ecology Not known.
Culture Characters Not known.

Known only from the type specimen.

Tomentella asperula (Karsten) Höhnel & Litsch. 1906
Sitzungsber. Kaiserl. Akad. Wiss., Math.-Naturwiss. Kl., Abt. 1, 115:1570.
Basionym *Hypochnus asperulus* Karsten, Bidrag Kännedom Finlands Natur Folk 48:441, 1889.
Distribution Canada: AB, ON.
United States: FL, ID, LA, MI, MN.
Hosts *Juniperus* sp., *J. virginiana, Populus* sp., *P. balsamifera, Thuja* sp.
Ecology Not known.
Culture Characters Not known.
References 202, 214, 242, 360, 362, 406.

Tomentella atrorubra (Peck) Bourd. & Galzin 1924
Bull. Soc. Mycol. France 40:134.
Basionym *Zygodesmus atroruber* Peck, Bot. Gaz. (Crawfordsville) 6:277, 1881.
Synonym *Hypochnus atroruber* (Peck) Burt 1916.
Distribution Canada: NS, ON. United States: KY, MA, MD, MI, NC, NH, NJ, NY.
Hosts *Abies balsamea, Acer* sp., *Castanea* sp., *C. dentata, Cedrus* sp., *Picea* sp., *Pinus* sp., *P. strobus, Populus* sp., *Quercus* sp., *Tsuga* sp., *T. canadensis*, angiosperm.
Ecology Humus.
Culture Characters Not known.
References 108, 183, 242, 249, 362, 448.

Tomentella avellanea (Burt) Bourd. & Galzin 1924
Bull. Soc. Mycol. France 40:153.
Basionym *Hypochnus avellaneus* Burt, Ann. Missouri Bot. Gard. 3:225, 1916.
Distribution Canada: BC, ON.
United States: AZ, CA, NY, TX, WA.
Hosts *Abies* sp., *Picea* sp., *Pinus* sp., *Pseudotsuga menziesii, Thuja* sp., *T. plicata*.
Ecology Humus.
Culture Characters Not known.
References 183, 242, 362, 448.

Tomentella bicolor (Atk. & Burt) Bourd. & Galzin 1924
Bull. Soc. Mycol. France 40:132.
Basionym *Hypochnus bicolor* Atk. & Burt, Ann. Missouri Bot. Gard. 3: 229, 1916.
Distribution Canada: ON.
United States: NC, NY, OH, PA.
Hosts *Acer* sp., angiosperm.

Ecology Not known.
Culture Characters Not known.
Reference 362.

Tomentella botryoides (Schw.) Bourd. & Galzin 1924 Bull. Soc. Mycol. France 40:159.
Basionym *Thelephora botryoides* Schw., Schriften Naturf. Ges. Leipzig 1:109, 1822.
Distribution Canada: NS, ON. United States: AL, AZ, FL, IN, LA, MA, MD, MI, MN, MS, NC, NH, NJ, NM, NY, OH, PA, SC, TN.
Hosts *Acer* sp., *Acer rubrum* var. *rubrum, Betula* sp., *Castanea* sp., *Fagus* sp., *Liquidambar* sp., *Pinus* sp., *P. ponderosa, P. strobus, Populus* sp., *P. tremuloides, Polyporus* sp., *Quercus* sp., *Q. rubra* var. *borealis, ?Sabal* sp., *Salix* sp., *Thuja* sp., *Tsuga* sp., *Ulmus* sp., angiosperm, gymnosperm.
Ecology Not known.
Culture Characters Not known.
References 116, 183, 197, 214, 216, 242, 249, 362, 406, 615.

Tomentella bresadolae (Brinkm.) Bourd. & Galzin 1924 Bull. Soc. Mycol. France 40:155.
Basionym *Hypochnus bresadolae* Brinkm. in Bresadola, Ann. Mycol. 1:108, 1903.
Distribution Canada: NS, ON. United States: AZ, KY, NC, NM, NY, OH, TN.
Hosts *Abies* sp., *A. balsamea, A. concolor, Acer* sp., *Juniperus virginiana, Picea* sp., *Pinus* sp., *P. strobus, P. ponderosa.*
Ecology Not known.
Culture Characters Not known.
References 183, 208, 242, 362, 448.

Tomentella brevispina (Bourd. & Galzin) Larsen 1970 Mycologia 62:136.
Basionym *Tomentella spongiosa* var. *brevispina* Bourd. & Galzin, Bull. Soc. Mycol. France 40:154, 1924.
Distribution Canada: ON. United States: AZ, CO.
Hosts *Populus* sp., *Thuja* sp., *T. occidentalis.*
Ecology Not known.
Culture Characters Not known.
References 242, 362, 402.

Tomentella brunneorufa Larsen 1974 Mycologia Memoir 4:37.
Distribution United States: MD, NC.
Hosts *Liriodendron* sp., *Pinus* sp., *Robinia* sp.
Ecology Not known.
Culture Characters Not known.

Known only from the holotype and two paratype specimens.

Tomentella bryophila (Pers.) Larsen 1974 Mycologia Memoir 4:51.
Basionym *Sporotrichum bryophilum* Pers., Myc. Eur. 1:78, 1822.
Synonyms *Tomentella pallidofulva* (Peck) Litsch. 1939, *Tomentella subferruginea* (Burt) Donk 1933. Both fide Larsen (362).
Distribution Canada: BC, MB, ON. United States: AZ, CO, DC, FL, GA, ID, IL, IN, LA, MA, MD, MI, MN, MT, NC, NM, NY, OH, PA, TN, VT.
Hosts *Abies* sp., *A. concolor, A. lasiocarpa* var. *arizonica, Acer* sp., *Alnus* sp., *A. incana, A. tenuifolia, Amelanchier* sp., *Betula* sp., *B. alleghaniensis, B. papyrifera, Castanea* sp., *Fagus* sp., *Fomes* sp., *Picea* sp., *P. engelmannii, Pinus* sp., *P. ponderosa, Populus* sp., *P. tremuloides, Quercus* sp., *Q. hypoleucoides, Q. rubra* var. *borealis, Robinia* sp., *Thuja* sp., *T. occidentalis, Tsuga* sp., angiosperm.
Ecology Not known.
Culture Characters Not known.
References 116, 183, 197, 208, 214, 216, 242, 362, 406, 448.

Tomentella caerulea (Bres.) Höhnel & Litsch. 1907 Sitzungsber. Kaiserl. Akad. Wiss., Math.-Naturwiss. Kl., Abt. 1, 116:831.
Basionym *Hypochnus caeruleus* Bres., Ann. Mycol. 1:109, 1903.
Synonyms *Tomentella papillata* Höhnel & Litsch. 1908, fide Larsen (362), *Hypochnus cervinus* Burt 1916, fide Larsen (362), *Tomentella cervina* (Burt) Bourd. & Galzin 1924.
Distribution Canada: AB, BC, MB, NS, ON. United States: AZ, CO, KY, MD, MI, MN, MT, NM, NS, WA.
Hosts *Acer* sp., *A. macrophyllum, Alnus* sp., *Betula* sp., *Fraxinus nigra, Pinus* sp., *P. ponderosa, Populus* sp., *P. tremuloides, P. trichocarpa, Poria* sp., *Prosopis juliflora, Quercus* sp., *Q. gambelii, Thuja* sp., *T. occidentalis, Tsuga* sp., angiosperm.
Ecology Not known.
Culture Characters Not known.
References 133, 183, 207, 208, 214, 216, 242, 249, 360, 362, 402, 406.

Tomentella calcicola (Bourd. & Galzin) Larsen 1967 Taxon 16:511.
Basionym *Caldesiella ferruginosa* var. *calcicola* Bourd. & Galzin, Hym. France, 471, 1928.
Synonym *Caldesiella calcicola* (Bourd. & Galzin) M. Christiansen 1960.
Distribution Canada: AB, BC. United States: CO, IA, ID, MT, NY, WA.
Hosts *Acer* sp., *Alnus* sp., *A. incana, A. tenuifolia, Picea* sp., *Populus* sp., *P. tremuloides, P. trichocarpa.*
Ecology Not known.
Culture Characters Not known.
References 183, 210, 242, 362, 402, 406, 448.

Tomentella carbonaria Larsen 1975 Beih. Nova Hedwigia 51:172.
Distribution United States: NM.
Host *Pinus contorta.*

Ecology Charred pine.
Culture Characters Not known.

Known only from the type specimen.

Tomentella chlorina (Massee) G.H. Cunn. 1953
Proc. Linn. Soc. New South Wales 77:279.
Basionym *Hypochnus chlorinus* Massee, Bull. Misc. Inform. Kew 158, 1901.
Synonyms *Sistotrema viride* Alb. & Schw. 1805, although this is the oldest epithet for this fungus it cannot be transferred to *Tomentella* because there is already a *T. viridis* (below), fide Larsen (362), *Caldesiella viridis* (Alb. & Schw.) Pat. 1900, *Tomentella viridis* (Berk.) G.H. Cunn. 1963, illegitimate because the basionym *Thelephora viridis* Berk. 1860 is a later homonym of *Thelephora viridis* Preuss 1851, see Larsen (362).
Distribution Canada: BC.
United States: AZ, CA, ID, MO, NM, NY, PA, VT.
Hosts *Abies* sp., *A. concolor, Acer* sp., *Alnus oblongifolia, Fraxinus velutina, Juglans* sp., *J. major, Larix* sp., *L. occidentalis, Picea* sp., *P. engelmannii, Pinus* sp., *P. ayacahuite, P. edulis, P. ponderosa, P. strobiformis, Platanus* sp., *Populus* sp., *P. tremuloides, Quercus* sp., *Q. hypoleucoides, Q. tomentella, Q. tomentosa*, gymnosperm.
Ecology Not known.
Culture Characters Not known.
References 93, 130, 183, 197, 208, 216, 362, 406, 448.

Tomentella cinerascens (Karsten) Höhnel & Litsch. 1906
Sitzungsber. Kaiserl. Akad. Wiss., Math.-Naturwiss. Kl. Abt. 1, 115:1570.
Basionym *Hypochnus cinerascens* Karsten, Meddeland. Soc. Fauna Fl. Fenn. 16:2, 1888.
Distribution Canada: BC, ON.
United States: AZ, MI, MT, NY, SD, WA.
Hosts *Abies concolor, Acer* sp., *Alnus* sp., *A. incana, A. tenuifolia, Betula* sp., *Castanea dentata, Fagus* sp., *F. grandifloia, Picea* sp., *Populus* sp., *Tilia* sp., *T. americana, Tsuga* sp., *T. canadensis.*
Ecology Not known.
Culture Characters Not known.
References 183, 216, 242, 362, 406, 448.

Tomentella clavigera Svrcek 1960
As Litsch. ex Svrcek, Sydowia 14:192.
Distribution United States: MN, NJ.
Hosts *Pinus* sp., *Populus* sp., *P. balsamifera.*
Ecology Not known.
Culture Characters Not known.
References 214, 362.

Tomentella crinalis (Fr.) Larsen 1967
Taxon 16:511.
Basionym *Hydnum crinale* Fr., Epicrisis, 516, 1838.
Synonym *Caldesiella ferruginosa* (Fr.) Sacc. 1877, fide Larsen (359).
Distribution Canada: AB, MB, NS, ON.
United States: AZ, CA, CO, IA, ID, IN, KY, LA, MD, ME, MN, MT, NY, SD, WA.
Hosts *Abies* sp., *A. concolor, A. lasiocarpa* var. *arizonica, Acer* sp., *Alnus* sp., *A. tenuifolia, Betula alleghaniensis, Fraxinus* sp., *F. nigra, Pinus* sp., *P. contorta, Populus* sp., *P. tremuloides, Quercus* sp., *Tsuga* sp., *T. canadensis*, angiosperm.
Ecology Not known.
Culture Characters Not known.
References 183, 214, 216, 242, 362, 406, 463.

Tomentella duemmeri (Wakef.) Larsen 1974
Mycologia Memoir 4:41.
Basionym *Caldesiella duemmeri* Wakef., Bull. Misc. Inform. Kew 3:73, 1916.
Distribution United States: FL.
Hosts Not known.
Ecology Not known.
Culture Characters Not known.
Reference 362.

Tomentella ellisii (Sacc.) Jülich & Stalpers 1980
Verh. Kon. Ned. Akad. Wetensch., Afd. Natuurk., Tweed Sect., 74:236.
Basionym *Zygodesmus ellisii* Sacc., Sylloge Fung. 4:808, 1886.
Synonyms *Tomentella microspora* (Karsten) Höhnel & Litsch. 1906, see Larsen (362) under *Zygodesmus ochraceus* Sacc. which Jülich and Stalpers (326) list in synonymy under *T. ellisii, Tomentella ochracea* (Sacc.) Larsen 1974, fide Jülich and Stalpers (326), a later homonym of *Zygodesmus ochracea* Corda 1835, *Tomentella sparsa* (Burt) Bourd. & Galzin 1924, see Larsen (362) as a synonym under *T. ochracea.*
Distribution Canada: BC, ON, PQ.
United States: AZ, ID, MN, NH, NJ, NY.
Hosts *Abies* sp., *A. balsamea, Acer* sp., *Alnus* sp., *Betula* sp., *B. alleghaniensis, B. occidentalis, Castanea* sp., *Larix* sp., *Picea* sp., *P. engelmannii, Pinus* sp., *P. ponderosa, Populus* sp., *Pseudotsuga* sp., *P. menziesii, Spiraea* sp., *Thuja* sp., *T. occidentalis, T. plicata, Tsuga* sp., angiosperm.
Ecology Not known.
Culture Characters Not known.
References 214, 216, 242, 360, 362, 448.

Tomentella epigaea (Burt) Larsen 1965
Canad. J. Bot. 43:1493.
Basionym *Hypochnus epigaeus* Burt, Ann. Missouri Bot. Gard. 3:226, 1916.
Distribution Canada: ON.
United States: MA, NH, NY.
Host *Acer* sp.
Ecology Not known.
Culture Characters Not known.
Reference 362.

Tomentella ferruginea (Pers.) Pat. 1887
Hym. Eur., 154.
Basionym *Corticium ferruginea* Pers., Obs. Mycol. 2:18, 1799.
Synonyms *Grandinia coriaria* Peck 1873, *Tomentella coriaria* (Peck) Bourd. & Galzin 1924, *Grandinia rudis* Peck 1878, *Tomentella fusca* (Pers.) Schröter 1888. All fide Larsen (362).
Distribution Canada: AB-NT, BC, MB, NB, NF, NS, ON. United States: AZ, CO, FL, IA, ID, KY, MA, MD, MI, MN, MT, NH, NM, NY, OH, SC, TN, VA, VT, WA, WI, WV.
Hosts *Abies* sp., *A. concolor, Acer* sp., *A. saccharum, Alnus* sp., *Betula* sp., *B. alleghaniensis, B. papyrifera, Fagus* sp., *Fraxinus nigra, Juglans* sp., *Juniperus* sp., *Liriodendron* sp., *Picea* sp., *Pinus* sp., *P. contorta, P. ponderosa, P. taeda, Platanus* sp., *P. wrightii, Populus* sp., *P. tremuloides, P. trichocarpa, Pseudotsuga menziesii, Quercus* sp., *Q. gambelii, Salix* sp., *Tilia* sp., *T. americana, Thuja plicata, Tsuga heterophylla, Yucca* sp., angiosperm, gymnosperm.
Ecology Leather scraps.
Culture Characters Not known.
References 116, 119, 183, 197, 208, 214, 216, 242, 249, 362, 402, 406.

Tomentella ferruginella Bourd. & Galzin 1924
Bull. Soc. Mycol. France 40:157.
Distribution Canada: AB. United States: LA, NM.
Hosts *Populus* sp., *P. tremuloides*.
Ecology Not known.
Culture Characters Not known.
References 208, 362.

Tomentella fraseri Larsen 1975
Beih. Nova Hedwigia 51:173.
Distribution Canada: AB.
Host *Populus* sp.
Ecology Not known.
Culture Characters Not known.

Known only from the type specimen.

Tomentella fuscoferruginosa (Bres.) Litsch. 1941
Ann. Mycol. 39:377.
Basionym *Hypochnus fuscoferruginosus* Bres., Ann. Mycol. 1:109, 1903.
Distribution Canada: ON.
United States: GA, NC, NY, TN, WA.
Hosts *Abies* sp., *A. balsamea, Acer* sp., *Betula* sp., *Pinus* sp., *Populus* sp., *Quercus* sp.
Ecology Not known.
Culture Characters Not known.
References 242, 360, 362, 406.

Tomentella galzinii Bourd. 1924
Bull. Soc. Mycol. France 40:143.
Distribution Canada: ON.
United States: AZ, LA, NY, PA.
Hosts *Betula* sp., *Fagus* sp., *Populus* sp., *Quercus* sp., *Taxus* sp., angiosperm.
Ecology Not known.
Culture Characters Not known.
References 362, 615.

Tomentella griseoumbrina Litsch. 1936
In Lundell and Nannfeldt, Fungi Exs. Suec. No. 357, Uppsala.
Distribution Canada: ON.
United States: AZ, MN, NY, OH.
Hosts *Abies balsamea, Larix* sp., *Pinus* sp., *Populus* sp., *Tsuga* sp., *T. canadensis*, gymnosperm.
Ecology Not known.
Culture Characters Not known.
References 214, 242, 358, 362.

Tomentella griseoviolacea Litsch. 1939
Glasn. Skopsk. Naucn. Drustva 20:20.
Distribution Canada: ON, PQ.
United States: AZ, FL, MD, NY, TN.
Hosts *Abies* sp., *Liriodendron* sp., *Populus* sp., *Quercus* sp., *Tilia* sp., *T. americana, Tsuga* sp., *T. canadensis*.
Ecology Bark.
Culture Characters Not known.
References 242, 360, 362, 608.

Tomentella kentuckiensis Larsen 1974
Mycologia Memoir 4:100.
Distribution United States: KY.
Hosts Not known.
Ecology Not known.
Culture Characters Not known.

Known only from the type specimen.

Tomentella kootenaiensis Larsen 1975
Beih. Nova Hedwigia 51:174.
Distribution Canada: AB, BC.
Hosts *Populus* sp., angiosperm, gymnosperm.
Ecology Not known.
Culture Characters Not known.

Known only from the type specimens.

Tomentella lapida (Pers.) Stalpers 1984
Studies in Mycol. (Baarn) 24:65.
Basionym *Sporotrichum lapidum* Pers., Mycol. Eur. 1:78, 1822.
Synonyms *Tomentella spinifera* (Burt) M. Christiansen 1960, fide Larsen (362), *Tomentella spongiosa* var. *spinifera* (Burt) Bourd. & Galzin 1924, *Tomentella trachychaetes* (Ellis & Ev.) Larsen 1968, fide Larsen (362), and Stalpers (589), *Tomentella violaceofusca* (Sacc.) Larsen 1974, fide Stalpers (589).
Distribution Canada: AB, MB, ON.
United States: AZ, FL, GA, LA, MA, MD, ME, MI, MN, MT, NC, NJ, NM, NY, SC, VT, WY.

Hosts *Abies* sp., *A. balsamea, Acer* sp., *Betula* sp., *Fraxinus* sp., *Larix occidentalis, Magnolia* sp., ?*Picea* sp., *P. glauca, Pinus* sp., *P. ponderosa, Populus* sp., *Quercus* sp., *Thuja* sp., *Trichoglossum* sp.
Ecology Not known.
Culture Characters Not known.
References 116, 183, 197, 214, 362, 448, 615.

Tomentella lateritia Pat. 1894
J. Bot. (Morot) 8:221.
Distribution Canada: ON. United States: AZ, CO, IL, MA, MI, MT, NC, NH, NM, NY, PA, TN.
Hosts *Acer* sp., *Betula* sp., *Fagus* sp., *Picea* sp., *Pinus* sp., *P. engelmannii, P. ponderosa, Platanus* sp., *Populus* sp., *P. grandidentata, P. tremuloides, Prunus serotina, Pseudotsuga* sp., *P. menziesii, Quercus* sp., *Robinia* sp., *Taxus* sp., *Tsuga* sp., *T. canadensis, Ulmus* sp.
Ecology Not known.
Culture Characters Not known.
References 183, 197, 208, 216, 242, 362, 402, 406, 448.

Tomentella molybdaea Bourd. & Galzin 1924
Bull. Soc. Mycol. France 40:142.
Distribution Canada: ON.
United States: CO, MT, NM, NY.
Hosts *Abies* sp., *A. balsamea, A. concolor, Pinus* sp., *P. banksiana, Populus* sp., *Quercus gambelii.*
Ecology Not known.
Culture Characters Not known.
References 208, 242, 360, 362, 400, 402.

Tomentella muricata (Ellis & Ev.) Wakef. 1960
Mycologia 52:924.
Basionym *Zygodesmus muricatus* Ellis & Ev., Bull. Torrey Bot. Club 11:17, 1884.
Distribution United States: FL, NJ, NY.
Hosts *Magnolia* sp., *Pinus* sp., *Quercus* sp.
Ecology Not known.
Culture Characters Not known.
References 183, 362.

Tomentella neobourdotii Larsen 1968
Mycologia 60:1179.
Distribution Canada: BC, NT, ON, PQ. United States: AZ, KY, MA, MI, MT, NC, NM, NY, TN, WI.
Hosts *Abies* sp., *Acer* sp., *Alnus* sp., *Betula* sp., *B. alleghaniensis, Carpinus* sp., *Fagus* sp., *Larix* sp., *Pinus* sp., *Polyporus* sp., *Populus* sp., *P. tremuloides, Quercus* sp., *Salix* sp., *Thuja* sp., *Tilia* sp., *T. americana, Tsuga* sp.
Ecology Not known.
Culture Characters Not known.
References 208, 216, 242, 362, 406.

Tomentella nitellina Bourd. & Galzin 1924
Bull. Soc. Mycol. France 40:151.
Distribution Canada: BC, NS, ON.
United States: NC, WA.
Hosts ?*Abies* sp., *A. balsamea, Picea* sp., *P. sitchensis, Populus* sp., *P. balsamifera, Thuja occidentalis, T. plicata.*
Ecology Twig.
Culture Characters Not known.
References 242, 327, 362, 448.

Tomentella olivascens (Berk. & Curtis) Bourd. & Galzin 1924
Bull. Soc. Mycol. France 40: 132.
Basionym *Zygodesmus olivascens* Berk. & Curtis, Grevillea 3:145, 1875.
Distribution Canada: NS, ON. United States: CT, MA, MD, NC, NY, OH, PA, SC, TN.
Hosts *Acer* sp., *Betula* sp., *Castanea* sp., *Fagus* sp., *Ganoderma* sp., *Hydnellum* sp., *Picea* sp., *Pinus* sp., *P. strobus, Populus* sp., *P. tremuloides, Prunus* sp., *Quercus* sp., *Tsuga* sp., *T. canadensis.*
Ecology Decayed wood.
Culture Characters Not known.
References 183, 242, 249, 362, 448.

Tomentella pilatii Litsch. 1933
Bull. Soc. Mycol. France 49:72.
Distribution United States: NM.
Host *Populus tremuloides.*
Ecology Not known.
Culture Characters Not known.
Reference 208.

Tomentella pilosa (Burt) Bourd. & Galzin 1924
Bull. Soc. Mycol. France 40:151.
Basionym *Hypochnus pilosus* Burt, Ann. Missouri Bot. Gard. 3:221, 1916.
Distribution Canada: MB, ON.
United States: AZ, CO, KY, LA, MD, MI, MN, MT, NC, NM, NY, PA, WA, WI.
Hosts *Abies* sp., *A. balsamea, Acer* sp., *Betula* sp., *Liriodendron* sp., *Pinus* sp., *P. ponderosa, P. strobus, Platanus* sp., *P. wrightii, Populus* sp., *P. grandidentata, P. tremuloides, Pseudotsuga* sp., *P. menziesii, Quercus* sp., *Q. alba, Q. gambelii, Tilia* sp., *Ulmus* sp., angiosperm, gymnosperm.
Ecology Not known.
Culture Characters Not known.
References 93, 116, 183, 197, 208, 214, 216, 242, 362, 399, 402, 406.

Tomentella pirolae (Ellis & Halsted) Larsen 1968
New York State Univ., Coll. Forestry, Syracuse, Publ. 93:105.
Basionym *Zygodesmus pirolae* Ellis & Halsted, J. Mycol. 6:34, 1890.
Distribution United States: NJ.
Host *Pyrola* sp.
Ecology Not known.
Culture Characters Not known.

Known only from the type specimen.

Tomentella punicea (Alb. & Schw.) Schröter 1888
In F. Cohn, Kryptogamen-Flora Schlesien 3 (1):420.
Basionym *Thelephora punicea* Alb. & Schw., Conspectus Fungorum, 278, 1805, *Zygodesmus granulosus* Peck 1905, fide Larsen (359).
Distribution Canada: AB, NS, ON.
United States: AL, AZ, FL, IL, KY, LA, MA, MI, MN, NC, NH, NJ, NY, OH, PA, TN.
Hosts *Abies balsamea, Acer* sp., *Betula* sp., *Fraxinus* sp., *Pinus* sp., *Populus* sp., *P. tremuloides, Quercus* sp., *Robinia* sp., *Thuja* sp., *Tsuga* sp., *Ulmus* sp., *Zea mays*, gymnosperm.
Ecology Not known.
Culture Characters Not known.
References 214, 242, 249, 362, 406.

Tomentella purpurea Wakef. 1966
Trans. Brit. Mycol. Soc. 49:361.
Distribution United States: FL.
Hosts Not known.
Ecology Not known.
Culture Characters Not known.
Reference 362.

Tomentella ramosissima (Berk. & Curtis) Wakef. 1960
Mycologia 52:927.
Basionym *Zygodesmus ramosissimus* Berk. & Curtis, Grevillea 3:145, 1875.
Synonym *Tomentella fuliginea* (Burt) Bourd. & Galzin 1924, fide Larsen (362).
Distribution Canada: AB, BC, NF, ON, PQ.
United States: AZ, CO, ID, MN, MI, MS, MT, NC, NH, NM, NY, SC, WA.
Hosts *Abies* sp., *A. concolor, Acer* sp., *Alnus* sp., *A. incana, Betula* sp., *Larix* sp., *L. occidentalis, Picea* sp., *P. engelmannii, P. mariana, Pinus* sp., *P. banksiana, P. ponderosa, P. strobus, Populus* sp., *P. tremuloides, Pseudotsuga* sp., *Quercus* sp., *Q. gambelii, Q. rubra* var. *borealis, Thuja* sp., *T. plicata, Tilia* sp., *T. americana.*
Ecology Not known.
Culture Characters Not known.
References 183, 197, 208, 214, 216, 242, 362, 398, 402, 406, 448.

Tomentella rubiginosa (Bres.) Maire 1906
Ann. Mycol. 4:335.
Basionym *Hypochnus rubiginosus* Bres., Atti Imp. Regia Accad. Rovereto, Ser. III, 3:116, 1897.
Distribution Canada: MB, NS, ON.
United States: AR, AZ, FL, IA, IL, IN, KY, LA, MD, MI, NH, NY, OH, OR, PA, VA, WV.
Hosts *Carex* sp., *Carya (= Hicoria)* sp., *Cymophyllus fraseri, Fagus* sp., *Fraxinus* sp., *Liriodendron* sp., *Picea glauca, P. rubens, Pinus* sp., *Populus* sp., *Prunus* sp., *Quercus* sp., *Q. alba, Q. hypoleucoides, Rhododendron* sp., *Thuja* sp., *T. occidentalis, Tilia* sp., *T. americana.*
Ecology Dead leaves and wood; soil.
Culture Characters Not known.
References 183, 216, 242, 249, 362, 406, 448.

Tomentella ruttneri Litsch. 1933
Bull. Soc. Mycol. France 49:67.
Distribution Canada: AB, NF, ON. United States: AR, AZ, MD, MI, MN, NJ, NM, NY, OH, TN, VT.
Hosts *Abies* sp., *A. balsamea, Acer* sp., *A. saccharum, Alnus* sp., *Betula* sp., *B. alleghaniensis, Fagus* sp., *F. grandifolia, Fraxinus* sp., *F. nigra, ?Juglans* sp., *Magnolia* sp., *Picea* sp., *Pinus* sp., *Polyporus* sp., *Populus* sp., *Thuja* sp., *Tilia* sp., *T. americana, Tsuga* sp., gymnosperm.
Ecology Not known.
Culture Characters Not known.
References 183, 214, 242, 362, 448.

Tomentella subalpina Larsen 1972
Mycologia 64:444.
Distribution Canada: AB.
Hosts Gymnosperm.
Ecology Not known.
Culture Characters Not known.
Reference 362.

Tomentella subcinerascens Litsch. 1939
Oesterr. Bot. Z. 88:133.
Distribution United States: NC.
Host *Acer* sp.
Ecology Not known.
Culture Characters Not known.
Reference 362.

Tomentella subclavigera Litsch. 1933
Bull. Soc. Mycol. France 49:57.
Distribution Canada: ON. United States: MT, NH.
Hosts *Betula* sp., *Thuja* sp., *T. occidentalis.*
Ecology Not known.
Culture Characters Not known.
References 242, 362.

Tomentella sublilacina (Ellis & Holway) Wakef. 1960
Mycologia 52:931.
Basionym *Zygodesmus sublilacinus* Ellis & Holway in J.C. Arthur et al., Bull. Geol. Nat. Hist. Surv., Minn. 3:34, 1887.
Distribution Canada: BC, NF, NS, ON, PQ.
United States: AZ, JD, KY, MA, MD, MI, MN, NC, NM, NY, OR, TN, WA, WV.
Hosts *Abies* sp., *A. balsamea, Acer* sp., *Alnus* sp., *Betula* sp., *B. alleghaniensis, B. papyrifera, Castanea* sp., *Fagus* sp., *F. grandifolia, Fomes* sp., *Ganoderma* sp., *Hypoxylon* sp., *Juglans* sp., *Lycopodium* sp., *Picea* sp., *P. glauca, P. mariana, Pinus* sp., *P. ponderosa, P. strobus, Polyporus* sp., *Populus* sp., *Pseudotsuga* sp., *P. menziesii, Prunus* sp., *Quercus* sp., *Thuja* sp., *T. occidentalis, T. plicata, Tsuga* sp., *T. heterophylla*, moss.

Ecology Not known.
Culture Characters Not known.
References 183, 208, 214, 242, 362, 448.

Tomentella subtestacea Bourd. & Galzin 1924
Bull. Soc. Mycol. France 40:144.
Distribution United States: AZ, CO.
Hosts *Abies balsamea, Quercus gambelii*, gymnosperm.
Ecology Not known.
Culture Characters Not known.
References 214, 362, 400, 402.

Tomentella subvinosa (Burt) Bourd. & Galzin 1924
Bull. Soc. Mycol. France 40:146.
Basionym *Hypochnus subvinosus* Burt, Ann. Missouri Bot. Gard. 3:231, 1916.
Distribution Canada: ON, PQ.
United States: AZ, MN, NH, NJ, NY, WA.
Hosts *Acer* sp., *Betula* sp., *Fagus* sp., *Picea* sp., *P. engelmannii, Populus* sp., *Sassafras* sp., *Tilia* sp.
Ecology Not known.
Culture Characters Not known.
References 133, 242, 362, 448.

Tomentella terrestris (Berk. & Br.) Larsen 1974
Mycologia Memoir 4:105.
Basionym *Zygodesmus terrestris* Berk. & Br., Ann. Mag. Nat. Hist., Ser. V, 7:130, 1881.
Synonyms *Tomentella badiofusca* Bourd. & Galzin 1928, *Tomentella umbrinella* Bourd. & Galzin 1924. Both fide Larsen (362).
Distribution Canada: BC, NS, ON.
United States: AZ, FL, ID, MI, MT, NC, NM, NY.
Hosts *Abies* sp., *A. balsamea, Acer* sp., *Betula* sp., *Castanea* sp., *Picea* sp., *P. engelmannii, Pinus* sp., *P. ponderosa, Populus* sp., *P. balsamifera, Quercus* sp., *Thuja* sp., *T. occidentalis, T. plicata, Tsuga* sp., *T. canadensis*, gymnosperm.
Ecology Not known.
Culture Characters Not known.
References 197, 208, 216, 242, 358, 362, 406, 448.

Tomentella umbrinospora Larsen 1968
New York State Univ., Coll. Forestry, Syracuse, Publ. 93:61.
Basionym *Zygodesmus rubiginosus* Peck, New York State Mus. Rep. 30:58, 1879, a *nom. nov.* was proposed by Larsen because there already was a *Tomentella rubiginosa* (Bres.) Maire 1906.
Distribution Canada: ON. United States: AL, AZ, FL, KY, LA, NC, NY, OH, TN, VT.
Hosts *Acer* sp., *A. saccharum, Hypoxylon* sp., *Pinus* sp., *Populus* sp., *Quercus* sp., *Salix* sp., *Tilia* sp., *Ulmus* sp.
Ecology Not known.
Culture Characters Not known.
References 183, 242, 362, 406.

Tomentella viridescens (Bres. & Torrend) Bourd. & Galzin 1928
Hym. France, 477.
Basionym *Hypochnus viridescens* Bres. & Torrend, Brotéria, Sér. Bot. 11:85, 1913.
Distribution Canada: ON.
United States: AZ, NM, NY.
Hosts *Abies* sp., *Acer* sp., *Juglans* sp., *Pinus* sp., *Platanus* sp., *Populus* sp.
Ecology Not known.
Culture Characters Not known.
References 242, 362, 406.

Tomentella viridula Bourd. & Galzin 1924
Bull. Soc. Mycol. France 40:144.
Distribution Canada: AB.
United States: AZ, NM, NY.
Hosts *Acer* sp., *Pinus* sp., *Platanus* sp., *Populus* sp., gymnosperm.
Ecology Not known.
Culture Characters Not known.
References 358, 362.

TOMENTELLASTRUM Svrcek 1958 Thelephoraceae

Tomentellastrum badium (Link) Larsen 1981
Nova Hedwigia 35:5.
Basionym *Sporotrichum badium* Link, Ges. Naturf. Freunde Berlin Mag. 7:25, 1816.
Synonyms *Thelephora floridana* Ellis & Ev. 1886, fide Larsen (362), *Tomentellastrum floridanum* (Ellis & Ev.) Larsen 1974, *Tomentella fimbriata* M. Christiansen 1960, fide Larsen (362), *Tomentella atroviolacea* Litsch. 1933, fide Larsen (362) and Stalpers (587).
Distribution Canada: ON, YT.
United States: AZ, CO, FL, ID, MN, NM, NY.
Hosts *Alnus tennuifolia, Betula alleghaniensis, Fagus* sp., *Picea glauca, Pinus ponderosa, Populus* sp., *P. tremuloides*, angiosperm.
Ecology Buried wood.
Culture Characters Not known.
References 197, 208, 214, 216, 358, 402, 406, 448.

Tomentellastrum brunneofirmum (Larsen) Larsen 1974
Mycologia Memoir 4:110.
Basionym *Tomentella brunneofirma* Larsen, Canad. J. Bot. 45:1300, 1967.
Distribution United States: AZ, CO, NM.
Hosts *Populus* sp., *P. tremuloides, Quercus* sp., *Robinia neomexicana.*
Ecology Not known.
Culture Characters Not known.
References 208, 359, 402.

Tomentellastrum fuscocinereum (Pers.) Svrcek 1958
Ceská Mykol. 12:69.
Basionym *Thelephora fuscocinerea* Pers., Mycol. Eur. 1:114, 1822.

Synonyms *Tomentellastrum alutaceo-umbrinum* (Bres.) Larsen 1974, fide Jülich and Stalpers (326), *Tomentella macrospora* Höhnel & Litsch. 1906, fide Larsen (362).
Distribution Canada: Locality not specified. United States: MN, NY.
Hosts Not known.
Ecology Duff.
Culture Characters Not known.
References 214, 359, 364.

Tomentellastrum montanensis (Larsen) Larsen 1974. Mycologia Memoir 4:120.
Basionym *Tomentella montanensis* Larsen, Canad. J. Bot. 45:1304, 1967.
Distribution Canada: AB. United States: MT, NM.
Hosts *Alnus* sp., *Populus* sp., *P. tremuloides*.
Ecology Not known.
Culture Characters Not known.
References 183, 208, 242, 364, 406.

TOMENTELLINA Höhnel & Litsch. 1906 Thelephoraceae

Tomentellina fibrosa (Berk. & Curtis) Larsen 1974 Mycologia Memoir 4:115.
Basionym *Zygodesmus fibrosus* Berk. & Curtis, Grevillea 3:145, 1873.
Synonyms *Kneiffiella fibrosa* (Berk. & Curtis) Larsen 1968, *Hypochnus canadensis* Burt 1916, fide Larsen (359), *Tomentella ferruginosa* (Höhnel & Litsch.) Sacc. & Trotter 1912, fide Larsen (362).
Distribution Canada: AB, BC, MB, NF, NT, ON, PQ, YT. United States: AK, AL, AZ, CO, ID, ME, MI, MN, MT, NH, NM, NY, SC, UT, VT, WA, WY.
Hosts *Betula alleghaniensis, Larix* sp., *L. occidentalis, Picea glauca, Pinus* sp., *P. aristata* var. *longaeva, P. banksiana, P. contorta* var. *contorta, P. ponderosa, P. strobus, Populus* sp. *P. balsamifera, P. tremuloides, Pseudotsuga menziesii, Quercus gambelii, Salix* sp., *Thuja plicata, T. occidentalis, Tsuga canadensis*, angiosperm, gymnosperm.
Ecology Decayed leaves; organic debris.
Culture Characters Not known.
References 116, 183, 197, 214, 216, 242, 359, 398, 402, 406, 407, 448.

TOMENTELLOPSIS Hjort. 1970 Thelephoraceae

Tomentellopsis bresadoliana (Sacc. & Trotter) Jülich & Stalpers 1980
Verh. Kon. Ned. Akad. Wetensch., Afd. Natuurk., Tweed Sect. 74:255.
Basionym *Corticium bresadolianum* Sacc. & Trotter, Sylloge Fung. 21:866, 1912.
Distribution United States: CO.
Host *Pinus ponderosa*.
Ecology Not known.
Culture Characters Not known.
Reference 402.

Tomentellopsis echinospora (Ellis) Hjort. 1970 Svensk Bot. Tidskr. 64:426.
Basionym *Corticium echinosporum* Ellis, Bull. Torrey Bot. Club 8:64, 1881.
Synonyms *Tomentella echinospora* (Ellis) Bourd. & Galzin 1924, *Hypochnus pennsylvanicus* Overh. 1929, fide Larsen (359).
Distribution Canada: MB. United States: ID, NJ, PA, WA.
Hosts *Carya* sp., *Pinus* sp., *P. banksiana, Populus* spp.
Ecology Bark.
Culture Characters Not known.
References 116, 119, 359, 514.

Tomentellopsis pallido-aurantiaca (Gilbn. & Budington) K.H. Larsson 1993, *ined.*
Basionym *Trechispora pallido-aurantiaca* Gilbn. & Budington, Mycologia 62:677, 1970.
Distribution Canada: AB, NT, NTM. United States: AZ, NM, NY.
Hosts *Picea mariana, Pinus banksiana, P. ponderosa, P. sylvestris, Populus* sp., *P. tremuloides*.
Ecology Not known.
Culture Characters Not known.
References 197, 205, 208, 242, 395, 406.

Tomentellopsis pusilla Hjort. 1974 Svensk Bot. Tidskr. 68:53.
Synonym *Trechispora variseptata* Burdsall & Gilbn. 1982, fide K.H. Larsson (in litt. 1992).
Distribution United States: AZ.
Host *Fraxinus velutina*.
Ecology Not known.
Culture Characters Not known.

Known in North America from only one specimen, the type of *T. variseptata*.

Tomentellopsis zygodesmoides (Ellis) Hjort. 1974 Svensk Bot. Tidskr. 68:55.
Basionym *Thelephora zygodesmoides* Ellis, North Amer. Fungi No. 715, Newfield, 1882.
Synonym *Tomentella zygodesmoides* (Ellis) Höhnel & Litsch. 1907.
Distribution Canada: AB-NT, ON. United States: NJ.
Hosts *Pinus* sp., *Populus tremuloides*.
Ecology Bark.
Culture Characters Not known.
References 242, 359.

TRECHISPORA Karsten 1890 Sistotremaceae

Trechispora alnicola (Bourd. & Galzin) Liberta 1966 Taxon 15:318.
Basionym *Grandinia alnicola* Bourd. & Galzin, Bull.

Soc. Mycol. France 30:254, 1914.
Anamorph Blastoconidia (395).
Distribution Canada: ON.
United States: IA, IL, IN, NJ, NY, OH, PA, WI.
Hosts *Poa pratensis, Populus* sp.
Ecology Associated with yellow ring disease of *Poa pratensis* (646).
Culture Characters Not known.
References 183, 242, 395, 463, 646.

Trechispora amianthina (Bourd. & Galzin) Liberta 1966
Taxon 15:318.
Basionym *Corticium amianthinum* Bourd. & Galzin, Bull. Soc. Mycol. France 27:260, 1911.
Synonyms *Corticium crustulinum* Burt 1926, a later homonym of *Corticium crustulinum* Bres. 1920, fide Liberta (395).
Distribution United States: AZ, IA, NM.
Host *Pinus ponderosa*.
Ecology Decayed wood; soil.
Culture Characters
2.3.7.32.36.(39).46-47.(54).(55) (Ref. 588).
References 197, 208, 395.

Trechispora cohaerens (Schw.) Jülich & Stalpers 1980
Verh. Kon. Ned. Akad. Wetensch., Afd. Natuurk., Tweede Sect. 74:257.
Basionym *Sporotrichum cohaerens* Schw., Trans. Amer. Philos. Soc., N.S., 4:272, 1832.
Synonyms *Corticium confine* Bourd. & Galzin 1911, fide Jülich and Stalpers (326), *Trechispora confinis* (Bourd. & Galzin) Liberta 1966, *Leptosporomyces juniperinus* Gilbn. & Lindsey 1978, fide K.H. Larsson (in litt. 1992, a synonym of *confinis*).
Distribution Canada: ON, PQ. United States: AZ, CO, MI, MN, NM, PA, RI, VT, WI.
Hosts *Abies concolor, Agrostis tenuis, Alnus oblongifolia, Chlorococcum* sp., *Festuca rubra, Juniperus deppeana, Picea* sp., *P. mariana, P. engelmannii, P. resinosa, Populus* sp., *P. tremuloides, Quercus* sp., *Tsuga canadensis*, annual meadow-grass, fungus, moss, angiosperm, gymnosperm.
Ecology Decaying leaves and needles; decaying wood; dead wood; dead fallen tree; associated with circular patches on turf; associated with an algae; associated with a white rot.
Culture Characters
2.3c.26.32.36.38.47.54.55 (Ref. 474).
References 93, 208, 214, 216, 242, 310, 395, 402, 406, 474.

Most references use the epithet *confinis*; two (402, 474) use *cohaerens*. K.H. Larsson, Göteborg, after intensive study of this genus, concluded (pers. comm. 1992) that *confinis* and *cohaerens* are two distinct species. It appears that the references refer to one species, but we await the publication of Larsson's studies before deciding which epithet applies to the North American reports.

Trechispora farinacea (Pers.:Fr.) Liberta 1966
Taxon 15:318.
Basionym *Hydnum farinaceum* Pers., Synop. Meth. Fung., 562, 1801.
Synonyms *Grandinia farinacea* (Pers.) Bourd. & Galzin 1914, *Trechispora sphaerospora* (Maire) Parm. 1968, fide Liberta (395).
Anamorph Arthroconidia (395).
Distribution Canada: AB, MB, NS, NTM, ON, PQ.
United States: AZ, CA, CO, FL, IA, ID, IL, IN, MD, MI, MN, MS, MT, NC, NC-TN, NM, NY, OH, UT, VT, WA, WI.
Hosts *Abies balsamea, A. fraseri, A. magnifica, Acer* sp., *A. negundo, A. saccharum, Betula* sp., *B. alleghaniensis, B. papyrifera, Celtis* sp., *Fraxinus velutina, Juniperus deppeana, J. virginiana, Liquidambar styraciflua, Malus domestica, Phoenix canariensis, Picea* sp., *P. glauca, P. engelmannii, P. pungens, Pinus aristata* var. *longaeva, P. cembroides, P. contorta* var. *contorta, P. engelmannii, P. ponderosa, P. resinosa, P. sylvestris, Platanus wrightii, Populus* sp., *P. tremuloides, Pseudotsuga menziesii, Quercus* sp., *Q. gambelii, Rhododendron?* sp., *R. catawbiense, Tsuga* sp., *T. canadensis, T. mertensiana, Ulmus* sp., angiosperm, gymnosperm.
Ecology Bark; decayed wood; brown rotted wood; twig; branch; dead branch; slash; dead fallen tree; logs; associated with a white rot.
Culture Characters
2.3c.31d.32.36.38.(46).47.(48).54.55 (Ref. 474).
References 119, 128 as '*farinosa*' and 128 as '*shaerospora*', 130, 183, 197, 202, 204, 208, 212, 214, 216, 242, 249, 327, 388, 390, 395, 400, 402, 406, 407, 463, 474.

Trechispora fastidiosa (Pers.:Fr.) Liberta 1966
Taxon 15:318.
Basionym *Merisma fastidiosum* Pers., Comment. fungis clavaeformibus .... Wolf, Lipsiae, 97, 1797.
Synonym *Grandinia membranacea* Peck & G.W. Clinton 1879, fide Gilbertson (189).
Distribution Canada: ON.
United States: AZ, CA, NY, RI.
Hosts *Populus* sp., *Quercus hypoleucoides, Salix gooddingii, Tsuga mertensiana*.
Ecology Much decayed wood.
Culture Characters Not known.
References 130, 189, 216, 395, 406.

Trechispora filia (Bres.) Liberta 1966
Taxon 15:318.
Basionym *Corticium filia* Bres., Ann. Mycol. 6:43, 1908.
Distribution Canada: NT.
Host *Pinus contorta*.
Ecology Not known.
Culture Characters Not known.
Reference 242.

Trechispora invisitata (Jackson) Liberta 1966 Taxon 15:318.
Basionym *Corticium invistatum* Jackson, Canad. J. Res., C, 26:155, 1948.
Anamorph Blastoconidia (395).
Distribution Canada: ON.
Hosts Not known.
Ecology Duff, debris.
Culture Characters Not known.

Known only from the type specimen, which was redescribed by Liberta (395).

Trechispora lunata (Romell) Jülich 1975 Persoonia 8:293.
Basionym *Grandinia lunata* Romell in Bourd. & Galzin, Hym. France, 410, 1928.
Synonym *Athelopsis lunata* (Romell) Parm. 1968.
Distribution United States: AZ.
Host *Pseudotsuga menziesii.*
Ecology Not known.
Culture Characters Not known.
Reference 216.

Trechispora microspora (Karsten) Liberta 1966 Taxon 15:319.
Basionym *Grandinia microspora* Karsten, Bidrag Kännedom Finlands Natur Folk 48:365, 1889.
Synonyms *Athelia microspora* (Karsten) Gilbn. 1974, *Corticium subnullum* Burt 1926, fide Liberta (395).
Distribution Canada: BC, ON, PQ. United States: CO, OH, PA, RI.
Host *Populus* sp.
Ecology Bark of decaying tree.
Culture Characters Not known.
References 93, 393, 395.

The name *Grandinia microspora* Karsten has been applied to at least two fungi. In a monograph of *Trechispora*, Liberta (395), having seen a lectotype specimen, described the basidiospores as having "irregularly scattered obtuse warts". However, Eriksson's drawing in Hjortstam et al. (294) seems to have much smaller ornamentation on the spore wall than the warts described by Liberta and may be a second fungus. In addition, Gilbertson and colleagues (197, 216, 402, 447) in checklists and floras applied the name *Athelia microspora* to a smooth spored fungus. These records are cited under *A. microspora* sensu Gilbn.

Trechispora mollusca (Pers.:Fr.) Liberta 1973 Canad. J. Bot. 51:1878.
Basionym *Boletus molluscus* Pers., Synop. Meth. Fung., 547, 1801.
Synonyms *Cristella mollusca* (Pers.:Fr.) Donk 1967, *Polyporus candidissimus* Schw. 1832, fide Donk (158) and Liberta (395), *Trechispora candidissima* (Schw.) Bondartsev & Singer 1941, *Phlebiella candidissima* (Schw.) W.B. Cooke 1952, *Poria candidissima* (Schw.) Cooke 1886.
Distribution Canada: BC, MB, NB, NF, NS, NTM, ON, PE, PQ. United States: AZ, CA, CO, ID, LA, MA, MN, MO, MT, NC, NM, NY, OR, PA, SC, TN, VA, WA.
Hosts *Abies* sp., *A. amabilis, A. balsamea, A. concolor, A. lasiocarpa, A. lasiocarpa* var. *arizonica, A. magnifica, Acer* sp., *A. macrophyllum, Alnus* sp., *A. rubra, Betula* sp., *B. alleghaniensis, Fomes fomentarius, Larix* sp., *Magnolia* sp., *Picea* sp., *P. mariana. P. sitchensis, Pinus edulis, P. jefferyi, P. monticola, P. ponderosa, P. ponderosa* var. *jeffreyi, Populus* sp., *P. tremuloides, Pseudotsuga menziesii, Quercus* sp., *Q. garryana, Thuja* sp., *T. plicata, Tsuga* sp., *T. canadensis, T. heterophylla, Umbellularia californica*, angiosperm, gymnosperm.
Ecology Causes a minor decay, mostly associated with wounds on *Umbellaria californica*; basidiomes produced on decaying wood; branch; slash; logs; associated with a white rot.
Culture Characters Not known.
References 119, 120, 128, 130, 183, 197, 208, 214, 217, 242, 249, 274, 327, 395, 402, 406, 416.

Only four of the references (130, 327, 395, 402) use the epithet *mollusca* for this fungus. And we presume that they refer to the poroid *candidissima*-type of fungus. K.H. Larsson (pers. comm. 1992) is of the opinion that *candidissima* is distinct from *mollusca*. We await the publication of Larsson's studies before revising the above synonymy.

Trechispora praefocata (Bourd. & Galzin) Liberta 1966 Taxon 15:319.
Basionym *Corticium sphaerosporum* R. Maire subsp. *praefocatum* Bourd. & Galzin, Hym. France, 233, 1928.
Distribution United States: AZ, NM.
Host *Pinus ponderosa.*
Ecology Not known.
Culture Characters Not known.
References 197, 204, 216.

Trechispora regularis (Murrill) Liberta 1973 Canad. J. Bot. 51:1878.
Basionym *Poria regularis* Murrill, Mycologia 12:87, 1920.
Distribution United States: FL, GA, IA, LA, NC, PA, SC, TX.
Hosts *Liquidambar styraciflua*, angiosperm.
Ecology Associated with a white rot.
Culture Characters 1.(2).3c.16.26.31d.32.36.(38).40.47.54 (Ref. 474).
References 395, 416, 474.

Trechispora stellulata (Bourd. & Galzin) Liberta 1966 Taxon 15:319.
Basionym *Corticium stellulatum* Bourd. & Galzin, Bull. Soc. Mycol. France 27:263, 1911.
Distribution Canada: PQ.
Host *Picea* sp.

Ecology Not known.
Culture Characters Not known.
References 242, 395.

Trechispora subsphaerospora (Litsch.) Liberta 1973
Canad. J. Bot. 51:1887.
Basionym *Corticium subsphaerosporum* Litsch. in Keissler, Nat. Hist. Juan Fernandes and Easter Island 2:549, 1928.
Distribution Canada: PQ. United States: NY.
Hosts Not known.
Ecology Not known.
Culture Characters Not known.
Reference 395.

TREMELLA Pers. 1801 Tremellaceae
Synonym is *Naematelia* Fr. 1822, fide Donk (157).

Tremella aspera Coker 1920
J. Elisha Mitchell Sci. Soc. 35:141.
Distribution United States: NC.
Host *Quercus* sp.
Ecology Stump.
Culture Characters Not known.

Known only from the type specimen.

Tremella aurantia Schw.:Fr. 1822
Schriften Naturf. Ges. Leipzig 1:114.
Distribution Canada: BC.
United States: CA, IA, LA, NC, NJ, OR, SC, TX.
Hosts *Stereum hirsutum* complex, but reported on *Alnus rubra*, angiosperm, less commonly gymnosperm.
Ecology Apparently mycoparasitic, fide R. J. Bandoni (in litt.).
Culture Characters Not known.
References 16, 116, 242, 419, 443, 562.

Tremella auricularia Möller 1895
Protobasidiomyceten. In A.F.W. Schimper (Ed.), Botanischen Mittheilungen aus den Tropen 8:113. G. Fischer, Jena.
Distribution United States: NC.
Host *Ligustrum sinense*.
Ecology Not known.
Culture Characters Not known.
References 23, 107.

Closely related to *T. foliacea*, fide Bandoni and Oberwinkler (23).

Tremella carneoalba Coker 1920
J. Elisha Mitchell Sci. Soc. 35:146.
Distribution United States: GA, NC, OH.
Hosts *Carpinus* sp., *Ligustrum* sp., *Robinia* sp., angiosperm, probably *Mimosa* sp.
Ecology Decayed wood; twigs on bushes and trees; fallen branch.
Culture Characters Not known.
References 107, 410, 498.

Tremella coalescens Olive 1951
Mycologia 43:678.
Distribution United States: LA.
Host *Quercus* sp.
Ecology Corticated limb.
Culture Characters Not known.

Known only from the type specimen, see Lowy (419).

Tremella concrescens (Schw.:Fr.) Burt 1921
Ann. Missouri Bot. Gard. 8:362.
Basionym *Peziza concrescens* Schw., Schriften Naturf. Ges. Leipzig 1:118, 1822.
Synonyms *Dacrymyces pellucidus* Schw. 1832, *Corticium tremellinum* Berk. & Rav. 1873. Both fide Burt (90).
Distribution Canada: PQ. United States: AL, GA, IA, LA, MI, MN, MO, NC, PA, VT, ?WI.
Hosts Not known.
Ecology Rotten wood; ground; soil.
Culture Characters Not known.
References 90, 443, 533.

A *Sebacina*, fide Wells (625:320).

Tremella diaporthicola Ginns & Lefebvre, *nom. nov.*
Basionym *Sebacina globospora* Whelden, Rhodora 37:126, 1935.
Distribution United States: KY.
Host *Diaporthe* sp.
Ecology Presumably mycoparasitic; basidiomes emerging from ostioles of perithecia on twigs of *Fraxinus* sp.
Culture Characters Not known.

Known only from the type specimen. McGuire (451), after studying the type, concluded "To be referred to *Tremella*". The apparent mycoparasitic habit of this fungus and McGuire's conclusion after morphological study are sufficient evidence to transfer the name to *Tremella*. A new epithet is necessary because there already is a *T. globospora* D. Reid 1970.

Tremella encephala Pers.:Fr. 1801
Synop. Meth. Fung. 623.
Synonyms *Naematelia encephala* (Pers.) Fr. 1822, *Tremella encephaliformis* Willd. 1788, fide Bandoni (16), *Naematelia encephaliformis* (Willd.) Coker 1920.
Distribution Canada: AB-NT, BC, NS, ON, PE, PQ. United States: MA, NC, NH, NY, OR, PA, VT, WI.
Hosts *Stereum sanguinolentum*, but typically reported on *Abies* sp., *A. balsamea*, *Pinus* sp., *P. contorta*, *Pseudotsuga menziesii*, *Quercus* sp., angiosperm, gymnosperm.

Ecology Mycoparasitic on *Stereum sanguinolentum*, see Jülich (323), but basidiomes often appear to be on bark or wood.
Culture Characters Not known.
References 16, 90, 107, 116, 242, 249, 332, 443, 496, 532.

Tremella exigua Desm. 1847
Ann. Sci. Nat. Bot., Sér. III, 8:191.
Synonym *Tremella atrovirens* (Fr.) Sacc. 1888, fide Bandoni (16).
Distribution Canada: NS.
Hosts Pyrenomycetes, *Cucurbitaria* sp., *Salix* sp.
Ecology Old perithecia, ascostromata, etc., fide Bandoni (in litt.), perhaps mycoparasitic, see Jülich (323).
Culture Characters Not known.
References 116, 249.

Tremella flavidula Lloyd 1924
Mycol. Writ. 7:1276.
Distribution United States: MA.
Host *Viburnum cassinoides*.
Ecology Live twig.
Culture Characters Not known.

Known only from the type specimen. Bandoni (15) redescribed this specimen and suggested it was related to European specimens labelled *T. lutescens*.

Tremella foliacea Pers.:Fr. 1799
Obs. Mycol. 2:98.
Synonyms *Tremella frondosa* Fr. 1822, fide Wojewoda (648) and see Donk (157:249), *Tremella fimbriata* Pers.:Fr. 1799, fide Donk (157).
Distribution Canada: BC, NS, PQ. United States: AZ, GA, ?IA, ID, IN, LA, NC, NEW ENGLAND, OR.
Hosts *Abies balsamea*, *A. lasiocarpa*, *Acer rubrum*, *Alnus incana*, *Pinus strobus*, *Pseudotsuga menziesii*, *Quercus* sp., *Q. alba*, *Q. arizonica*, *Q. garryana*, *Q. hypoleucoides*, angiosperm, gymnosperm.
Ecology Dead twigs and branches; log; stump; uprooted stump.
Culture Characters Not known.
References 107, 116, 183, 216, 242, 249, 335 as *T. foliosa*, 410, 419, 443, 498, 501, 641.

Tremella fuciformis Berk. 1856
Hooker's J. Bot. Kew Gard. Misc. 8:277.
Distribution United States: AL, IA, LA, NC, SC.
Host *Quercus* sp.
Ecology Dead wood; limb; fallen branch; decaying log.
Culture Characters Not known.
References 90, 107, 444, 499.

Tremella globospora D. Reid 1970
Trans. Brit. Mycol. Soc. 55:414.
Distribution Canada: BC.
Hosts *Diaporthe* sp., but typically reported on angiosperms and gymnosperms.
Ecology Mycoparasitic, see Jülich (323).
Culture Characters Not known.
Reference 54.

Tremella grandibasidia Olive 1958
J. Elisha Mitchell Sci. Soc. 74:41.
Distribution United States: NC.
Host ?*Mimosa* sp.
Ecology Bark of fallen branch.
Culture Characters Not known.

Known only from the type specimen. Detailed description and illustrations in Olive (495).

Tremella mesenterica Retz.:Fr. 1769
Kongl. Vetensk. Acad. Handl. 30:249.
Synonym *Tremella lutescens* Pers.:Fr. 1798, fide Wong et al. (650).
Distribution Canada: AB, AB-NT, BC, MB, NF, NFL, NS, ON, PQ, YT. United States: AZ, CA, CO, GA, IA, ID, LA, MA, MI, NC, NY, OH, OR, WA, WI.
Hosts Associated with *Peniophora aurantiaca*, *P. cinerea*-like species, and *P. incarnata*, but typically reported as lignicolous on angiosperms and gymnosperms.
Ecology Apparently mycoparasitic, see Wong et al. (650).
Culture Characters Only a partial code is known .60 (Ref. 18).
References 18, 96, 107, 116, 119, 120, 183, 216, 242, 249, 334, 335, 406, 419, 443, 448, 465, 497, 498, 499, 532, 641, 650.

Tremella moriformis J.E. Smith & Sowerby:Fr. 1812
In J.E. Smith, English Botany 34:tab. 2446.
Synonym *Tremella colorata* Peck 1873, fide Coker (107).
Distribution United States: LA, NC, NY, SC.
Hosts *Fraxinus* sp., angiosperm.
Ecology Corticated dead branches.
Culture Characters Not known.
References 107, 419, 502.

Tremella mycetophiloides Y. Kobayasi 1939
Tokyo Bunr. Daig. Sci. Rep. 4:13.
Synonym *Tremella mycophaga* G.W. Martin 1940, fide Bandoni and Ginns (21).
Distribution Canada: ON, PQ. United States: NY.
Host *Aleurodiscus amorphus*.
Ecology Mycoparasitic.
Culture Characters Not known.
References 116, 437, 443.

The reports from Georgia and Louisiana were based upon specimens of *T. obscura*, see Martin (443).

Tremella obscura (Olive) M. Christiansen 1954
Friesia 5:62.
Basionym *Tremella mycophaga* var. *obscura* Olive, Mycologia 38:540, 1946.

Distribution United States: GA, LA.
Hosts *Dacryomitra stipitata, Dacrymyces* sp., *D. deliquescens, D. minor*, but reported on decorticated angiosperm and cedar.
Ecology Mycoparasitic within basidiomes.
Culture Characters Not known.
References 419, 497, 498, 499.

Tremella polyporina D. Reid 1970
Trans. Brit. Mycol. Soc. 55:416.
Distribution Canada: BC. United States: NY.
Host *Tyromyces lacteus.*
Ecology Presumably mycoparasitic; basidiomes lining the tubes of *T. lacteus.*
Culture Characters Not known.
References 345, 578.

Tremella reticulata (Berk.) Farlow 1908
Rhodora 10:9.
Basionym *Corticium tremellinum* var. *reticulatum* Berk., Grevillea 1:180, 1873.
Synonyms *Corticium reticulatum* (Berk.) Cooke 1891, a later homonym of *Corticium reticulatum* (Fr.) Fr. 1874, *Tremella vesicaria* Fr. sensu N. Amer. Aucts., fide Donk (157:334), *Tremella clavarioides* Lloyd 1908, fide Burt (90) and Bandoni (15), *Tremella sparassoidea* Lloyd 1919, fide Burt (90) and Bandoni (15).
Distribution Canada: MB, ON, PQ. United States: IA, LA, MI, MN, NC, NY, OH, PA, WI, VT.
Hosts Angiosperm.
Ecology Rotten wood; rotten stumps; soil; grassy woodland pasture; ground in shady woods.
Culture Characters Not known.
References 15, 90, 107, 419, 443, 510, 525, 532, 593.

The New York report (525), labelled *Tremella vesicaria*, was redetermined as *T. reticulata* by Donk (157).

Tremella rufobrunnea Olive 1948
Mycologia 40: 591.
Distribution United States: LA.
Hosts Angiosperm.
Ecology Dead corticated wood; dead wood.
Culture Characters Not known.
References 419, 499.

Tremella simplex Jackson & G.W. Martin 1940
Mycologia 32:687.
Distribution Canada: ON, PQ.
Host *Aleurodiscus thujae.*
Ecology Mycoparasitic.
Culture Characters Not known.
References 244, 437, 443.

The reports (207, 216) of *T. simplex* on *Dacrymyces minor* fruiting on *Prosopis juliflora*, and "parasitic on basidiocarps of other wood-inhabiting fungi" were probably misidentifications because the hosts for *T. simplex* do not include these species.

Tremella subanomala Coker 1920
J. Elisha Mitchell Sci. Soc. 35:148.
Distribution Canada: ON.
United States: IA, MI, MN, NY, NC, OR, TN, VT.
Hosts *Acer saccharinum, Alnus* sp., *Quercus macrocarpa*, angiosperm.
Ecology Twigs.
Culture Characters Not known.
References 107, 443, 562.

Perhaps the same as *Tremella indecorata* Sommerf., fide Bandoni and Oberwinkler (23). *Tremella indecorata* is a European species mycoparasitic on *Diatrype* sp., see Jülich (323).

Tremella tremelloides (Berk.) Massee 1889
J. Mycol. 1889:184.
Basionym *Sparassis tremelloides* Berk., Grevillea 2:6, 1873.
Synonyms *Naematelia aurantia* sensu Burt, *Naematelia quercina* Coker 1920. Both fide Bandoni (16).
Distribution United States: IA, LA, NC, NH, NJ, SC, TN, TX.
Hosts *Quercus* sp., angiosperm.
Ecology Fallen twig; wood in a woodpile.
Culture Characters Not known.
References 16, 90, 107.

The type of *Sparassis tremelloides* is "apparently an immature specimen of *Tremella reticulata*", fide Burdsall and Miller (71).

Tremella tubercularia Berk. 1860
Outlines British Fung., 288.
Distribution Canada: ON, PQ. United States: AZ, IA, KY, NEW ENGLAND, OH, TN, WI.
Hosts *Diaporthe* sp., *Eutypella* sp. and similar Pyrenomycetes; *Quercus* sp.
Ecology Presumably mycoparasitic on stromata.
Culture Characters Not known.
References 216, 443.

Sensu Martin (443) is *T. globospora* D. Reid, fide Wojewoda (648:185).

Tremella virens Schw. 1822
Schriften Naturf. Ges. Leipzig 4:115.
Distribution United States: NC.
Hosts *Cornus* sp., *Quercus* sp.
Ecology Corticated rotting branch; dead limbs.
Culture Characters Not known.
Reference 107.

Tremella wrightii Berk. & Curtis 1869
J. Linn. Soc., Bot. 10:341.
Distribution United States: FL.
Hosts Not known.
Ecology Not known.
Culture Characters Not known.
Reference 15.

TREMELLOSCYPHA D. Reid 1979 Sebacinaceae

Tremelloscypha gelatinosa (Murrill) Oberw. & Wells 1982
Mycologia 74:325.
Basionym *Eichleriella gelatinosa* Murrill, Ann. Missouri Bot. Gard. 2:748, 1915.
Distribution United States: FL.
Hosts Not known.
Ecology Decaying wood.
Culture Characters Not known.
Reference 629.

TREMISCUS (Pers.) Lév. 1846 Hyaloriaceae
Synonym is *Phlogiotis* Quél. 1886, fide Donk (157).

Tremiscus helvelloides (DC.:Fr.) Donk 1958
Taxon 7:164.
Basionym *Tremella helvelloides* DC., Flora France, Ed. 3, 2:93, 1805.
Synonyms *Phlogiotis helvelloides* (DC.) G.W. Martin 1936, *Gyrocephala rufa* (Jacq.) Bref. 1888, fide Donk (157).
Distribution Canada: BC, MB, NS, ON, PQ.
United States: CA, ID, NY, MI, OR, WA.
Hosts Gymnosperm.
Ecology Needles and rotten wood; very rotten wood; litter; saprophytic on ground.
Culture Characters Not known.
References 133, 119, 249, 335, 443, 532, 557.

TUBULICIUM Oberw. 1965 Tubulicrinaceae

Tubulicium capitatum (D.P. Rogers & Boquiren) Burdsall & Nakasone 1983
Mycotaxon 17:265.
Basionym *Epithele capitata* D.P. Rogers & Boquiren, Mycologia 63:942, 1971.
Synonym *Dextrinocystis capitata* (D.P. Rogers & Boquiren) Gilbn. & Blackwell 1988.
Distribution United States: FL, LA.
Hosts *Bambusa* sp., *Phoenix canariensis, Sabal palmetto.*
Ecology Dead culms of bamboo; dead leaves of standing palm; associated with a uniform white rot.
Culture Characters Not known.
References 75, 203.

Type species for *Dextrinocystis* Gilbn. & Blackwell.

Tubulicium vermiferum (Bourd.) Jülich 1979
As Oberw. ex Jülich, Persoonia 10:335.
Basionym *Peniophora vermifera* Bourd., Rev. Sci. Bourbonnais Centr. France 23:13, 1910.
Synonyms *Epithele vermifera* (Bourd.) Boquiren 1971, *Xenasma vermiferum* (Bourd.) Liberta 1960.
Distribution Canada: BC. United States: CA, FL.
Hosts *Antidesma platyphyllum, Calluna vulgaris, Cibotium chamissoi, Citharexylum fruticosum, Cocos nucifera, Dysoxylum spectabile, Erica arborea, E. cinerea, Eugenia axillaris, Gaultheria shallon, Livistonia* sp., *Metrosideros lucida, M. umbellata, Olearia furfuracea, Osmaronia cerasiformis, Pseudopanax crassifolium, Salix cinerea, Senecio rotundifolius, Vaccinium* sp., *Vitex lucens*, angiosperm.
Ecology Not known.
Culture Characters Not known.
References 49, 116, 384, 634.

TUBULICRINIS Donk 1956 Tubulicrinaceae

Tubulicrinis accedens (Bourd. & Galzin) Donk 1956
Fungus 26:14.
Basionym *Peniophora glebulosa* subsp. *accedens* Bourd. & Galzin, Bull. Soc. Mycol. France 28:386, 1913.
Synonym *Peniophora accedens* (Bourd. & Galzin) Wakef. & A. Pearson 1918.
Distribution Canada: BC, ON, PQ.
United States: CO, IA, IL, IN, MA, OR, RI.
Hosts *Abies balsamea, Picea* sp., *Pinus* sp., *P. strobus, Pseudotsuga menziesii, Tsuga canadensis*, angiosperm, gymnosperm.
Ecology Decayed wood.
Culture Characters Not known.
References 116, 242, 267, 387, 390, 584, 634.

Tubulicrinis angustus (D.P. Rogers & Weresub) Donk 1956
Fungus 26:14.
Basionym *Peniophora angusta* D.P. Rogers & Weresub, Canad. J. Bot. 31:764, 1953.
Distribution Canada: BC, ON, PQ.
United States: MA, OR, RI, WI.
Hosts *Fagus* sp., *Picea* sp., *P. mariana, Pinus contorta, P. strobus, Quercus* sp., *Tsuga canadensis*, angiosperm, gymnosperm.
Ecology Not known.
Culture Characters Not known.
References 242, 388, 584, 634, 635.

Tubulicrinis borealis Eriksson 1958
Symb. Bot. Upsal. 16 (1):79.
Distribution Canada: BC. United States: OR.
Hosts *Pinus contorta, Pseudotsuga menziesii.*
Ecology Not known.
Culture Characters
1.3c.(7).(26).32.36.38.47.55.58.61 (Ref. 259);
2.3c.8.13.(16).26.32.36.38.47.55.58 (Ref. 474).
References 255, 474.

Tubulicrinis calothrix (Pat.) Donk 1956
Fungus 26:14.
Basionym *Corticium calothrix* Pat., Cat. Rais. Pl. Cell. Tunis. 59, 1897.

Synonyms *Peniophora calothrix* (Pat.) D.P. Rogers & Jackson 1943, ?*Peniophora delectans* Overh. 1934, fide Rogers and Jackson (567) and Weresub (634), ?*Peniophora pirina* (Bourd. & Galzin) Bourd. & Galzin 1928, fide Rogers and Jackson (567) and Weresub (634).
Distribution Canada: AB, AB-NT, BC, ON, PQ. United States: AZ, CO, IA, IL, LA, MA, NM, NY, OR, PA, TN, WI.
Hosts *Abies* sp., *Picea* sp., *Pinus* sp., *P. cembroides, P. contorta, P. leiophylla, P. leiophylla* var. *chihuahuana, P. ponderosa, P. strobus, Pseudotsuga menziesii, Quercus gambelii, Tsuga canadensis*, angiosperm, gymnosperm.
Ecology Slash; logs; associated with a white rot.
Culture Characters
2a.3c.(7).(26).32.36.38.47.54.(55).58.61 (Ref. 259).
References 116, 183, 197, 204, 216, 242, 267, 388, 398, 402, 491, 584, 615, 634, 635.

Tubulicrinis chaetophorus (Höhnel) Donk 1956
Fungus 26:14.
Basionym *Hypochnus chaetophorus* Höhnel, Sitzungsber. Kaiserl. Akad. Wiss., Math.-Naturwiss. Kl., Abt. 1, 111:1007, 1902.
Synonyms *Peniophora chaetophora* (Höhnel) Höhnel & Litsch. 1907, *Peniophora dissoluta* Overh. 1934, fide Rogers and Jackson (567) and Weresub (634).
Distribution Canada: BC, ON, PQ.
United States: AZ, CA, ID, MA, NM, NY, OR, PA.
Hosts *Abies magnifica, Picea* sp., *Pinus monticola, P. ponderosa, Populus tremuloides, Thuja plicata, Tsuga mertensiana*, gymnosperm.
Ecology Rotting wood.
Culture Characters Not known.
References 120, 128, 130, 183, 197, 204, 216, 242, 584, 634, 635.

Tubulicrinis confusus Larsson & Hjort. 1986
Mycotaxon 26:437.
Distribution Canada: ON.
Hosts Not known.
Ecology Not known.
Culture Characters Not known.

Known only from the type specimens from Canada, Norway and Sweden.

Tubulicrinis corticioides (Overh.) Ginns & Lefebvre, *comb. nov.*
Basionym *Odontia corticioides* Overh., Mycologia 22:239, 1930.
Distribution United States: CO.
Hosts Gymnosperm.
Ecology Not known.
Culture Characters Not known.

Known only from the type specimen, which was redescribed by Gilbertson (190). A member of *Peniophora* sect. *tubuliferae*, very similar to *P. propinqua* and *P. juniperina*, fide Gilbertson (190). We agree with Gilbertson and transfer the epithet to *Tubulicrinis* where the latter species are now placed.

Tubulicrinis effugiens (Bourd. & Galzin) Oberw. 1965
Z. Pilzk. 31:35.
Basionym *Peniophora effugiens* Bourd. & Galzin, Bull. Soc. Mycol. France 28:386, 1913.
Distribution United States: CO.
Host *Populus angustifolia.*
Ecology Associated with a white rot.
Culture Characters Not known.
Reference 402.

Tubulicrinis globisporus Larsson & Hjort. 1978
Mycotaxon 7:123.
Distribution Canada: BC.
Host *Abies lasiocarpa.*
Ecology Fallen trunk.
Culture Characters Not known.

Known only from the three Swedish and one Canadian specimens.

Tubulicrinis gracillimus (D.P. Rogers & Jackson) G.H. Cunn. 1963
New Zealand Dept. Sci. Ind. Res. Bull. 145:141.
Basionym *Peniophora gracillima* Ellis & Ev. ex D.P. Rogers & Jackson, Farlowia 1:317, 1943.
Synonyms *Corticium glebulosum* (Fr.) Bres. 1898, a nom. conf., fide Rogers and Jackson (567), *Peniophora glebulosa* (Bres.) Sacc. & Syd. 1902, *Tubulicrinis glebulosus* (Bres.) Donk 1956.
Distribution Canada: BC, MB, NS, ON, PQ. United States: AZ, CA, CO, ID, IL, ME, MI, MT, NC, NC-TN, NE, NH, NJ, NM, NY, OR, VT, WA, WI.
Hosts *Abies* sp., *A. balsamea, A. lasiocarpa* var. *lasiocarpa, A. magnifica, Acer* sp., *A. saccharum, Alnus* spp., *Betula alleghaniensis, Fagus grandifolia, Juniperus* sp., *Picea* sp., *P. glauca, Pinus banksiana, P. contorta, P. edulis, P. ponderosa, Populus* sp., *Prunus* sp., *Pseudotsuga menziesii, Salix* sp., *Thuja plicata, Tsuga canadensis, T. heterophylla*, angiosperm, gymnosperm.
Ecology Bark; rarely bark; decaying wood; twig; slash; bark or between loose bark and wood of logs; associated with a white rot.
Culture Characters
2.3c.26.32.36.38.47.54.55.58 (Ref. 474);
1.3.7.34.36.38.46.55.64 (sibling species A);
2a.(2b).26.32.36.38.47.55.58.61 (sibling species B);
2a.(2b).26.32.36.38.47.54.58.61 (sibling species on angiosperm substrates).
References 92, 116, 119, 120, 128, 133, 183, 197, 204, 208, 216, 249, 255, 267, 327, 334, 388, 402, 474, 584, 635.

The 3 sibling species described by Hallenberg (259), known only from Sweden, are incompatible with each other.

Tubulicrinis hamatus (Jackson) Donk 1956 Fungus 26:14.
Basionym *Peniophora hamata* Jackson, Canad. J. Res., C, 26:133, 1948.
Distribution Canada: ON, PQ. United States: AZ.
Hosts *Abies balsamea, Picea* sp., *Pseudotsuga menziesii, Thuja occidentalis*, gymnosperm.
Ecology Not known.
Culture Characters Not known.
References 116, 216, 242, 302, 584, 635.

Tubulicrinis hirtellus (Bourd. & Galzin) Eriksson 1958 Symb. Bot. Upsal. 16 (1):82.
Basionym *Peniophora hirtella* Bourd. & Galzin, Bull. Soc. Mycol. France 28:386, 1913.
Distribution Canada: ON.
Hosts Gymnosperm.
Ecology Not known.
Culture Characters Not known.
Reference 584.

Tubulicrinis inornatus (Jackson & D.P. Rogers) Donk 1956 Fungus 26:14.
Basionym *Peniophora inornata* Jackson & D.P. Rogers, Canad. J. Res., C, 26: 139, 1948.
Distribution Canada: ON. United States: OR.
Hosts *Abies balsamea, Pinus* sp., *P. contorta, P. strobus, Pseudotsuga menziesii*, gymnosperm.
Ecology Not known.
Culture Characters Not known.
References 116, 242, 302, 584, 635.

Tubulicrinis juniperinus (Bourd. & Galzin) Donk 1956 Fungus 26:14.
Basionym *Peniophora glebulosa* subsp. *juniperina* Bourd. & Galzin, Bull. Soc. Mycol. France 28:386, 1913.
Synonym *Peniophora juniperina* (Bourd. & Galzin) Bourd. & Galzin 1928.
Distribution Canada: BC, ON, PQ. United States: ID, MA, OR.
Hosts *Picea* sp., gymnosperm.
Ecology Not known.
Culture Characters Not known.
References 242, 584, 634.

Tubulicrinis medius (Bourd. & Galzin) Oberw. 1966 Z. Pilzk. 31:26.
Basionym *Peniophora media* Bourd. & Galzin, Bull. Soc. Mycol. France 28:385, 1913.
Distribution Canada: YT. United States: CO, FL, LA.
Hosts *Juniperus virginiana, J. phoenicea, Picea glauca, Pinus edulis, P. ponderosa, Quercus gambelii.*
Ecology Decorticated branches; associated with a white rot.
Culture Characters 2a.(2b).3c.(25).26.32.36.38.47.55.58.61 (Ref. 259).
References 201, 202, 242, 402.

Tubulicrinis prominens (Jackson & Dearden) Donk 1956 Fungus 26:14.
Basionym *Peniophora prominens* Jackson & Dearden, Mycologia 43:57, 1951.
Distribution United States: AZ, ID, NM.
Hosts *Pinus monticola, P. ponderosa.*
Ecology Rotting wood.
Culture Characters Not known.
References 197, 204, 216, 309, 635.

Listed as a questionable synonym of *Hyphodontia (Grandinia) barba-jovis* by Jülich and Stalpers (326).

Tubulicrinis propinquus (Bourd. & Galzin) Donk 1956 Fungus 26:14.
Basionym *Peniophora cretacea* subsp. *propinqua* Bourd. & Galzin, Hym. France, 288, 1928.
Synonym *Peniophora propinqua* (Bourd. & Galzin) Laurila 1939.
Distribution Canada: MB, ON, PQ. United States: MA, MN, NM, NY.
Hosts *Abies concolor, Picea* sp., *P. glauca*, gymnosperm.
Ecology Not known.
Culture Characters Not known.
References 183, 208, 214, 242, 584.

Tubulicrinis regificus (Jackson & Dearden) Donk 1956 Fungus 26:14.
Basionym *Peniophora regifica* Jackson & Dearden, Mycologia 43:57, 1951.
Distribution United States: OR.
Hosts Gymnosperm.
Ecology Not known.
Culture Characters Not known.

Known only from the type specimen, which was redescribed by Weresub (635).

Tubulicrinis sceptriferus (Jackson & Weresub) Donk 1956 Fungus 26:14.
Basionym *Peniophora sceptrifera* Jackson & Weresub, Canad. J. Bot. 31:772, 1953.
Distribution Canada: ON. United States: WI.
Hosts *Tsuga canadensis*, gymnosperm, leafy liverwort.
Ecology Rotten wood.
Culture Characters Not known.
References 388, 584, 634, 635.

Tubulicrinis subulatus (Bourd. & Galzin) Donk 1956 Fungus 26:14.
Basionym *Peniophora glebulosa* subsp. *subulata* Bourd. & Galzin, Bull. Soc. Mycol. France 28:386, 1913.
Synonym *Peniophora subulata* (Bourd. & Galzin) Donk 1931.
Distribution Canada: AB, AB-NT, BC, ON, PQ. United States: AZ, CA, CO, ID, IL, MA, MT, NM, NY, OH, OR, RI, WA.

Hosts *Abies balsamea, A. concolor, A. magnifica, Picea* sp., *P. pungens, Pinus banksiana, P. contorta, P. engelmannii, P. ponderosa, P. resinosa, P. strobus*, angiosperm, gymnosperm.
Ecology Slash; associated with a white rot.
Culture Characters
2a.2b.3c.13.32.36.38.47.55.58.61 (Ref. 259, old spruce forest type);
2a.2b.3c.26.32.36.38.47.55.60.61 (Ref. 259, mixed forest type).
References 120, 128, 183, 197, 204, 208, 216, 242, 255, 267, 402, 584.

The two forest types described by Hallenberg (259), known only from Sweden and Norway, were incompatible.

TULASNELLA Schröter 1888 Tulasnellaceae
Synonym is *Gloeotulasnella* Höhnel & Litsch. 1908, fide Olive (506) and Donk (157).

Tulasnella aggregata (Olive) Olive 1957
Mycologia 49:677.
Basionym *Gloeotulasnella aggregata* Olive, Bull. Torrey Bot. Club 80:40, 1953.
Distribution United States: NC.
Hosts Not known.
Ecology Rotting, decorticated wood.
Culture Characters Not known.

Known only from the type specimen (503).

Tulasnella albida Bourd. & Galzin 1928
Hym. France, 59.
Distribution United States: NC.
Host *Quercus* sp.
Ecology Underside of rotten decorticated log.
Culture Characters Not known.
Reference 495.

Perhaps only a variant of *T. araneosa*, fide Olive (506).

Tulasnella allantospora Wakef. & A. Pearson 1923
Trans. Brit. Mycol. Soc. 8:220.
Distribution Canada: ON, PQ. United States: AZ, IA, MA, MO, NC, NM, NY, OH.
Hosts *Pinus ponderosa, Quercus* sp., *Q. hypoleucoides, Q. macrocarpa*, angiosperm, rarely gymnosperm.
Ecology Not known.
Culture Characters Not known.
References 183, 197, 216, 443, 506, 532, 561.

Tulasnella araneosa Bourd. & Galzin 1924
Bull. Soc. Mycol. France 39:265.
Distribution Canada: ?ON. United States: IA, NC.
Hosts *Alnus* sp., *Fraxinus* sp., *Prunus* sp., *Tilia* sp., angiosperm.
Ecology Not known.
Culture Characters Not known.
References 443, 506, 561.

Tulasnella bifrons Bourd. & Galzin 1924
Bull. Soc. Mycol. France 39:264.
Distribution Canada: ON.
United States: IA, MA, MO, NC, OH, OR.
Hosts *Acer rubrum* var. *rubrum, Alnus* sp., *Betula* sp., *Pinus* sp., *Quercus* sp., *Ulmus* sp., polypore.
Ecology Underside of moist rotted wood; hymenium of decaying conk.
Culture Characters Not known.
References 183, 443, 495, 561.

Perhaps only a variant of *T. araneosa*, fide Olive (506).

Tulasnella calospora (Boudier) Juel 1897
Bih. Kongl. Svenska Vetensk.-Akad. Handl. 23:23.
Basionym *Prototremella calospora* Boudier, J. Bot. (Morot) 10:85, 1896.
Synonym *Gloeotulasnella calospora* (Boudier) D.P. Rogers 1933.
Anamorph *Epulorhiza repens* (N. Bernard) R.T. Moore 1987 (≡ *Rhizoctonia repens* N. Bernard 1909).
Distribution Canada: AB, ON.
United States: IA, ME, NC, NH, OR, WI.
Hosts *Calopogon* sp., *Goodyera* sp., *Habenaria* sp., *Pinus contorta, Platanthera obtusata, Pogonia* sp., *Quercus* sp., *Spiranthes* sp., angiosperm, gymnosperm, mosses, polypore.
Ecology Mycorrhizal with species of the Orchidaceae; basidiomes produced on old conk; dead mosses.
Culture Characters Not known.
References 93, 124, 138, 183, 388, 443, 496, 561.

Tulasnella caroliniana (Olive) Olive 1957
Mycologia 49:676.
Basionym *Gloeotulasnella caroliniana* Olive, Bull. Torrey Bot. Club 80:41, 1953.
Distribution United States: NC.
Hosts Angiosperm.
Ecology Decorticated limbs.
Culture Characters Not known.
Reference 503.

Tulasnella conidiata Olive 1957
Mycologia 49:675.
Distribution United States: NC.
Hosts Not known.
Ecology Old soggy piece of heartwood.
Culture Characters Not known.
References 495.

Known only from the type specimen. In the reference the name appears as *Gloeotulasnella cystidiocarpa*, but that name is not validly published (506).

Tulasnella curvispora Donk 1966
Persoonia 4:263.
Synonym *Tulasnella rutilans* sensu D.P. Rogers 1933, fide Donk (157).
Distribution Canada: ON. United States: IA.
Hosts *Quercus* sp., angiosperm.
Ecology Decorticated wood.
Culture Characters Not known.
References 443, 561.

Tulasnella cystidiophora Höhnel & Litsch. 1906
Sitzungsber. Kaiserl. Akad. Wiss., Math.-Naturwiss. Kl., Abt. 1, 115 (1):1557.
Synonym *Gloeotulasnella cystidiophora* (Höhnel & Litsch.) Höhnel & Litsch. 1908.
Distribution Canada: ON. United States: IA, MA, NC.
Hosts Angiosperm.
Ecology Decorticated wood.
Culture Characters Not known.
References 443, 501, 506.

Tulasnella eichleriana Bres. 1903
Ann. Mycol. 1:113.
Distribution Canada: MB, ON.
United States: IA, ID, MA, NH, NY, PA, WA.
Hosts *Betula* sp., *B. papyrifera, Populus* sp., *Quercus* sp., *Salix* sp., angiosperm.
Ecology Rotting wood and bark.
Culture Characters Not known.
References 88, 116, 183, 516, 560.

Tulasnella fuscoviolacea Bres. 1900
Fungi Tridentini 2:98.
Distribution Canada: AB-NT, ON, PE. United States: CA, IA, ID, MN, NH, NY, OH, PA, WA.
Hosts *Abies* sp., *A. balsamea, A. magnifica, Picea* sp., *P. rubens, Pinus albicaulis, P. contorta, P. strobus, Tsuga heterophylla*, angiosperm, gymnosperm.
Ecology Bark; rotten wood; pine cone.
Culture Characters Not known.
References 88, 120, 183, 242, 443, 506, 511, 516, 561.

Tulasnella hyalina Höhnel & Litsch. 1908
Sitzungsber. Kaiserl. Akad. Wiss., Math-Naturwiss. Kl., Abt. I, 117 (1):34.
Synonym *Gloeotulasnella metachroa* Bourd. & Galzin 1924, fide Olive (506).
Distribution United States: IA, NC.
Hosts Angiosperm, gymnosperm.
Ecology Not known.
Culture Characters Not known.
References 443, 506.

Tulasnella lactea Bourd. & Galzin 1924
Bull. Soc. Mycol. France 38:263.
Distribution United States: IA, MO.
Hosts *Crataegus* sp., *Malus* sp., *Quercus* sp., *Salix* sp., angiosperm.
Ecology Not known.
Culture Characters Not known.
References 443, 506, 560, 561.

Tulasnella papillata (Olive) Olive 1957
Mycologia 49:676.
Basionym *Gloeotulasnella papillata* Olive, Bull. Torrey Bot. Club 81:335, 1954.
Distribution United States: NC.
Host *Quercus* sp.
Ecology Decorticated wood.
Culture Characters Not known.

Known only from the type specimen.

Tulasnella pinicola Bres. 1903
Ann. Mycol. 1:114.
Synonyms *Gloeotulasnella pinicola* (Bres.) D.P. Rogers 1933, *Tulasnella tulasnei* (Pat.) Juel sensu D.P. Rogers 1932.
Distribution Canada: ON. United States: GA, IA, LA, MN, NC, NY, OH, TN, VT, WY.
Hosts *Betula nigra, Carya* sp., *Cytisus* sp., *Liriodendron tulipifera, Mimosa* sp., *Quercus* sp., *Q. rubra*, angiosperm, gymnosperm, polypore
Ecology Decaying bark and wood; decorticated wood; old polypores.
Culture Characters Not known.
References 443, 497, 498, 499, 506, 527, 560, 561.

Tulasnella pruinosa Bourd. & Galzin 1924
Bull. Soc. Mycol. France 39:264.
Distribution Canada: ON, PQ.
United States: IA, MT, NC, NH, NY, OH, WA, WI.
Hosts *Acer* sp., *Betula nigra, Carpinus caroliniana, Castanea* sp., *Fagus* sp., *Picea* sp., *P. engelmannii, Quercus* sp., *Salix* sp., *Ulmus* sp., angiosperm.
Ecology Not known.
Culture Characters Not known.
References 133 as "*pruinata*", 183, 443, 448, 506, 532, 561.

Tulasnella rogersii (Olive) Olive 1957
Mycologia 49:677.
Basionym *Gloeotulasnella rogersii* Olive, Mycologia 43: 689, 1951.
Distribution United States: LA.
Hosts Angiosperm.
Ecology Corticated and decorticated wood.
Culture Characters Not known.

Known only from the holotype and one paratype specimen.

Tulasnella species A
Anamorph *Epulorhiza anaticula* (Currah) Currah, Canad. J. Bot. 68:1174, 1990 (≡ *Rhizoctonia anaticula* Currah, Canad. J. Bot. 65:2474-2476, 1987).

Distribution Canada: AB.
Hosts *Calypso bulbosa, Coeloglossum viride, Platanthera dilatata, P. hyperborea, P. obtusata.*
Ecology Endophyte, isolated from healthy root and tuber tissue.
Culture Characters Not known, but see Currah (loc. cit.).

Produces monilioid cells in culture which are distinct from those of the over 30 known orchid endophytic rhizoctonias.

Tulasnella traumatica (Bourd. & Galzin) Olive 1957
Mycologia 49:677.
Basionym *Gloeotulasnella traumatica* Bourd. & Galzin, Bull. Soc. Mycol. France 25:32, 1909.
Synonym *Gloeotulasnella opalea* D.P. Rogers 1933, fide Olive (496).
Distribution Canada: ON. United States: IA, NC.
Hosts *Platanus* sp., angiosperm.
Ecology Large fallen limb; underside of old logs.
Culture Characters Not known.
References 443, 496.

Tulasnella tremelloides Wakef. & A. Pearson 1918
Trans. Brit. Mycol. Soc. 6:70.
Synonym *Gloeotulasnella tremelloides* (Wakef. & A. Pearson) D.P. Rogers 1933.
Distribution United States: IA.
Hosts Not known.
Ecology Wood and plant litter.
Culture Characters Not known.
References 443, 506.

Tulasnella violacea (Johan-Olsen) Juel 1897
Bih. Kongl. Svenska Vetensk.-Akad. Handl. 23:22.
Basionym *Pachysterigma violacea* Johan-Olson in J.O. Brefeld, Untersuch. Gesammtgeb. Mykol. 8 (3):6, 1888.
Distribution Canada: NB, NS, ON.
United States: CA, GA, IA, NC, NEW ENGLAND.
Hosts *Abies balsamea, Quercus* sp., angiosperm, herbs.
Ecology Bark; limb; fallen limb; sticks.
Culture Characters Not known.
References 130, 242, 249, 443, 496, 498, 506, 561.

Tulasnella violea (Quél.) Bourd. & Galzin 1909
Bull. Soc. Mycol. France 25:32.
Basionym *Hypochnus violeus* Quél., Assoc. Franç. Avancem. Sci. (La Rochelle) 11:401, 1883.
Distribution Canada: AB, AB-NT, BC, MB, NS, ON. United States: AZ, CO, CT, GA, IA, ID, IL, MA, ME, MO, NC, NH, NM, NY, OH, OR, PA, SC, VT, WA, WI.
Hosts *Acer* sp., *Acer rubrum* var. *rubrum, Alnus* sp., *Betula* sp., *Carya* sp., *C. ovata, Fagus* sp., *F. grandifolia, Fraxinus velutina, Holodiscus discolor, Liriodendron tulipifera, Nyssa* sp., *N. sylvatica, Picea pungens, Pinus banksiana, P. contorta, P. ponderosa, Polyporus* sp., *Populus* sp., *Quercus* sp., *Q. hypoleucoides, Salix* sp., angiosperm, lichens, polypores.
Ecology Bark; very rotten wood; limb; fallen branches; old conks; humus.
Culture Characters Not known.
References 88, 116, 119, 183, 197, 216, 242, 249, 402, 443, 448, 498, 503, 506, 512, 516, 560, 561.

TYLOSPORA Donk 1960 — Atheliaceae

Tylospora asterophora (Bonorden) Donk 1960
Taxon 9:220.
Basionym *Hypochnus asterophora* Bonorden, Handb. Allg. Mykol., 160, 1851.
Distribution Canada: ON, PQ. United States: OR.
Host *Picea* sp.
Ecology Not known.
Culture Characters Not known.
References 242, 312.

Tylospora fibrillosa (Burt) Donk 1960
Taxon 9:220.
Basionym *Hypochnus fibrillosus* Burt, Ann. Missouri Bot. Gard. 3:238, 1916.
Distribution Canada: ON, PQ.
Hosts *Picea* sp., *Pinus banksiana*
Ecology Mycorrhizal, at least, with species of *Picea* (602).
Culture Characters Not known.
References 242, 312.

UTHATOBASIDIUM Donk 1956 — Ceratobasidiaceae

Uthatobasidium fusisporum (Schröter) Donk 1958
Fungus 28:22.
Basionym *Hypochnus fusispora* Schröter in F. Cohn, Kryptogamen-Flora Schlesien 3 (1):416, 1888.
Synonyms *?Hypochnus flavescens* Bonorden 1851, *Botryobasidium flavescens* (Bonorden) sensu D.P. Rogers 1935, fide Donk (153), *Pellicularia flavescens* (Bonorden) sensu D.P. Rogers 1943 pro parte, see Talbot (599:391).
Distribution Canada: BC, MB, ON, PQ. United States: CA, FL, IA, MA, ME, MO, NH, NY, PA.
Hosts *Abies balsamea, A. magnifica, Acer negundo, A. spicatum, Betula* sp., *Carpinus caroliniana, Carya* sp., *Juglans cinerea, Phytolacca decandra, Picea* sp., *P. glauca, Pinus strobus, Populus* sp., *?P. balsamifera,*

Populus tremuloides, Salix sp., *Ulmus americana, Ustulina vulgaris*, angiosperm.
Ecology Rotten wood; decaying wood.
Culture Characters
2.6.(21).(24).(25).26.32.36.38.42.(54).(55) (Ref. 588).
References 116, 120, 128, 183, 406, 563, 565.

There is uncertainty over the identity of *Hypochnus flavescens*, fide Jülich and Stalpers (326), and Talbot (599). Talbot has equated it with *Pellicularia flavescens* sensu D.P. Rogers and we follow Talbot. Unfortunately, Rogers (565) had a broad species concept, see Talbot, which included *U. ochraceum*, thus some of Rogers' records are of *U. ochraceum*.

Uthatobasidium ochraceum (Massee) Donk 1958
Fungus 28:23.
Basionym *Coniophora ochracea* Massee, J. Linn. Soc., Bot. 25: 137, 1889.
Synonym *Botryobasidium ochraceum* (Massee) Donk in D.P. Rogers 1935.
Distribution Canada: BC. United States: MI.
Host *Picea glauca.*
Ecology Bark and dead wood.
Culture Characters Not known.
References 242, 563.

VARARIA Karsten 1903 Lachnocladiaceae
Synonym is *Asterostromella* Höhnel & Litsch. 1907, fide Rogers and Jackson (567).

Vararia athabascensis Gilbn. 1970
Madroño 20:282.
Distribution Canada: AB. United States: AZ, UT.
Hosts *Picea engelmannii, Pinus aristata* var. *longaeva, P. contorta, Populus* sp., *P. tremuloides.*
Ecology Small twigs and branches on the ground; fallen large branches and logs; associated with a white rot.
Culture Characters
2.3.8.15.36.38.44.55.58.61 (Ref. 45);
2.3c.15(p).31g.32.36.38.46-47.(50).54.55.58 (Ref. 474).
References 45, 194, 216, 406, 407, 448, 474.

Vararia fibra Welden 1965
Mycologia 57:507.
Distribution United States: AZ.
Hosts *Juniperus deppeana, Quercus* sp.
Ecology Dead fallen trees; associated with a white rot.
Culture Characters Not known.
Reference 212.

Vararia gomezii Boidin & Lanquetin 1975
Bull. Soc. Mycol. France 91:462.
Distribution United States: FL, LA.
Hosts *Juniperus phoenicea, J. virginiana.*
Ecology Dead branches on live tree; associated with a white rot.
Culture Characters Not known.
References 201, 202.

Vararia investiens (Schw.) Karsten 1898
Kritisk öfversigt Finlands Basidsvampar 3:32.
Basionym *Radulum investiens* Schw., Trans. Amer. Philos. Soc., N.S., 4:165, 1832.
Synonyms *Corticium investiens* (Schw.) Bres. 1897, *Corticium alutarium* Berk. & Curtis 1873, fide Burt (93), and Rogers and Jackson (567).
Distribution Canada: NB, NF, NS, ON, PQ. United States: CO, CT, DC, IA, ID, IN, MA, MD, ME, MI, MN, MT, NC, NH, NY, PA, RI, VT, WA, WI, WY.
Hosts *Abies balsamea, A. grandis, A. lasiocarpa, Acer rubrum, A. rubrum* var. *rubrum, A. saccharum, A. spicatum, Alnus* sp., *Betula alleghaniensis, Castanea dentata, Fagus* sp., *F. grandifolia, Carya (Hicoria) alba, Larix* sp., *L. occidentalis, Liriodendron tulipifera, Paxistima myrsinites, Picea* sp., *P. engelmannii, P. pungens, P. rubens, P. rubra, Pinus* sp., *P. albicaulis, P. contorta, P. monticola, P. strobus, Populus* sp., *P. tremuloides, Prunus* sp., *Quercus* sp., *Q. alba, Q. borealis, Q. rubra, Q. rubra* var. *borealis, Rhododendron canadense, Tsuga canadensis*, angiosperm, gymnosperm.
Ecology Fallen twigs and branches; logs; associated with a white rot.
Culture Characters
2a.3c.15.25d.32.37.38.43.55.58.61 (Ref. 45);
2.3c.8v.15p.32.37.38.44-45.(53).54.55.58 (Ref. 474).
References 45, 93, 183, 193, 214, 227, 242, 249, 402, 406, 435, 448, 474, 618.

Burt's (93: 285) citation "Pennsylvania: Michener, type of *Corticium alutarium* (in Curtis Herb., 6349)" effectively designated a lectotype for *C. alutarium*. The Curtis herbarium is at FH. Hjortstam (288) did not find the "type" of *C. alutarium* at K.

Vararia pectinata (Burt) D.P. Rogers & Jackson 1943
Farlowia 1:298.
Basionym *Corticium pectinatum* Burt, Ann. Missouri Bot. Gard. 13:286, 1926.
Distribution United States: FL, MD.
Hosts *Vitis* sp., angiosperm.
Ecology Dead branches; fallen wood; associated with a white rot.
Culture Characters
2a.(2b).3c.9.15.22.25d.32.37.38.43.54.58.61 (Ref. 45).
References 93, 193, 618.

Vararia peniophoroides (Burt) D.P. Rogers & Jackson 1943
Farlowia 1:294.
Basionym *Hypochnus peniophoroides* Burt, Ann. Missouri Bot. Gard. 3:234. 1916.
Distribution United States: LA.
Hosts Not known.
Ecology Not known.
Culture Characters Not known.

Known from only the holotype from Jamacia and the Louisiana paratype.

Vararia phyllophila (Massee) D.P. Rogers & Jackson 1943 Farlowia 1:323.

Basionym *Peniophora phyllophila* Massee, J. Linn. Soc., Bot. 25:150, 1889.

Distribution United States: FL, ID, NM, SC.

Hosts *Pinus monticola, Populus* sp., *P. tremuloides, Quercus* sp., angiosperm, gymnosperm.

Ecology Dried leaves; fallen leaves and limbs; wood on ground.

Culture Characters Not known.

References 92, 183, 193, 208, 406, 618.

Vararia racemosa (Burt) D.P. Rogers & Jackson 1943 Farlowia 1:299.

Basionym *Corticium racemosa* Burt, Ann. Missouri Bot. Gard. 13:287, 1926.

Distribution Canada: AB, BC.
United States: AZ, CO, ID, MT, UT, WA.

Hosts *Abies* sp., *A. grandis, A. lasiocarpa, A. lasiocarpa* var. *arizonica, A. lasiocarpa* var. *lasiocarpa, Larix occidentalis, Picea* sp., *P. engelmannii, Pinus contorta, Populus* sp., *P. tremuloides, Pseudotsuga menziesii, Thuja plicata*, angiosperm, gymnosperm.

Ecology Decaying wood and bark; slash; logs; associated with a white rot.

Culture Characters
2.4.7.34.36.38.(39).43-44.(50).54.55 (Ref. 474).

References 93, 116, 193, 204, 216, 242, 402, 406, 448, 474.

Vararia sphaericospora Gilbn. 1965 Pap. Michigan Acad. Sci. 50:176.

Distribution United States: AL, FL, GA, LA.

Hosts *Acer* sp., *Liquidambar styraciflua, Pinus* sp., angiosperm.

Ecology Bark; root; associated with a uniform white rot.

Culture Characters
2.3c.15.25d.34.36.38.42.54.59.61 (Ref. 45);
2.3c.7.34.36.38.41-42.(51).54.55.60 (Ref. 474).

References 193, 203, 474.

Vararia thujae Boidin 1981 Naturaliste Canad. 108:201.

Distribution Canada: ON, PQ.

Hosts *Thuja* sp., *T. occidentalis.*

Ecology Bark; live and dead twigs.

Culture Characters Not known.

Known only from the holotype and one paratype specimen.

Vararia tropica Welden 1965 Mycologia 57:516.

Distribution United States: AZ.

Hosts *Prosopis juliflora, P. velutina.*

Ecology Associated with a white rot.

Culture Characters
2.6.15.25d.32.36.(39).45.64 (Ref. 45);
2.6.7.32.36.38.47.54 (Ref. 474).

References 207, 474.

VELUTICEPS Pat. 1894 Stereaceae

Synonym is *Columnocystis* Pouzar 1959, fide Hjortstam and Tellería (297), and Nakasone (473).

Veluticeps abietina (Pers.:Fr.) Hjort. & Tellería 1990 Mycotaxon 37:54.

Basionym *Thelephora abietina* Pers., Synop. Meth. Fung. 573, 1801.

Synonyms *Stereum abietinum* (Pers.) Fr. 1838, *Columnocystis abietina* (Pers.) Pouzar 1959, *Hymenochaete abnormis* Peck 1889, fide Burt (89), Lentz (379) and Nakasone (473).

Distribution Canada: AB, BC, MB, NF, NS, NT, NTM, ON, PQ, YT.
United States: AK, AZ, CA, CO, ID, MI, MT, NH, NY, OR, PA, TN, UT, VT, WA, WI, WY.

Hosts *Abies* sp., *A. amabilis, A. balsamea, A. grandis, A. lasiocarpa, A. lasiocarpa* var. *arizonica, A. lasiocarpa* var. *lasiocarpa, A. magnifica, A. procera, Larix* sp., *L. occidentalis, Picea* sp., *P. engelmannii, P. glauca, P. mariana, P. rubens, P. sitchensis, Pinus* sp., *P. banksiana, P. contorta, P. murrayana, P. ponderosa, Pseudotsuga menziesii, Thuja plicata, Tsuga* sp., *T. canadensis, T. heterophylla, T. mertensiana*, gymnosperm.

Ecology One of the four most important brown heart rot fungi in *Tsuga heterophylla*; causes minor trunk rot in *Abies* species and a minor root rot in *Picea mariana*; basidiomes produced on bark; wood and logs; decorticated log; fence timber; associated with a brown rot.

Culture Characters
1.3c.8.9.27.34.37.39.47.50.55 (Refs. 473 and 474);
1.3c.9.13.27.34.37.39.47.50.55 (Ref. 473);
1.3i.8.34.37.39.47.53.55.57.66 (Ref. 104);
1.3.(4).(5).7.(8).34.37.39.47.53.55 (Ref. 487).

References 89, 97, 104, 116, 130, 183, 197, 199, 216, 249, 274, 334, 379, 402, 448, 473, 474, 484, 511, 519.

Nakasone (473) segregated *V. fimbriata* from the *V. abietina* complex, thus some of the western North American reports of *V. abietina* may have included *V. fimbriata.*

Veluticeps ambigua (Peck) Hjort. & Tellería 1990 Mycotaxon 37:54.

Basionym *Stereum ambiguum* Peck, Annual Rep. New York State Mus. 47:145, 1894.

Synonym *Columnocystis ambigua* (Peck) Pouzar 1959.

Distribution Canada: ON. United States: ID, ME, MT, NC, NH, NY, PA, TN, VT.

Hosts *Abies* sp., *A. fraseri, A. lasiocarpa, Picea* sp., *P. glauca, P. rubens, Pinus* sp., ?*P. strobus, Tsuga* sp., *T. canadensis*, gymnosperms.

Ecology Bark; moss-covered wood; cut wood; butt; log; associated with a brown rot.
Culture Characters
1.3r.7.32.37.39.47.55.57.66 (Ref. 104).
References 59, 89, 104, 183, 199, 327, 448, 473, 511.

Veluticeps berkeleyi Cooke 1880
Grevillea 8:149.
Basionym *Nom. nov.* for *Hymenochaete veluticeps* Berk. & Curtis, J. Linn. Soc., Bot. 10:333, 1869.
Synonym *Veluticeps fusca* C.J. Humphrey & W.H. Long in Burt 1926, fide Gilbertson et al. (215).
Distribution United States: AZ, NM, SD, WA.
Hosts *Abies concolor, Pinus* sp., *P. engelmannii, P. ponderosa.*
Ecology Causes a heart rot in live *Pinus ponderosa* and a sapwood rot of down timber; basidiomes produced on charred wood; trunks; stumps; logs; associated with a brown cubical rot.
Culture Characters
1.3c.8.9.21.34.37.38.(39).47.(50).(51).(53).55.60 (Ref. 474).
References 93, 197, 199, 208, 215, 216, 274, 473, 516.

Veluticeps fimbriata (Ellis & Ev.) Nakasone 1990
Mycologia 82:634.
Basionym *Hymenochaete fimbriata* Ellis & Ev., J. Mycol. 1:149, 1885.
Synonyms *Hymenochaete rugispora* Ellis & Ev. 1890, fide Chamuris (104), and Nakasone (473), *Stereum rugisporum* (Ellis & Ev.) Burt 1920.
Distribution Canada: BC.
United States: AK, AZ, CA, CO, ID, OR, WA, WY.
Hosts *Abies* sp., *A. concolor, A. lasiocarpa, A. magnifica, Larix* sp., *Picea* sp., *Picea sitchensis, Pinus* sp., *P. contorta, P. contorta* var. *latifolia, P. monticola, Picea engelmannii, Tsuga mertensiana*, gymnosperm.
Ecology Associated with a brown rot.
Culture Characteristics:
2.3r.8.9.31c.34.37.39.46.50.55 (Ref. 473);
2.3r.9.13.31c.34.37.39.46.50.55 (Ref. 473).
References 89, 97, 116, 120, 121, 128, 133, 402, 473, 514.

Veluticeps pimeriensis (Gilbn.) Hjort. & Tellería 1990
Mycotaxon 37:54.
Basionym *Columnocystis pimeriensis* Gilbn., Fungi That Decay Ponderosa Pine, 87, 1974.
Distribution United States: AZ, NM.
Host *Pinus ponderosa.*
Ecology Apparently associated with a brown cubical rot.
Culture Characters Not known.
References 197, 199, 216.

VESICULOMYCES Hagström 1977 Gloeocystidiellaceae

Vesiculomyces citrinum (Pers.) Hagström 1977
Bot. Not. 130:53.
Basionym *Thelephora citrina* Pers., Mycol. Eur. 1:136, 1822.
Synonyms *Gloeocystidiellum citrinum* (Pers.) Donk 1956, *Corticium radiosum* (Fr.) Fr., 1838, fide Eriksson and Ryvarden (173), and McKay and Lentz (452), *Gloeocystidiellum radiosum* (Fr.) Boidin 1957.
Distribution Canada: AB, AB-NT, BC, NB, NS, ON, PQ. United States: AK, AR, AZ, CA, CO, GA, ID, MA, MI, MO, MT, NC, NJ, NM, NY, PA, TN, VT, WA, WV.
Hosts *Abies amabilis, A. balsamea, A. concolor, A. fraseri, A. lasiocarpa, A. lasiocarpa* var. *arizonica, A. lasiocarpa* var. *lasiocarpa, A. magnifica, Alnus oblongifolia, Betula alleghaniensis, Fagus grandifolia, Picea* sp., *P. engelmannii, P. rubens, Pinus contorta, P. ponderosa, P. sylvestris, Populus* sp., *P. tremuloides, Pseudotsuga menziesii, Salix* sp., *S. nigra, Thuja plicata, Tsuga* sp., *T. canadensis*, angiosperm, gymnosperm.
Ecology A major root and butt rot fungus in *Abies lasiocarpa*; basidiomes produced on moss-covered wood and bark; rotten wood; tree base; stump; logs; associated with a white rot.
Culture Characters
2.6.15.26.32.36.38.47.(50).(52).54.55.58 (Ref. 474);
2.6.15.32.36.38.47.52.54.55-58 (Ref. 431);
2.6.15.32.36.39.46-47.55 (Ref. 452).
References 93, 97, 116, 120, 128, 133, 144, 183, 197, 204, 208, 216, 242, 249, 274, 327, 402, 406, 452, 474.

VUILLEMINIA Maire 1902 Vuilleminiaceae

Vuilleminia comedens (Nees:Fr.) Maire 1902
Bull. Soc. Mycol. France 18:81.
Basionym *Thelephora comedens* Nees, System Pilze Schwämme, 329, 1817.
Distribution Canada: YT.
Host *Betula glandulosa.*
Ecology Small (up to 1.5 cm diam), dead standing stems, associated with a white rot.
Culture Characters
2.3.7.32.36.38.43-45.54.60 (Ref. 588);
2.3.7.32.36.42-44.53.60 (compiled from Refs. 36 and 601).
Reference 243.

WAITEA Warcup & Talbot 1962 Ceratobasidiaceae

Waitea circinata Warcup & Talbot 1962
Trans. Brit. Mycol. Soc. 45:503.

Anamorphs *Moniliopsis oryzae* (Ryker & Gooch) R. Moore 1987 (≡ *Rhizoctonia oryzae* Ryker & Gooch 1938) and *Moniliopsis zeae* (Voorhees) R. Moore 1987 (≡ *Rhizoctonia zeae* Voorhees 1938, = *Rhizoctonia endophytica* Saksena & Vaartaja var. *filicata* Saksena & Vaartaja 1960, fide Oniki et al., Trans. Mycol. Soc. Japan 26:197, 1985).
Distribution Canada: SK. United States: AR, CA, FL, GA, ID, IN, LA, MS, NC, ND, SD, TX.
Hosts *Agrostis palustris, A. stolonifera, Avena sativa, Beta vulgaris, Bromus tectorum, Cucumis sativus, Daucus carota, Festuca arundinacea, F. pratensis, Glycine max, Helianthus annuus, Hordeum vulgare, Oryza sativa, Pennisetum glaucum, Phaseolus lunatus, Pinus banksiana, Raphanus sativus, Schizachyrium scoparium, Stenotaphrum secundatum, Triticum aestivum, Zea mays.*
Ecology Causes brown patch disease of turfgrasses; sclerotial rot of *Zea mays*; sheath spot and root rot of *Oryza* and other grasses; damping-off of pine seedlings.
Culture Characters Not known.
References 183, 449, 585.

Rhizoctonia oryzae in Farr et al. (183) is restricted to Poaceae, whereas *R. zeae* occurs on Poaceae as well as plants in several unrelated families.

WOLDMARIA W.B. Cooke 1961 — Cyphellopsidaceae

Woldmaria crocea (Karsten) W.B. Cooke 1961
Beih. Sydowia 4:29.
Basionym *Solenia crocea* Karsten, Hedwigia 23:88, 1884.
Distribution Canada: MB, NS, ON.
Hosts *Matteuccia struthiopteris, Onoclea struthiopteris, Pinus* sp., *Pteretis* sp., *Struthiopteris germanica.*
Ecology Not known.
Culture Characters Not known.
References 125, 249.

XENASMA Donk 1957 — Tubulicrinaceae

Xenasma praeteritum (Jackson) Donk 1957
Fungus 27:25.
Basionym *Peniophora praeterita* Jackson, Canad. J. Res., C, 28:533, 1950.
Distribution Canada: ON. United States: CT, IL, IN.
Hosts *Acer* sp., *Betula* sp., *Populus* sp., *Salix* sp., *Ulmus* sp., *U. americana*, angiosperm.
Ecology Decayed wood.
Culture Characters Not known.
References 242, 267, 306, 384, 387, 584.

Xenasma pruinosum (Pat.) Donk 1957
Fungus 27:25.
Basionym *Corticium pruinosum* Pat., Cat. rais. pl. cell. Tunisie, 60, 1897.
Synonym *Peniophora pruinosa* (Pat.) Jackson 1950.
Distribution United States: NY.
Hosts *Platanus occidentalis*, angiosperm.
Ecology Not known.
Culture Characters (2).3.13.(16).(24).(25).32.(36).38.43-45 (Ref. 588).
References 306, 384, 492, 584.

Xenasma pulverulentum (Litsch.) Donk 1957
Fungus 27:25.
Basionym *Corticium pulverulentum* Litsch., Oesterr. Bot. Z. 88:112, 1939.
Synonym *Peniophora pulverulenta* (Litsch.) Jackson 1950.
Distribution Canada: ON. United States: WI.
Hosts *Tilia americana, Tsuga canadensis*, angiosperm.
Ecology Decayed ?log.
Culture Characters 1.3.13.(16).32.36.38.47.(48).(54).(55) (Ref. 588).
References 116, 242, 306, 384, 388.

Xenasma rimicola (Karsten) Donk 1957
Fungus 27:26.
Basionym *Corticium rimicola* Karsten, Hedwigia 35:45, 1896.
Synonym *Peniophora rimicola* (Karsten) Höhnel & Litsch. 1906.
Distribution Canada: BC, ON, PQ. United States: MI, MO, OR, WA.
Hosts *Acer* sp., *A. macrophyllum, Alnus* sp., *Phellinus (Fomes) conchatus, Fraxinus* sp., *Picea* sp., *Populus* sp., *Quercus garryana, Rubus* sp., *Salix* sp., *Ulmus* sp., angiosperm.
Ecology Not known.
Culture Characters Not known.
References 116, 242, 306, 384, 492, 563, 584.

XENOLACHNE D.P. Rogers 1947 — Tremellaceae

Xenolachne flagellifera D.P. Rogers 1947
Mycologia 39:562.
Distribution United States: MA, OR.
Hosts *Hyaloscypha atomaria, Libocedrus decurrens, Pinus strobus, Pseudotsuga menziesii.*
Ecology Not known.
Culture Characters Not known.

Known only from the type specimens.

XENOSPERMA Oberw. 1965 — Tubulicrinaceae

Xenosperma ludibundum (D.P. Rogers & Liberta) Jülich 1979
As Oberw. ex Jülich, Persoonia 10:335.
Basionym *Xenasma ludibundum* D.P. Rogers & Liberta, Mycologia 52:902, 1960.
Distribution United States: MA, OR, RI.
Hosts *Chamaecyparis thyoides, Quercus* sp.

Ecology Bark and decayed wood.
Culture Characters Not known.
Reference 384.

Xenosperma murrillii Gilbn. & Blackwell 1987
Mycotaxon 28:400.
Distribution United States: FL.
Host *Juniperus virginiana.*
Ecology Not known.
Culture Characters Not known.

Known only from the type specimen.

XYLOBOLUS Karsten 1881 Stereaceae

Xylobolus frustulatus (Pers.:Fr.) Karsten 1881
Acta Soc. Fauna Fl. Fenn. 2:40.
Basionym *Thelephora frustulata* Pers., Synop. Meth. Fung. 577, 1801.
Synonyms *Stereum frustulosum* Fr. 1838, *Stereum frustulatum* (Pers.) Fuckel 1861, superfluous combination.
Distribution Canada: NS, ON.
United States: AL, AR, CT, DE, FL, GA, IA, IL, IN, KS, KY, LA, MA, MD, MN, MO, MS, NC, ND, NE, NJ, NY, OH, OK, OR, PA, RI, SC, SD, TN, TX, VA, VT, WI, WV.
Hosts *Betula alleghaniensis, Carya* sp., *Castanea* sp., *C. dentata, C. pumila, Liriodendron tulipifera, Pinus strobus, Pseudotsuga menziesii, Quercus* sp., *Q. alba, Q. fellos, Q. lyrata, Q. macrocarpa, Q. muhlenbergii, Q. palustris, Q. phellos, Q. rubra, Q. rubra* var. *borealis, Q. velutina*, angiosperm.
Ecology Causes a trunk rot in *Castanea* and *Quercus*; decay in treated Douglas-fir utility poles; basidiomes produced on dead wood; logs and stumps; trunk; associated with a white rot or a white pocket rot.
Culture Characters
1.3r.13.20.21.31g.32.(36).37.38.44-45.(53).54.(57) (Ref. 474);
1.3r.8.21.32.37.39a.44-46.47.54.57.66 (Ref. 104);
1.3r.13.32.37.(39).44-45.(53).54.57.66 (Ref. 609).
References 89, 96, 108, 116, 183, 214, 242, 249, 274, 333, 379, 474, 519, 609.

Fries (1838) and Karsten (1881) misspelled the epithet as 'frustulosus', see Chamuris (104). However, Lentz (379) treated *Stereum frustulosus* (Fr.) Fr. as a distinct epithet. We follow Chamuris and consider 'frustulosus' as a typographical error. Thus the combination listed in the above synonymy by Fuckel becomes a superfluous combination.

Xylobolus subpileatus (Berk. & Curtis) Boidin 1958
Rev. Mycol. (Paris) 23:341.
Basionym *Stereum subpileatum* Berk. & Curtis, Hooker's J. Bot. Kew Gard. Misc. 1:238, 1849.
Synonyms *Stereum frustulatum* var. *subpileatum* (Berk. & Curtis) Welden 1971, *Stereum insigne* Bres. 1891, *Stereum sepium* Burt 1920. Both fide Lentz (379).
Distribution Canada: NS, ON. United States: AL, AR, DE, FL, GA, IA, IL, KY, LA, MD, MO, MS, NC, NJ, NY, OH, OR, PA, SC, TN, TX, VA.
Hosts *Castanea* sp., *Celtis occidentalis, Fagus* sp., *Liquidambar* sp., *L. styraciflua, Malus* sp., *M. pumila, Platanus occidentalis, Quercus* sp., *Q. alba, Q. garryana, Q. laurifolia, Q. nuttallii, Q. phellos, Q. rubra, Q. stellata, Q. virginiana, Taxodium distichum, Ulmus* sp., *U. americana*, angiosperm.
Ecology Causes a minor decay in southern angiosperms; basidiomes produced on bark; old logs; underside of rotten log; trunks; dead stump.
Culture Characters
1.3r.8.21.32.37.39a.44-46.47.(53).54.57.66 (Ref. 104);
1.(2).3r.(20).31g.32.(36).37.39.45-47.50.51.54.(57) (Ref. 474).
References 89, 96, 104, 108, 127, 183, 242, 249, 274, 327, 347, 379, 474, 519.

YPSILONIDIUM Donk 1972 Ceratobasidiaceae

Ypsilonidium sterigmaticum (Bourd.) Donk 1972
Verh. Kon. Ned. Akad. Wetensch., Afd. Natuurk., Tweed Sect. C, 75:371.
Basionym *Corticium sterigmaticum* Bourd., Rev. Sci. Bourbonnais Centr. France 35:4, 1922.
Synonym *Ceratobasidium sterigmaticum* (Bourd.) D.P. Rogers 1935.
Distribution United States: IA.
Hosts Angiosperm.
Ecology Dead wood; underside of rotted log.
Culture Characters Not known.
References 443, 563.

EXCLUDED NAMES

(Names which are invalid, doubtful or misapplied to North American taxa)

Aleurodiscus helveolus Bres. 1925
Mycologia 17:71.

The type is on *Salix lasiandra* from Washington State. Rogers and Jackson (567) examined the type and placed the name in synonymy under ***Clavariadelphus fistulosus*** (Fr.) Corner var. ***contortus*** Corner.

Aleurodiscus tsugae Yasuda 1921
In Lloyd, Mycol. Writ. 6:1066.

Reported (116, 242, 375) from North America, but Ginns (244) redetermined the North American specimens to be either ***Aleurodiscus occidentalis*** or ***A. thujae***.

Auricularia scutellaeformis (Berk. & Curtis) Lloyd 1924
Mycol. Writ. 7:1275.
Basionym *Hirneola scutellaeformis* Berk. & Curtis, Grevillea 1:19, 1873.

Reported from Alabama (593) and the type is from Alabama. Lowy (417), apparently based solely upon Lloyd's description, excluded the name from *Auricularia*. The microscopic features of the type are unknown.

Ceratobasidium fibulatum Tu & Kimbrough 1978
Bot. Gaz. (Crawfordsville) 139:457.

The name is not validly published and should not be validated (236).

Coniophora harperi Burt 1917
Ann. Missouri Bot. Gard. 4:252.

Reported (96) from Georgia. Ginns (234), studied the type specimen, but could not satisfactorily place the epithet in a genus.

Corticium albido-carneum (Schw.) Massee 1890
J. Linn. Soc., Bot. 27:142.
Basionym *Thelephora albido-carnea* Schw., Trans. Amer. Philos. Soc., N.S., 4:169, 1832.

Reported (93) from Michigan, Pennsylvania (type specimen), and Virginia. However, the Michigan and Virginia collections were misidentified (245). Because the type specimen lacks basidia and basidiospores, Ginns (245) declared the name a *nomen dubium*.

Corticium alutaceum (Schrader) Bres. sensu Lyman 1907
Proc. Boston Soc. Nat. Hist. 33:160-167.

Reported on *Salix nigra* from Georgia (178) and from Tennessee (333). Lyman's concept is now considered to be ***Sistotrema biggsiae***, fide Nakasone (474) and Rogers (566), but the Georgia and Tennessee records may not be that fungus.

Corticium argentatum Burt 1926
Ann. Missouri Bot. Gard. 13:257.

Reported (93, 116, 183, 393) from Manitoba, Kansas and Nebraska on *Fraxinus pennsylvanica* and *Salix* sp. The type from Nebraska lacks basidia and spores (393). Gloeocystidia numerous, flexuous, 40-50 x 8-12 µm, but some 6-12 µm, spherical brown masses; spores uncertain, the few seen were hyaline, even, 4-6 x 3 µm, fide Burt (93). The species concept is uncertain.

Corticium aschistum Berk. 1873
Grevillea 2:3.

Presumably from South Carolina, nevertheless, the name is illegitimate because it is a later homonym of *Corticium aschistum* Berk. & Curtis 1858, from Nicaragua.

Corticium auberianum Mont. 1845
In La Sagra, Hist. Cuba 9 (2):226.

Burt (93) is the only author to have considered this fungus to be part of the mycota of the United States. Ginns (245) studied the ten collections Burt cited and concluded, because they represented at least six genera, that the name *C. auberianum* is not documented from the United States.

Corticium auriforme Berk. & Curtis 1873
Grevillea 1:166.

Presumably from South Carolina. The type is not at K, fide Hjortstam (288) and the species concept is uncertain.

Corticium basale Peck 1890
Annual Rep. New York State Mus. 43:69.

Known only from the original description. Peck stated that the type specimens were sterile.

Corticium calceum Fr. sensu Burt 1926
Ann. Missouri Bot. Gard. 13:203.

Burt cited specimens from a number of Provinces and States but they should be renamed. A *nomen confusum*, fide D. Rogers and Jackson (567).

Corticium centrifugum (Lév.) Bres. 1903
Ann. Mycol. 1:96.

Reported (93, 116, 120, 242, 249) from several localities on a variety of hosts. Probably misapplied as used in North America, see D. Rogers and Jackson (567). Burt's species concept was very broad, e.g., the Stoney Mountain, Manitoba collection is ***Fibulomyces canadensis*** (312), the Winnipeg, Manitoba collection is ***Athelia fibulata*** (312), the Maine collection is ***Leptosporomyces galzinii*** (312) the Pennsylvania collection is ***F. mutabilis*** (312), and the specimens named by Bresadola are heterogeneous (312). The name is illegitimate because it is a later homonym of *Corticium centrifugum* (Weinm.) Fr. 1875. Therefore, the application of this name in North America is uncertain.

Corticium cultum Burt 1926
Ann. Missouri Bot. Gard. 13: 231.

Reported in bark beetle tunnels in coniferous logs in Idaho. *Nomen confusum*, fide Liberta (393).

Corticium debile Berk. & Curtis ex Massee 1890
J. Linn. Soc., Bot. 27:131.

Burt (93) cited three collections from the United States. Two represent two different species, perhaps neither conspecific with the type; the third collection cited was not studied (245). Ginns (245) found part of the type, from Venezuela, to be in such poor condition it could not be adequately characterized. The name is not adequately documented from the United States.

Corticium deglubens Berk. & Curtis 1873
Grevillea 1:166.
Synonym *Sebacina deglubens* (Berk. & Curtis) Burt 1915.

The disposition of this epithet is uncertain. In the original description only one specimen is cited: "No. 4557. On Juniper. Alabama, Peters." Thus 4557 is the type and part is at the Royal Botanic Gardens (K), see Hjortstam (288), but a part is presumably in the Farlow Herbarium (FH) because Rogers and Jackson (567) acknowledged FH and they did not mention having seen specimens at K. Rogers and Jackson (567), after studying the type, concluded that the epithet was a synonym of *Sebacina incrustans*. However, Hjortstam (288), also having studied the type, placed the epithet in synonymy with "*Eichleriella deglubens* (Berk. & Curtis) Reid." Hjortstam's citation of "Curtis" as an author is a slip for the basionym for the *Eichleriella* combination is *Radulum deglubens* Berk. & Br., Ann. Mag. Nat. Hist., Ser IV, 15:32, 1875. The differing opinions, by very reputable mycologists, suggests that the type specimens of *C. deglubens* and *R. deglubens* should be carefully compared.

Corticium flavidum Berk. & Curtis 1873
Grevillea 1:178.

The holotype (Michener 4084 at K) from Pennsylvania is an anamorphic state, fide Hjortstam (288).

Corticium gracilis Copeland 1904
Ann. Mycol. 2:508.

The type from California could not be found, fide Cooke (125). Perhaps a synonym of *Rectipilus fasciculatus*, fide Cooke (125, 131). A *nomen confusum*.

Corticium hypopyrrhinum Berk. & Curtis 1873
Grevillea 1:179.

Presumably from South Carolina; the type (Ravenel 1704 at K) is a hyphomycete, fide Hjortstam (288).

Corticium improvisum Jackson 1950
Canad. J. Res., C, 28:720.

Known only from the type and one paratype specimen, both collected near Montreal, Quebec on the same day. The paratype from Isle Perrot (DAOM 30762) has the features described and illustrated by Jackson, except we did not find basidia. A pseudoparenchymatous tissue, about 1 cell thick, was seen but its connection to the basidia and the subicular hyphae is unclear. A variety of hyphae were present in mounts, as illustrated by Jackson, but they may be from different fungi. Imperfectly known but certainly not a *Corticium sensu stricto* and its generic affinities are uncertain.

Corticium koleroga (Cooke) Höhnel 1910
Sitzungsber. Kaiserl. Akad. Wiss., Math.-Naturwiss. Kl., Abt. 1, 119:395.
Basionym *Pellicularia koleroga* Cooke, Grevillea 4:116, 1876.

The type is a mixture and the name is a *nomen confusum* (Art. 76), fide Donk (151) and Talbot (599). The North American reports are presumed to be *P. koleroga* sensu D. Rogers and are listed under ***Ceratobasidium ochroleuca***.

Corticium lacteum Fr. sensu Burt 1926
Ann. Missouri Bot. Gard. 13: 212.

This is the only citation of *C. lacteum* which refers to specimens from Canada and the United States. Rogers and Jackson (567), following study of all these specimens, concluded they were misdetermined.

Corticium pelliculare (Karsten) Karsten 1896
Hedwigia 35:46.
Basionym *Corticium laeve* subsp. *pelliculare* Karsten, Bidrag Kännedom Finlands Natur Folk 48:411, 1889.

Reported (93, 116, 120, 128, 133) from a variety of localities but Jülich (312), having studied Karsten's specimens, concluded that it was not clear which fungus Karsten had described. Jülich designated the name a *nomen dubium*. Thus the North American specimens should be renamed.

Corticium peterga Burrill
Reported (7) as a thread blight and leaf spot on *Crataegus* sp. in Florida. We have not found the original description and presume it is an unpublished name.

Corticium rosae Burt 1926
Ann. Missouri Bot. Gard. 13: 239.

Reported from British Columbia on *Rosa* sp. Related to species of *Phlebia* (393), but imperfectly known.

Corticium rubicundum Burt 1926
Ann. Missouri Bot. Gard. 13: 235.

Reported (93, 133, 393) from Ontario, Colorado and Washington on bark of *Picea* sp., *Pinus* sp., *Tsuga canadensis*. Type was restudied (393) but generic placement not discussed.

Corticium rubrocanum Thümen 1876
Bull. Torrey Bot. Club 6:95, 1876 and Mycotheca Univ. Exs. 409, 1876.

Known (245) from Alabama, Louisiana, New Jersey and South Carolina on bark of dead twigs and small branches of *Quercus coccinea*. All collections are sterile, but the available features agree with *C. albido-carneum* (above) and *C. rubrocanum* was placed in synonymy.

Corticium scariosa Berk. & Curtis 1873
Grevillea 2:3.
Synonym *Sebacina scariosa* (Berk. & Curtis) Burt 1915.

Known only from the type specimen, pieces of which are at K and FH, from South Carolina. Burt (81) described the basidia as longitudinally septate, pyriform to subglobose, 12-15 x 9-12 μm but he did not find any spores. Hjortstam (288) observed basidiospores as numerous, subglobose, inamyloid, 10-12 μm, basidia not seen, gloeocystidia sulfo-negative, hyphae simple-septate. The species concept is uncertain.

Corticium spretum Burt 1926
Ann. Missouri Bot. Gard. 13:229.

Known only from the type specimen from Washington on *Fraxinus latifolia* (see 133, 183). Burt described the spores as "hyaline, even, 8-10 x 5-6 μ" but Liberta (393) did not find basidiospores on the type collection (FH - Suksdorf 962), and neither did we. Clamp connections are common on the hyphae and the base of the hymenial cells, but Burt stated "not nodose-septate". The thin-walled probasidia and narrow, branched dendrohyphidia indicate that fungus is a *Corticium* species but in the absence of basidiospores the species concept is uncertain.

Corticium subaurantiacum Peck 1890
Annual Rep. New York State Mus. 43:69.

Known only from the type specimen. The microscopic features need to be described.

Corticium subcontinuum Berk. & Curtis 1868
J. Linn. Soc., Bot. 10:337.

Reported (93) from Louisiana and Texas, but these specimens are not conspecific with the type, fide Ginns (245). The type from Cuba (Wright 537 at K) lacks basidia and spores but may be a species of *Duportella*, fide Hjortstam (288). The fungus is not documented from Canada or the United States.

Corticium subochraceum Bres. 1896
Hedwigia 35:290.

Ginns (245) concluded that all six collections cited by Burt (93) were misidentified, i.e., none is conspecific with the Brazilian type, and the six represent five or six species. This fungus is not accepted as part of the mycota. The name should not be confused with *Grandinia subochracea* Bres. 1894, which is ***Phlebia subochracea***.

Corticium sulphureum Fr. sensu Burt 1926
Ann. Missouri Bot. Gard. 13:177.

A smooth spored fungus which Burt stated "Not *Corticium sulphureum* Pers." (now named ***Phlebiella sulphurea***). Burt's description suggests he had a species of *Piloderma*.

Corticium vinaceum Burt 1926
Ann. Missouri Bot. Gard. 13:298.

Reported from Alabama and Louisiana on the underside of decaying coniferous planks. Generic disposition uncertain, fide Liberta (393).

Cyphella arachnoidea Peck 1891
Annual Rep. New York State Mus. 44:134.

Currently named ***Rimbachia arachnoidea*** (Peck) Redhead (547) and not treated here because the genus primarily occurs on mosses.

Cyphella cinereo-fusca (Schw.:Fr.) Sacc. 1881
Michelia 2:303.

Basionym *Peziza cinereo-fusca* Schw., Schriften Naturf. Ges. Leipzig 1:119, 1822.

Reported (80) from North Carolina on decorticated branches of *Cercis* sp. A specimen in the Schweinitz Herbarium (PH) lacked any cyphellaceous fungus and no specimen was found in Michener's herbarium (BPI), thus Cooke (125) "rejected" the name.

Cyphella perexigua Sacc. 1880
Michelia 2:136.

Reported (80) from South Carolina on decorticated branches. "Based on a specimen collected by Ravenel in Carolina and reported to be smaller than *C. eruciformis* and *C. cupuliformis* which it resembled" (125). Basidia and basidiospores have not been described (80). The species concept is uncertain.

Cyphella trachychaeta Ellis & Ev. 1888
J. Mycol. 4:73.
Synonym *Flagelloscypha trachychaeta* (Ellis & Ev.) W.B. Cooke 1962.

Cooke's (125) concept was very broad and Agerer (2) has segregated *Flagelloscypha langloisii* and *F. minutissima* from Cooke's synonymy. In addition, Agerer found the type specimen, from Louisiana, of *C. trachychaeta* to be in poor condition. Thus, the North American reports of *C. trachychaeta* in Burt (80), Conners (116) and Cooke (125) should be re-evaluated.

Cyphella tsugae Coker 1927
J. Elisha Mitchell Sci. Soc. 43:138.

Known only from the original description, where it was reported from North Carolina and "obviously ... from the bark itself and not from the [adjacent] mosses." Coker stated "Spores not found" and the description lacks the microscopic features necessary to place the fungus in a currently accepted cyphelloid genus.

Dacrymyces abietinus (Pers.) sensu N. Amer. Aucts.

Reported (51, 52, 90, 107, 119, 133, 402, 448, 498, 502, 579) from a variety of locations and on coniferous wood of several species. After studying the type specimen, McNabb (461) concluded the name had been misapplied to North American specimens. He thought that most records were probably of specimens of *D. variisporus*.

Dacrymyces aurantius (Schw.:Fr.) Lloyd 1913
Mycol. Writ. 4 (Letter 44:6).
Basionym *Tremella aurantia* Schw., Schriften Naturf. Ges. Leipzig 1:114, 1822.

Reported (107) from North Carolina and the type was from North Carolina. Coker was probably following Farlow (182) in assigning a dacrymycetaceous fungus to this name. Farlow, however, had misapplied the name *Tremella aurantia* Schw. to specimens of *D. palmatus*, fide Donk (157). Other specimens labelled *D. aurantius* are probably *D. chrysospermus*, see McNabb (461). The epithet is now applied to a *Tremella*. The combination in *Dacrymyces* is sometimes attributed to Farlow, but he only stated, tentatively, of *T. aurantia* "There can be no doubt, I think, that the fungus should be placed in *Dacrymyces*,...."

Dacrymyces conglobatus Peck 1880
Annual Rep. New York State Mus. 32:37.

Reported on bark of *Thuja occidentalis* in New York. We have not seen the illustration cited in the original description. However, Peck (Contributions to the Botany of the State of New York. Museum Bull. 1 (2): 27-28, plate 1, figs. 1-4, 1887) illustrated conidiophores with spinulose walls and terminal clusters of 6 to 8 spores. The spores are allantoid, about 12 μm long with three oil drops, the walls hyaline, smooth, thin. Apparently it is an anamorph of a Heterobasidiomycete. Peck placed his name in synonymy with '*Umbrophila rubella* Quél.' Kennedy (337) who had seen Peck's type specimen referred the name to *Craterocolla cerasi* (Tul.) Bref., Untersuchungen Gesammtgebiete Mykol. 7:98, 1888. Perhaps the same fungus was reported from Washington State by Cooke (133) under the unpublished combination "*Craterocolla rubella* Pers." All these names are poorly understood, fide Donk (157) and the status of the species in North America is uncertain.

Ditiola albizziae Coker 1920
J. Elisha Mitchell Sci. Soc. 35:180.

The type from North Carolina is in poor condition and the fungus could not be adequately characterized, fide McNabb (460). A *nomen dubium*.

Exidia bullata Lloyd 1924
Mycol. Writ. 7:1275.

From North Carolina with globose basidia, but spores not seen by Lloyd. No subsequent reports of this name have been found.

Exidia fragilis Olive 1954
Bull. Torrey Bot. Club 81:333.
Basionym *Nom. nov.* for *Seismosarca cartilaginea* Olive, Bull. Torrey Bot. Club 78:110, 1951, because there already was an *Exidia cartilaginea* Lundell & Neuhoff 1935.

Impossible to demonstrate basidia and the true affinities of the type from South Carolina are not known, fide Wells (623).

Gloeocystidiellum heimii Boidin 1966
Cahiers Maboké 4:10.

Reported from Minnesota (214), but the specimens were redetermined to be *G. karstenii* by Freeman (in litt. 1978).

Gloeocystidiellum radiosum (Fr.) Donk (sic)
Basionym *Thelephora radiosum* Fr., Obs. Mycol. 2:277, 1818.

We could not verify that Donk ever proposed the combination, although it appears in Gilbertson and Buddington's (204) report of Arizona fungi. Nevertheless, the identity of the Arizona fungus is uncertain. The epithet is generally accepted (171, 326) as a synonym of *Vesiculomyces citrinum*. However, in North America, Burt (93, index) did not comment on *citrinum* and treated specimens under *Corticium radiosum* (Fr.) Fr. Gilbertson and Budington (204) reported specimens under both epithets but did not explain the differences between the fungi.

Grandinia virescens Peck 1878
Annual Rep. New York State Mus. 30:47.

A *nomen confusum*, fide Gilbertson (189). The type specimen is contains two species.

Helicogloea sebacinoidea Olive 1948
Mycologia 40:588.

Nomen confusum, original description was based on *Exidia glandulosa* parasitized by the *Helicogloea*. Olive (502) renamed the parasite *H. parasitica* but that name is now a synonym of *H. longispora*.

Hericium fasciculare (Alb. & Schw.) Banker 1906
Mem. Torrey Bot. Club 12:123.
Basionym *Hydnum fasciculare* Alb. & Schw., Conspectus Fungorum, 269, 1805.

Reported (25) in Pennsylvania on decayed pine and fir trunks, but the identity of the specimen(s) is uncertain.

Hericium fimbriatum Banker 1906
Mem. Torrey Bot. Club 12:122.

The type from Pennsylvania at NY has spores "hyaline, smooth, cylindric, straight to slighty curved, 4-4.5 x 2-2.5 μ", fide Gilbertson (191); thus the fungus is not a *Hericium*.

Hydnum chrysodon Berk. & Curtis 1873
Grevillea 1:98.

Type from South Carolina is "unidentifiable," fide Gilbertson (192). A *nomen dubium*.

Hydnum conchiforme Sacc. 1888
Sylloge Fung. 6:458.
Basionym *Nom. nov.* for *Hydnum plumarium* Berk. & Curtis, Grevillea 1:97, 1873, a later homonym of *Hydnum plumarium* Berk. & Curtis, J. Linn. Soc., Bot. 10: 324, 1869.

Banker (26) declared the type from the Carolinas too scanty to be of much value. However the fungus is in the *Steccherinum queletii* group as *S. conchiforme* (Sacc.) Hjort. (289). We realized this too late to move this species to the treatment of *Steccherinum*.

Hydnum halei Berk. & Curtis 1873
Grevillea 1:98.

The type from Louisiana lacks any hymenium, fide Gilbertson (192). A *nomen dubium*.

Hydnum plumosa Duby 1830
Bot. Gallicum 2:779.

Duby wrote "?plumosum" and mentioned a *Sistotrema plumosum* Pers., Mycol. Eur. 2:201, which he cited as a synonym. Fries (Elenchus 1:141, 1828) stated "S. plumosum l.c. monstrosa progenies". However, Saccardo (Sylloge Fung. 6:475, 1888) credited the name to Duby. Berkeley (Grevillea 1:100, 1873) reported two collections from New England and Carolina. Bourdot and Galzin (50:741) mention the name but do not discuss a concept. The North American specimens should be re-examined.

Hyphoderma teutobergense (Brinkm.) Eriksson 1958
Symb. Bot. Upsal. 16 (1):100.

Reported (216) in Arizona, but the identity of these specimens is uncertain because Eriksson (167) misapplied the name to a fungus now called *H. sibiricum* (173:537). The type specimen is an *Athelia* (313).

Hypochnus fuscus Pers. sensu Burt 1916
Ann. Missouri Bot. Gard. 3:215.

Reported (108, 334) from Michigan and North Carolina. Burt's concept was a mixture, see Larsen (359). The identity of the specimens in these reports is uncertain.

Irpex cinnamomeus Fr. 1838
Epicrisis, 524.

The type from North America is ***Cerrena unicolor*** (Bull.:Fr.) Murrill (Polyporaceae), fide Maas Geesteranus (428).

Irpex concrescens Lloyd 1915
Mycol. Writ. 4 (Lett. 60:9).

The type from Alabama is *Poria ambigua* Bres. (Polyporaceae), fide Gilbertson (190), which is now a synonym of ***Oxyporus latemarginata*** (Mont.) Donk (217).

Irpex coriaceus Berk. & Rav. 1873
Grevillea 1:101.

The type from South Carolina may be *Cerenella farinacea* (Fr.) Murrill, which is now a synonym of ***Fuscocerrena portoricensis*** (Fr.) Ryv. (Polyporaceae), see Maas Geesteranus (428).

Irpex crassitatus Lloyd 1920
Mycol. Writ. 6:909.

The type from Iowa is ***Spongipellis pachyodon*** (Pers.:Fr.) Kotlaba & Pouzar (Polyporaceae), fide Maas Geesteranus (428).

Irpex farinaceus Fr. 1830
Linnaea 5:523.

A synonym of ***Fuscocerrena portoricensis***, fide Ryvarden (572).

Irpex flavus Klotzsch 1833
Linnea 8:488.

Originally reported from North America but never again recorded from this continent. A common species in the Southern Hemisphere. Ryvarden (570) believes the label with the type has been switched and the specimen was not from North America. Currently known as ***Flavodon flavus*** (Klotzsch) Ryv. (Steccherinaceae).

Irpex mollis Berk. & Curtis 1849
Hooker's J. Bot. Kew Gard. Misc. 1:236.

The type from South Carolina is ***Spongipellis pachyodon***, fide Maas Geesteranus (428).

Irpex nodulosus Peck 1888
Annual Rep. New York State Mus. 41:79.

The type, on bark of dead standing trunk of *Populus* sp. in New York, was redescribed by Gilbertson (189). After consulting Gilbertson's paper, Maas Geesteranus (428) concluded it was not a species of *Irpex*. The generic placement of this species is uncertain.

Irpex ochraceus Schw. 1832
Trans. Amer. Philos. Soc., N.S., 4:164.

The type from Pennsylvania was not found at PH, thus Maas Geesteranus (428) declared the name a *nomen dubium*.

Irpex paradoxus (Schrader:Fr.) Fr. 1838
Epicrisis, 522.

Reported (96) from Georgia. The name is now ***Schizopora paradoxa*** (Schrader:Fr.) Donk (Polyporaceae).

Irpex pityreus Berk. & Curtis 1873
Grevillea 1:102.

The type from Rhode Island is "a mere fragment", thus Maas Geesteranus (428) declared the name a ***nomen** dubium*.

Irpex schweinitzii Berk. & Curtis 1873
Grevillea 1:102.

The type from Alabama is probably a *Poria s.l.*, but could not be adequately characterized, fide Gilbertson (192). A *nomen dubium*.

Irpex viticola Cooke & Peck 1881
Annual Rep. New York State Mus. 34:43.

The type (K), on vine of *Vitis* sp. in from New York, was redescribed by Maas Geesteranus (428). He reported simple septate generatives, 3-5 μm diam; hyaline skeletals in the spines, to 5.5 μm diam, curved into the hymenium to form cystidia which become incrusted, no spores seen. Peck did not give any sizes and no microscopic features. Not an *Irpex*, fide Maas Geesteranus (428). A *nomen dubium*.

Kneiffia candidissima Berk. & Curtis 1849
Hooker's J. Bot. Kew Gard. Misc. 1:237.

The type from South Carolina is sterile and in poor condition, fide Gilbertson (192). A *nomen dubium*.

Kneiffia tessulata Berk. & Curtis 1873
Grevillea 1:147.

The type on *Quercus* sp. from South Carolina has distinctive hyphae but is sterile, fide Gilbertson (192). A *nomen dubium*.

Lachnella ciliata (Sauter) W.B. Cooke 1962
Beih. Sydowia 4:71.
Basionym *Cyphella ciliata* Sauter, Flora 28:134, 1845.

Reported (125) from Maine, Ontario, Oregon and Virginia. The type specimen cannot be found and the species concept is uncertain, fide Agerer (5). Because Cooke (125) listed *Cyphella villosa* var. *orthospora* Bourd. & Galzin as a synonym, his records of *L. ciliata* are listed herein under ***Flagelloscypha orthospora***.

Lachnella pinicola W.B. Cooke 1961
Beih. Sydowia 4:76.

The type, on a fallen branch of *Pinus* sp. in North Carolina, could not be found, thus Agerer (5) concluded that the species concept was not definitive but the fungus might be a *Henningsomyces*.

Lachnella punctiformis (Fr.) W.B. Cooke 1962
Beih. Sydowia 4:77.

Cooke's (125) concept of this species was in error, fide Agerer (2). The specimen Cooke reported from South Carolina should be re-examined.

Lachnella septentrionalis W.B. Cooke 1962
Beih. Sydowia 4:79.

The type is from Ontario and the name is now ***Leptoglossum septentrionalis*** (W.B. Cooke) Agerer (Tricholomataceae).

Leaia piperata Banker 1906
Mem. Torrey Bot. Club 12:175.

Reported from Iowa, Nebraska and New York. The name needs to be characterized microscopically.

Leaia stratosum (Berk.) Banker 1906
Mem. Torrey Bot. Club 12:177.
Basionym *Hydnum stratosum* Berk., London J. Bot. 4:307, 1845.

Reported (25) from Indiana, Ohio and New York on rotting logs. The identity of the specimens is uncertain.

Merulius albus Burt 1917
Ann. Missouri Bot. Gard. 4:3370.

The type is from Alabama. A synonym of *Poria cocos* F.A. Wolf, fide Ginns (226) which is now ***Wolfiporia cocos*** (F.A. Wolf) Ryv. & Gilbn. (Polyporaceae).

Merulius cupressi Schw. 1822
Schriften Naturf. Ges. Leipzig 1:92.

The type is an insect gall, fide Ginns (220).

Merulius dubius Burt 1917
Ann. Missouri Bot. Gard. 4:330.

The type from New York is *Poria sanguinolenta* (Alb. & Schw.) Cooke, fide Ginns (226), which is now ***Physisporinus sanguinolentus*** (Alb. & Schw.:Fr.) Pilát (Polyporaceae).

Merulius fugax Fr. sensu N. Amer. Aucts.

Reported (97, 116, 120, 128, 249, 512) from Alaska, California, Pennsylvania, British Columbia, Manitoba and Nova Scotia, on wood of a variety of conifers. This name has been applied to *Leucogyrophana mollusca* and *L. romellii*, and reports of *M. fugax* were probably a mixture.

Merulius lichenicola Burt 1917
Ann. Missouri Bot. Gard. 4:329.

The type from New York is *Poria mollusca* (Fr.) Cooke sensu Lowe (416), fide Ginns (226), which is now ***Ceriporiopsis mucida*** (Pers.:Fr.) Gilbn. & Ryv. (Polyporaceae), see Donk (160:159).

Merulius odoratus Schw. 1822
Schriften Naturf. Ges. Leipzig 1:92.

This epithet is currently classified as ***Cantharellus odoratus*** (Schw.) Fr. (Cantharellaceae), see Corner (135).

Merulius patellaeformis Berk. & Curtis 1872
Grevillea 1:72.

The type, from "Carolina", lacks hymenial detail, fide Ginns (222). A *nomen dubium*, fide Ginns (226).

Merulius roseus Schw. 1822
Schriften Naturf. Ges. Leipzig 1:92.

No specimens are known in North America (220), but Burt (79) mentioned a Schweintiz specimen that may be at Uppsala (UPS). The microscopic characters are unknown. All descriptions, e.g., Burt (79) and Corner (135), have been based upon the original description. Generally accepted as ***Cantharellus roseus*** (Schw.) Fr., see Corner (135).

Merulius sulcatus Peck 1878
Bot. Gaz. (Crawfordsville) 4:138.

No specimens have been found and the microscopic features are unknown (220). A *nomen dubium*, fide Ginns (220).

Merulius tomentosus Burt 1917
Ann. Missouri Bot. Gard. 4:334.

The type from British Columbia lacks features necessary to accurately characterize the species, fide Ginns (218). A *nomen dubium*.

Merulius wrightii Berk. 1872
Grevillea 1:69.

The type from Texas is undeterminable (222). A *nomen dubium*.

Odontia albo-miniata Berk. & Curtis 1849
Hooker's J. Bot. Kew Gard. Misc. 1:237.

The type on *Quercus* sp. from South Carolina "is a very striking and peculiar fungus, unfortunately sterile," fide Gilbertson (192). A *nomen dubium*.

Odontia fragilis Karsten
Reported by Burt (658) on *Quercus* sp. from Oregon. We have not found the original place of publication of the name. Perhaps a *lapsus calami* for *Radulum fragile* Karsten, Hedwigia 25:231, 1886. However, the collection from Oregon may not be that species.

Odontia papillosa (Fr.) Bres. 1897
Atti Imp. Regia Accad. Rovereto, Ser. III, 3:361.

Reported (116) on *Fagus grandifolia* in Nova Scotia. Because the name is of uncertain application, fide Eriksson and Ryvarden (172), the identity of the Nova Scotia specimen is uncertain.

Odontia subochracea Bres.
Reported (335) on *Umbellularia* sp. in Oregon. We have not found the original place of publication of the name. Apparently a *lapsus calami* for either *Grandinia subochracea* Bres. (herein *Phlebia subochracea*) or *Corticium subochraceum* Bres. The identity of the collection from Oregon in uncertain.

Odontia wrightii Berk. & Curtis

Reported (333) from Tennessee. We have not found the original place of publication of the name.

Peniophora admirabilis Burt 1926
Ann. Missouri Bot. Gard. 12:304.

Reported on *Ulmus* sp. in New York. A *nomen confusum*, fide Rogers and Jackson (567).

Peniophora albofarcta Burt 1926
Ann. Missouri Bot. Gard. 12:228.

Reported on rotten stump of *Citrus* sp. in Louisiana. Perhaps a *nomen confusum*, fide Liberta (392).

Peniophora argentea Ellis & Ev. ex Burt 1926
Ann. Missouri Bot. Gard. 12:346.

Reported on *Fraxinus* sp. in Louisiana. Perhaps a *nomen confusum*, fide Rogers and Jackson (567).

Peniophora isabellina Burt 1926
Ann. Missouri Bot. Gard. 12:253.

Reported by Burt from Alabama and Virginia on *Rubus* sp. and from Kentucky (333). Perhaps a *nomen confusum*, compare Liberta (392).

Peniophora obscura (Pers.) Bres.
Atti Imp. Regia Accad. Rovereto, Ser. III, 3:113.

Bresadola applied this name to specimens of ***Dendrophora versiformis***, fide Eriksson (164:69), and the specimens reported (96) from Georgia may be that fungus.

Peniophora viridis (Preuss) Bres. 1928
In Bourdot & Galzin, Hym. France, 295.
Basionym *Thelephora viridis* Preuss, Linnea 24:152, 1851.

Reported (116) from British Columbia on *Abies grandis*. Because the application of the name is in doubt, see Weresub (635), the identity of the Canadian specimen is uncertain.

Peniophora zonata Burt 1926
Ann. Missouri Bot. Gard. 12:245.

Reported on decayed conifer in Oregon. A *Phlebia*, fide Liberta (392).

Physalacria solida Clements & Clements 1941
In G. Baker, Bull. Torrey Bot. Club 68:287.

The name first appeared on an exsiccatae label no. 333 as "n. sp." in Clements's *Cryptogamae Formationum Coloradensium*, issued 1906, but the fungus was not described. Baker provided a brief description but the required Latin description was lacking, thus the name was not validly published. Described by Berthier as *Heterocephalacria solida* Berthier (see 32). The type specimen was subsequently interpreted (241) to be two fungi, ***Syzygospora solida*** (Berthier) Ginns (Syzygosporaceae) which was parasitic on ***Marasmius*** *pallidocephalus* Gilliam.

Platygloea fuscoatra Jackson & G.W. Martin 1940
Mycologia 32:691.

Reported (116, 437, 443) on ***Tsuga canadensis*** and coniferous wood in Ontario. The type is an immature dacrymycetaceous fungus parasitized by ***Platygloea*** *arrhytidae*, fide Bandoni (13). A ***nomen confusum***.

Podoscypha elegans (Meyer:Fr.) Pat. 1900
Essai taxonomique ..., 71.
Basionym *Thelephora elegans* Meyer, Primitiae Florae Essequeboensis, 305, 1818.
Synonym *Stereum elegans* (Meyer) Fr. 1838.

Reported (112, 613) from North and South Carolina, but the identity of the collections is uncertain. The name has been applied to different fungi. A ***nomen*** *confusum*, fide Reid (555).

Sebacina dendroidea (Berk. & Curtis) Lloyd 1915
Mycol. Writ. 4:538.
Basionym *Hymenochaete dendroidea* Berk. & Curtis, J. Linn. Soc., Bot. 14:69, 1873.

Reported (593) from Massachusetts, New York, Pennsylvania, and West Virginia. Not a ***Sebacina***, fide Lloyd (409). The generic disposition of this epithet is uncertain.

Sebacina sordida Olive 1944
J. Elisha Mitchell Sci. Soc. 60:21.

Reported on dead wood in North Carolina. Not validly published because the required Latin description was lacking. Wells (622) in 1957 proposed the new combination *Exidiopsis sordida* (Olive), which, being based upon an invalid name, is also invalid. Olive (508) validly published the name in 1958 (see p. 143, above). The poor condition of the type led Wells (625) to conclude "a dubious member" of *Exidiopsis*. Earlier reports (622) have been redetermined (625). Similar to *E. grisea* (625), presumably referring to *E. grisea* sensu N. Amer. Aucts. which is now *E. plumbescens*, fide Wells and Raitviir (631).

Seismosarca hydrophora Cooke
Reported from Iowa but the specimen is *Sebacina galzinii* Bres., fide Martin (443), which is now ***Bourdotia galzinii***.

Serpula erecta (Lloyd) W.B. Cooke 1957
Mycologia 49:216.
Basionym *Merulius erectus* Lloyd, Mycol. Writ. 6:1049, 1921.

The type from Minnesota on soil may be a species of *Aphelaria*, fide Ginns (226).

Serpula hexagonoides (Burt) W.B. Cooke 1957
Mycologia 49:217.
Basionym *Merulius hexagonoides* Burt, Ann. Missouri Bot. Gard. 4:351, 1917.

Although the type from California on charred wood of *Sequoia sempervirens* was redescribed, the generic disposition is uncertain, fide Ginns (218).

Serpula lacrimans var. *carbonaria* (Lloyd) W.B. Cooke 1957
Mycologia 49:209.
Basionym *Merulius carbonarius* Lloyd, Mycol. Writ. 6:963, 1920.

Apparently known only from the type specimens (Lloyd Herb. 9930 and 17510 at BPI) from Washington. The type has spores cylindrical to subfusiform, hyaline, smooth, 6.5-9 x 2-3 μm and hyphae thick-walled, amyloid, 3-8 μm diam. The generic disposition is uncertain, fide Ginns (226).

Steccherinum rawakense (Pers.) Banker 1912
Mycologia 4:312.
Basionym *Hydnum rawakense* Pers. in Gaudichaud in Freycin., Bot. Voy. Monde, 175, 1827.

The erroneous synonymy of *Steccherinum reniforme*, a North American name, under *S. rawakense*, see Maas Geesteranus (428), was responsible for the report (463) from the United States. Maas Geesteranus (428) gave the distribution of *S. rawakense* as Australasia.

Stereum calyculus Berk. & Curtis 1849
Hook. J. Bot. Kew Gard. Misc. 1:238.

The type is from South Carolina (see 553) and the name is now ***Craterellus calyculus*** (Berk. & Curtis) Burt.

Stereum cuneatum Lloyd 1916
Mycol. Writ. 4 (Letter 54:7).

Densely caespitose; arising from the ground in Florida. Subsequently shown to have cruciate basidia and treated as *Tremellodendron cuneatum* (Lloyd) Lloyd 1916.

Stereum fimbriatum Ellis 1877
Bull. Torrey Bot. Club 6:133.

The type from New Jersey has no hymenium, fide Burt (89). A *nomen dubium*.

Stereum grantii Lloyd 1924
Mycol. Writ. 7:1314.

Known only from the two United States collections cited (see 593) in the original description. The name is now a synonym of ***Aphelaria tuberosa*** (Grev.) Corner, fide Reid (553).

Stereum laetum Lloyd 1922
Mycol. Writ. 7:1157.

Apparently collected in Quebec. An illegitimate name because it is a later homonym of ***Stereum laetum*** Berk. 1853.

Stereum membranacea (Fr.) Fr. 1838
Epicrisis, 547.

The report (593) from Florida needs to be verified. The specimen may not be the same as the type, if one exists.

Stereum unicum Lloyd 1913
Mycol. Writ. 4 (Stip. Stereums:35).

The type from New York has no basidia, and is indeterminable, fide Burt (89) and see Reid (553). A *nomen dubium*.

Thelephora rosella Peck 1885
Annual Rep. New York State Mus. 35:136.

The type is from New York (see 553) and the name is now ***Xylocoremium flabelliforme*** (Schw.:Fr.) J.D. Rogers, Mycologia 76:914, 1984.

Thelephora subundulata Peck 1895
Bull. Torrey Bot. Club 22:492.

The type is from Delaware (see 553) and the name is now ***Craterellus subundulatus*** (Peck) Peck.

Tomentella höhnelii Skovsted 1950
Compt. Rend. Lab. Carlsb. sér. Physiol. 25:29.

Reported (249) from Nova Scotia but the specimen is of uncertain identity because Larsen (362) declared the name a *nomen dubium*.

Tomentella jaapii (Bres.) Bourd. & Galzin 1924
Bull. Soc. Mycol. France 40:155.

Reported (607) from Minnesota on *Betula* sp. but the specimen is *T. subvinosa*, fide Larsen (359:84 & 138).

Tomentella pannosa (Berk. & Curtis) Bourd. & Galzin 1924
Bull. Soc. Mycol. France 40:150.
Basionym *Zygodesmus pannosus* Berk. & Curtis, Grevillea 3:112, 1875.
Synonym *Hypochnus pannosus* (Berk. & Curtis) Burt 1916.

The type is from Pennsylvania, fide Larsen (359). Also reported (116, 133) from British Columbia, Manitoba and Washington. These specimens should be re-identified. A *nomen confusum*, fide Larsen (359, 362). Perhaps Michener 4380 should be designated lectotype.

Tomentella spongiosa (Schw.) Höhnel & Litsch. 1908
Oesterr. Bot. Z. 58:333.
Basionym *Thelephora spongiosa* Schw., Schriften Naturf. Ges. Leipzig 1:109, 1822.
Synonym *Hypochnus spongiosus* (Schw.) Burt 1916.

Reported from Michigan and Oregon (334, 335), and Nova Scotia (249) but the specimens are of uncertain identity. The type from Pennsylvania is an abnormally developed *Thelephora*, fide Larsen (359, 362).

Tomentella subfusca (Karsten) Höhnel & Litsch. 1906
Sitzungsber. Kaiserl. Akad. Wiss., Math.-Naturwiss. Cl., Abt. 1, 115:1572.
Basionym *Hypochnus subfuscus* Karsten, Bidrag Kännedom Finlands Natur Folk 37:163, 1882.

Reported (116) on *Fagus grandifolia* in Nova Scotia. The type could not be located and the species circumscription is uncertain, fide Larsen (362).

Tomentella testacea Bourd. & Galz. 1924
Bull. Soc. Mycol. France 40:146.

Reported (249) from Nova Scotia, but the specimen is of uncertain identity because the characters to be associated with this name are uncertain, see Larsen (362).

Tremella albida W. Hudson:Fr. 1778
Flora anglica, Ed. II, 1:565.

This name has been applied to several different fungi (159) and the identity of the fungus reported (333) from Tennessee is uncertain.

Tremella mycetophila Peck 1876
Annual Rep. New York State Mus. 28:53
Synonym *Christiansenia mycetophila* (Peck) Ginns & S. Sunhede 1978

Reported (241, 249) from New Brunswick, Nova Scotia, Ontario, Quebec, New Hampshire, New York, and Vermont. It is currently named ***Syzygospora mycetophila*** (Peck) Ginns. Because this fungus is parasitic on agarics and clearly not lignicolous it is not included above.

Tremella phyllachoroidea Sacc. 1904
Harriman Alaska Exped. 5. The Fungi of Alaska, 42, tab. 3, f. 1.

From Alaska on decaying fallen leaves of *Menziesia ferruginea*, see also Conners (116). Saccardo inserted a "?" before the epithet, perhaps because basidia and spores were not seen. He concluded "dubia tamen, quia sterilis."

Tremella pinicola Britz. 1891
Hym. Sudbayer 9:22, f. 19.

Reported (107) from North and South Carolina on *Pinus taeda*. The specimens should be renamed. The name is illegitimate because it is a later homonym of *T. pinicola* Peck 1886 (see ***Exidia pinicola***).

Tremella subcarnosa Peck 1879
Annual Rep. New York State Mus. 32:36.

Known only from the type specimen. The microscopic features need to be described.

Tremella subochracea Peck 1881
Annual Rep. New York State Mus. 34:43.

Known only from the type specimen. The microscopic features need to be described.

Tremella uliginosa Karsten 1883
Meddeland. Soc. Fauna Fl. Fenn. 9:111.

Reported (345) to be "parasitic on sterile septate mycelium" of a Basidiomycete sclerotium in British Columbia. The fungus is not a *Tremella*, and is the type species of *Tetragoniomyces* Oberw. & Bandoni (494).

Tremella vesicaria Bull. 1791
Histoire Champignons France, 224, tab. 427.

Reported from Pennsylvania (410, 510) and in New York (525). The latter was referred to *T. reticulata* by Donk (157). A *nomen dubium*, fide Donk (157).

Tremella ?viscosa Berk.
This name, with a question mark before the epithet, appeared in Conners (116) and was based upon a specimen from Manitoba. The name perhaps should have been *Tremella viscosa* (Pers.) Berk. & Br. 1854 (basionym *Corticium viscosum* Pers., Obs. Mycol. 2:18, 1799). Donk (157) concluded that the name had been applied to several fungi and was of uncertain application. The name is not adequately documented from North America.

REFERENCES

1. Agerer, R. 1973. *Rectipilus*, Eine neue Gattung cyphelloider Pilze. Persoonia 7: 389-436.
2. Agerer, R. 1975. *Flagelloscypha*, Studien an cyphelloiden Basidiomyceten. Sydowia 27: 131-264.
3. Agerer, R. 1979. *Flagelloscypha* sect. *Lachnelloscypha*. Persoonia 10: 337-346.
4. Agerer, R. 1979. A new combination in the genus *Flagelloscypha* and a contribution to the identity of *Cyphella peckii*. Mycotaxon 9: 464-468.
5. Agerer, R. 1983. Typusstudien an cyphelloiden Pilzen IV. *Lachnella* Fr. s.l. Mitt. Bot. Staatssamml. München 19: 163-334.
6. Ainsworth, G.C., F.K. Sparrow & A.S. Sussman (Eds.) 1973. *The Fungi. Volume 4B.* Academic Press, New York. 504 p.
7. Alfieri, S.A., K.R. Langdon, C. Wehlburg & J.W. Kimbrough. 1984. Index of plant diseases in Florida. Florida Dept. Agric. & Consumer Services, Plant Industry Bull. 11 (revised): 1-389.
8. Baker, G.E. 1936. A study of the genus *Helicogloea*. Ann. Missouri Bot. Gard. 23: 69-128.
9. Baker, G.E. 1941. Studies on the genus *Physalacria*. Bull. Torrey Bot. Club 68: 265-288.
10. Baker, G.E. 1946. Addenda to the genera *Helicogloea* and *Physalacria*. Mycologia 38: 630-638.
11. Baker, K.F. 1970. Types of *Rhizoctonia* diseases and their occurrence. Pp. 125-148. *In: Rhizoctonia solani, biology and pathology.* Ed., J.R. Parmeter. Univ. California Press, Berkeley. 255 p.
12. Bandoni, R.J. 1955. A new species of *Helicobasidium*. Mycologia 47: 918-919.
13. Bandoni, R.J. 1956. A preliminary survey of the genus *Platygloea*. Mycologia 48: 821-840.
14. Bandoni, R.J. 1957. The spores and basidia of *Sirobasidium*. Mycologia 49: 250-255.
15. Bandoni, R.J. 1958. Some tremellaceous fungi in the C.G. Lloyd Collection. Lloydia 21: 137-151.
16. Bandoni, R.J. 1961. The genus *Naematelia*. Amer. Midl. Naturalist 66: 319-328.
17. Bandoni, R.J. 1963. *Dacrymyces ovisporus* from British Columbia. Mycologia 55: 360-361.
18. Bandoni, R.J. 1963. Conjugation in *Tremella mesenterica*. Canad. J. Bot. 41: 467-474.
19. Bandoni, R.J. 1973. Epistolae mycologicae II. Species of *Platygloea* from British Columbia. Syesis 6: 229-232.
20. Bandoni, R.J. 1984. The Tremellales and Auriculariales: an alternative classification. Trans. Mycol. Soc. Japan 25: 489-530.
21. Bandoni, R.J. & J. Ginns. 1993. On some species of *Tremella* associated with Corticiaceae. Trans. Mycol. Soc. Japan 34: 21-36.
22. Bandoni, R.J. & B.N. Johri. 1975. Observations on the genus *Exobasidiellum*. Canad. J. Bot. 53: 2561-2564.
23. Bandoni, R.J. & F. Oberwinkler. 1983. On some species of *Tremella* described by Alfred Möller. Mycologia 75: 854-863.
24. Baniecki, J.F. & H.E. Bloss. 1969. The basidial stage of *Phymatotrichum omnivorum*. Mycologia 61: 1054.
25. Banker, H.J. 1906. A contribution to a revision of the North American Hydnaceae. Mem. Torrey Bot. Club 12: 99-194.
26. Banker, H.J. 1912. Type studies in the Hydnaceae II. The genus *Steccherinum*. Mycologia 4: 309-318.
27. Barr, M.E. & H.E. Bigelow. 1968. *Martindalia* unmasked. Mycologia 60: 456-457.
28. Barrett, M.F. 1910. Three common species of *Auricularia*. Mycologia 2: 12-18.
29. Basham, J.T. 1959. Studies in forest pathology XX. Investigations of the pathological deterioration in killed balsam fir. Canad. J. Bot. 37: 291-326.
30. Basham, J.T. 1966. Heart rot of jack pine in Ontario I. The occurrence of Basidiomycetes and microfungi in defective and normal heartwood of living jack pine. Canad. J. Bot. 44: 275-295.
31. Basham, J.T., J. Hudak, D. Lachance, L.P. Magasi & M.A. Stillwell. 1976. Balsam fir death and deterioration in eastern Canada following girdling. Canad. J. Forest Res. 6: 406-414.
32. Berthier, J. 1985. Les Physalacriaceae du Globe. Biblio. Mycol. 98: 1-128.
33. Bigelow, H.E. 1966. Contributions to the fungus flora of northeastern North America IV. Rhodora 68: 175-191.
34. Bodman, M.C. 1949. The genus *Heterochaete* in the United States. Mycologia 42: 527-536.
35. Bodman, M.C. 1952. A taxonomic study of the genus *Heterochaete*. Lloydia 15: 193-233.
36. Boidin, J. 1958. Essai Biotaxonomique sur les Hydnés résupinés et les Corticiés. Rev. Mycol. Mém. 6: 1-387.
37. Boidin, J. 1966. Basidiomycètes Corticiaceae de la République Centrafricaine 1. Le genre *Gloeocystidiellum* Donk. Cahiers de la Maboké 4: 5-17.
38. Boidin, J. 1966. Basidiomycètes Auriscalpiaceae de la République Centrafricaine. Cahiers de la Maboké 4: 18-25.
39. Boidin, J. 1981. Nouvelles espèces de Lachnocladiaceae du Canada (Basidiomycetes). Naturaliste Canad. 108: 199-203.
40. Boidin, J. & P. Lanquetin. 1980. Contribution a l'étude du genre *Dichostereum* Pilát (Basidiomycetes Lachnocladiaceae). Bull. Soc. Mycol. France 96: 381-406.
41. Boidin, J. & P. Lanquetin. 1983. Basidiomycètes Aphyllophorales epitheloides étales. Mycotaxon 16: 461-499.
42. Boidin, J. & P. Lanquetin. 1984. Le genre *Amylostereum* (Basidiomycetes) intercompatibilités partielles entre espèces allopatriques. Bull. Soc. Mycol. France 100: 211-236.
43. Boidin, J. & P. Lanquetin. 1984. Répertoire des données utils pour effectuer les tests d'intercompatibilité chez les Basidiomycètes. Crypt. Mycol. 5: 193-245.
44. Boidin, J. & P. Lanquetin. 1987. Le genre *Scytinostroma* Donk (Basidiomycetes, Lachnocladiaceae). Biblio. Mycol. 114: 1-130.
45. Boidin, J., P. Lanquetin, P. Terra & C. Gomez. 1976. *Vararia* subg. *Vararia* (Basidiomycetes, Lachnocladiaceae). Bull. Soc. Mycol. France 92: 247-277.
46. Boidin, J., P. Lanquetin, G. Gilles, F. Candoussau & R. Hugueney. 1985. Contribution à la connaissance des Aleurodiscoideae à spores amyloides. Bull. Soc. Mycol. France 101: 333-367.

47. Boidin, J. & M. des Pomeys. 1961. Hétérobasidiomycètes saprophytes et homobasidiomycètes résupinés IX. Bull. Soc. Mycol. France 77: 237-261.

48. Boidin, J., P. Terra & P. Lanquetin. 1968. Contribution à la connaissance des charactères mycéliens et sexuels des genres *Aleurodiscus*, *Dendrothele*, *Laeticorticium* et *Vuilleminia*. Bull. Soc. Mycol. France 84: 53-84.

49. Boquiren, D.T. 1971. The genus *Epithele*. Mycologia 63: 937-957.

50. Bourdot, H. & A. Galzin. 1928. *Hyménomycètes de France.* M. Bry, Sceaux. 764 p.

51. Brasfield, T.W. 1938. The Dacrymycetaceae of temperate North America. Amer. Midl. Naturalist 20: 211-235.

52. Brasfield, T.W. 1940. Notes on the Dacrymycetaceae. Lloydia 3: 105-108.

53. Bresadola, G. 1925. New species of fungi. Mycologia 17: 68-77.

54. Brough, S.G. 1974. *Tremella globospora*, in the field and in culture. Canad. J. Bot. 52: 1853-1859.

55. Brough, S. & R.J. Bandoni. 1975. Epistolae mycologicae VI. Occurrence of *Dacryonaema* in British Columbia. Syesis 8: 301-303.

56. Brown, C.A. 1935. Morphology and biology of some species of *Odontia*. Bot. Gaz. (Crawfordsville) 96: 640-675.

57. Budington, A.B. & R.L. Gilbertson. 1973. Some southwestern lignicolous Hymenomycetes of special interest. Southw. Naturalist 17: 409-422.

58. Burdsall, H.H. 1969. Stephanocysts: unique structures in the Basidiomycetes. Mycologia 61: 915-923.

59. Burdsall, H.H. 1971. Notes on some lignicolous Basidiomycetes of the southeastern United States. J. Elisha Mitchell Sci. Soc. 87: 239-245.

60. Burdsall, H.H. 1975. A new species of *Laeticorticium* (Aphyllophorales, Corticiaceae) from the southern Appalachians. J. Elisha Mitchell Sci. Soc. 91: 243-245.

61. Burdsall, H.H. 1975. Taxonomic and distributional notes on Corticiaceae (Homobasidiomycetes, Aphyllophorales) of the southern Appalachians. Pp. 265-286. *In: Distributional history of the Biota of the southern Appalachians. Part IV. Algae and Fungi.* Eds., B.C. Parker and M.K. Roane. Univ. Virginia Press, Charlottesville. 416 p.

62. Burdsall, H.H. 1979. Review of: *Basidiomycetes that decay aspen in North America*. Mycologia 71: 229-231.

62a. Burdsall, H.H. 1981. The taxonomy of *Sporotrichum pruinosum* and *Sporotrichum pulverulentum*/ *Phanerochaete chrysosporium.* Mycologia 73: 765-680.

63. Burdsall, H.H. 1984. The genus *Candelabrochaete* (Corticiaceae) in North America and a note on *Peniophora mexicana*. Mycotaxon 19: 389-395.

64. Burdsall, H.H. 1985. A contribution to the taxonomy of the genus *Phanerochaete* (Corticiaceae, Aphyllophorales). Mycologia Memoir 10: 1-165.

65. Burdsall, H.H. & W.E. Eslyn. 1974. A new *Phanerochaete* with a *Chrysosporium* imperfect state. Mycotaxon 1: 123-133.

66. Burdsall, H.H. & R.L. Gilbertson. 1974. Three new species of *Phanerochaete* (Aphyllophorales, Corticiaceae). Mycologia 66: 780-790.

67. Burdsall, H.H. & R.L. Gilbertson. 1982. New species of Corticiaceae (Basidiomycotina, Aphyllophorales) from Arizona. Mycotaxon 15: 333-340.

68. Burdsall, H.H., H.C. Hoch, M.G. Boosalis & E.C. Setliff. 1980. *Laetisaria arvalis* (Aphyllophorales, Corticiaceae): a possible biological control agent for *Rhizoctonia solani* and *Pythium* species. Mycologia 72: 728-736.

69. Burdsall, H.H. & M.J. Larsen. 1974. *Lazulinospora*, a new genus of Corticiaceae and a note on *Tomentella atrocyanea*. Mycologia 66: 96-100.

70. Burdsall, H.H. & F.F. Lombard. 1976. The genus *Gloeodontia* in North America. Mem. New York Bot. Gard. 28: 16-31.

71. Burdsall, H.H. & O.K. Miller. 1988. Type studies and nomenclatural considerations in the genus *Sparassis*. Mycotaxon 31: 199-206.

72. Burdsall, H.H. & O.K. Miller. 1988. Neotypification of *Sparassis crispa*. Mycotaxon 31: 591-593.

73. Burdsall, H.H. & K.K. Nakasone. 1978. Taxonomy of *Phanerochaete chrysorhizon* and *Hydnum omnivorum*. Mycotaxon 7: 10-22.

74. Burdsall, H.H. & K.K. Nakasone. 1981. New or little known lignicolous Aphyllophorales (Basidiomycotina) from southeastern United States. Mycologia 73: 454-476.

75. Burdsall, H.H. & K.K. Nakasone. 1983. Species of effused Aphyllophorales (Basidiomycotina) from the southeastern United States. Mycotaxon 17: 253-268.

76. Burdsall, H.H., K.K. Nakasone & G.W. Freeman. 1981. New species of *Gloeocystidiellum* (Corticiaceae) from the southeastern United States. Systematic Bot. 6: 422-434.

77. Burpee, L. 1982. Isolates of *Rhizoctonia cerealis* cause damping-off of cereal seedlings. Canad. J. Plant Path. 4: 304.

78. Burt, E.A. 1914. The Thelephoraceae of North America I. Ann. Missouri Bot. Gard. 1: 185-228.

79. Burt, E.A. 1914. The Thelephoraceae of North America II. *Craterellus*. Ann. Missouri Bot. Gard. 1: 327-350.

80. Burt, E.A. 1914. The Thelephoraceae of North America III. *Craterellus borealis* and *Cyphella*. Ann. Missouri Bot. Gard. 1: 357-382.

81. Burt, E.A. 1915. The Thelephoraceae of North America V. *Tremellodendron, Eicheriella* and *Sebacina*. Ann. Missouri Bot. Gard. 2: 731-770.

82. Burt, E.A. 1916. The Thelephoraceae of North America VI. *Hypochnus*. Ann. Missouri Bot. Gard. 3: 203-241.

83. Burt, E.A. 1917. *Merulius* in North America. Ann. Missouri Bot. Gard. 4: 305-362.

84. Burt, E.A. 1918. Corticiums causing Pellicularia disease of the coffee plant, Hypochnose of pomaceous fruits and Rhizoctonia disease. Ann. Missouri Bot. Gard. 5: 119-132.

85. Burt, E.A. 1918. The Thelephoraceae of North America X. *Hymenochaete*. Ann. Missouri Bot. Gard. 5: 301-372.

86. Burt, E.A. 1919. *Merulius* in North America, supplementary notes. Ann. Missouri Bot. Gard. 6: 143-145.

87. Burt, E.A. 1919. *Protomerulius farlowii* Burt, *n. sp.* Ann. Missouri Bot. Gard. 6: 175-177.

88. Burt, E.A. 1919. The Thelephoraceae of North America XI. *Tulasnella, Veluticeps, Mycobonia, Epithele*, and *Lachnocladium*. Ann. Missouri Bot. Gard. 6: 253-280.

89. Burt, E.A. 1920. The Thelephoraceae of North America XII. *Stereum*. Ann. Missouri Bot. Gard. 7: 81-248.

90. Burt, E.A. 1921. Some North American Tremellaceae, Dacrymycetaceae, and Auriculariaceae. Ann. Missouri Bot. Gard. 8: 361-396.

91. Burt, E.A. 1924. The Thelephoraceae of North America XIII. *Cladoderris, Hypolyssus, Cymatella, Skepperia, Cytidia, Solenia, Matruchotia, Microstroma, Protocoronospora*, and *Asterostroma*. Ann. Missouri Bot. Gard. 11: 1-36.

92. Burt, E.A. 1925. The Thelephoraceae of North America XIV. *Peniophora*. Ann. Missouri Bot. Gard. 12: 213-357.

93. Burt, E.A. 1926. The Thelephoraceae of North America XV. *Corticium*. Ann. Missouri Bot. Gard. 13: 173-358.

94. Cahill, J.V., N.R. O'Neill and P.H. Dernoeden. 1982. Variations among isolates of *Corticium fuciforme* causing red thread disease of turf grasses. Phytopathology 72: 932.

95. Campbell, W.A. & R.W. Davidson. 1940. Top rot in glaze-damaged black cherry and sugar maple on the Allegheny Plateau. J. Forest. (Washington) 38: 963-965.

96. Campbell, W.A., J.H. Miller & G.E. Thompson. 1950. Notes on some wood-decaying fungi of Georgia. Pl. Dis. Reporter 34: 128-134.

97. Cash, E.K. 1953. A check list of Alaskan fungi. Pl. Dis. Reporter, Suppl. 219: 1-70.

98. Cauchon, R. & G.B. Ouellette. 1966. Cultural characters of *Phlebia atkinsoniana* W.B. Cooke. Canad. J. Bot. 44: 527-528.

99. Chamuris, G.P. 1984. Nomenclature adjustments in *Stereum* and *Cylindrobasidium* according to the Sydney Code. Mycotaxon 20: 587-588.

100. Chamuris, G.P. 1985. On distinguishing *Stereum gausapatum* from the "*S. hirsutum* complex". Mycotaxon 22: 1-12.

101. Chamuris, G.P. 1986. The *Cystostereum pini-canadense* complex in North America. Mycologia 78: 380-390.

101a. Chamuris, G.P. 1986. Modification of Nobles' species code for identification of Basidiomycete cultures. Mycotaxon 25: 159-160.

102. Chamuris, G.P. 1987. Notes on stereoid fungi I. The genus *Dendrophora* stat. nov. and *Peniophora malenconii* subsp. *americana, subsp. nov.* ("*Stereum heterosporum*"). Mycotaxon 28: 540-552.

103. Chamuris, G.P. 1987. The population structure of *Peniophora rufa* in an aspen plantation. Mycologia 79: 451-457.

104. Chamuris, G.P. 1988. The non-stipitate stereoid fungi in the northeastern United States and adjacent Canada. Mycologia Memoir 14: 1-247.

105. Clark, W.S., M.J. Richardson & R. Watling. 1983. Calyptella root rot - a new fungal disease of tomatoes. Plant Pathology 32: 95-99.

106. Clements, M.A. 1988. Orchid mycorrhizal associations. Lindleyana 3: 73-86.

107. Coker, W.C. 1920. Notes on the lower Basidiomycetes of North Carolina. J. Elisha Mitchell Sci. Soc. 35: 113-182.

108. Coker, W.C. 1921. Notes on the Thelephoraceae of North Carolina. J. Elisha Mitchell Sci. Soc. 36: 146-196.

109. Coker, W.C. 1926. Further notes on Hydnums. J. Elisha Mitchell Sci. Soc. 41: 270-286.

110. Coker, W.C. 1927. New or noteworthy Basidiomycetes. J. Elisha Mitchell Sci. Soc. 43: 129-145.

111. Coker, W.C. 1928. Notes on Basidiomycetes. J. Elisha Mitchell Sci. Soc. 43: 233-242.

112. Coker, W.C. 1929. Notes on fungi. J. Elisha Mitchell Sci. Soc. 45: 164-178.

113. Coker, W.C. 1930. Notes on fungi, with a description of a new species of *Ditiola*. J. Elisha Mitchell Sci. Soc. 46: 117-120.

114. Coker, W.C. 1948. Notes on some higher fungi. J. Elisha Mitchell Sci. Soc. 64: 135-146.

115. Coker, W.C. & A. Beers. 1951. *The stipitate Hydnums of the eastern United States.* Univ. N. Carolina Press, Chapel Hill. 86 p.

116. Conners, I.L. 1967. An annotated index of plant diseases in Canada. Canad. Dept. Agric. Res. Br. Publ. 1251: 1-381.

117. Cooke, W.B. 1951. The genus *Cytidia*. Mycologia 43: 196-210.

118. Cooke, W.B. 1952. Western fungi-II. Mycologia 44: 245-261.

119. Cooke, W.B. 1955. Fungi, lichens and mosses in relation to vascular plant communities in eastern Washington and adjacent Idaho. Ecol. Monogr. 25: 119-180.

120. Cooke, W.B. 1955. Fungi of Mount Shasta (1936-1951). Sydowia 9: 94-215.

121. Cooke, W.B. 1955. Subalpine fungi and snowbanks. Ecology 36: 124-130.

122. Cooke, W.B. 1956. The genus *Phlebia*. Mycologia 48: 386-405.

123. Cooke, W.B. 1957. The genera *Serpula* and *Meruliporia*. Mycologia 49: 197-225.

124. Cooke, W.B. 1957. The Porotheleaceae: *Porotheleum*. Mycologia 49: 680-693.

125. Cooke, W.B. 1961. The cyphellaceous fungi, a study in the Porotheleaceae. Beih. Sydowia 4: 1-144.

126. Cooke, W.B. 1961. The genus *Schizophyllum*. Mycologia 53: 575-599.

127. Cooke, W.B. 1973. Back-yard fungi. Ohio J. Sci. 73: 88-96.

128. Cooke, W.B. 1974. Some Aphyllophorales of the southern Cascade Mountains. Trans. Mycol. Soc. Japan 15: 324-340.

129. Cooke, W.B. 1976. On *Cyphellopsis anomala* (Persoon ex Fries) Donk. Mem. New York Bot. Gard. 28: 32-37.

130. Cooke, W.B. 1985. Fungi of Lassen Volcanic National Park. Univ. California at Davis, Institute of Ecology Tech. Rept. 21: 1-251.

131. Cooke, W.B. 1989. The cyphelloid fungi of Ohio. Mem. New York Bot. Gard. 49: 158-172.

132. Cooke, W.B. & R. Pomerleau. 1964. IX International Botanical Congress - Field Trip No.16 - Fungi. Mycologia 56: 607-618.

133. Cooke, W.B. & C.G. Shaw. 1953. The Suksdorf fungus collections. Res. Stud. State Coll. Wash. 21: 3-56.

134. Corner, E.J.H. 1950. A monograph of *Clavaria* and allied genera. Ann. Bot. Mem. 1: 1-740.

135. Corner, E.J.H. 1966. *A monograph of cantharelloid fungi*. Oxford University Press, London. 255 p.

136. Corner, E.J.H. 1968. A monograph of *Thelephora* (Basidiomycetes). Beih. Nova Hedwigia 27: 1-110.

137. Corner, E.J.H. 1970. Supplement to " A monograph of *Clavaria* and allied genera." Beih. Nova Hedwigia 33: 1-299.

138. Currah, R.S., L. Sigler & S. Hambleton. 1987. New records and new taxa of fungi from the mycorrhizae of terrestrial orchids of Alberta. Canad. J. Bot. 65: 2473-2482.

138a. Currah, R.S., E.A. Smreciu & S. Hambleton. 1990. Mycorrhizae and mycorrhizal fungi of boreal species of *Platanthera* and *Coeloglossum* (Orchidaceae). Canad. J. Bot. 68:1171-1181.

139. Dale, J.L. 1978. Atypical symptoms of *Rhizoctonia* infection on *Zoysia*. Pl. Dis. Reporter 62: 645-647.

140. Danielson, R.M. & M. Pruden. 1989. The ectomycorrhizal status of urban spruce. Mycologia 81: 335-341.

141. Davidson, A.G. & D.E. Etheridge. 1963. Infection of balsam fir, *Abies balsamea* (L.) Mill., by *Stereum sanguinolentum* (Alb. and Schw. ex Fr.) Fr. Canad. J. Bot. 41: 759-765.

142. Davidson, R.W. 1934. *Stereum gaustapatum*, cause of heart rot of oaks. Phytopathology 24: 831-832.

143. Davidson, R.W., W.A. Campbell & R.C. Lorenz. 1941. Association of *Stereum murrayi* with heart rot and cankers of living hardwoods. Phytopathology 31: 82-87.

144. Davidson, R.W. & T.E. Hinds. 1958. Unusual fungi associated with decay in some forest trees in Colorado. Phytopathology 48: 216-218.

145. Davidson, R.W., P.L. Lentz, & H.H. McKay. 1960. The fungus causing pecky cypress. Mycologia 52: 260-279.

146. Davidson, R.W. & F.F. Lombard. 1953. Large-brown-spored house-rot fungi in the United States. Mycologia 45: 88-100.

147. Demoulin, V. 1985. *Stereum fasciatum* (Schw.) Fr. and *S. lobatum* (Kunze: Fr.) Fr.: two distinct species. Mycotaxon 23: 207-217.

148. Dodd, J.L. 1972. The genus *Clavicorona*. Mycologia 64: 737-773.

149. Domanski, S. 1976. *Resinicium bicolor* in Poland. Mem. New York Bot. Gard. 28: 58-66.

150. Domanski, S. & A. Orlicz. 1969. Studium nad grzybem wieloporowatym *Irpex lacteus* (Fr. ex Fr.)Fr. Acta Mycol. 5: 149-159.

151. Donk, M.A. 1954. Notes on resupinate Hymenomycetes-I. On *Pellicularia* Cooke. Reinwardtia 2: 425-434.

152. Donk, M.A. 1957. Notes on resupinate Hymenomycetes-IV. Fungus 27: 1-29.

153. Donk, M.A. 1957. Notes on resupinate Hymenomycetes-V. Fungus 28: 16-36.

154. Donk, M.A. 1957. The generic names proposed for Hymenomycetes-VII. Taxon 6: 68-85.

155. Donk, M.A. 1959. Notes on 'Cyphellaceae'-I. Persoonia 1: 25-110.

156. Donk, M.A. 1962. Notes on 'Cyphellaceae'-II. Persoonia 2: 331-348.

157. Donk, M.A. 1966. Check list of European Hymenomycetous Heterobasidiae. Persoonia 4: 145-335.

158. Donk, M.A. 1967. Notes on European Polypores-II. Notes on *Poria*. Persoonia 5: 47-130.

159. Donk, M.A. 1974. Check list of European Hymenomycetous Heterobasidiae. Supplement and corrections. Persoonia 8: 33-50.

160. Donk, M.A. 1974. Check list of European Polypores. Verh. Kon. Ned. Akad. Wetensch., Afd. Natuurk., Ser. 2, 62: 1-469.

161. Doty, M.S. 1947. *Clavicorona*, a new genus among the clavarioid fungi. Lloydia 10: 38-44.

162. Duncan, E.G. 1972. Microevolution in *Auricularia polytricha*. Mycologia 64: 394-404.

163. Duncan, E.G. & J.A. MacDonald. 1967. Micro-evolution in *Auricularia auricula*. Mycologia 59: 803-818.

164. Eriksson, J. 1950. *Peniophora* Cke. sect. *coloratae* Bourd. & Galz. A taxonomical study with special reference to the Swedish species. Symb. Bot. Upsal. 10 (5): 1-76.

165. Eriksson, J. 1953. *Odontia breviseta* (Karst.) John Erikss. n. comb. Pp. 22. *In: Fungi exsiccati Suecici. Nos. 2101-2200.* Eds., S. Lundell & J.A. Nannfeldt.

166. Eriksson, J. 1958. Studies in Corticiaceae (*Botryohypochnus* Donk, *Botryobasidium* Donk, and *Gloeocystidiellum* Donk). Svensk Bot. Tidskr. 52: 1-17.

167. Eriksson, J. 1958. Studies in the Heterobasidiomycetes and Homobasidiomycetes - Aphyllophorales of Muddus National Park in North Sweden. Symb. Bot. Upsal. 16 (1): 1-172.

168. Eriksson, J. & K. Hjortstam. 1969. Studies in the *Botryobasidium vagum* complex (Corticiaceae). Friesia 9: 10-17.

169. Eriksson, J., K. Hjortstam & L. Ryvarden. 1978. The Corticiaceae of North Europe 5: 887-1047. Fungiflora, Oslo.

170. Eriksson, J., K. Hjortstam & L. Ryvarden. 1981. The Corticiaceae of North Europe 6: 1048-1276. Fungiflora, Oslo.

171. Eriksson, J., K. Hjortstam & L. Ryvarden. 1984. The Corticiaceae of North Europe 7: 1279-1449. Fungiflora, Oslo.

172. Eriksson, J. & L. Ryvarden. 1973. The Corticiaceae of North Europe 2: 60-286. Fungiflora, Oslo.

173. Eriksson, J. & L. Ryvarden. 1975. The Corticiaceae of North Europe 3: 287-546. Fungiflora, Oslo.

174. Eriksson, J. & L. Ryvarden. 1976. The Corticiaceae of North Europe 4: 547-886. Fungiflora, Oslo.

175. Eriksson, J., S. Sunhede & A.-E. Torkelsen. 1977. *Femsjonia peziziformis* (Dacrymycetales) in North Europe. Bot. Not. 130: 241-247.

176. Eslyn, W.E. 1981. Ability of isolates of *Confertobasidium olivaceoalbum* to stain and decay wood. Canad. J. Forest Res. 11: 497-501.

177. Eslyn, W.E., T.L. Highley & F.F. Lombard. 1985. Longevity of untreated wood in use above ground. Forest Products J. 35: 28-35.

178. Eslyn, W.E. & F.F. Lombard. 1984. Fungi associated with decayed wood in stored willow and cottonwood logs. Mycologia 76: 548-550.

179. Eslyn, W.E. & K.K. Nakasone. 1984. Fifteen little-known wood-products-inhabiting Hymenomycetes. Mater. U. Organ. 19: 201-240.

180. Etheridge, D.E. 1961. Factors affecting branch infection in aspen. Canad. J. Bot. 39: 799-816.

181. Evans, D., D.P. Lowe & R.S. Hunt. 1978. Annotated check list of forest insects and diseases of the Yukon Territory. Canad. Forestry Serv., Victoria, Report BC-X-169: 1-31.

182. Farlow, W.G. 1883. Notes on the cryptogamic flora of the White Mountains [New Hampshire]. Appalachia 3: 232-251.

183. Farr, D.F., G.F. Bills, G.P. Chamuris & A.Y. Rossman. 1989. *Fungi on plants and plant products in the United States.* APS Press, St. Paul. 1252 p.

184. Filer. T.H. 1966. Red thread found on bermudagrass. Pl. Dis. Reporter 50: 525-526.

185. Freeman, G.W. & R.H. Petersen. 1979. Studies in the genus *Scytinostromella*. Mycologia 71: 85-91.

186. Fries, N. 1941. Über die Sexualitàt einiger Hydnaceen. Bot. Not. 1941: 285-300.

187. Froidevaux, L. 1975. Identification of some Douglas-fir mycorrhizae. European J. For. Path. 2: 212-216.

188. Froidevaux, L., R. Amiet, M. Jaquenoud-Steinlin. 1978. Les Hyménomycètes résupinés mycorrhiziques dans le bois pourri. Schweiz. Z. Pilzk. 56: 9-14.

189. Gilbertson, R.L. 1962. Resupinate hydnaceous fungi of North America I. Type studies of species described by Peck. Mycologia 54: 658-677.

190. Gilbertson, R.L. 1963. Resupinate hydnaceous fungi of North America II. Type studies of species described by Bresadola, Overholts and Lloyd. Pap. Michigan Acad. Sci. 48: 137-149.

191. Gilbertson, R.L. 1964. Resupinate hydnaceous fungi of North America III. Additional type studies. Pap. Michigan Acad. Sci. 49: 15-25.

192. Gilbertson, R.L. 1965. Resupinate hydnaceous fungi of North America V. Type studies of species described by Berkeley and Curtis. Mycologia 57: 845-871.

193. Gilbertson, R.L. 1965. Some species of *Vararia* from temperate North America. Pap. Michigan Acad. Sci. 50: 161-184.

194. Gilbertson, R.L. 1970. A new *Vararia* from western North America. Madrono 20:282-287.

195. Gilbertson, R.L. 1971. Phylogenetic relationships of Hymenomycetes with resupinate hydnaceous basidiocarps. Pp. 275-307. *In: Evolution in the higher Basidiomycetes.* Ed., R.L Petersen. Univ. Tennessee Press, Knoxville. 562 p.

196. Gilbertson, R.L. 1973. Notes on some corticioid lignicolous fungi associated with snowbanks in southern Arizona. Persoonia 7: 171-182.

197. Gilbertson, R.L. 1974. *Fungi that decay ponderosa pine.* Univ. Arizona Press, Tucson. 197 p.

198. Gilbertson, R.L. 1980. Wood-rotting fungi of North America. Mycologia 72: 1-49.

199. Gilbertson, R.L. 1981. North American wood-rotting fungi that cause brown rots. Mycotaxon 12: 372-416.

200. Gilbertson, R.L. & M. Blackwell. 1984. Two new Basidiomycetes on living live oak in the southeast and Gulf Coast region. Mycotaxon 22: 85-93.

201. Gilbertson, R.L. & M. Blackwell. 1985. Notes on wood-rotting fungi on junipers on the Gulf Coast region. Mycotaxon 24: 325-348.

202. Gilbertson, R.L. & M. Blackwell. 1987. Notes on wood-rotting fungi on junipers on the Gulf Coast region II. Mycotaxon 28: 369-402.

203. Gilbertson, R.L. & M. Blackwell. 1988. Some new or unusual corticioid fungi from the Gulf Coast region.

Mycotaxon 33: 375-386.
204. Gilbertson, R.L. & A.B. Budington. 1970. New records of Arizona wood-rotting fungi. J. Arizona Acad. Sci. 6: 91-97.
205. Gilbertson, R.L. & A.B. Budington. 1970. Three new species of wood-rotting fungi in the Corticiaceae. Mycologia 62: 673-678.
206. Gilbertson, R.L. & H.H. Burdsall. 1975. *Peniophora tamaricicola* in North America. Mycotaxon 2: 143-150.
207. Gilbertson, R.L., H.H. Burdsall & E.R. Canfield. 1976. Fungi that decay mesquite in southern Arizona. Mycotaxon 3: 487-551.
208. Gilbertson, R.L., H.H. Burdsall & M.J. Larsen. 1975. Notes on wood-rotting Hymenomycetes in New Mexico. Southw. Naturalist 19: 347-360.
209. Gilbertson, R.L., E.R. Canfield & G.B. Cummins. 1972. Notes on fungi from the L.N. Gooding Research Natural Area. Arizona Acad. Sci. 7: 129-138.
210. Gilbertson, R.L. & M.J. Larsen. 1965. Resupinate hydnaceous fungi of North America IV. Some western species of special interest. Bull. Torrey Bot. Club 92: 51-61.
211. Gilbertson, R.L. & J.P. Lindsey. 1975. Basidiomycetes that decay junipers in Arizona. Great Basin Naturalist 35: 288-304.
212. Gilbertson, R.L. & J.P. Lindsey. 1978. Basidiomycetes that decay junipers in Arizona II. Great Basin Naturalist 38: 42-48.
213. Gilbertson, R.L. & J.P. Lindsey. 1989. North American species of *Amylocorticium* (Aphyllophorales, Corticiaceae), a genus of brown rot fungi. Mem. New York Bot. Gard. 49: 138-146.
214. Gilbertson, R.L. & F.F. Lombard. 1976. Wood-rotting Basidiomycetes - Itasca State Park annotated list. J. Minnesota Acad. Sci. 42: 25-31.
215. Gilbertson, R.L., F.F. Lombard & T.E. Hinds. 1968. *Veluticeps berkeleyi* and its decay of pine in North America. Mycologia 60: 29-41.
216. Gilbertson, R.L., K.J. Martin & J.P. Lindsey. 1974. Annotated check list and host index for Arizona wood-rotting fungi. Univ. Arizona Agric. Exp. Sta. Bull. 209: 1-48.
217. Gilbertson, R.L. & L. Ryvarden. 1986-1987. *North American polypores*. Volumes 1 & 2. Fungiflora, Oslo.
218. Ginns, J. 1968. The genus *Merulius* I. Species proposed by Burt. Mycologia 60: 1211-1231.
219. Ginns, J. 1969. The genus *Merulius* II. Species of *Merulius* and *Phlebia* proposed by Lloyd. Mycologia 61: 357-372.
220. Ginns, J. 1970. The genus *Merulius* III. Species of *Merulius* and *Phlebia* proposed by Schweinitz and Peck. Mycologia 62: 238-255.
221. Ginns, J. 1970. Taxonomy of *Plicatura nivea* (Aphyllophorales). Canad. J. Bot. 48: 1039-1043.
222. Ginns, J. 1971. The genus *Merulius* IV. Species proposed by Berkeley, by Berkeley and Curtis, and by Berkeley and Broome. Mycologia 63: 219-236.
223. Ginns, J. 1971. The genus *Merulius* V. Taxa proposed by Bresadola, Bourdot & Galzin, Hennings, Rick, and others. Mycologia 63: 800-818.
224. Ginns, J. 1973. *Coniophora*: study of 22 type specimens. Canad. J. Bot. 51: 249-259.
225. Ginns, J. 1974. *Schizophyllum commune*. Fungi Canadenses No. 42. National Mycological Herbarium, Agriculture Canada, Ottawa.
226. Ginns, J. 1976. *Merulius*: s.s. and s.l., taxonomic disposition and identification of species. Canad. J. Bot. 54: 100-167.
227. Ginns, J. 1976. *Exidia recisa*. Fungi Canadenses No. 86. National Mycological Herbarium, Agriculture Canada, Ottawa.
228. Ginns, J. 1978. *Leucogyrophana* (Aphyllophorales): identification of species. Canad. J. Bot. 56: 1953-1973.
229. Ginns, J. 1979. The genus *Ramaricium* (Gomphaceae). Bot. Not. 132: 93-102.
230. Ginns, J. 1982. *Athelopsis glaucina*. Fungi Canadenses No. 231. National Mycological Herbarium, Agriculture Canada, Ottawa.
231. Ginns, J. 1982. *Cerinomyces ceraceus sp. nov.* and the similar *C. grandinioides* and *C. lagerheimii*. Canad. J. Bot. 60: 519-524.
232. Ginns, J. 1982. *Hyphoderma sibiricum*. Fungi Canadenses No. 230. National Mycological Herbarium, Agriculture Canada, Ottawa.
233. Ginns, J. 1982. *Steccherinum oreophilum*: cultural characters and occurrence in North America. Mycologia 74: 20-25.
234. Ginns, J. 1982. A monograph of the genus *Coniophora* (Aphyllophorales, Basidiomycetes). Opera Bot. 61: 1-61.
235. Ginns, J. 1982. The wood-inhabiting fungus, *Aleurodiscus dendroideus sp. nov.*, and the distinctions between *A. grantii* and *A. amorphus*. Canad. Field-Naturalist 96: 131-148.
236. Ginns, J. 1983. "*Ceratobasidium fibulatum*" an invalid name. Mycotaxon 18: 439-442.
237. Ginns, J. 1984. *Helicogloea farinacea*. Fungi Canadenses No. 286. National Mycological Herbarium, Agriculture Canada, Ottawa.
238. Ginns, J. 1984. *Hericium coralloides* N. Amer. Auct. (= *H. americanum sp. nov.*) and the European *H. alpestre* and *H. coralloides*. Mycotaxon 20: 39-43.
239. Ginns, J. 1985. *Hericium* in North America: cultural characteristics and mating behavior. Canad. J. Bot. 63: 1551-1563.
240. Ginns, J. 1986. The genus *Dentipellis* (Hericiaceae). Windahlia 16: 35-45.
241. Ginns, J. 1986. The genus *Syzygospora* (Heterobasidiomycetes: Syzogosporaceae). Mycologia 78: 619-636.
242. Ginns, J. 1986. Compendium of plant disease and decay fungi in Canada 1960-1980. Canad. Dept. Agric. Publ. 1813: 1-416.
243. Ginns, J. 1989. Descriptions and notes for some unusual North American corticioid fungi (Corticiaceae: Aphyllophorales). Mem. New York Bot. Gard. 49: 129-137.
244. Ginns, J. 1990. *Aleurodiscus occidentalis sp. nov., A. thujae sp. nov.*, and the *A. tsugae* complex in North America. Mycologia 82: 753-758.
245. Ginns, J. 1992. Reevaluation of reports of 15 uncommon species of *Corticium* from Canada and the United States. Mycotaxon 44: 197-217.
246. Ginns, J. & J. Clark. 1989. *Crustoderma longicystidia* associated with decay of lumber in British Columbia, and the cultural features of *C. dryina*. Mycologia 81: 920-925.
247. Ginns, J. & D. Malloch. 1977. *Halocyphina*, a marine Basidiomycete (Aphyllophorales). Mycologia 69: 53-58.
248. Ginns, J. & L.K. Weresub. 1976. Sclerotium-producing species of *Leucogyrophana* (Aphyllophorales). Mem. New York Bot. Gard. 28: 86-97.
249. Gourley, C.O. 1983. An annotated index of the fungi of Nova Scotia. Nova Scotian Inst. Sci. Proc. 32: 75-294.
250. Greuter, W. 1988. International code of botanical nomenclature. Regnum Veg. 118: 1-328.
251. Gross, H.L. 1964. The Echinodontiaceae. Mycopathol. Mycol. Appl. 24: 1-26.
252. Guerrero, R.T. 1978. Estimacion del numero de factores de incompatibilidad en la poblacion de *Auricularia fuscosuccinea*. Revista Fac. Agron. Univ. Nac. La Plata 54: 25-37.
253. Hallenberg, N. 1981. *Phlebia centrifuga* Karst. (Corticiaceae,

Basidiomycetes) - compatibility between specimens from Sweden and Canada. Windahlia 11 (= Götesborgs Svampklubb 1981): 33-37.

254. Hallenberg, N. 1983. Cultural studies in *Hypochnicium* (Coriticiaceae, Basiomycetes). Mycotaxon 16: 565-571.

255. Hallenberg, N. 1984. Compatibility between species of Corticiaceae s.l. (Basidiomycetes) from Europe and North America. Mycotaxon 21: 335-388.

256. Hallenberg, N. 1984. A taxonomic analysis of the *Sistotrema brinkmannii* complex (Corticiaceae, Basidiomycetes). Mycotaxon 21: 389-411.

257. Hallenberg, N. 1985. Compatibility between species of Corticiaceae s.l. (Basidiomycetes) from Europe and Canada II. Mycotaxon 24: 437-443.

258. Hallenberg, N. 1985. *The Lachnocladiaceae and Coniophoraceae of North Europe.* Fungiflora, Oslo. 96 p.

259. Hallenberg, N. 1986. Cultural studies in *Tubulicrinis* and *Xenasmatella* (Corticiaceae, Basidiomycetes). Mycotaxon 27: 361-375.

260. Hallenberg, N. 1987. Culture studies in Corticiaceae (Basidiomycetes) II. Windahlia 17: 43-47.

261. Hallenberg, N. 1989. Culture studies in Corticiaceae (Basidiomycetes) III. Windahlia 18: 25-30.

262. Hallenberg, N. & A. Bernicchia. 1987. Cultural studies in *Fibricium* (Corticiaceae, Basidiomycetes). Mycotaxon 30: 203-208.

263. Hallenberg, N. & T. Hallingbäck. 1974. Interfertility and polarity in *Amylostereum laevigatum* (Fr.) Boid. and *Peniophora lycii* (Pers.) Höhn. & Litsch. Göteborgs Svampklubb Årsskrift 1974: 18-21.

264. Hallenberg, N., K. Hjortstam & L. Ryvarden. 1985. *Pirex* genus nova (Basidiomycetes, Corticiaceae). Mycotaxon 24: 287-291.

265. Harmsen, L. 1954. Om *Polyporus caesius* og *Ditola radicata* som tømmersvampe. Bot. Tidsskr. 51: 117-123.

266. Harmsen, L. 1960. Taxonomic and cultural studies on brown spored species of the genus *Merulius*. Friesia 6: 233-277.

267. Harris, R.J. 1966. The genus *Peniophora* in Illinois. Trans. Illinois State Acad. Sci. 59: 254-274.

268. Harrison, K.A. 1973. The genus *Hericium* in North America. Michigan Bot. 12: 177-194.

269. Harrison, K.A. 1985. *Creolophus* in North America. Mycologia 76: 1121-1123.

270. Hassan, F. Kasim & A. David. 1983. Studies on the cultural characteristics of 16 species of *Hyphodontia* Eriksson and *Chaetoporellus* Bond. & Sing. ex Sing. Sydowia 36: 139-149.

271. Hawksworth, D.L. 1991. The fungal dimension of biodiversity: magnitude, significance, and conservation. Mycol. Res. 95: 641-655.

272. Hawksworth, D.L., B.C. Sutton & G.C. Ainsworth. 1983. *Dictionary of the Fungi.* 7th ed. Commonwealth Mycological Inst., Kew. 445 p.

273. Hennon, P.E. 1990. Fungi on *Chamaecyparis nootkatensis*. Mycologia 82: 59-66.

274. Hepting, G.H. 1971. Diseases of forest and shade trees of the United States. U.S. Dept. Agr. Handbook 386: 1-658.

275. Hesler, L.R. 1955. Notes on south Appalachian fungi XII. J. Tennessee Acad. Sci. 30: 212-221.

276. Highley, T.L. & T.K. Kirk. 1979. Mechanisms of wood decay and the unique features of heartrots. Phytopathology 69: 1151-1157.

277. Hjortstam, K. 1980. Notes on Corticiaceae (Basidiomycetes) VII. Mycotaxon 11: 430-434.

278. Hjortstam, K. 1981. Notes on Aphyllophorales (Basidiomycetes) from Omberg, Ostergotland, Sweden. Windahlia 11 (=Götesborgs Swampklubb 1981): 39-55.

279. Hjortstam, K. 1983. Studies in the genus *Hyphodontia* (Basidiomycetes) I: *Hyphodontia* John Erikss. sectio *Hyphodontia*. Mycotaxon 17: 550-554.

280. Hjortstam, K. 1983. Studies in tropical Corticiaceae (Basidiomycetes) V. Mycotaxon 17: 555-572.

281. Hjortstam, K. 1983. Notes on Corticiaceae (Basidiomycetes) XII. Mycotaxon 17: 577-584.

282. Hjortstam, K. 1984. Corticiaceous fungi of northern Europe - Checklist of the species in the nordic countries. Windahlia 14: 1-29.

283. Hjortstam, K. 1986. A contribution to the knowledge of *Pellidiscus pallidus*. Windahlia 15: 59-61.

284. Hjortstam, K. 1986. *Hypochnicium subrigescens*, a new species of northern Europe. Windahlia 16: 69-71.

285. Hjortstam, K. 1986. Notes on Corticiaceae (Basidiomycetes) XIV. Mycotaxon 25: 273-277.

286. Hjortstam, K. 1987. A check-list to genera and species of corticioid fungi (Hymenomycetes). Windahlia 17: 55-85.

287. Hjortstam, K. 1987. Studies in tropical Corticiaceae (Basidiomycetes) VII. Mycotaxon 28: 19-37.

288. Hjortstam, K. 1989. Corticioid fungi described by M.J. Berkeley. Kew Bull. 44: 301-315.

289. Hjortstam, K. 1990. Corticioid fungi described by M.J. Berkeley II. Species from Cuba. Mycotaxon 39: 415-423.

290. Hjortstam, K. & E. Hogholen. 1980. Notes on Corticiaceae (Basidiomycetes) VI. *Rogersella eburnea nov. spec.* Mycotaxon 10: 265-268.

291. Hjortstam, K. & K.-H. Larsson. 1977. Notes on Corticiaceae (Basidiomycetes). Mycotaxon 5: 475-480.

292. Hjortstam, K. & K.-H. Larsson 1978. Notes on Corticiaceae (Basidiomycetes) II. Mycotaxon 7: 117-124.

293. Hjortstam, K. & K.-H. Larsson. 1987. Additions to *Phlebiella* (Corticiaceae, Basidiomycetes) with notes on *Xenasma* and *Sistotrema*. Mycotaxon 29: 315-319.

294. Hjortstam, K., K.-H. Larsson, & L. Ryvarden. 1988. The Corticiaceae of North Europe 8: 1450-1631. Fungiflora, Oslo.

295. Hjortstam, K. & L. Ryvarden. 1979. Notes on Corticiaceae (Basidiomycetes) IV. Mycotaxon 9: 505-519.

296. Hjortstam, K. & L. Ryvarden. 1979. Notes on Corticiaceae (Basidiomycetes) V. Mycotaxon 10: 201-209.

297. Hjortstam, K. & M.T. Telleria. 1990. *Columnocystis*, a synonym of *Veluticeps*. Mycotaxon 37: 53-56.

298. Höhnel, F. & V. Litschauer. 1906. Beiträge zur Kenntnis der Corticieen. Akad. Wiss. Wien Sitzungsber., Math.-Naturwiss. Kl., Abt. 1, 115: 1549-1620.

299. Holmgren, P.K., N.H. Holmgren & L.C. Barnett. 1990. Index herbariorum, Part 1. Eighth ed. Regnum Veg. 120: 1-693.

300. Hughes, S.J. 1951. Studies on micro-fungi VII. *Allescheriella crocea, Oidium simile,* and *Pellicularia pruinata*. Mycological Papers (Kew) 41: 1-17.

301. Hung, Ching-Yuan & K. Wells. 1975. Genetic control of compatibility in *Myxarium nucleatum*. Mycologia 67: 1181-1187.

302. Jackson, H.S. 1948. Studies of Canadian Thelephoraceae I. Some new species of *Peniophora*. Canad. J. Res., C, 26: 128-139.

303. Jackson, H.S. 1948. Studies of Canadian Thelephoraceae II. Some new species of *Corticium*. Canad. J. Res., C, 26: 143-157.

304. Jackson, H.S. 1949. Studies of Canadian Thelephoraceae IV. *Corticium anceps* in North America. Canad. J. Res., C, 27: 241-252.

305. Jackson, H.S. 1950. Studies of Canadian Thelephoraceae V. Two new species of *Aleurodiscus* on conifers. Canad. J. Res., C, 28: 63-77.

306. Jackson, H.S. 1950. Studies of Canadian Thelephoraceae VI. The *Peniophora rimicola* group. Canad. J. Res., C, 28: 525-534.

307. Jackson, H.S. 1950. Studies of Canadian Thelephoraceae VII. Some new species of *Corticium*, section *Athele*. Canad. J. Res., C, 28: 716-725.

308. Jackson, H.S. & E.R. Dearden. 1949. Studies of Canadian Thelephoraceae III. Some new species from British Columbia. Canad. J. Res., C, 27: 147-156.

309. Jackson, H.S. & E.R. Dearden. 1951. Studies of North American Thelephoraceae I. Some new western species of *Peniophora*. Mycologia 43: 54-61.

310. Jackson, N. 1973. Superficial invasion of turf by two Basidiomycetes. J. Sports Turf Res. Inst. No. 48: 20-23.

311. Jahn, H. 1971. Stereoide Pilze in Europa (Stereaceae Pil. emend. Parm. u.a., *Hymenochaete*). Westfäl. Pilzbriefe 8: 69-176.

312. Jülich, W. 1972. Monographie der Athelieae (Corticiaceae, Basidiomycetes). Beih. Willdenowia 7: 1-283.

313. Jülich, W. 1973. Studien an resupinaten Basidiomyceten-II. Persoonia 7: 381-388.

314. Jülich, W. 1974. On *Scotoderma* and *Phlyctibasidium*, two new genera of lower Basidiomycetes with resupinate basidiocarps. Verh. Kon. Ned. Akad. Wetensch., Afd. Natuurk., Tweed Sect., 77: 149-156.

315. Jülich, W. 1974. The genera of the Hyphodermoideae (Corticiaceae). Persoonia 8: 59-97.

316. Jülich, W. 1975. On *Cerocorticium* P. Henn., a genus described from Java. Persoonia 8: 217-220.

317. Jülich, W. 1975. Studies in resupinate Basidiomycetes III. Persoonia 8: 291-305.

318. Jülich, W. 1976. Studies in resupinate Basidiomycetes IV. Persoonia 8: 431-442.

319. Jülich, W. 1976. Studies in hydnoid fungi-I. On some genera with hyphal pegs. Persoonia 8: 447-458.

320. Jülich, W. 1978. A new lichenized *Athelia* from Florida. Persoonia 10: 149-151.

321. Jülich, W. 1981. Higher taxa of Basidiomycetes. Biblio. Mycol. 85: 1-485.

322. Jülich, W. 1982. Studies in resupinate Basidiomycetes VII. Int. J. Myc. Lich. 1: 27-37.

323. Jülich, W. 1983. Parasitic Heterobasidiomycetes on other fungi. Int. J. Myc. Lich. 1: 189-203.

324. Jülich, W. 1984. *Galzinia geminispora* Olive new to Europe. Persoonia 12: 189-191.

325. Jülich, W. 1984. Die Nichtblätterpilze, Gallertpilze und Bauchpilze. Pp. 1-626. *In: Kleine Kryptogamenflora II b/1.* Ed., H. Gams. G. Fischer, Stuttgart.

326. Jülich, W. & J.A. Stalpers. 1980. The resupinate non-poroid Aphyllophorales of the temperate northern hemisphere. Verh. Kon. Ned. Akad. Wetensch., Afd. Natuurk., Ser. 2, 74: 1-335.

327. Jung, H.S. 1987. Wood-rotting Aphyllophorales of the southern Appalachian spruce-fir forest. Biblio. Mycol. 119: 1-260.

328. Kallio, P. 1980. Some observations on the fungi of the central Quebec-Labrador Peninsula. Centre for Northern Studies & McGill Univ., Montreal, Subarctic Res. Paper 30: 1-16.

329. Kaplan, J.D. & N. Jackson. 1982. Variations in growth and pathogenicity of fungi associated with red thread disease of turf grasses. Phytopathology 72: 262.

330. Kataria, H.R. & G.M. Hoffmann. 1988. A critical review of plant pathogenic species of *Ceratobasidium* Rogers. Z. Pflanzenkrankh. Pflanzenschutz. 95: 81-107.

331. Kauffman, C.H. 1908. Unreported Michigan fungi for 1907, with an outline of the Gasteromycetes of the State. Ann. Report, Mich. Acad. Sci. 10: 63-84.

332. Kauffman, C.H. 1915. The fungi of North Elba. New York State Mus. Bull. 179: 80-104.

333. Kauffman, C.H. 1917. Tennessee and Kentucky fungi. Mycologia 9: 159-166.

334. Kauffman, C.H. [1929]. A study of the fungous flora of the Lake Superior region of Michigan, with some new species. Pap. Michigan Acad. Sci. 9: 169-218.

335. Kauffman, C.H. [1930]. The fungous flora of the Siskiyou mountains in southern Oregon. Pap. Michigan Acad. Sci. 11: 151-210.

336. Keller, J. 1986. Ultrastructure des parois sporiques de quelques Aphyllophorales. Mycol. Helvetica 2: 1-34.

337. Kennedy, L.L. 1958. The genus *Dacrymyces*. Mycologia 50: 896-915.

338. Kennedy, L.L. 1964. The genus *Ditola*. Mycologia 56: 298-308.

339. Kennedy, L.L. 1972. Basidiocarp development in *Calocera cornea*. Canad. J. Bot. 50: 413-417.

340. Kimmey, J.W. & J.A. Stevenson. 1957. A forest disease survey of Alaska. Pl. Dis. Reporter 247: 87-98.

341. Kirk, T.K. 1988. Lignin degradation by *Phanerochaete chrysoporium*. ISI Atlas of science: Biochemistry 1 (1): 71-76.

342. Kirk, T.K. & R.L. Farrell. 1987. Enzymatic "combustion": the microbial degradation of lignin. Ann. Rev. Microbiol. 41: 465-505.

343. Kirk, T.K. & K.-E. Eriksson. 1989. Roles for biotechnology in manufacture. Pp. 23-28. *In: World Pulp & Paper Technology 1990.* Ed. F. Roberts. Sterling Publishing Group, London.

344. Kobayasi, Y., N. Hiratsuka, K. Aoshima, R.P. Korf, M. Soneda, K. Tubaki, & J. Sugiyama. 1967. Mycological studies of the Alaskan Arctic. Ann. Rep. Inst. Fermentation, Osaka 3: 1-138.

345. Koske, R.E. 1972. Two unusual tremellas from British Columbia. Canad. J. Bot. 50: 2565-2567.

346. Kotlaba, F. 1953. Nebezpecny parasit jabloni-*Sarcodontia crocea* (Schweinitz) *c. n.* Ceská Mykol. 7: 117-123.

347. Kotlaba, F. 1985. A remarkable *Stereum: S. subpileatum* (Aphyllophorales), its ecology and distribution special regard to Czechoslovakia. Ceská Mykol. 39: 193-204.

348. Kropp, B.R. & J.M. Trappe. 1982. Ectomycorrhizal fungi of *Tsuga heterophylla*. Mycologia 74: 479-488.

349. Laflamme, G. & M. Lortie. 1973. Micro-organismes dans les tissus colorés et cariés du peuplier faux tremble. Canad. J. Forest Res. 3: 155-160.

350. Lair, E.D. 1946. Smooth patch, a bark disease of oak. J. Elisha Mitchell Sci. Soc. 62: 212-220.

351. Lamar, R.T. & D.M. Dietrich. 1990. In situ depletion of pentachlorophenol from contaminated soil by *Phanerochaete* spp. Appl. Environ. Microbiol. 56: 3093-3100.

352. Lamar, R.T., J. Glaser & T.K. Kirk. 1990. Fate of pentachlorophenol (PCP) in sterile soils inoculated with the white-rot Basidiomycete *Phanerochaete chrysosporium*: mineralization, volatilization and depletion of PCP. Soil Biol. Biochem. 22: 433-440.

353. Lanquetin, P. 1973. Intérêt des caractères culturaux et des tests d'interfertilité dans l'étude des *Vararia* Karst. subg. *Dichostereum* (Pilát) Boid. (Basidiomycetes). Compt. Rend. Hebd. Séances Acad. Sci., D, 276: 1677-1680.

354. Lanquetin, P. 1973. Interfertilités et polarités chez les *Scytinostroma* sans boucles (Basidiomycetes, Lachnocladiaceae). Naturaliste Canad. 100: 33-49.

355. Lanquetin, P. 1973. Utilisation des cultures dans la systematique des *Vararia* Karst. subg. *Dichostereum* (Pilát) Boid. (Basidiomycetes, Lachnocladiaceae). Bull. Mens. Soc. Linn. Lyon 42: 167-192.

356. Larsen, M.J. 1964. *Hyphodontia alutacea* in North America. Canad. J. Bot. 42: 1167-1172.
357. Larsen, M.J. 1965. *Tomentella* and related genera in North America I. Studies of nomenclatural types of species of *Hypochnus* described by Burt. Canad. J. Bot. 43: 1485-1510.
358. Larsen, M.J. 1967. *Tomentella* and related genera in North America V. New North American records of tomentelloid fungi. Mycopathol. Mycol. Appl. 32: 37-67.
359. Larsen, M.J. 1968. Tomentelloid fungi of North America. New York State Univ., Coll. Forestry, Syracuse, Publ. 93: 1-157.
360. Larsen, M.J. 1970. *Tomentella* and related genera in North America VI. Some synonymy and additional new records. Mycologia 62: 256-271.
361. Larsen, M.J. 1971. The genus *Pseudotomentella* (Basidiomycetes, Thelephoraceae s. str.). Nova Hedwigia 22: 599-619.
362. Larsen, M.J. 1974. A contribution to the taxonomy of the genus *Tomentella*. Mycologia Memoir 4: 1-145.
363. Larsen, M.J. 1974. Some notes on *Pseudotomentella*. Mycologia 66: 165-168.
364. Larsen, M.J. 1981. The genus *Tomentellastrum* (Aphyllophorales, Thelephoraceae s. str.). Nova Hedwigia 35: 1-16.
365. Larsen, M.J. 1984. Notes on laeticorticioid fungi. Mycologia 76: 353-355.
366. Larsen, M.J. 1986. *Lindtneria thujatsugina sp. nov.* (Stephanosporales, Stephanosporaceae) and notes on other resupinate basidiomycetes with ornamented basidiospores. Nova Hedwigia 43: 255-267.
367. Larsen, M.J. & R.L. Gilbertson. 1974. New taxa of *Laeticorticium* (Aphyllophorales, Corticiaceae). Canad. J. Bot. 52: 687-690.
368. Larsen, M.J. & R.L. Gilbertson. 1977. Studies in *Laeticorticium* (Aphyllophorales, Corticiaceae) and related genera. Norw. J. Bot. 24: 99-121.
369. Larsen, M.J. & R.L. Gilbertson. 1978. *Laeticorticium lombardiae* (Aphyllophorales, Corticiaceae): a newly recognized segregate from the *L. roseum* complex. Mycologia 70: 206-208.
370. Larsen, M.J., M.F. Jurgensen & A.E. Harvey. 1981. *Athelia epiphylla* associated with colonization of subalpine fir foliage under psychrophilic conditions. Mycologia 72: 1195-1202.
371. Larsen, M.J. & K.K. Nakasone. 1984. Additional new taxa on *Laeticorticium* (Aphyllophorales, Corticiaceae). Mycologia 76: 528-532.
372. Larsen, M.J. & B. Zak. 1978. *Byssoporia gen. nov.*: taxonomy of the mycorrhizal fungus *Poria terrestris*. Canad. J. Bot. 56: 1122-1129.
373. Lavallee, A. 1969. Incidence des microorganismes dans le bois de l'érable à sucre et du bouleau jaune. Phytoprotection 50: 16-22.
374. Leger, J.C. 1981. Les *Hymenochaete* a elements hymeniens pinnatifides. Mycotaxon 13: 241-256.
375. Lemke, P.A. 1964. The genus *Aleurodiscus* (sensu stricto) in North America. Canad. J. Bot. 42: 213-282.
376. Lemke, P.A. 1964. The genus *Aleurodiscus* (sensu lato) in North America. Canad. J. Bot. 42: 723-768.
377. Lemke, P.A. 1965. *Dendrothele* (1907) vs. *Aleurcorticium* (1963). Persoonia 3: 365-367.
377a. Lentz, P.L. 1942. The genus *Thelephora* in Iowa. Iowa Acad. Sci. 49: 175-184.
378. Lentz, P.L. 1947. Some species of *Cyphella, Solenia* and *Porotheleum*. Proc. Iowa Acad. Sci. 54: 141-154.
379. Lentz, P.L. 1955. *Stereum* and allied genera of fungi in the Upper Mississippi Valley. U.S.D.A. Monograph 24: 1-74.
380. Lentz, P.L. 1967. Delineations of forest fungi, several species of Deuteromycetes and a newly described *Botryobasidium*. Mycopathol. Mycol. Appl. 32: 1-25.
381. Lentz, P.L. & H.H. Burdsall. 1973. *Scytinostroma galactinum* as a pathogen of woody plants. Mycopathol. Mycol. Appl. 49: 289-305.
382. Lentz, P.L. & H.H. McKay. 1970. Sclerotial formation in *Corticium olivascens* and other Hymenomycetes. Mycopathol. Mycol. Appl. 40: 1-13.
383. Lentz, P.L. & H.H. McKay. 1976. Basidiocarp and culture descriptions of *Hyphoderma* (Corticiaceae) in the delta region. Mem. New York Bot. Gard. 28: 141-162.
384. Liberta, A.E. 1960. A taxonomic analysis of section *Athele* of the genus *Corticium* I. Genus *Xenasma*. Mycologia 52: 884-914.
385. Liberta, A.E. 1961. A taxonomic analysis of section *Athele* of the genus *Corticium* II. Mycologia 53: 443-450.
386. Liberta, A.E. 1962. The genus *Paullicorticium* (Thelephoraceae). Brittonia 14: 219-223.
387. Liberta, A.E. 1964. Notes on Illinois resupinate Hymenomycetes. Mycologia 56: 249-252.
388. Liberta, A.E. 1965. Notes on Wisconsin resupinate Basidiomycetes. Mycologia 57: 459-464.
389. Liberta, A.E. 1965. A new species of *Xenasma*. Mycologia 57: 967-968.
390. Liberta, A.E. 1966. Notes on Colorado resupinate Basidiomycetes. Trans. Illinois State Acad. Sci. 59: 392-393.
391. Liberta, A.E. 1966. Resupinate Hymenomycetes from Gaspé and adjacent counties. Mycologia 58: 927-933.
392. Liberta, A.E. 1968. Descriptions of the nomenclatural types of *Peniophoras* described by Burt. Mycologia 60: 827-857.
393. Liberta, A.E. 1969. Descriptions of the nomenclatural types of *Corticiums* described by Burt. Nova Hedwigia 18: 215-233.
394. Liberta, A.E. 1971. Notes on Illinois and Wisconsin resupinate Basidiomycetes. Trans. Illinois State Acad. Sci. 64: 306.
395. Liberta, A.E. 1973. The genus *Trechispora* (Basidiomycetes, Corticiaceae). Canad. J. Bot. 51: 1871-1892.
396. Linder, D.H. 1933. The genus *Schizophyllum* I. Species of the western hemisphere. Amer. J. Bot. 20: 552-564.
397. Linder, D.H. 1942. A contribution towards a monograph of the genus *Oidium* (Fungi Imperfecti). Lloydia 5: 165-207.
398. Lindsey, J.P. 1985. Basidiomycetes that decay gambel oak in southwestern Colorado. Mycotaxon 22: 327-362.
399. Lindsey, J.P. 1986. Basidiomycetes that decay gambel oak in southwestern Colorado: II. Mycotaxon 25: 67-83.
400. Lindsey, J.P. 1986. Basidiomycetes that decay gambel oak in southwestern Colorado: III. Mycotaxon 27: 325-345.
401. Lindsey, J.P. 1987. A new species of *Aleurodiscus* from Colorado. Mycotaxon 30: 433-437.
402. Lindsey, J.P. 1988. Annotated check-list with host data and decay characteristics for Colorado wood rotting Basidiomycotina. Mycotaxon 33: 265-278.
403. Lindsey, J.P. & R.L. Gilbertson. 1975. Wood-inhabiting Homobasidiomycetes on saguaro in Arizona. Mycotaxon 2: 83-103.
404. Lindsey, J.P. & R.L. Gilbertson. 1977. A new *Steccherinum* (Aphyllophorales, Steccherinaceae) on quaking aspen. Mycologia 69: 193-197.
405. Lindsey, J.P. & R.L. Gilbertson. 1977. New species of corticioid fungi on quaking aspen. Mycotaxon 5: 311-319.
406. Lindsey, J.P. & R.L. Gilbertson. 1978. Basidiomycetes that decay aspen in North America. Biblio. Mycol. 63: 1-406.
407. Lindsey, J.P. & R.L. Gilbertson. 1983. Notes on basidiomycetes that decay bristlecone pine. Mycotaxon 18: 541-559.

408. Lipps, P.E. & L.J. Herr. 1982. Etiology of *Rhizoctonia cerealis* in sharp eyespot of wheat. Phytopathology 72: 1574-1577.
409. Lloyd, C.G. 1920. Mycological Note 62. Mycol. Writ. 6: 917.
410. Lloyd, C.G. 1921. Mycological Note 65. Mycol. Writ. 6: 1029-1101.
411. Lloyd, C.G. 1922. Mycological Note 67. Mycol. Writ. 7: 1137-1168.
412. Lloyd, C.G. 1924. Mycological Note 72. Mycol. Writ. 7: 1269-1300.
413. Loman, A.A. 1962. The influence of temperature on the location and development of decay fungi in lodgepole pine logging slash. Canad. J. Bot. 40: 1545-1559.
414. Loman, A.A. & G.D. Paul. 1963. Decay of lodgepole pine in two foothills sections of the boreal forest in Alberta. Forest. Chron. 39: 422-435.
415. Lombard, F.F., H.H. Burdsall & R.L. Gilbertson. 1975. Taxonomy of *Corticium chrysocreas* and *Phlebia livida*. Mycologia 67: 495-510.
416. Lowe, J.L. 1966. Polyporaceae of North America - The genus *Poria*. New York State Univ., Coll. Forestry, Syracuse, Publ. 90: 1-183.
417. Lowy, B. 1952. The genus *Auricularia*. Mycologia 44: 656-692.
418. Lowy, B. 1954. A new *Dacrymyces*. Bull. Torrey Bot. Club 81: 300-303.
419. Lowy, B. 1955. Illustrations and keys to the tremellaceous fungi of Louisiana. Lloydia 18: 149-181.
420. Lowy, B. 1956. A note on *Sirobasidium*. Mycologia 48: 324-327.
421. Lowy, B. 1957. A new *Exidia*. Mycologia 49: 899-902.
422. Lowy, B. 1971. Tremellales. Flora Neotrop. Monogr. 6: 1-153.
423. Lowy, B. & A.L. Welden. 1959. Synopsis of Louisiana polypores. Amer. Midl. Naturalist 61: 329-349.
424. Luck-Allen, E.R. 1960. The genus *Heterochaetella*. Canad. J. Bot. 38: 559-569.
425. Luck-Allen, E.R. 1963. The genus *Basidiodendron*. Canad. J. Bot. 41: 1025-1052.
426. Maas Geesteranus, R.A. 1962. Hyphal structure in *Hydnums*. Persoonia 2: 377-405.
427. Maas Geesteranus, R.A. 1971. Hydnaceous fungi of the eastern old world. Verh. Kon. Ned. Akad. Wetensch., Afd. Natuurk., Tweede Sect. 60(3): 1-171.
428. Maas Geesteranus, R.A. 1974. Studies in the genera *Irpex* and *Steccherinum*. Persoonia 7: 443-591.
429. Maas Geesteranus, R.A. & P. Lanquetin. 1975. Observations sur quelques champignons hydnoïdes de l'Afrique. Persoonia 8: 145-165.
430. Maekawa, N. 1990. Redefinition of *Botryobasidium pruinatum* (Corticiaceae, Aphyllophorales) and its relationship to *B. asperulum* and *B. laeve*. Trans. Mycol. Soc. Japan 31: 467-478.
431. Maekawa, N., I. Arita & Y. Hayashi. 1982. Corticiaceae in Japan 1. Three species of the genus *Gloeocystidiellum* previously unrecorded from Japan. Rept. Tottori Mycol. Inst. (Japan) 20: 33-41.
432. Martin, G.W. 1932. The genus *Protodontia*. Mycologia 24: 508-511.
433. Martin, G.W. 1934. The genus *Stypella*. Stud. Nat. Hist. Iowa Univ. 2: 143-150.
434. Martin, G.W. 1934. Three new Heterobasidiomycetes. Mycologia 26: 261-265.
435. Martin, G.W. 1937. Notes on Iowa fungi VII. Proc. Iowa Acad. Sci. 44: 45-53.
436. Martin, G.W. 1939. Notes on Iowa fungi VIII. Proc. Iowa Acad. Sci. 46: 89-95.
437. Martin, G.W. 1940. Some Heterobasidiomycetes from eastern Canada. Mycologia 32: 683-695.
438. Martin, G.W. 1942. New or noteworthy tropical fungi-II. Lloydia 5: 158-164.
439. Martin, G.W. 1943. Notes on Iowa fungi X. Proc. Iowa Acad. Sci. 50: 165-169.
440. Martin, G.W. 1944. The Tremellales of north central United States and adjacent Canada. Stud. Nat. Hist. Iowa Univ. 18: 1-88.
441. Martin, G.W. 1949. The genus *Ceracea* Cragin. Mycologia 41: 77-86.
442. Martin, G.W. 1952. Notes on Iowa fungi XII. Proc. Iowa Acad. Sci. 59: 111-118.
443. Martin, G.W. 1952. Revision of the north central Tremellales. Stud. Nat. Hist. Iowa Univ. 19: 1-123.
444. Martin, G.W. 1954. Notes on Iowa fungi XIII. Proc. Iowa. Acad. Sci. 61: 138-140.
445. Martin, G.W. 1960. Notes on Iowa fungi XIV. Proc. Iowa Acad. Sci. 67: 139-144.
446. Martin, K.J. & R.L. Gilbertson. 1976. Cultural and other morphological studies of *Sparassis radicata* and related species. Mycologia 68: 622-639.
447. Martin, K.J. & R.L. Gilbertson. 1977. Synopsis of wood-rotting fungi on spruce in North America I. Mycotaxon 6: 43-77.
448. Martin, K.J. & R.L. Gilbertson. 1980. Synopsis of wood-rotting fungi on spruce in North America III. Mycotaxon 10: 479-501.
449. Martin, S.B. & L.T. Lucas. 1984. Characterization and pathogenicity of *Rhizoctonia* and binucleate *Rhizoctonia*-like fungi from turfgrasses in North Carolina. Phytopathology 74: 170-171.
450. Maxwell, M.B. 1954. Studies of Canadian Thelephoraceae XI. Conidium production in the Thelephoraceae. Canad. J. Bot. 32: 259-280.
451. McGuire, J.M. 1941. The species of *Sebacina* (Tremellales) of temperate North America. Lloydia 4: 1-43.
452. McKay, H.H. & P.L. Lentz. 1960. Descriptions of some fungi associated with forest tree decay in Colorado. Mycopathol. Mycol. Appl. 13: 265-286.
453. McKeen, C.G. 1952. Studies of Canadian Thelephoraceae IX. A cultural and taxonomic study of three species of *Peniophora*. Canad. J. Bot. 30: 764-787.
454. McNabb, R.F.R. 1964. Taxonomic studies in the Dacrymycetaceae I. *Cerinomyces* Martin. New Zealand J. Bot. 2: 415-424.
455. McNabb, R.F.R. 1965. Some auriculariaceous fungi from the British Isles. Trans. Brit. Mycol. Soc. 48: 187-192.
456. McNabb, R.F.R. 1965. Taxonomic studies in the Dacrymycetaceae II. *Calocera* (Fries) Fries. New Zealand J. Bot. 3: 31-58.
457. McNabb, R.F.R. 1965. Taxonomic studies in the Dacrymycetaceae III. *Dacryopinax* Martin. New Zealand J. Bot. 3: 59-72.
458. McNabb, R.F.R. 1965. Taxonomic studies in the Dacrymycetaceae IV. *Guepiniopsis* Patouillard. New Zealand J. Bot. 3: 159-169.
459. McNabb, R.F.R. 1965. Taxonomic studies in the Dacrymycetaceae V. *Heterotextus* Lloyd. New Zealand J. Bot. 3: 215-222.
460. McNabb, R.F.R. 1966. Taxonomic studies in the Dacrymycetaceae VII. *Ditiola* Fries. New Zealand J. Bot. 4: 546-558.
461. McNabb, R.F.R. 1973. Taxonomic studies in the Dacrymycetaceae VIII. *Dacrymyces* Nees ex Fries. New Zealand J. Bot. 11: 461-524.

462. Mielke, J.L. & R.W. Davidson. 1947. Notes on some western wood-decay fungi. Pl. Dis. Reporter 31: 27-30.

463. Miller, L.W. & J.S. Boyle. 1943. The Hydnaceae of Iowa. Stud. Nat. Hist. Iowa Univ. 18: 1-92.

464. Miller, O.K. & R.L. Gilbertson. 1969. Notes on Homobasidiomycetes from northern Canada and Alaska. Mycologia 61: 840-844.

465. Millspaugh, C.F. & L.W. Nuttall. 1923. Flora of Santa Catalina Island (California). Field Mus. Nat. Hist. Bot. Ser. 5: 1-413.

466. Molnar, A.C. & A.G. Davidson. 1974. *Annual report of the forest insect and disease survey, 1973.* Canad. Forestry Service, Ottawa. 101 p.

467. Mordue, J.E.M. 1974. *Corticium rolfsii.* CMI Descript. Path. Fungi Bact. 410. Commonw. Mycol. Inst., Kew.

468. Morris, B.B. & J.J. Motta. 1982. Isolation of an apparent temperature-sensitive cell cycle mutant of *Schizophyllum commune*. Mycologia 74: 412-422.

469. Nakasone, K.K. 1982. Cultural and morphological studies of *Gloeocystidiellum porosum* and *Gloeocystidium clavuligerum*. Mycotaxon 14: 316-324.

470. Nakasone, K.K. 1983. Cultural and morphological studies on *Cystostereum australe* (Corticiaceae), a new species from southeastern U.S.A. and Costa Rica. Mycotaxon 17: 269-274.

471. Nakasone, K.K. 1984. Taxonomy of *Crustoderma* (Aphyllophorales, Corticiaceae). Mycologia 76: 40-50.

472. Nakasone, K.K. 1985. Additional species of *Crustoderma*. Mycotaxon 22: 415-418.

473. Nakasone, K.K. 1990. Taxonomic study of *Veluticeps* (Aphyllophorales). Mycologia 82: 622-641.

474. Nakasone, K.K. 1990. Cultural studies and identification of wood-inhabiting Corticiaceae and selected Hymenomycetes from North America. Mycologia Memoir 15: 1-412.

474a. Nakasone, K.K. & H.H. Burdsall. 1984. *Merulius*, a synonym of *Phlebia.* Mycotaxon 21: 241-246.

475. Nakasone, K.K., H.H. Burdsall & L.A. Noll. 1982. Species of *Phlebia* section *Leptocystidiophlebia* (Aphyllophorales, Corticiaceae) in North America. Mycotaxon 14: 3-12.

476. Nakasone, K.K. & W.E. Eslyn. 1981. A new species, *Phlebia brevispora*, a cause of internal decay in utility poles. Mycologia 73: 803-810.

477. Nakasone, K.K. & R.L. Gilbertson. 1978. Cultural and other studies of fungi that decay ocotillo in Arizona. Mycologia 70: 266-299.

478. Nakasone, K.K. & R.L. Gilbertson. 1982. Three brown-rot fungi in the Corticiaceae. Mycologia 74: 599-606.

479. Nakasone, K.K. & J.A. Micales. 1988. *Scytinostroma galactinum* species complex in the United States. Mycologia 80: 546-559.

480. Nobles, M.K. 1935. Conidial formation, mutation and hybridization in *Peniophora allescheri*. Mycologia 27: 286-301.

481. Nobles, M.K. 1937. First Canadian record of *Aleurodiscus subcruentatus*. Mycologia 29: 387-391.

482. Nobles, M.K. 1937. Production of conidia by *Corticium incrustans*. Mycologia 29: 557-566.

483. Nobles, M.K. 1942. Secondary spores in *Corticium effuscatum*. Canad. J. Res., C, 20: 347-357.

484. Nobles, M.K. 1948. Studies in forest pathology VI. Identification of cultures of wood-rotting fungi. Canad. J. Res., C, 26: 281-431.

485. Nobles, M.K. 1953. Studies in wood-inhabiting Hymenomycetes I. *Odontia bicolor*. Canad. J. Bot. 31: 745-749.

486. Nobles, M.K. 1956. Studies in wood-inhabiting Hymenomycetes III. *Stereum pini* and species of *Peniophora* sect. *Colorate* on conifers in Canada. Canad. J. Bot. 34: 104-130.

487. Nobles, M.K. 1965. Identification of cultures of wood-inhabiting Hymenomycetes. Canad. J. Bot. 43: 1097-1139.

488. Nobles, M.K. 1967. Conspecificity of *Basidioradulum (Radulum) radula* and *Corticium hydnans*. Mycologia 59: 192-211.

489. Nobles, M.K., R. Macrae & B.P. Tomlin. 1957. Results of interfertility tests on some species of Hymenomycetes. Canad. J. Bot. 35: 377-387.

490. Nobles, M.K. & V.J. Nordin. 1955. Studies in wood-inhabiting Hymenomycetes II. *Corticium vellereum* Ellis and Cragin. Canad. J. Bot. 33: 105-112.

491. Oberwinkler, F. 1965. Die Gattung *Tubulicrinis* Donk s.l. (Corticiaceae). Z. Pilzk. 31: 12-48.

492. Oberwinkler, F. 1965. Primitive Basidiomyceten. Revision einiger Formenkreise von Basidienpilzen mit plastischer Basidie. Sydowia 19: 1-72.

493. Oberwinkler, F. 1977. Species- and generic-concepts in the Corticiaceae. Pp. 332-348. *In*: H. Clémençon (Ed.). *The species conmcept in Hymenomycetes.* Biblio. Mycol. 61. 444 p.

494. Oberwinkler, F. & R.J. Bandoni. 1981. *Tetragoniomyces gen. nov.* and Tetragoniomycetaceae *fam. nov.* (Tremellales). Canad. J. Bot. 59: 1034-1040.

495. Olive, L.S. 1944. New or rare Heterobasidiomycetes from North Carolina I. J. Elisha Mitchell Sci. Soc. 60: 17-26.

496. Olive, L.S. 1946. New or rare Heterobasidiomycetes from North Carolina II. J. Elisha Mitchell Sci. Soc. 62: 65-71.

497. Olive, L.S. 1946. Some taxonomic notes on the higher fungi. Mycologia 38: 534-547.

498. Olive, L.S. 1947. Notes on the Tremellales of Georgia. Mycologia 39: 90-108.

499. Olive, L.S. 1948. Taxonomic notes on Louisiana fungi-II Tremellales. Mycologia 40: 586-604.

500. Olive, L.S. 1950. A new genus of the Tremellales from Louisiana. Mycologia 42: 385-390.

501. Olive, L.S. 1951. New or noteworthy species of Tremellales from the southern Appalachians. Bull. Torrey Bot. Club 78: 103-112.

502. Olive, L.S. 1951. Taxonomic notes on Louisiana fungi III. Additions to the Tremellales. Mycologia 43: 677-690.

503. Olive, L.S. 1953. New or noteworthy species of Tremellales from the southern Appalachians II. Bull. Torrey Bot. Club 80: 33-42.

504. Olive, L.S. 1954. New or noteworthy species of Tremellales from the southern Appalachians III. Bull. Torrey Bot. Club 81: 329-339.

505. Olive, L.S. 1954. Two species of *Galzinia* from the southern Appalachians. Mycologia 46: 794-799.

506. Olive, L.S. 1957. Tulasnellaceae of Tahiti. A revision of the family. Mycologia 49: 663-679.

507. Olive, L.S. 1957. Two new genera of the Ceratobasidiaceae and their phylogenetic significance. Amer. J. Bot. 44: 429-435.

508. Olive, L.S. 1958. Latin diagnoses of earlier described species of Tremellales. J. Elisha Mitchell Sci. Soc. 74: 41.

509. Overholts, L.O. 1919. Some Colorado fungi. Mycologia 11: 245-258.

510. Overholts, L.O. 1920. Some mycological notes for 1919. Mycologia 12: 135-142.

511. Overholts, L.O. 1921. Some New Hampshire fungi. Mycologia 13: 24-37.

512. Overholts, L.O. 1922. Mycological notes for 1920. Bull. Torrey Bot. Club 49: 163-173.

513. Overholts, L.O. 1924. Mycological notes for 1921-22. Mycologia 16: 233-239.

514. Overholts, L.O. 1929. Mycological notes for 1926-27. Mycologia 21: 274-287.

515. Overholts, L.O. 1930. Mycological notes for 1928-29. Mycologia 22: 232-246.

516. Overholts, L.O. 1933. Mycological notes for 1930-32. Mycologia 25: 418-430.

517. Overholts, L.O. 1934. Mycological notes for 1933. Mycologia 26: 502-515.

518. Overholts, L.O. 1938. Mycological notes for 1934-35. Mycologia 30: 269-279.

519. Overholts, L.O. 1939. The genus *Stereum* in Pennsylvania. Bull. Torrey Bot. Club 66: 515-537.

520. Overholts, L.O. 1940. Mycological notes for 1936-38. Mycologia 32: 251-263.

521. Parker, A.K. & A.L.S. Johnson. 1960. Decay associated with logging injury to spruce and balsam in the Prince George region of British Columbia. Forest. Chron. 36: 30-45.

522. Parmasto, E. 1968. *Conspectus Systematis Corticiacearum.* Inst. Zool. Bot. Acad. Sci. SSR Estonia, Tartu. 261 p.

523. Parmasto, E. 1986. On the origin of the Hymenomycetes (What are corticioid fungi?). Windahlia 16: 3-19.

524. Parmeter, J.R. & H.S. Whitney. 1970. Taxonomy and nomenclature of the imperfect state. Pp. 7-19. *In: Rhizoctonia solani, biology and pathology.* Ed., J.R. Parmeter. Univ. California Press, Berkeley. 255 p.

525. Peck, C.H. 1879. Report of the Botanist. Ann. Rep. New York State Mus. 28: 31-88.

526. Peck, C.H. 1903. Report of the State botanist 1902. New York State Mus. Bull. 67: 1-194.

527. Petersen, R.H. 1968. Hyphal spore germination and asexual spore production in *Gloeotulasnella pinicola*. Mycopathol. Mycol. Appl. 35: 145-149.

528. Petersen, R.H. 1971. A new genus segregated from *Kavinia* Pilát. Ceská Mykol. 25: 129-134.

529. Petersen, R.H. 1976. Cultural characteristics of *Auriscalpium* and *Gloiodon*. Mycotaxon 3: 358-362.

530. Petersen, R.H. 1980. *Gloeomucro* and a note on *Physalacria concinna*. Mycologia 72: 301-311.

531. Peterson, R.L. & R.S. Currah. 1990. Synthesis of mycorrhizae between protocorms of *Goodyera repens* (Orchidaceae) and *Ceratobasidium cereale*. Canad. J. Bot. 68: 1117-1125.

532. Pomerleau, R. 1980. *Flore des champignons au Québec et régions limitrophes.* Les Editions la Presse, Montréal. 652 p.

533. Pomerleau, R. 1984. *Supplément à la flore des champignons au Québec.* Les Editions la Presse, Montréal. 88 p.

534. Pomerleau, R. & J. Brunel. 1939. Inventaire descriptif de la flore mycologique de Québec IV. Naturaliste Canad. 66: 28-32.

535. Pouzar, Z. 1958. New genera of higher fungi II. Ceská Mykol. 12: 31-36.

536. Pouzar, Z. 1964. *Stereum subtomentosum sp. nov.* and its taxonomic relations. Ceská Mykol. 18: 147-156.

537. Pouzar, Z. 1982. The problem of the correct name of *Vararia granulosa* (Lachnocladiaceae). Ceská Mykol. 36: 72-76.

538. Pouzar, Z. 1982. Taxonomic studies in resupinate fungi I. Ceská Mykol. 36: 141-145.

539. Prentice, R.M. & A.G. Davidson. 1965. *Annual report of the forest insect and disease survey, 1964.* Canad. Dept. Forestry, Ottawa. 141 p.

540. Punja, Z.K., R.G. Grogan, & G.C. Adams. 1982. Influence of environment, and the isolate, on basidiocarp formation, development, and structure in *Athelia (Sclerotium) rolfsii*. Mycologia 74: 917-926.

541. Punter, D., J. Reid & A.A. Hopkin. 1984. Notes on sclerotium-forming fungi from *Zizania aquatica* (wildrice) and other hosts. Mycologia 76: 722-732.

542. Punugu, A., M.T. Dunn & A.L. Welden. 1980. The peniophoroid fungi of the West Indies. Mycotaxon 10: 428-454.

543. Quack, W., T. Anke, F. Oberwinkler, B.M. Giannettii & W. Steglich. 1978. Antibiotics from Basidiomycetes V. Merulidial, a new antibiotic from the Basidiomycete, *Merulius tremellosus* Fr. J. Antibiotics 31: 737-741.

544. Raper, C.A. 1978. Sexuality and breeding. Pp. 83-117. *In: The biology and cultivation of edible mushrooms.* Eds., S.T. Chang and W.A. Hayes. Academic Press, New York. 819 p.

545. Redhead, S.A. 1973. Epistolae mycologicae I. Some cyphelloid basidiomycetes from British Columbia. Syesis 6: 221-227.

546. Redhead, S.A. 1981. Agaricales on wetland monocotyledoneae in Canada. Canad. J. Bot. 59: 574-589.

547. Redhead, S.A. 1984. *Arrhenia* and *Rimbachia*, expanded generic concepts, and a reevaluation of *Leptoglossum* with emphasis on musicicolous North American taxa. Canad. J. Bot. 62: 865-892.

548. Redhead, S.A. 1989. A biogeographical overview of the Canadian mushroom flora. Canad. J. Bot. 67: 3003-3062.

549. Redhead, S.A. & D.A. Reid. 1983. *Craterellus humphreyi*, an unusual *Stereopsis* from western North America. Canad. J. Bot. 61: 3088-3090.

550. Redhead, S.A. & J.A. Traquair. 1981. *Calyptella capula.* Fungi Canadenses No. 202. National Mycological Herbarium, Agriculture Canada, Ottawa.

551. Reeves, F. & A.L. Welden. 1967. West Indian species of *Hymenochaete*. Mycologia 59: 1034-1049.

552. Reid, D.A. 1959. The genus *Cymatoderma* Jungh. (*Cladoderris*). Kew Bull. 13: 518-530.

553. Reid, D.A. 1962. Notes on fungi which have been referred to the Thelephoraceae sensu lato. Persoonia 2: 109-170.

554. Reid, D.A. 1964. Notes on some fungi of Michigan-I 'Cyphellaceae'. Persoonia 3: 97-154.

555. Reid, D.A. 1965. A monograph of the stipitate stereoid fungi. Beih. Nova Hedwigia 18: 1-382.

556. Reid, D.A. 1970. New or interesting records of British Hymenomycetes-IV. Trans. Brit. Mycol. Soc. 55: 413-441.

557. Reid, D.A. 1973. New or interesting records of British Hymenomycetes-V. Persoonia 7: 293-303.

558. Reid, D.A. 1974. A monograph of the British Dacrymycetales. Trans. Br. Mycol. Soc. 62: 433-494.

559. Richard, J. & T.L. Highley. 1988. Biocontrol of decay or pathogenic fungi in wood and trees. Pp. 9-15. *In: 2nd Internat. Trichoderma/Gliocladium workshop.* Ed. A. Gear. Trichoderma Newsletter No. 4. H. Doubleday Res. Assoc., Ryton.

560. Rogers, D.P. 1932. A cytological study of *Tulasnella*. Bot. Gaz. (Crawfordsville) 94: 86-105.

561. Rogers, D.P. 1933. A taxonomic review of the Tulasnellaceae. Ann. Mycol. 31: 181-208.

562. Rogers, D.P. 1933. Some noteworthy fungi from Iowa. Stud. Nat. His. Iowa Univ. 15: 9-29.

563. Rogers, D.P. 1935. Notes on the lower Basidiomycetes. Stud. Nat. His. Iowa Univ. 17: 1-43.

564. Rogers, D.P. 1936. Basidial proliferation through clamp-formation in a new *Sebacina*. Mycologia 28: 347-362.

565. Rogers, D.P. 1943. The genus *Pellicularia* (Thelephoraceae). Farlowia 1: 95-118.

566. Rogers, D.P. 1944. The genera *Trechispora* and *Galzinia* (Thelephoraceae). Mycologia 36: 70-103.

567. Rogers, D.P. & H.S. Jackson. 1943. Notes on the synonymy of some North American Thelephoraceae and other resupinates. Farlowia 1: 263-328.

568. Rogers, D.P. & G.W. Martin. 1955. The genus *Mylittopsis*. Mycologia 47: 891-894.

569. Ryvarden, L. 1972. *Radulodon*, a new genus in the Corticiaceae (Basidiomycetes). Canad. J. Bot. 50: 2073-2076.

570. Ryvarden, L. 1976. Type-studies in the Polyporaceae 4. Species described by J.F. Klotzsch. Mem. New York Bot. Gard. 28: 199-207.

571. Ryvarden, L. 1978. A study of *Hydnum subcrinale* and *Odontia laxa*. Norw. J. Bot. 25: 293-296.

572. Ryvarden, L. 1982. *Fuscocerrena*, a new genus in the Polyporaceae. Trans. Brit. Mycol. Soc. 79: 279-281.

573. Ryvarden, L. 1982. The genus *Hydnochaete* Bres. (Hymenochaetaceae). Mycotaxon 15: 425-447.

574. Sanders, P.L., L.L. Burpee & H. Cole. 1978. Preliminary studies on binucleate turfgrass pathogens that resemble *Rhizoctonia solani*. Phytopathology 68: 145-148.

575. Schlechte, G. 1981. Diagnose des Bleiglanzerregers *Chondrostereum purpureum* (Pers. ex Fr.) Pouz. 1959 in Reinkultur. Gartenbauwissenschaft 46: 59-64.

576. Schweinitz, L.D. 1832. Synopsis fungorum in America boreali. Trans. Amer. Philos. Soc., Ser. II, 4: 273.

577. Seifert, K. 1983. Decay of wood by the Dacrymycetales. Mycologia 75: 1011-1018.

578. Setliff, E.C. 1982. *Tremella polyporina* from New York State. Canad. J. Bot. 60: 1028-1029.

579. Shaw, C.G. & W.B. Cooke. 1953. Reliquiae Suksdorfiana. Fungi collected by Wilhelm N. Suksdorf 1882-1927. Wash. State Agric. Exp. Station Circular 217: 1-29.

580. Shields, J.K. 1969. Microflora of eastern Canadian wood chip piles. Mycologia 61: 1165-1168.

581. Siepmann, R. 1969. Artdiagnose einiger holzzerstörender Hymenomyceten an Hand von Reinkulturen II. Nova Hedwigia 18: 183-201.

582. Siepmann, R. 1970. Artdiagnose einiger holzzerstörender Hymenomyceten an Hand von Reinkulturen III. Nova Hedwigia 20: 833-863.

583. Siepmann, R. 1971. Artdiagnose einiger Holzzerstörender Hymenomyceten an Hand von Reinkulturen IV. Nova Hedwigia 21: 843-875.

584. Slysh, A.R. 1960. The genus *Peniophora* in New York State and adjacent regions. New York State Univ., Coll. Forestry, Syracuse, Publ. 83: 1-95.

585. Smith, J.D., N. Jackson and A.R. Woolhouse. 1989. *Fungal diseases of amenity turf grasses*. E. & F. Spon, New York. 401 p.

586. Sprague, R. 1950. *Diseases of cereals and grasses in North America*. Roland Press, New York. 538 p.

587. Stalpers, J.A. 1975. Notes on *Sporotrichum* 1. On the segregate genus *Alytosporium*. Rev. Mycol. 29: 97-101.

588. Stalpers, J.A. 1978. Identification of wood-inhabiting Aphyllophorales in pure culture. Studies in Mycology, Baarn 16: 1-248.

589. Stalpers, J.A. 1984. A revision of the genus *Sporotrichum*. Studies in Mycol., Baarn 24: 1-105.

590. Stalpers, J.A. 1985. Type studies of the species of *Corticium* described by G.H. Cunningham. New Zealand J. Bot. 23: 301-310.

591. Stalpers, J.A. 1988. *Auriculariopsis* and the Schizophyllales. Persoonia 13: 495-504.

592. Stalpers, J.A. & W.M. Loerakker. 1982. *Laetisaria* and *Limonomyces* species (Corticiaceae) causing pink diseases in turf grasses. Canad. J. Bot. 60: 529-537.

593. Stevenson, J.A. & E.K. Cash. 1936. The new fungus names proposed by C.G. Lloyd. Lloyd Library Bull. 35, Mycol. Ser. 8: 1-209.

594. Stillwell, M.A. 1966. Woodwasps (Siricidae) in conifers and the associated fungus, *Stereum chailletii*, in eastern Canada. Forest Sci. 12: 121-128.

595. Stillwell, M.A. & D.J. Kelly. 1964. Fungus deterioration of balsam fir killed by spruce budworm in northwestern New Brunswick. Forest. Chron. 40: 482-487.

596. Strid, A. 1986. Tremellaceous fungi with small, spiny fruitbodies. Windahlia 16: 99-112.

597. Sumstine, D.R. 1941. Notes on some new or interesting fungi. Mycologia 33: 17-22.

598. Talbot, P.H.B. 1958. Studies of some South African resupinate Hymenomycetes II. Bothalia 7 (1): 131-187.

599. Talbot, P.H.B. 1965. Studies of '*Pellicularia*' and associated genera of Hymenomycetes. Persoonia 3: 371-406.

600. Talbot, P.H.B. 1970. Taxonomy and nomenclature of the perfect state. Pp. 20-31. *In: Rhizoctonia solani, biology and pathology.* Ed., J.R. Parmeter. Univ. California Press, Berkeley. 255 p.

601. Tassinari, M. 1956. Recherches sur la polarité de vingt-trois Théléphoracées. Compt. Rend. Hebd. Séances Acad. Sci. 242: 2661-2662.

602. Taylor, A.F.S. & I.J. Alexander. 1991. Ectomycorrhizal synthesis with *Tylospora fibrillosa*, a member of the Corticiaceae. Mycological Research 95: 381-384.

603. Thomas, G.P., D.E. Etheridge & G. Paul. 1960. Fungi and decay in aspen and balsam poplar in the boreal forest region, Alberta. Canad. J. Bot. 38: 459-466.

604. Thorn, R.G. & G.L. Barron. 1986. *Nematoctonus* and the tribe *Resupinatae* in Ontario, Canada. Mycotaxon 25: 321-453.

605. Tu, C.C., D.A. Roberts & J.W. Kimbrough. 1969. Hyphal fusion, nuclear condition, and perfect stages of three species of *Rhizoctonia*. Mycologia 61: 775-783.

606. de Vries, B.W.L. 1987. Some new corticioid taxa. Mycotaxon 28: 77-90.

607. Wakefield, E.M. 1960. Some species of *Tomentella* from North America. Mycologia 52: 919-933.

608. Wakefield, E.M. 1966. Some extra-european species of *Tomentella*. Trans. Brit. Mycol. Soc. 49: 357-362.

609. Wang, C.J.K. & R.A. Zabel (Eds.). 1990. *Identification manual for fungi from utility poles in the eastern United States.* American Type Culture Collection, Rockville. 356 p.

610. Weaver, M.G. & R.L. Shaffer. 1969. Records of the higher fungi of Minnesota. J. Minn. Acad. Sci. 35: 123-128.

611. Weaver, M.G. & R.L. Shaffer. 1972. Higher fungi of Minnesota II. J. Minn. Acad. Sci. 38: 46-52.

612. Welden, A.L. 1954. Some tropical American stipitate *Stereums*. Bull. Torrey Bot. Club 81: 422-439.

613. Welden, A.L. 1958. A contribution toward a monograph of *Cotylidia* (Thelephoraceae). Lloydia 21: 38-44.

614. Welden, A.L. 1958. Notes and brief articles - two unrecognized species of *Cytidia*. Mycologia 50: 304-306.

615. Welden, A.L. 1958. Prodromus fungi Ludovicianae. J. Tennessee Acad. Sci. 33: 252-257.

616. Welden, A.L. 1960. Prodromus fungorum Ludovicianorum II. J. Tennessee Acad. Sci. 35: 231-237.

617. Welden, A.L. 1960. The genus *Cymatoderma* (Thelephoraceae) in the Americas. Mycologia 52: 856-876.

618. Welden, A.L. 1965. West Indian species of *Vararia* with notes on extralimital species. Mycologia 57: 502-520.

619. Welden, A.L. 1966. West Indian species of *Asterostroma*, with notes on extralimital species. Amer. Midl. Naturalist 76: 222-229.

620. Welden, A.L. 1975. *Lopharia*. Mycologia 67: 530-551.

621. Welden, A.L. & J.W. Bennett. 1973. The cultural characteristics and mating-type behavior in *Podoscypha multizonata* and *P. ravenelii*. Mycologia 65: 203-207.

622. Wells, K. 1957. Studies of some Tremellaceae. Lloydia 20: 43-65.

623. Wells, K. 1958. Studies of some Tremellaceae II. The genus *Ductifera*. Mycologia 50: 407-416.

624. Wells, K. 1959. Studies of some Tremellaceae III. The genus *Bourdotia*. Mycologia 51: 541-563.

625. Wells, K. 1962. Studies of some Tremellaceae IV. *Exidiopsis*. Mycologia 53: 317-370.

626. Wells, K. 1964. The basidia of *Exidia nucleata* I. Ultrastructure. Mycologia 56: 327-341.

627. Wells, K. 1975. Studies of some Tremellaceae V. A new genus *Efibulobasidium*. Mycologia 67: 147-156.

628. Wells, K. 1987. Comparative morphology, intracompatibility, and intercompatibility of several species of *Exidiopsis* (Exidiaceae). Mycologia 79: 274-288.

629. Wells, K. & F. Oberwinkler. 1982. *Tremelloscypha gelatinosa*, a species of a new family Sebacinaceae. Mycologia 74: 325-331.

630. Wells, K. & A. Raitviir. 1975. The species of *Bourdotia* and *Basidiodendron* (Tremellaceae) of the U.S.S.R. Mycologia 67: 904-922.

631. Wells, K. & A. Raitviir. 1977. The species of *Exidiopsis* (Tremellaceae) of the U.S.S.R. Mycologia 69: 987-1007.

632. Wells, K. & A. Raitviir. 1980. The species of *Eichleriella* (Tremellaceae) of the U.S.S.R. Mycologia 72: 564-577.

633. Wells, K. & G. Wong. 1985. Interfertility and comparative morphological studies of *Exidiopsis plumbescens* from the West Coast. Mycologia 77: 285-299.

634. Weresub, L.K. 1953. Studies of Canadian Thelephoraceae X. Some species of *Peniophora*, section *Tubuliferae*. Canad. J. Bot. 31: 760-778.

635. Weresub, L.K. 1961. Typification and synonomy of *Peniophora* species sect. *Tubuliferae* (Corticiaceae). Canad. J. Bot. 39: 1453-1495.

636. Weresub, L.K. 1974. *Amylocorticium canadense*. Fungi Canadenses 45. National Mycological Herbarium, Agriculture Canada, Ottawa.

637. Weresub, L.K. 1976. *Amylocorticium cebennense*. Fungi Canadenses 87. National Mycological Herbarium, Agriculture Canada, Ottawa.

638. Weresub, L.K. 1977. *Amylocorticium subincarnatum*. Fungi Canadenses 91. National Mycological Herbarium, Agriculture Canada, Ottawa.

639. Weresub, L.K. & S. Gibson. 1960. "*Stereum pini*" in North America. Canad. J. Bot. 38: 833-867.

640. Weresub, L.K. & W.I. Illman. 1980. *Corticium centrifugum* sensu Butler, re-isolated from fisheye rot of stored apples. Canad. J. Bot. 58: 137-146.

641. Whelden, R.M. 1934. Cytological studies in the Tremellaceae I. *Tremella*. Mycologia 26: 415-435.

642. Whelden, R.M. 1935. Cytological studies in the Tremellaceae II. *Exidia*. Mycologia 27: 41-57.

643. Whelden, R.M. 1935. Cytological studies of the Tremellaceae III. *Sebacina*. Mycologia 27: 503-520.

644. White, L.T. 1951. Studies of Canadian Thelephoraceae VIII. *Corticium galactinum* (Fr.) Burt. Canad. J. Bot. 29: 279-296.

645. Whitney, R.D. & D.T. Myren. 1978. Root-rotting fungi associated with mortality of conifer saplings in northern Ontario. Canad. J. Forest Res. 8: 17-22.

646. Wilkinson, H.T. 1987. Association of *Trechispora alnicola* with yellow ring disease of *Poa pratensis*. Canad. J. Bot. 65: 150-153.

647. Wilson, A.D. 1990. The genetics of sexual incompatibility in the Indian Paint Fungus, *Echinodontium tinctorium*. Mycologia 82: 332-341.

648. Wojewoda. W. 1977. *Grzyby (Mycota). Vol. 8.* Polska Akad. Nauk, Inst. Bot., Warszawa. 334 p.

649. Wong, G.J. & K. Wells. 1987. Comparative morphology, compatibility, and interfertility of *Auricularia cornea, A. polytricha*, and *A. tenuis*. Mycologia 79: 847-856.

650. Wong, G.J., K. Wells & R.J. Bandoni. 1985. Interfertility and comparative morphological studies of *Tremella mesenterica*. Mycologia 77: 36-49.

651. Wright, J.E. & J.R. Deschamps. 1972. Basidiomicetos xilófagos de los Bosques Andinopatagónicos. Revista Investiga. Apropecuarias, INTA, Buenos Aires, Ser. 5, Pathologia Vegetal 9: 111-195.

652. Young, C.S. & J.H. Andrews. 1990. Inhibition of pseudothecial development of *Venturia inaequalis* by the Basidiomycete *Athelia bombacina* in apple leaf litter. Phytopathology 80: 536-542.

653. Young, C.S., H.H. Burdsall & J.H. Andrews. 1987. A new species of *Athelia*? Mycol. Soc. Amer. Newsletter 38 (1): 56-57.

654. Zabel, R.A., F.F. Lombard, C.J.K. Wang & F. Terracina. 1985. Fungi associated with decay in treated southern pine utility poles in the eastern United States. Wood Fiber Sci. 17: 75-91.

655. Zak, B. 1969. Characterization and classification of mycorrhizae of Douglas-fir I. *Pseudotsuga menziesii* + *Poria terrestris* (blue- and orange-staining strains). Canad. J. Bot. 47: 1833-1840.

656. Zak, B. 1976. Pure culture synthesis of pacific madrone ectendomycorrhizae. Mycologia 68: 362-369.

657. Zak, B. & M.J. Larsen. 1978. Characterization and classification of mycorrhizae of Douglas-fir III. *Pseudotsuga menziesii* + *Byssoporia (Poria) terrestre* vars. *lilacinorosea, parksii* and *sublutea*. Canad. J. Bot. 56: 1416-1424.

658. Zeller, S.M. 1922. Contributions to our knowledge of Oregon fungi-1. Mycologia 14: 173-199.

659. Zycha, H. & H. Knopf. 1966. Cultural characteristics of some fungi which cause red-stain of *Picea abies*. Lloydia 29: 136-145.

GENUS SPECIES INDEX

(Teleomorph names of recognized taxa are in bold)

SPECIES INDEX

(Teleomorph names of recognized taxa are in bold)